Cougar

THE LION *by Virginia Bennett*

She waits in the deep, dense forest
Lurking in the shadows where the sun is defied
Lapping water from an ice-encrusted stream
She is stealth wrapped up in a tawny hide.

She hears more by instinct than by listening
Her paws like radar upon the glistening shale
And, she's keenly aware, when you are two miles away,
Of your horse as he plods up the trail.

She has ample time to consider her options
Whether scientists believe she can reason or not.
She could stay where she's at, undetected,
Or head back up the slope at a trot.

Instead, she crosses your path when you're almost upon her,
Like a dancing sunbeam teasing a child.
Leaving her track in the trail just to inform you. . . .
You've been that close to something that wild.

(November 1997; from *In the Company of Horses*)

Cougar
Ecology and Conservation

Edited by Maurice Hornocker
and Sharon Negri

The University of Chicago Press
Chicago and London

MAURICE HORNOCKER is director of the Selway Institute.
SHARON NEGRI is the director and founder of WildFutures.

Frontispiece: "Lapping water from an ice-encrusted stream" (photo by
Maurice Hornocker); "The Lion" (Poem by Virginia Bennett from *In the
Company of Horses*).

The University of Chicago Press, Chicago 60637
The University of Chicago Press, Ltd., London
© 2010 by Maurice Hornocker and Sharon Negri
All rights reserved. Published 2009
Printed in the United States of America

18 17 16 15 14 13 12 11 10 09   1 2 3 4 5

ISBN-13: 978-0-226-35344-9 (cloth)
ISBN-10: 0-226-35344-3 (cloth)

Library of Congress Cataloging-in-Publication Data

Cougar : ecology and conservation / edited by Maurice Hornocker and
Sharon Negri.
        p. cm.
  Includes bibliographical references and index.
   ISBN-13: 978-0-226-35344-9 (cloth : alk. paper)
   ISBN-10: 0-226-35344-3 (cloth : alk. paper) 1. Puma—
Conservation. 2. Wildlife conservation. 3. Wildlife management.
I. Hornocker, Maurice G. II. Negri, Sharon.
   QL737.C23C6775 2010
   599.75'24—dc22

The University of Chicago Press and the volume editors gratefully
acknowledge the Thaw Charitable Trust, the Laura Moore Cunningham
Foundation, the Richard King Mellon Foundation, and the Wildlife
Conservation Society for their generous contributions toward publication
of this work.

♾ The paper used in this publication meets the minimum requirements of
the American National Standard for Information Sciences—Permanence
of Paper for Printed Library Materials, ANSI Z39.48-1992. The text of
this book is printed on 30% recycled paper.

Contents

Range Map   vii

Foreword
Alan Rabinowitz   viii

Preface
Maurice Hornocker and Sharon Negri   x

Acknowledgments   xiii

Part I. Research and Management Come of Age

The Bitterroot Tom
Maurice Hornocker   3

1   To Save a Mountain Lion: Evolving Philosophy
of Nature and Cougars
R. Bruce Gill   5

2   The Emerging Cougar Chronicle
Harley Shaw   17

3   Lessons and Insights from Evolution, Taxonomy,
and Conservation Genetics
Melanie Culver   27

4   Cougar Management in North America
United States: Charles R. Anderson Jr. and Frederick Lindzey
Canada: Kyle H. Knopff, Martin G. Jalkotzy, and Mark S. Boyce   41

Part II. Populations

Tracking for a Living
Kerry Murphy   57

5   Cougar Population Dynamics
Howard Quigley and Maurice Hornocker   59

6   What We Know about Pumas in Latin America
John W. Laundré and Lucina Hernández   76

7   The World's Southernmost Pumas in Patagonia
and the Southern Andes
Susan Walker and Andrés Novaro   91

Part III. Cougars and Their Prey

Notes from the Field
Linda L. Sweanor   103

8   Behavior and Social Organization of a Solitary Carnivore
Kenneth A. Logan and Linda L. Sweanor   105

9   Diet and Prey Selection of a Perfect Predator
Kerry Murphy and Toni K. Ruth   118

10   Cougar-Prey Relationships
Toni K. Ruth and Kerry Murphy   138

11   Competition with Other Carnivores for Prey
Toni K. Ruth and Kerry Murphy   163

Part IV. Conservation and Coexisting with People

Death of a Towncat
Harley Shaw   175

12   A Focal Species for Conservation Planning
Paul Beier   177

13   Cougar-Human Interactions
Linda L. Sweanor and Kenneth A. Logan   190

14   People, Politics, and Cougar Management
David J. Mattson and Susan G. Clark   206

15   Cougar Conservation: The Growing Role of Citizens and Government
Sharon Negri and Howard Quigley   221

16   Pressing Business
Maurice Hornocker   235

Appendix 1: Genetics Techniques Primer   248

Appendix 2: Cougar Harvest in the United States   252

Appendix 3: Groups Participating in Cougar Management   254

Appendix 4: Cougar Litigation Summary, a Partial Listing   260

Appendix 5: Summary of Cougar Ballot Initiatives in the United States   264

References   266
Contributors   299
Index   303

Color plates follow page 114

Range Map

North and South America Contemporary Cougar Population

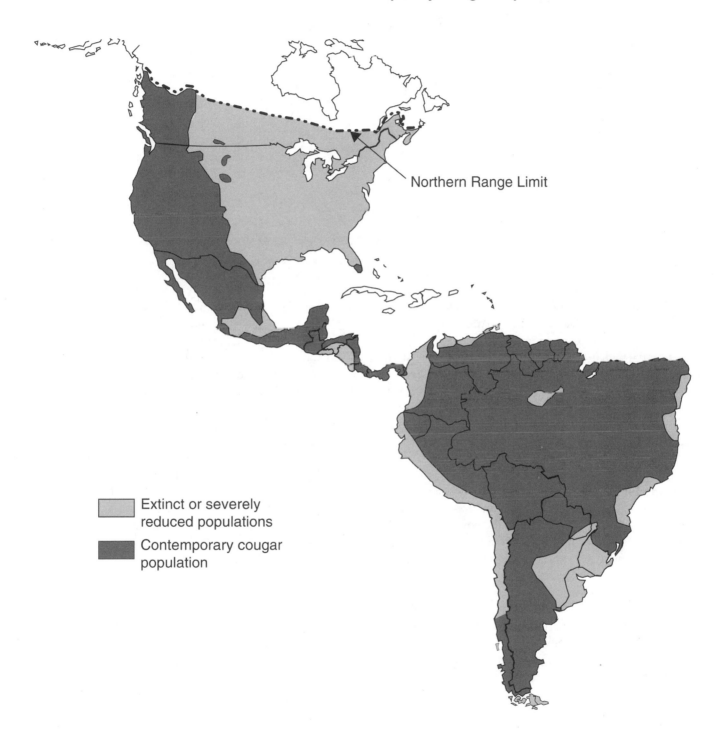

Northern Range Limit

Extinct or severely reduced populations

Contemporary cougar population

Foreword

Alan Rabinowitz, President, Panthera

THE BOY suppressed a shiver, brought on partly by the chill of the December weather and the start of the summer solstice in his birthplace of Cuzco, the Inca capital high in the Andes. He'd been there since sunrise, standing beside four other boys of noble blood waiting for the start of the *Capac Raymi* ceremony. This was the day he would leave behind his youthful freedom and the comfort of his mother's bosom to take on the serious responsibilities of manhood. Suddenly, a deep booming sound reverberated through the courtyard, breaking his reverie.

Four Inca warriors, dressed as pumas, emerged from the shadows into the early morning light. The warriors pounded the drums of the Sun God, Inti, to produce "the voice of the puma." Tingling with excitement, the boy remembered the story his father told him of how the land of the Inca was formed from a great puma laying astride the high mountains, imbuing his people with the strength and courage to conquer other tribes. As the warriors reached them, he and the other young aristocrats were handed heads of pumas complete with golden ear spools. Each of them understood that he had now passed through that invisible door into adult Inca society and royalty. And they realized, more than ever before, that high up in these snow-clad Peruvian Andes, their protector, the dominant feline icon of Inca civilization, is the majestic puma.

Spanning almost every available habitat from northern British Columbia to Patagonia and from the Atlantic to the Pacific, the second largest felid in the New World has an extensive geographic distribution, matched only by that of humans. It is known by many names. Most North Americans refer to this cat either as the cougar, a name originating from the Tupí language of Brazil; the puma, from the Quechua language of Peru; or the lion, a misnomer from the early explorers and Spanish conquerors who mistakenly thought that the first skins they saw were those of female African lions. Occurring sympatrically with the jaguar throughout much of the neotropical portion of its range, and with body characteristics similar to both larger and smaller cat species, the puma has often had its place in the pantheon of indigenous cultures subverted by the more fearsome jaguar. Exceptions to this were in those habitats or elevations where the puma was dominant, such as in the arid deserts or the high Andes.

As befell most of the world's larger cats, it was not long before this six- to seven-foot-long, 150-pound-or-more top carnivore ran afoul of the early European settlers who thronged to the New World. Prime habitat was lost and native prey declined as large areas of range were increasingly filled with domestic livestock, which were easy pickings for the cat. Suddenly, the shy, elusive puma, revered as a "ghost cat" by some indigenous New World cultures, went from deity to villain, an animal now portrayed as an evil, sneaky, bloodthirsty miscreant that threatened human livelihoods and well-being. One account of exploration in the northern Rockies included the following warning, attributed to a British traveler, Lord Southesk, in 1875: "When all is dark and silent, the insidious puma glides in, and the sleeper knows but a short awakening before its fangs are buried in his throat" (Etling 2001).

Not surprisingly, the true behavior of these cats was best known to the men who spent much of their time trying to kill them. In his 1948 book *Hunting American Lions*, Frank Hibben describes being followed or tracked by pumas repeatedly, noting that it was "seemingly without purpose and from an urge of sheer curiosity." He describes the animal he spent so many years hunting as a "fascinating, terrifying, and sometimes loveable cat" (1948, 225). Still, persecution of the puma continued unabated due to pervasive fear and misunderstanding about the true nature of this animal.

The earliest recorded bounty on pumas was in the 1500s, when Jesuit priests in Baja California offered natives one bull for each animal killed. By the late 1800s and early 1900s, eastern United States puma populations had been eradicated or severely reduced, while the distribution of western populations had greatly diminished. In 1915, ranchers lobbied Congress to set up a predator elimination program. The Animal Damage Control Act was passed in 1931, and over the next three to four decades, government trappers killed more than 7,000 pumas. Between 1907 and 1978, the U.S. Fish and Wildlife Service estimated a minimum of 66,665 cougars killed. Between 1965 and 1973, the status of the cougar changed from predator to game animal in many western states. In 1990, the sport kill in eleven western U.S. states recorded legal kills of 1,875 cougars. A 1990 referendum in California gave full protection to the cougar (Sunquist and Sunquist 2002).

Scientists now realize that large carnivores are important components of healthy natural systems, often helping structure and shape ecological relationships among the myriad species with which they coexist. Several of those interactions are examined by the authors of this book. The question now is whether large carnivore species, particularly the wide-ranging, solitary large cats, will be allowed to survive in the face of increasing human population pressures and increasing human-wildlife conflicts. In approaching an uncertain future, much can be gained from closely examining the past. If there are lessons to be learned from how populations of large cats have managed to survive—even thrive—into the twenty-first century, there is perhaps no better species to focus on than the puma. Capable of enduring the harsh desert environments of northern Mexico or the southwestern United States, sharing habitat and prey with their larger relative the jaguar throughout much of Central and South America, and managing to live alongside major residential developments in parts of California, the puma has displayed a resiliency unmatched by most other big cat species.

My first sighting of a wild puma was, in fact, my first encounter with any large wild cat and occurred several years before I decided to devote my life to big cat conservation. As such, the significance of that encounter went unrecognized at the time. In 1976, while hiking with a friend in the Florida Everglades, I rounded a bend as a large tawny-colored cat that I recognized immediately as a puma stepped out into the trail twenty feet ahead of me. Initially, it seemed as surprised as I was by the encounter. Before my mind could process any thought other than how lucky I was to meet this elusive species, it disappeared into the brush as quickly and soundlessly as it had appeared. Only later would I learn that I had just met one of the few remaining individuals of the Florida panther, the most critically endangered puma population still surviving. Thought to have been extinct for several decades, it had been rediscovered only three years earlier. And at the same time as my sighting, the U.S. Fish and Wildlife Service was in the process of appointing a recovery team to save this subspecies.

Ironically, following that first puma sighting, I would scarcely see another in more than three decades of working to conserve large cats—capturing and studying jaguars, tigers, and leopards. I have tracked pumas myriad times through the jungles of Central America, up the slopes of the Teton Range in Wyoming, and over the rocky precipices of the River of No Return Wilderness Area in Idaho. Always the cat evaded me. I would catch sight of a wild puma only once more, a brief glimpse of hindquarters as the cat ran across an old timber road deep in the jungle of Belize.

Working closely with governments of developing countries on conservation efforts, I find it intriguing that no high-level official has ever expressed the belief that the nation would be better off if it lost or eradicated all its large cat species; more often, the reverse sentiment is expressed. These same officials ask repeatedly for guidelines on mitigating human-wildlife conflict or for examples of how wealthier countries balance a conservation agenda with development. Unfortunately, successful scenarios from developed countries are sorely lacking. This book is part of an effort to shift that paradigm. Interested citizens and the United States government, at the state and federal levels, have the opportunity to manage the survival of the puma properly and set an example of how people and large cats can live together.

The book could not be more timely. Written and edited by the world's leaders in cougar biology and conservation, it is the most comprehensive guide to date on the history, behavior, biology, and conservation of one of the world's most magnificent and adaptable large carnivore species. Latin America, often underexposed in cougar discussion, receives substantial treatment. New findings from work in genetics are presented. The knowledge collected provides a rich resource for those who wish to understand the intricate role a large carnivore plays in the natural world. And, more important, it provides a foundation for all those wishing to help shape a future in which this top predator, this elusive, iconic feline, forever endures.

Preface

Maurice Hornocker and Sharon Negri

All genuine knowledge originates in direct experience.
—Mao Tse-tung

T HESE FEW WORDS set the tone for this book. All of the contributing authors are professionals who have devoted their lives to the science and conservation of cougars—they truly have "walked the walk" and write with the insight of direct experience. We believe that this, more than anything, lends strength and credibility to these collected works.

Cougar has been a long time in the making. The editors have more than sixty-five years combined experience with *Puma concolor,* the cat of one color. Our involvement has been in biological research and sociopolitics, in management and conservation, and with cultural concerns. We have not always agreed with each other, nor are we in accord with all the assertions of the contributing authors. Divergent voices are represented here. We did agree to let the authors' experience and views be heard and therefore did not edit their opinions or conclusions to fit some common standpoint. We understood from the beginning that the authors have different perspectives and may not necessarily agree with the conclusions reached by others. These different perspectives share the world that the cougar lives in and reflect the diverse views surrounding this controversial cat and its management and conservation.

It seems fitting that at this stage in our careers, as a biologist and a conservationist, we should work together to compile as complete a work on this species as possible. To accomplish this, we agreed to approach professionals who have devoted their careers to creating a greater understanding of this carnivore. The enthusiastic response by this distinguished group of authors has been gratifying. All had full schedules, yet they all agreed to devote their personal time to this project.

The result is the most complete work yet on this species. It is aimed at researchers and managers, academics, and everyone else with an interest not only in cougars but in all wildlife. The intended audience is broad and so, too, is public interest in the cougar; perceptions and emotions surrounding this cat run the gamut. Ecologists, conservationists, ranchers, hunters, wildlife managers, and average citizens all have voiced opinions about this large carnivore. We recognize that we cannot include every perspective or everything that is known about the cougar. There are topics covered in more depth and detail elsewhere. The authors cite these works, and we urge readers to follow up and read further.

The book consists of four parts, sixteen contributed chapters, and five appendices. We open each part with a close-up and personal look into the world of the scientist, aiming to offer a window into their direct experience as they face the many logistics, challenges, and rewards of learning the ways of cougars. Frontline field work to gather core knowledge is the essential basis for managing and conserving a species, but with this cryptic species, collecting the data is hardly a straightforward affair. Some chapters open with introductory material on the basic techniques and principles of specialized fields, such as genetics or modeling the effects of predation, aiming to deepen nonspecialists' understanding of how data are applied and what models can tell us.

Beginning with a philosophical treatment of human involvement with wildlife, Chapter 1 summarizes the philosophical literature about our relationship with nature, how this relationship shaped the history of cougar management, and what it might portend for the future of cougar conservation. Chapter 2 traces the evolution of our knowledge of *Puma concolor*. Early research is chronicled and placed in perspective with the evolved methodology and technology of current times, bringing us to where we are now. The modern field of genetics is treated in Chapter 3. Both molecular genetics and population genetics are discussed along with taxonomy and evolution. The importance of this relatively new field in conservation is stressed.

Rounding out Part I is a historical and contemporary view of cougar management in the United States and Canada. In Chapter 4, the authors review the phases of management from the bounty era, extermination of eastern cougar populations, and intensified hunting and trapping during the early part of the twentieth century to an era of more sustained sport hunting and maintaining viable cougar populations. They provide a broad overview of the history of management, the development of sport hunting programs in terms of sex-age restrictions and apportionment of permits, and the impact of cougar hunting on livestock and other hunted species. Of potential threats to the species, sport hunting is the most visible and easily fixed by simply banning it, but it may be the least important in the long term when compared to the more serious threat of habitat loss. The chapter concludes with a description from Canadian researchers of how northern cougars are managed and conserved differently than in the United States.

Populations are the focus of Part II. Cougar population dynamics are addressed in Chapter 5. Discussion covers population characteristics, including reproduction, mortality, survival, longevity, rates of population increase, and population estimation and survey methods. Particular attention is paid to diagnosing the degree to which cougar populations are regulated by food, habitat, and social organization. Perspectives are presented on the current status of cougar populations in western North America, their potential responses to habitat loss, and their expansion into former range, particularly in the Midwest region of the United States.

Chapters 6 and 7 explore our knowledge of cougars (pumas) in Central and South America. Because formal studies and records have been limited in Latin America until recent years, available information is general for some countries and more detailed for others. While there is still much we do not know about the puma, we are conscious of the importance of these southern populations: recent work places the center of genetic diversity for the species firmly in the south, where habitats and predation patterns vary even more radically than do those of the north.

Chapter 6 provides an overview of the taxonomic status of pumas in Latin America followed by an analysis of their current status and abundance. Attention is given to our knowledge of the ecology and behavior of pumas in this region, their role as predators, and their interaction with other species. The authors present what is known about conservation efforts for this species and what the future holds for pumas in Latin America. Following up, Chapter 7 treats the world's southernmost cougars in Patagonia and the southern Andes. The current situation of cougars in this region is discussed— their role in ecosystems where they are the only top predator, relying mostly on introduced prey, and the research, management, and conservation issues in the region.

Part III is about cougars and their prey. Chapter 8 focuses on behavior and social organization. In particular, the authors discuss how and why cougars interact with one another in a population and with their environment. In addition to exploring the role behavior plays in population dynamics, the authors consider behavior and social organization in the context of cougar management and conservation. Chapters 9, 10, and 11 treat food habits, cougar-prey relationships, and cougar-carnivore interactions. The authors examine how human factors, climate variations, and interaction with other carnivores such as wolves, bears, and coyotes affect cougar-prey relations. The latest theories on cougar predation and prey selection and recent advances in techniques to determine the rate of predation are included.

The contributions in Part IV address cougar conservation, governance, and the conflicts arising when a large predator coexists with people. Chapter 12 explores conservation planning to accommodate how cougars select and use habitats, which depends upon vegetation, topography, the presence of human infrastructure, and other factors. It summarizes various selection patterns and examines isolated cougar populations, such as those in south Florida, on Vancouver Island, British Columbia, in Washington's Olympic Mountains, and in the Black Hills of South Dakota. The status of these populations is described with special reference to gene flow, potential for extinction, and the beginning of recolonization.

Chapter 13 looks at cougar-human interactions, summarizing information on documented cougar attacks and threatening encounters that have occurred in North America. It also takes a closer look at cougar behavioral research conducted in two different environments: a remote New Mexico mountain range where cougars have minimal contact with humans, and a state park in California with annual visitation of 500,000 people. Using this information as well as findings from other research, the authors present suggestions for reducing the probability of negative cougar-human encounters and identify areas for future research.

People respond to cougars and the issues of cougar management and conservation in a variety of ways. In Chapter 14, the authors review the various schematics that have been developed for describing how people orient to and value nature and wildlife, and ways in which these worldviews frame how we think of cougars. How do we organize ourselves in response to one another as participants in the discussion? How do the venues of ballot initiatives, litigation, media, informal relations with managers, and shared culture affect cougar management and policy development? In addition, this chapter explores models of management and governance, especially those employed by agency managers, and looks at how those various models play out to affect social and other policy dynamics.

Chapter 15 explores the roles of citizens and government in cougar conservation. The authors ask the question: "What can we do today to ensure that one hundred years from now cougars are still living on the landscape?" They discuss both past and current citizen involvement in conserving cougars, and provide examples of progressive government agency programs and collaborative projects that have promising models for addressing the most egregious threats to cougars. These authors stress the importance of education and provide six essential principles to guide agencies and citizens' groups in conserving cougars and their habitats both now and in the future.

The book concludes with observations accumulated over the senior editor's career of more than forty-five years. Opinions are offered on our knowledge of cougar biology and ecology, on hunting, and on cougar-human interactions.

Research is suggested to answer questions of both practical and theoretical importance. The point is made that cultural considerations have been—and can still be—powerful tools in the conservation of wildlife.

The cougar belongs to that special group of big and potentially dangerous cats that humankind has feared and revered since the beginning of recorded history. The tiger has been prominent in the culture and art of the great Asian civilizations. The lion has been used as a symbol of royalty throughout the ages. The jaguar and cougar figured prominently in pre-Columbian civilizations in Mexico and Central America.

At the same time, these big cat populations have declined drastically because of human activities—direct killing and habitat destruction. In recent years, however, worldwide awareness of the plight of these magnificent species has developed. Reserves have been created, protective measures have been put in place, and some populations have responded. Cougars are now at the forefront of conservation efforts in some regions. In many situations conservationists face a dilemma—human welfare and safety must be stressed, while at the same time, cougar populations in nonconflict wildland areas should be maintained or even enhanced. But we believe real progress is being made.

These matters are addressed in detail by several authors in this book. We hope readers gain a better understanding, not only of the cougar itself, but also of the issues important to its perpetuity. And we hope more and more people choose to become engaged in efforts to conserve what Wallace Stegner (1984) called "this species of such evolved beauty and precise function" to keep it with us for always.

Acknowledgments

S O MANY PEOPLE have contributed to this project. First and foremost, we gratefully acknowledge all the authors for participating—their devotion to science has been inspiring, and their patience during the many phases of this project was always appreciated. Some authors were especially gracious with their time, anonymously reviewing other chapters, and some also donated photographs. Their individual acknowledgments follow ours.

We would like to thank the following individuals who were instrumental in making this book possible: Gene Thaw and Sherry Thompson of the Thaw Charitable Trust, Harry Bettis of the Laura Moore Cunningham Foundation, Mike Watson and Prosser Mellon of the Richard King Mellon Foundation, and John Robinson of the Wildlife Conservation Society. We would also like to thank Harley Shaw and Dave Brown for their early review of the manuscript. Their comments were extremely helpful, as was Linda Sweanor's help with the photographs and Lori Arakaki's sharp attention to detail in the reference section. We benefited greatly from Kristi Negri's valuable advice on the publishing process and skillful editing during the early stages. Special acknowledgment goes to Sally Antrobus, whose deft editing skills and advice significantly improved the book.

Harley Shaw thanks Kevin Hansen for his Chapter 2 review; Jenny Lisignoli, who conducted the search for the Hibben photos; and Linda Sweanor, who edited and contributed to the table on cougar characteristics.

Melanie Culver acknowledges John Norton for skillfully drawing the maps and other figures in Chapter 3, and Sharon Negri for time spent editing the many details in it. She thanks Lisa Haynes, Ron Thompson, Brad McRae, Harley Shaw, Linda Sweanor, Ken Logan, John Laundré, Carlos Driscoll, Melody Roelke, and Warren Johnson for stimulating discussions about cougars over the past decade. Melanie especially thanks Steve O'Brien for his mentorship during many of the years she spent on cougar research.

Financial support for the work of Susan Walker and Andrés Novaro was provided by the Argentine Research Council (CONICET) and Science Agency (IM40 program, SETCIP) and the Wildlife Conservation Society (WCS). For sharing their years of knowledge about carnivore management in Patagonia, Susan and Andrés wish to thank M. Funes of WCS Argentina, N. Soto of the Agricultural Agency in southern Chile, and M. Failla, S. Rivera, and B. Alegre of the wildlife agencies of Río Negro, Chubut, and Santa Cruz provinces in southern Argentina, respectively. Others to whom the authors are indebted for sharing information about research on pumas in the region are A. Travaini, M. Pessino, S. Montanelli, and R. Baldi. Discussions with M. Anz of Los Remolinos ranch and Maurice Hornocker of WCS enlightened their thinking about Patagonian pumas. They thank Maurice Hornocker and Sharon Negri for inviting them to participate in this book.

Linda Sweanor's cooperators on her New Mexico cougar research included the Hornocker Wildlife Institute, New Mexico Department of Game and Fish, the United States Army at White Sands Missile Range, and the San Andres National Wildlife Refuge. The California research was supported by California State Parks, California Department of Fish and Game, and private donations to the Wildlife Health Center at the University of California at Davis. Those directly involved in the capture and tracking of cougars in these studies included J. F. Smith, B. R. Spreadbury, Toni K. Ruth, J. L. Cashman, J. R. Augustine, J. W. Bauer, and

B. Millsap; the work was successful because of their efforts. The research team acknowledges Walter Boyce for providing the opportunity to study cougars in California and better understand the conservation challenges that lie ahead. Ken Logan and Linda Sweanor thank David Mattson for providing information and insight on managing for human safety in cougar range. They offer special thanks to their mentor, Maurice Hornocker, for initiating the New Mexico research, for giving the researchers the experience of a lifetime, and for his support and encouragement over the years.

Regarding their chapter on social organization, Ken Logan and Linda Sweanor want to express their gratitude to the professional leaders who helped them in their development—especially Maurice Hornocker, John and Frank Craighead, David Mech, and George Schaller—whose own accomplishments inspired and nurtured Ken and Linda's curiosity about the ecology of carnivores in their wild environments. They applaud all the ecologists with whom they worked while directly investigating wild cougars, pointing out that these individuals' dedicated work and the ideas that spring forth from studying the animals in wild habitats are what move cougar science and management beyond the conventional.

Kerry Murphy and Toni K. Ruth thank Ken Logan, who graciously reviewed Chapter 9 and provided helpful comments. They greatly appreciate reviews of Chapter 10 by John Laundré, Ken Logan, and Harley Shaw, which tremendously improved the text. Dennis Murray and Jason Husseman provided helpful comments on the Chapter 10 figures. They thank Mark Hurley and Chuck Schwartz for discussions about recent work and the concepts of limitation and regulation. Ken Logan and Harley Shaw also reviewed Chapter 11. For help in obtaining literature during their writing from remote locations, the authors' heartfelt thanks go to the staff at Interlibrary Loan, Teton County Library, Wyoming.

Paul Beier thanks Mark Dowling, Gary Koehler, and Pete Singleton for helpful responses to his queries, and Rick Hopkins and Ken Logan for reviewing a draft of Chapter 12.

David Mattson and Susan Clark note that several people and organizations were instrumental in making Chapter 14 possible and improving its quality. They owe a great debt of gratitude to Harold Lasswell and his successors in the policy sciences for the intellectual underpinnings of their work. Martin Nie and Steve Kellert provided helpful and thought-provoking reviews at a busy time in their schedules. The USGS Southwest Biological Science Center (SBSC) and Yale School of Forestry and Environmental Studies provided monetary support in the form of the authors' salaries. Mark Sogge of SBSC provided important encouragement and the benefits of a broad vision. For the many lessons and insights offered, they thank all the people with whom they have interacted over the years in the arena of cougar management.

Sharon Negri and Howard Quigley thank all who contributed in various ways to Chapter 15: Paul Beier, Chris Belden, Deanna Dawn, Marianne Dugan, Toni Frohoff, Marilyn Gilbert, Kevin Hansen, Rick Hopkins, Kyle Knopff, Kathy Lang, Martha Lonner, Chris Papouchis, Nina Rogozen, D. J. Schubert, Harley Shaw, Mindy Toomay, Corinna Wainwright, and Neva Welton. They greatly appreciate the backing of Melanie Lambert of the Summerlee Foundation, who has generously provided support for numerous research and conservation efforts throughout the cougar's range.

Maurice Hornocker thanks his many students and colleagues, too numerous to list, who have contributed to his thinking over the years. He is grateful to Sally Antrobus, Fred Lindzey, Toni Ruth, Harley Shaw, and Tony Wright for their reviews, comments, and helpful suggestions.

Finally, we all owe special thanks to the University of Chicago Press. In particular, our editor, Christie Henry, showed incredible patience and kindly helped us navigate the publishing process. Few other projects call for such wide-ranging collaborative effort. The book includes multiple standpoints, brings together work conducted over several decades, and covers an enormous geographic range. We appreciate Christie's support and enthusiasm throughout its gestation period.

Part I

Research and Management
Come of Age

The Bitterroot Tom

Maurice Hornocker

THE CALL came late on a cold Montana December evening. Two houndsmen—Floyd Partney and Bill Zeiler—had treed a male cougar some thirty-five miles southeast of Missoula in the Bitterroot River drainage and were calling for me to come and tranquilize him. I quickly assembled my makeshift equipment—tranquilizing gear, climbing spurs, ropes—and drove off into snowy darkness.

With John Craighead's encouragement and help in securing funding, I had begun an exploratory study of cougars in western Montana in the winter of 1962. My aim was to learn if cougars could successfully be captured, tranquilized, marked, and released for future study. This had never been done before and all capturing and handling techniques had to be worked out. I had gotten the word out to cougar hunters that I would pay $50 for each cougar treed and held there until I arrived. The Partney-Zeiler team had captured the first cougar—a mature female—a week earlier. This was the second opportunity in my fledgling study.

"We've got a big tom for you!" Bill exclaimed as I arrived at their hunting truck parked on the snowy logging road. "Leave your snowshoes. It's too steep to use them," he directed. Shouldering our packs, we started the climb to the treed cougar, where Floyd waited with the hounds on a precipitous snow-covered slope. It was well over an hour of hard climbing in the darkness before we sighted the fire Floyd had built for light and warmth near the base of a huge ponderosa pine.

Our flashlight beams revealed a sight I'll never forget. Lying calmly, almost peacefully, on a horizontal limb of the old pine was the most beautiful cat I'd ever seen. The prime male returned our stares with his own expressionless, noncommittal stare. Quickly, I assembled a tranquilizing dart and fired it into his exposed hip. At the impact he came alive, literally bolting upward in the towering pine. At about seventy feet the pine forked; it was here the big male stopped climbing and draped himself comfortably in the fork. As he showed signs of becoming tranquilized, I started my ascent. The aim was to reach him and lower him to the ground on a rope. But midway up the massive and partially dead tree, I had second thoughts. *This is crazy* crossed my mind—*it's dark, the wind is howling, you are a novice tree climber, and a partially (if at all) tranquilized mountain lion awaits you seventy feet above the ground.*

Good sense prevailed and I retreated down the swaying old pine. "We'll wait until daylight," I announced to the others. So we dragged in logs and limbs, stoked the fire, and huddled in the smoke, half dozing the night away.

At first light, I checked the cougar. He had obviously recovered from the tranquilization and had come down to the original limb, where he now balefully eyed us once again. I quickly snapped a few photographs and then fired another dart into his hip. To my dismay, he once again climbed upward to the high fork. This time, in daylight, I was a bit more confident as I cautiously climbed, setting the spurs deeply into the pine's thick bark.

"He looks drugged!" Floyd yelled from the ground. I readied my lowering rope as I climbed the last few feet to the big cat above me. His head was away from me, presenting his haunches with his thick-based tail and huge hind limbs. I reached for one of his hind legs to slip the noose over his foot. As I grasped his leg, the cat—only partially affected by the drug—came alive. Letting loose a deafening growl, he swung his huge head around the tree trunk not three feet from my face. Eyes blazing and fangs bared, he lurched his body, attempting to move from the fork. As he did so, his tail whipped across my chest. Instinctively, I grabbed that thick tail and with all my might pulled the big, now unbalanced, cat from the tree, literally launching him past my face and into space. I watched the big lion sail through the air as if in slow motion and fall into the deep, cushioning snow, where he skidded harmlessly down the steep slope.

Suddenly, I was shaking so hard I had to cling to the trunk with all my strength. Floyd and Bill had reacted quickly and within minutes had the big male secured with restraining ropes. Regaining my composure, I slowly made my way down. As my feet touched terra firma, I thought: *Will it be this way every time? What a way to make a living!*

We brought the big male to our holding facilities at the University of Montana, where I conducted a series of experiments with him over the next month, developing handling techniques. After learning a great deal from him, I returned him to his old territory and released him. We had learned to know each other and, after bolting from his cage and running a hundred yards up the old logging road, he paused and looked back. As he turned and disappeared, I whispered: "Good-bye, old friend. Thanks and good luck."

We captured and processed twelve more cougars that winter, and my objective of developing techniques was achieved. None of the ensuing captures provided the challenge of the Bitterroot male, and procedures, while not routine, became a bit more predictable.

Two years later, the big Bitterroot tom was killed by a hunter not far from where we first captured him. He still wore one of the ear tags I had attached, and his ear tattoo was clearly legible. He weighed 180 pounds. When I learned of his death I felt a real loss—he truly was a research pioneer.

Chapter 1 To Save a Mountain Lion: Evolving Philosophy of Nature and Cougars

R. Bruce Gill

N O ONE KNOWS the specifics of the first encounter between humans and cougars, but it could easily have gone this way: The air was cold and sharp on a bright autumn morning, perhaps in Alberta, likely some 10,000 to 20,000 years ago. A herd of deer grazed unconcerned and unaware that two starkly different predators stalked them from opposite directions. From the south, a large, sleek, strong, stealthy cat crept slowly and deliberately, employing every bit of available cover. From the north, a group of two-legged predators cautiously positioned themselves among boulders adjacent to the travel path of the deer. Compared to the cat, the two-legged hunters were slow and clumsy; armed with atlatls, stone axes, knives, and extraordinary intelligence and cooperation, however, they were the equal of any prey or predator (Merchant 2002).

The deer wandered to within ten meters of the humans, who patiently waited for the opportune moment to strike. Meanwhile, the cougar had approached to within twenty meters and, with tail twitching, waited like the humans until all the deer were feeding. Then both predators struck simultaneously. The lead human arose quickly and launched a spear at the nearest deer. The cat leaped, covering ten meters in the first bound. Both struck their prey with deadly force, immediately discovering the competition. At that moment, the first human immigrants to America met their most widespread predator (Young and Goldman 1946a). With that first encounter, a philosophical relationship was inaugurated that would vacillate between reverence and warfare.

First Philosophy

The beliefs and attitudes that the first immigrants brought with them to the New World are not known because no historical records exist. But one has only to view the cave-art paintings of their Western European contemporaries in Lascaux, France, for a sense of the force that large animals and awe of the natural environment exerted in their art and religion (Bataille 1955; Hadingham 1979). Petroglyphs in North America provide us with a similar window on early worldviews (Figure 1.1).

Figure 1.1 A dramatic rendering of a cougar is among the petroglyphs in Petrified Forest National Park, Arizona. Discovered in 1934, the petroglyph, dated between 1050 and 1250 AD, is the work of the Anasazi or Ancestral Puebloans. Photo courtesy of T. Scott Williams, Petrified Forest National Park, NPS.

Spiritualism and Connectedness

European immigrants arriving in North America in the sixteenth century encountered stories, legends, myths, and ceremonies of the Native Americans that revealed much about the natural philosophies of the natives' earliest ancestors. Among these early people, all experiences with nature were embedded in a metaphysical dimension. Their theology of nature was both physical and spiritual. They considered themselves as but one part of nature rather than apart from it, as later immigrants would believe (Hughes 1996).

All things, both animate and inanimate, were possessed of a spirit, and all spirits were interconnected. Individual spirits emanated from the Great Spirit, who was the creator of the universe and greater in both power and influence than the sum of the individual spirits (Brown 1997; Kracht 2000). The Lakota Sioux holy man John Lame Deer explained the concept this way:

Nothing is so small and unimportant but it has a spirit given it by Wakan Tanka. Tunkan is what you might call a stone god, but he is also a part of the Great Spirit. The gods are separate beings, but they are all united in Wakan Tanka. It is hard to understand—something like the Holy Trinity. You can't explain it except by going back to the *circles within circles* idea, the spirit splitting itself up into stones, trees, tiny insects even, making them all *wakan* his ever presence. And in turn all these myriad of things which makes up the universe flowing back to their source, united in one Grandfather Spirit. (Erdoes 1976, 102–3)

Spirits could be beneficial or detrimental, depending upon circumstances. Balance within the spirit world, therefore, was critical because spiritual harmony was necessary for the survival of all life. To maintain this spiritual balance, Native Americans developed elaborate ceremonies of prayer and sacrifice to placate the spirits of those others who were necessarily taken to sustain their own. Pre-Columbian Americans also believed that spiritual power was hierarchical, with the sun possessing more power than the earth and the eagle being stronger than the buffalo (Boas 1930).

Philosophical Pillars

The Native American understanding of nature rested on four broad pillars. First, all was sacred and everything in nature was inherently spiritual. Second, all was interrelated and one could not act upon one element of the environment without affecting all others. Third, all elements of creation shared a spiritual kinship with Mother Earth. Native Americans regarded themselves as part of the land, not apart from it. Fourth, people were obliged spiritually and ethically to respect Mother Earth and her inhabitants. They had to act righteously to preserve and maintain the complex physical and spiritual balance of nature (Calicott 1982; Booth and Harvey 1990; Jostad et al. 1996).

A Predator's Place

As a result of their reverence for nature, Indians espoused a conservation philosophy consisting of two simple tenets: take only what you need and use all of what you take. They enforced this through customs and taboos that functioned like modern game laws. The Utes would not kill the gray jay because they believed the bird's raucous call helped hunters detect predators. The Gosiutes of Nevada had a custom of waiting twelve years between antelope drives to allow the herds to replenish. Although members of several tribes hunted predators to inherit their power, no predator was hunted with the intent to exterminate the entire species (Kracht 2000).

Since predators possessed more power than their prey, frequently they were invoked as spiritual guardians to ward off enemies, sickness, and disaster (Hultkrantz 1981; Hughes 1996). Several tribes, particularly those of the southwestern United States, regarded the cougar as an icon of power, protection, and friendship. The cougar was regarded so reverently by some native peoples in southern California that even in modern times some refused to kill the cats even to protect livestock from depredations (Tinsley 1987; Bolgiano 2001).

Although reverence for the cougar was virtually universal among Native Americans, the ceremonial specifics varied. Pueblo people accorded the cougar the top of the hierarchy of the beast gods. Because it was such a superb hunter, it possessed great power and was a patron deity of hunters and warriors (Saunders 1998). Early Pueblos carved two elaborate stone lion statues (Figure 1.2) at the Cochiti Pueblo to invoke the cougar spirit's hunting prowess (Saunders 1998). Among the Zunis, the cougar was considered the master of all prey gods. Prey gods guarded the Zunis from threats to the north, south, east, and west, as well as from the earth and sky. The cougar guarded the north god, a direction of great spiritual significance (Cushing 1883). The Navaho celebrated the cougar way of hunting in their rituals. Cougars were believed to possess powers that greatly augmented success among Navaho hunters.

The Cheyenne tell a traditional story of a woman who had lost her child. While mourning, she wandered into the woods and came across a den of motherless cougar kittens. She nursed the kittens until they could survive on their own. In gratitude, they returned the favor by bringing her a share of their kills. From this legend came the belief that the cougar was both provider and friend (Seger 1905).

Figure 1.2 The Shrine of the Stone Lions, near the ruins of Yapashenye on Portrero de la Vaca in what is now Bandelier National Monument, are among the largest relics of antiquity found in New Mexico. Governor L. Bradford Prince brought them to public attention in *The Stone Lions of Cochiti* (Prince 1903). Photo by David E. Brown.

Figure 1.3 South American Indians used three-strand throwing devices called bolas for hunting the puma (Young and Goldman 1946b).

Despite its strong spiritual context, the relationship between early Americans and cougars was not entirely peaceful; occasionally, it was also lethal (Bolgiano 2001). Cougars killed humans for food and in self-defense. Their method was simple, consistent, and effective: stalk, wait, and ambush. Humans killed cougars for protection and to gain the power that could only be obtained by killing one. The methods people used to capture and kill cougars were diverse, ingenious, and deadly. The Incas, for example, conducted circular drives in which as many as 30,000 individuals formed a large circle several miles in diameter. Gradually, they closed the circle, working steadily inward. All predators thus encircled were killed (Young and Goldman 1946a).

Central American natives waited in ambush at night for cougars that they attracted with instruments made from hollowed bone or branches. The instruments imitated the calls of distressed prey of jaguars—the original "varmint calls" (Tinsley 1987). Some South American Indians used bolas, a three-strand throwing device tipped with weighted balls (Figure 1.3). When successfully cast, it entangled the limbs of a cougar and immobilized the animal. Regardless of how cougars were killed, they were taken ceremoniously with great deference and reverence so that the cougar's spirit would reciprocate by empowering and protecting the bearer of its icons (Saunders 1998). But the world of the cougar changed dramatically when European immigrants began to invade America beginning in the fifteenth century.

Feckless Philosophy

The second wave of Americans brought a new religion based upon homocentric ideas of dominion and private property ownership. The perspective of these immigrants was feckless compared to the reverential attitude of their predecessors. The new arrivals brought domesticated animals and food plants to lessen their dependence upon nature. Both their religious and natural philosophies taught them they had the right to use, alter, or destroy anything in nature that impeded progress. Likewise, anything that satisfied their amusements was fair game (Kline 1997). The entire belief and value structure of the new settlers was decidedly antinature.

Colonization

The colonists saturated their views of nature with intense moral connotations that assigned natural objects to the absolute ethical categories of good or bad (Kleese 2002). Wilderness was considered bad for several reasons. It was feared because it was populated by strange, wild predators that preyed upon domestic livestock and, occasionally, humans. Wilderness was inconvenient because it obstructed travel and the expansion of agriculture; forests quickly reinvaded clearings unless vigorously and regularly removed. Wild places were considered untidy because they were not well kept like the manicured landscapes of Europe. So the colonists attacked wilderness with both vengeance and evangelism (Davis 1996; Kleese 2002; D. Foster et al. 2004). Their perspective toward nature was born not only from greed, arrogance, and ignorance but also from the considerable grief and agony the wilderness caused them (Taylor 1995). Conquering the wilderness, including the cougar, was as much about security as it was about dominance.

First to fall were the old-growth forests of the East. Prior to sixteenth-century colonization, old-growth forests occupied as much as 950 million acres of land (Davis 1996). Under the onslaught of saw, axe, and plow, deforestation occurred so rapidly that, by 1800, residents of New York fretted about a fuel wood shortage in the Hudson River valley (Taylor 1995). By the mid-1800s, from 50 to 75 percent of the eastern landscape consisted of open agricultural land, exceeding 90 percent in some locales (Foster et al. 2004).

As forests gave way to farm lots, wildlife dwindled in response partly to vanishing habitat and partly to direct slaughter from shooting, trapping, and poisoning. Of the forest animals, predators, including cougars, were singularly despised (Kleese 2002). Faced with rampant habitat loss and unrestricted hunting, white-tailed deer numbers plummeted. So rapidly did deer populations decline that hunting seasons were closed as early as 1639. Nonetheless, over most of the eastern range of the white-tailed deer, market hunting continued relentlessly and between 1755 and 1773 accounted for the exportation of 600,000 deer hides from Savannah, Georgia, alone (Demarais et al. 2002).

As populations of deer and other prey species began to decline, cougar populations also began to wane. Human persecution, however, was the final nail in the eastern cougar's coffin. Throughout the eastern United States, every settler owned a gun and every predator was a target. Opportunistic killing alone probably would not have doomed the cougar. But when bounties were established, a cadre of professional killers emerged who specialized in predator hunting. Some made their entire living from bounties collected by killing predators. The bounty hunters used diverse methods to kill their quarry. Pit traps, steel traps, guns, and poisons all were employed with varying success (Young and Goldman 1946a).

Encircling drives were adapted from the Amerindians and were employed frequently by communities throughout the East to rid areas of vermin. Approximately two hundred individuals would form a large circle up to thirty miles in diameter and gradually close it inward, using guns, bells, dogs, fires, and other disturbances to drive animals inward. On one particular drive, 41 cougars, 109 wolves, 112 foxes, 114 "mountain cats" (lynx and bobcats), 17 black bears, 12 wolverines, 3 fishers, an otter, and a grizzly bear were dispatched (Danz 1999). By the mid- to late 1800s, the combination of habitat loss, prey depletion, and dogged human persecution had taken its toll. Cougars were exterminated from areas east of the Mississippi River except for small, isolated remnant populations in Florida and, perhaps, Louisiana (Cardoza and Langlois 2002).

Settlement

As civilization expanded westward, so did the carnage. The plight of cougars and other carnivores was dictated by a sequence of events. First, as the European immigrants expanded westward, they clashed increasingly and violently with the American Indians. As these clashes increased in frequency and violence, white settlers urged the U.S. Army to confine Native Americans to reservations. After Colonel George A. Custer's death in 1876 at the Little Bighorn River in Montana, demands to confine Native Americans to reservations rose to a clamorous din (M. Wilson 2002). A strategy was developed to hasten their confinement. One part of the strategy called for the elimination of the bison to make the native tribes dependent upon beef provided by the Bureau of Indian Affairs.

Market hunters, already bountifully engaged in provisioning settlers and railroad workers with meat, were encouraged to slaughter bison without restraint. They killed bison by the tens of thousands, often taking only hides; additional thousands were killed by sport hunters. In less than one hundred years, bison declined from millions to near extinction (Garreston 1938; Haines 1970).

The demise of the bison affected large predators in at least three significant ways. First, it greatly reduced the number of available prey, thus reducing predator numbers and forcing survivors to find alternative prey. Second, it left a vacant niche that was almost immediately filled with rangeland livestock, bringing large predators into direct conflict with human commerce. Third, it increased both predator depredations on domestic livestock, especially by wolves, and the scope and intensity of government predator control programs.

Loss of the bison probably did not greatly affect cougars directly. They were distributed only sparsely throughout the Great Plains, primarily inhabiting the wooded stream bottoms and brushy feeder gullies (Young and Goldman 1946a). It was a nexus of events, including the discovery of gold and other precious minerals in the West, the growth of the western market hunting industry, the completion of the transcontinental railway, and the arrival of rangeland livestock that put western cougars and man on a collision path (Trefethen 1975; Robinson 2005).

At first, most of the pioneering settlement of the West occurred across the Great Plains, but with the discovery of gold in California in 1848 and in Colorado in 1859, hordes of hopeful miners invaded the foothills and mountains. Events that occurred during the settlement of the East were repeated. Forests were felled and game was depleted to meet the needs of the expanding human populations. By the early 1900s, deer, elk, pronghorn, and bighorn sheep were rare across most of their former range (Kie and Czech 2000; Gill 2001a; Robinson 2005; Heffelfinger 2006).

Predator Wars

Wholesale slaughter of the bison and other wild game provided predators, especially wolves and coyotes, with a surfeit of carrion, initially prompting a sharp increase in their numbers (McIntyre 1995; Robinson 2005). Once the carrion supply was exhausted, predators switched to domestic livestock. Because wolves were the most problematic, they were marked for extermination. As in the East, bounties were placed on their heads to encourage hunters to take as many as possible. Although bounties resulted in countless predator deaths, even this system probably would not have eliminated predators completely. It took a combination of bounties, traps, ready access to cheap and effective poisons, and the formation of a government predator control program to doom both the wolf and the grizzly bear (Young 1946a; Danz 1999; Robinson 2005).

From 1840 through 1860, domestic livestock were uncommon in the West (Voight, 1976; Holechek et al. 2004; Hess 1992). Upon completion of the transcontinental railway system in 1879, however, national and European markets became accessible, stimulating rapid growth in western livestock numbers. By 1870, there were an estimated 4.7 million cattle in the seventeen western states, and by 1884 cattle numbers peaked between 35 and 40 million head. The western sheep industry meanwhile also was expanding. Numbers of range sheep grew briskly between 1880 and 1890 and peaked around 1910 (Holechek et al. 2004).

The westward movement of settlers was gradual during the early years of the nineteenth century. Much of the West remained sparsely settled until the discovery of gold at Sutter's Mill near Sacramento, California, in January 1848. That discovery spawned a mass westward migration along the Oregon-California Trail. Although the original lure for settlers was the offer of free land in Oregon, following the discovery of gold in California more than 200,000 people went to the gold fields (Dary 2004).

Then, in 1859, gold was discovered at Cripple Creek in south-central Colorado, bringing an additional 60,000 to 100,000 immigrants. Many travelers stopped along the way and established farmsteads and ranches. As miners surged into mountain valleys, they demanded fresh meat. Initially, buffalo supplied by hordes of market hunters filled the demand. However, by the mid-1860s too few bison were left to meet the demand. Market hunters began to turn to other sources of wild game. Roads proliferated into the mountains to access mining communities. Game was rapidly depleted from the plains and market hunters concentrated on mountain populations of deer, elk, and bighorn sheep. Market hunting intensified between 1880 and 1900 and caused numbers of big game animals to plummet throughout the West.

As livestock replaced bison all across western ranges, ranchers formed grazing associations to protect their grazing monopolies and enhance their political influence. The first of these associations, the Colorado Stock Growers, was formed in 1867. Soon other states followed the Colorado example (Voight 1976). As their political power grew, the livestock associations successfully lobbied territorial and state governments to take over predator bounty programs. By 1914, state and territorial governments were spending more than $1 million per year to fund bounty programs (Young 1946a; Robinson 2005). Bounty programs were expanded to include grizzly bears, coyotes, and cougars shortly thereafter (Young 1946a; Young and Goldman 1946a).

In 1907, the federal government entered the picture. The newly fledged U.S. Forest Service agreed to control wolves on national forest lands in exchange for an agreement from the livestock industry to accept grazing fees. Federal predator control authority was transferred from the Forest Service to the Bureau of Biological Survey in 1914. The following year, Congress allocated the Bureau $125,000 to assist in organizing predator control operations on forests and other public lands (Cameron 1929; Young 1946a). The Bureau began assembling a staff of professional hunters and trappers who were expected to devote themselves full time to prosecuting the war on predators. The combined effects of bounties, poisons, and government agents drove both the grizzly and the wolf to near extinction (McIntyre 1995; Robinson 2005).

Cougars fared better than wolves and grizzlies for at least three reasons. First, they were not easily attracted to poisoned baits. Second, they were much harder to trap because they seldom scavenged. Third, they were mostly solitary and widely dispersed, making their control uneconomical. However, as the Bureau increased its staff of predator control agents, eventually they recruited professional hunters who specialized in cougar hunting.

Scattered throughout the West during the late 1800s and early 1900s were colorful characters who devoted their lives to hunting cougars (Figure 1.4). They were expert at training and using dogs to pursue the elusive cats. A handful of legendary cougar hunters were responsible for the deaths of nearly a thousand cougars each during their hunting careers. More important, they demonstrated how effectively a single hunter with dogs and determination could catch and kill cougars. They set the example for the government cougar hunters who would soon follow (Danz 1999).

By the early 1900s, the efforts of the professional hunters and a cadre of recreational hunters had significantly diminished cougar numbers, reducing their range to nearly half of the original North American distribution (Logan and Sweanor 2000; See Range Map, p. vii). Nonetheless, by the 1930s the Bureau of Biological Survey (since 1940 the U.S. Biological Survey) employed more than two hundred professional cougar hunters to protect livestock from depredations (Tinsley 1987).

Figure 1.4 Jay Bruce was a notable bounty hunter in California. Born in 1881, he made his first lion hunt in 1915 and killed more than seven hundred lions over the next thirty years. He was publicly credited with promoting California's deer population and helping protect livestock in the mountains. Photo courtesy of California Department of Fish and Game.

Forked Philosophy

At the dawn of the twentieth century, large predators in America were gripped in a struggle to survive (Dunlap 1988). Forests had been laid bare, rangelands were overgrazed, soils were gullied, and wild game was rare (Merchant 2002). Early on, the voices of Thoreau, Emerson, Catlin, and a few others protested the profligate waste. Although their protests did not curb the excessive destruction, they slowly stirred the collective public conscience and sowed the seeds for the nascent environmental movement (Kline 1997).

Other voices, primarily a newly emerging group of sport hunters, began to protest the unalleviated destruction of wildlife. They organized into sportsmen's clubs to promote ethical hunting of desirable game species. As species after species dwindled under relentless pressure from market hunters, sportsmen's clubs began to buy key hunting areas and wildlife habitats. Later, they successfully lobbied state legislatures to pass restrictive laws establishing shortened hunting seasons to protect females with young and limit the numbers of animals that could be taken. Numbers of both game animals and predators slowly began to rise (Trefethen 1975).

The Preservation Movement

By the early 1900s, the movement to save remnants of wildlands and wildlife had coalesced around two competing philosophies—*preservation* and *conservation*. Preservationists sought to preserve nature *from* people, while conservation sought to preserve nature *for* people. The primary value of nature preservation, according to preservationists, resided in amenities such as spiritual inspiration, scenic beauty, and psychological renewal. Conservationists proposed that nature's primary benefits were found in useful commodities such as timber, water, livestock forage, and minerals.

John Muir was the foremost leader of the preservationist movement. The son of a fundamentalist Scottish minister, Muir became disenchanted with the Christian concept of "soul-less" nature. The nature that inspired him was pervaded with spirit and soul. He often referred to wilderness as a "cathedral" and a "temple." Muir was both throwback and harbinger. His belief in the spirituality of nature evoked the perspectives of the Native Americans who preceded him, yet he also believed that inorganic and organic nature were functionally interconnected, anticipating the emergence of the scientific discipline of ecology (Fox 1981; Merchant 2002).

From the onset, Muir struggled with the contradiction that continues to plague the preservation movement even today. Although preservationists sought to protect nature from humans, they nevertheless needed public support to enact legislation creating nature preserves. Protecting nature from people alienated and abated public support.

Preservationists unexpectedly found themselves allied with a strange but powerful bedfellow. Near the turn of the century, the railroads were beginning to realize the economic potential of an emerging tourist industry. Railroad owners favored the preservation of sites with exceptional scenic beauty because they attracted tourists. By providing tourists with transportation, lodging, and food, the railroaders stood to make a handsome profit. Indeed, the number of visitors to newly created western parklands swelled from 69,000 in 1908 to 335,000 in 1915 (Kline 1997).

Three enduring legacies of the preservation movement are national parks, wilderness areas, and urban, county, and state parks and open space. Initially, the establishment of the national park system, beginning with Yellowstone National Park in 1872, had mixed results for cougars and other predators. "Good" wildlife, living in natural surrounds, was a powerful tourist attraction. In order to hasten the recovery of populations of bison, elk, deer, and pronghorn, professional hunters were enlisted to annihilate "bad" wildlife, such as the cougar and other large predators.

Two events combined to change both public attitudes and policies toward predators in national parks. First, freed from significant predation, big game populations exploded. During the winter of 1908–9, thousands of elk starved to death in Yellowstone and Jackson Hole, Wyoming, as heavy snows obscured already impoverished forage supplies. The saga was repeated in the winters of 1916–17 and 1919–20. In response, scientific and public pressure began to mount for a change in policy that protected predators along with other wildlife.

The first protests originated among members of the scientific community. In 1925, Dr. Charles C. Adams of the New York State Museum published an article in the *Journal of Mammalogy* in which he declared, "Without question our National Parks should be one of our main sanctuaries for predacious mammals, and these parks should be of sufficient size to insure the safety and perpetuity of such mammals" (1925, 90). Finally, in 1936, the National Park Service relented to public pressure and implemented policies that abandoned predator control except in circumstances where it was necessary to protect public safety (Kline 1997; Gottlieb 2004; Robinson 2005).

As tourist visits to western national parks increased, the public image of predators slowly began to metamorphose. In Yellowstone National Park, for example, hotels began to proliferate at the park boundary. In their early days, it was customary for hotel managers to discard refuse and garbage in areas adjacent to the hotels. Bears were attracted to the garbage and, in turn, attracted tourists. Eventually, it became aesthetically and hygienically necessary to centralize the garbage dumps. Bear visits became localized and predictable to the extent that the Park Service constructed bleachers adjacent to the garbage dump to accommodate viewers, and the rangers presented scheduled lectures on bear behavior and biology (Schullery 1992; Gill 2002). As the public image of bears and other predators in national parks began to improve, they became tourist attractions in their own right.

The preservation movement directly benefited cougars in at least three important aspects: (1) national and state parks provided places of protection from hunting, (2) as the touring public began to develop knowledge and positive experiences with cougars and other predators, the public image of predators began to improve, and (3) parks and wilderness areas supported abundant and diverse prey that were important to maintain entire populations of cougars.

The Conservation Movement

Conservationists espoused a competing philosophy and employed a different tactic. To conservationists, natural resources were commodities, some of which were essential to the nation's continued economic and political progress (Kline 1997).

Presidents Benjamin Harrison and Grover Cleveland inaugurated the conservation movement following the 1891 enactment of legislation empowering the president to designate public forest preserves. Harrison and Cleveland set aside a combined 35 million acres of public forests and declared them off limits to commercial exploitation. But neither articulated an overarching conservation philosophy intended to perpetuate the reserves. That task fell to President Theodore Roosevelt and the first chief of the U.S. Forest Service, Gifford Pinchot.

During Theodore Roosevelt's two terms as president, he expanded the forest reserves to 172 million acres. In addition, he created fifty-one national wildlife refuges and eighteen national monuments, some of which would ultimately become new national parks. Shortly after becoming chief of the Forest Service, Pinchot declared, "The object of our forest policy is not to preserve the forests because they are beautiful . . . or because they are refuges for the wild creatures of wilderness. The forests are to be used by man. Every other consideration comes secondary" (Kline 1997, 58). Conservation, according to Pinchot, was the limited (wise) use of renewable natural resources to assure an unending supply for current and future generations of people (Kline 1997; Merchant 2002).

Pinchot and his supporters campaigned actively to ensconce conservation philosophy as the fundamental policy for all public land management. By the end of Theodore Roosevelt's administration, conservation was firmly established as the paramount natural resource management paradigm, thereby sowing the seeds of the enduring conflict between the competing standpoints of conservation and preservation.

The period 1935–60 has been called the Golden Age of Conservation. President Franklin D. Roosevelt employed conservation as a cornerstone in his campaign to help America emerge from the Great Depression. Roosevelt created a plethora of new conservation agencies, including the Bureau of Land Management (BLM), Soil Conservation Service, Tennessee Valley Authority, and Civilian Conservation Corps, to put the unemployed to work on forest, range, and water restoration and reclamation projects (Gottlieb 2004).

Over time, the relationships between private resource user, public resource agency, and legislative resource overseer developed into iron triangles of mutual interests, so called because once set, they endured with the rigidity of iron. That rigidity often effectively excluded dissenting public voices from the policy-making process (Mosher 1982; Gill 1996a; Gottlieb 2004; for details, see Chapter 14).

No more rigid and impenetrable iron triangle was forged than that between the livestock industry, Congress, and Bureau of Biological Survey. During the 1920s, professional mammalogists, concerned with the Bureau's poisoning campaign to exterminate large predators, began to object for philosophical and economic reasons (Adams 1925; Dice 1925). These professionals persuaded the American Society of Mammalogists to appoint a committee to evaluate the Bureau's predator control program. The Bureau claimed, with little evidence, that each mountain lion cost ranchers $1,000 annually in damages. The mammalogists countered, also with little evidence, that predators provided economic benefits to ranchers by controlling wild animals that competed with livestock for forage.

In 1928, the committee published its results in the *Journal of Mammalogy*. In fact, it issued two reports because committee members could not agree on many issues. The first report, signed by all committee members, was ambivalent and simply urged that predators should be preserved because they had scientific, economic, and educational values, and should be protected in national parks and "isolated parts" of the public domain where conflicts with livestock were likely to be minimal (Bailey et al. 1928).

The second report was signed by the committee's three university scientists and accused the Bureau and its livestock constituency of conducting all-out war on predators in an attempt to exterminate entire species, a war they claimed was unjustified. The report pointed out that the Bureau and the ranchers had strong incentives to inflate losses from predators. It concluded by recommending the abrogation of extermination and replacing it with a policy that focused on the control of individual predators causing problems in specific locales (Adams et al. 1928).

The Bureau and the livestock industry mounted a public relations campaign to refute the recommendations of the second report. First, the Bureau sought (unsuccessfully) to eliminate the word "extermination" from both its rhetoric and its publications. Next, it persuaded the Congressional Agricultural Committees to hold hearings to garner support for legislation to increase predator control appropriations and strengthen the authority of the secretary of agriculture to control livestock predators unilaterally on public and private land. Then the Bureau sent Stanley P. Young of its West Coast office on a public relations blitz to discredit the second report and its authors. His efforts deflated the Bureau's critics and promoted passage of the Animal Damage Control Act of 1930, which codified and solidified the enduring iron triangle of the federal animal damage control program (Dunlap 1988).

State wildlife agencies went the federal government one better. Not only did they adopt conservation as the cornerstone of wildlife management policy but they also funded wildlife management almost entirely with hunting and angling license fees and associated excise taxes. This had the effect of establishing "diamond triangle" relationships that were even more impenetrable to concerned outsiders than iron triangle relationships. The formation of iron and diamond triangle relationships cemented conservationism as the dominant environmental management paradigm throughout the ensuing seven decades. It would also set the stage to resurrect the preservation-conservation conflict.

At first, conservation abetted the war on large predators. State wildlife professionals believed that predators suppressed the abundance of desirable game. Thus they actively supported both federal and state initiatives for vigorous control of cougars, wolves, and bears. Gradually, however, public attitudes toward large predators became more favorable, and state wildlife agencies adapted. By the 1960s, the legal status of cougars and bears in most states had changed from unprotected varmints to game animals (Cougar Management Guidelines Working Group 2005).

No one exemplified the transformation of public attitudes toward predators more clearly than Aldo Leopold. The onset of Leopold's professional career found him squarely in the conservationist camp, but as he neared the end of his career, he would migrate into the preservationist camp. Leopold was educated as a professional forester and first employed by the U.S. Forest Service. In 1915, while still a Forest Service employee, he supported government-sponsored predator control programs (Leopold 1991).

By 1923, however, Leopold's thinking about the role of predators and humankind's relationship to nature was undergoing profound transformation. He argued that far from being "dead," the entire earth was a living organism that we humans were *morally* obliged to protect. Near the twilight of his career, Leopold penned his now famous essay "The Land Ethic," precursor and inspiration to the environmental movement. According to the land ethic, "predators are inherent members of the community," and no special interest "has the right to exterminate them for the sake of a benefit, real or fancied to itself" (1949, 211–12). It was time to "quit thinking about decent land-use as solely an economic problem. . . . A thing is right when it tends to preserve the integrity, stability, and beauty of the biotic community. It is wrong when it tends otherwise" (224–25).

Factious Philosophy

Late in the 1920s, as the science of ecology emerged, ecologists began to unravel the fascinating interrelationships of the complex, ever-renewing web of life. It became increasingly clear to ecologists that humans could no longer consider themselves apart from nature. For better or for worse, we were an integral part of natural ecological processes, affecting and affected by their interactions. Most people either ignored or were ignorant of this reality until Rachel Carson published her epic *Silent Spring* (Carson 1962).

Environmental Revolution

Silent Spring did not abruptly change public attitudes as much as catalyze existing attitudes so that they coalesced (Brooks 1972; Quaratiello 2004; Gottlieb 2004). Propelled by the twin engines of industrialization and urbanization, American attitudes toward nature had been transforming for decades from utility to appreciation, from consumption to protection (Kellert and Westervelt 1982; Decker et al. 2001). *Silent Spring* galvanized many people with latent nature-protection attitudes into overt environmental activism.

Silent Spring impacted iron triangle relationships in at least three important ways. It heightened public awareness of problems incurred from profligate application of pesticides, an issue that government, science, and industry had so far either overlooked or ignored. It revealed the undemocratic alliance among government, science, and industry that collectively made critical public-interest decisions without involving those affected by the decisions. And it undermined public confidence in the objectivity, credibility, and authority of science as the unqualified foundation for natural resource policy making (Smith 2001). It also sowed the seeds for a backlash that would take nearly two decades to develop.

Following the publication of *Silent Spring*, new members flocked in droves to established nongovernmental organizations like the Sierra Club, the National Wildlife Federation, and the Wilderness Society, and a handful of newly emerging environmental organizations such as the Environmental Defense Fund and Fund for Animals (Decker et al. 2001). Increased membership brought increased influence, and collectively these organizations launched a populist political campaign that produced an astonishing array of environmental legislation.

Of particular importance was the National Environmental Policy Act (NEPA) of 1969. NEPA required government agencies and private sector businesses who were proposing projects that were to be conducted on federal lands or funded with federal money to evaluate their potential environmental effects before the projects could proceed. It also provided for open public review of and comment on each impact statement. The Environmental Protection Agency (EPA) was created to implement the Act, bypassing established federal agencies that might try to delay or weaken the implementation the provisions of NEPA.

In 1973, Congress enacted the third iteration of the Endangered Species Act (ESA), which set up criteria for identifying species that were threatened or in danger of extinction. It included provisions for species recovery and the protection of critical habitats (Dunlap 1988; Kline 1997; Gottlieb 2004).

Seldom were cougars the object of federal protectionist legislation. Clean water and air, wilderness preservation, and endangered species were the legislative objects. With the exception of the ESA, all of the other provisions aided cougars primarily by protecting, enhancing, and expanding key habitats and by providing average citizens entrée, however tenuous, into the natural resource policy-making processes. For the Florida panther, the ESA was and is the gossamer tether preventing its obliteration. The Florida panther was listed as endangered in 1973, and today, after decades of research and restorative efforts, still barely averts the oblivion that was inevitable without endangered species designation and protection (Maehr 1997a; for more on the Florida panther, see Chapter 12).

Buoyed by early success, environmentalists pushed to change nature philosophy and environmental policy. They neglected to check the barometer of public opinion, particularly in economically stressed areas of the West, where the burst of new legislation triggered first a backlash and then a full-scale counterattack. Although environmentalism tapped a wave of public support for nature protection and restoration, it dominated rather than replaced conservationism. Those who made a living from the wildland products were subdued, but hardly defeated. Ineluctably, the stage was set for a counter-revolution.

Counter-Revolution

The first shot fired in the counter-revolution was rather innocuous. In 1972, President Richard Nixon signed an executive order that banned the use of predator poisons on federal lands (Robinson 2005). Nixon was approaching reelection in a hostile political environment. Mindful of the tremendous popularity of environmentalism that had followed from *Silent Spring*, he sought support among the new wave of environmental voters. The support turned out to be superfluous (he won reelection in a landslide vote), and the ban was short-lived. When Nixon resigned from the presidency in 1974, Vice President Gerald Ford amended Nixon's executive order and relaxed poisoning restrictions by allowing for experimental uses of poisons. Then on January 27, 1982, President Ronald Regan rescinded both executive orders, removing all poisoning restrictions.

The anti-environmental movement turned into a full-blown skirmish in 1979. State Senator Deane Rhodes introduced a bill into the Nevada legislature that called for the federal government to hand over all forty-eight million acres of BLM lands within the state to Nevada and launched the "sagebrush rebellion." Soon, the states of Utah, Idaho, Wyoming, Arizona, and Alaska followed suit with similar land transfer resolutions (Helvarg 1997). The ultimate aim, at least of some sagebrush rebels, was to return these lands to private ownership via state land sales.

Almost as quickly as it ignited, the sagebrush rebellion fizzled. The Reagan administration quickly appointed conservatives to key natural resource posts in the new administration. The sagebrush rebels anticipated staunch support from the conservative appointees. Instead, the very interests that had encouraged the rebellion in the first place undermined them for fear of a political backlash against selling public lands to private interests (Helvarg 1997).

Although the battle was lost, the war was hardly over. In the mid-1980s, anti-environmental interests began to organize a full-scale countermovement designed to roll back or overturn legislation put in place earlier by environmental activism. The movement was called the "wise use movement" and was billed as a grassroots, populist uprising against regulatory excess that stifled public land management (Echeverria and Eby 1995). Particularly galling to wise users was the ESA (Luoma 1992; Maughan and Nilson 1993; Tokar 1995). Wise use groups focused intently on neutering the ESA. For the first time in decades, both the executive branch and the majority party in the Congress were allied behind an antiregulatory agenda.

Wise use advocates regarded the protection of endangered carnivores, particularly grizzly bears and wolves, as a potent symbol of federal regulatory excess on public lands. It has been said that the controversy over reintroducing wolves into Yellowstone National Park had less to do with wolves themselves than with what wolf reintroduction symbolized. Wolf reintroduction took center stage in natural resource policy debates because of a convergence of three contentious social issues: inequitable access to political power, conflicting interpretations of the extent of private property rights, and contrasting philosophies about the relationship of humans to the natural environment. Polarization between wise use advocates and environmentalists along each of these dimensions created a caustic natural resource policy and management milieu that permeated every discussion of carnivore conservation (Wilson 1997).

Stalemated in their attempts to effect environmental policy change at the federal level, environmentalists shifted the battlefield to the states. As of today, twenty-seven states permit some type of petition process whereby citizens can make law through direct democracy. Environmental laws, particularly laws affecting open space preservation and wildlife management, have become increasingly popular (see Chapters 14, 15).

In 1990, California citizens proposed to voters an initiative that banned cougar hunting throughout the entire state. Soon thereafter cougar advocates in Oregon and Washington followed California's lead by enacting legislation that banned the use of dogs to hunt cougars. No one knows for sure how each of these initiatives has impacted cougar populations (see Chapter 4), but California cougars have likely increased following the elimination of sport hunting. On the other hand, cougar populations in Oregon and Washington may have declined. Wildlife agencies in both Oregon and Washington responded to bans on hunting cougars with dogs by dramatically increasing the number of available cougar hunting licenses. As a result, hunting mortality of cougars increased, especially among females and young cougars (Beausoleil et al. 2003; Whittaker 2005; see Chapter 4).

It is likely that cougars have been aided somewhat by citizen initiatives that banned the use of leghold traps to take wildlife, and they have benefited substantially from initiatives and other legislation that preserved open space (see Appendix 5 for a partial list of initiatives related to cougars). Expanded open space has increased habitat for prey species and protected travel corridors that promote genetic interchange among populations.

The success of citizen-initiated environmental legislation, like the success of federal environmental legislation, provoked an anti-environmental backlash. Several initiatives have surfaced to repeal previous environmental legislation outright, while others aimed to disarm wildlife protection initiatives by granting state wildlife agencies unilateral authority to develop and implement wildlife policy (Minnis 1998; see Chapters 14, 15).

In effect, what we now face is a perpetuating values war in which both sides win battles, but neither side wins the war. Each time administrations change or the power of Congress shifts from one party to the other, environmental policy shifts with it. Neither environmentalists nor wise users will end the values war until both acknowledge these fundamental conflicting values and find ways to resolve them. And that will require more dialogue and less legislation and litigation.

Yet while the politics of environmentalism was stalemated, the perspectives of environmentalism were spreading. Nowhere was this fundamental change in values more evident than in public attitudes toward large carnivores. Wolves and grizzly bears, once considered worthless varmints, have climbed the list of the most favored of American mammals (Kellert 1985; Bright and Manfredo 1996; Kellert and Smith 2000). Along the Front Range of Colorado, despite a period of chronic conflict between cougars and people, nearly 80 percent of the public still expressed positive attitudes toward cougars. Although the public supported the general notion that authorities ought to take steps to control the number of cougars coming into residential areas along the Front Range, lethal control was not an acceptable control method unless the cougar had killed or injured a pet or person (Zinn and Manfredo 1996; Baron 2004).

The California ban on cougar hunting was upheld by citizen initiative in 1990. Asked by the National Rifle Association (NRA) to repeal the ban in 1996, California voters refused, despite conflicts between agriculture and cougars, cougar depredations on pets, and even human death and injury from cougar attacks. Citizen initiatives in Oregon and Washington that banned the use of dogs to hunt cougars affirmed that Californians were not unique in their support for cougars (Minnis 1998; see Chapter 15 and Appendix 5). Ironically, as the century ended, the public held the cougar in higher regard than the politicians who wrangled over its fate (Riley and Decker 2000; Orren 1997).

Future Philosophy

Two major environmental issues loom on the horizon of the twenty-first century: global climate change and unrelenting human population growth. Sometime in 1999, the world's human population reached the six billion mark. Even though the trend in birth rates appears to be declining, population inertia promises to propel that number ever upward until around 2050, when human numbers are expected to stabilize at between nine billion and twelve billion people (M. Wilson 2002; Brown 2006).

Dark Clouds Gathering

The combination of demographic inertia and ecological constraints will force humans through a bottleneck that threatens not only human existence but also the existence of hundreds of other species of plants and animals (M. Wilson 2002). No group of species faces a more insecure future than large carnivorous mammals (Gittleman et al. 2001). As humans invade wildlands, they dissect habitats, degrading their capacity to sustain populations of cougars, wolves, bears, lions, tigers, and other species.

Examples of the fragmenting effects of human expansion can be found in Arizona, California, Colorado, Montana, and New Mexico, with urban populations increasingly expanding into adjacent foothills and canyonlands—ideal cougar habitat (Best 2005). Frequent contact with humans, pets, and livestock increases conflicts between resident humans and resident cougars, during which cougars often lose. As human populations grow, both habitat fragmentation and cougar-human conflicts are likely to increase, amplifying the difficulty of cougar preservation (Torres et al. 1996; Sweanor et al. 2000).

Global climate change complicates the already inscrutable environmental problems posed by human population growth. Coastal cougar populations will face habitat inundations from rising sea levels. Inland populations increasingly will face isolation as migration corridors disappear and quality of extant habitats declines because of climate- and human-caused habitat conversion. Genetic diversity of isolated cougar populations will gradually decline, further complicating their conservation (Kurz and Sampson 1991; Burkett 2001; Maehr et al. 2002; Lovejoy and Hannah 2005; Root and Schneider 2005).

The environmental movement found itself forced to confront the challenges of human population growth and global warming at precisely the wrong time. Much of the old movement was mired in malaise from the grueling political stalemate with the wise use movement. One of the most effective strategies the wise use movement employed against environmentalism was to assert repeatedly that environmentalists were elitists with little concern for ordinary citizens. It was effective because it was partially true. Forced to confront a bewildering array of policy initiatives that sought to undermine environmental protection legislation, most mainstream environmental groups focused their attention on Washington, D.C. In the process, some mainstream environmental organizations strayed from their grassroots values (Shabecoff 2000; Gottlieb 2004).

Sunlight Peeking Through

Recent events suggest that environmentalism is moving to reinvent and reinvigorate itself by returning to those grassroots values of its founding. First, environmentalists have launched an explosion of scientific activity to document and safeguard against the pending effects of climate change and human expansion. Biodiversity inventories have expanded to document worldwide trends of plant and animal species. Genetic material of a growing number of species is being stored in genetic banks to allow for future cloning and restoration of species that become extinct before they can be preserved (Holt et al. 1996; Hold et al. 2004).

Elsewhere, experiments are under way to restore habitats and ecosystems already impaired by human cultural encroachments (van Andel and Aronson 2006). Large carnivores, in the United States and elsewhere, are priorities for restoration because large carnivores require vast expanses of natural habitat to maintain viable populations (Gittleman et al. 2001; Maehr et al. 2001; Fascione et al. 2004; Clark et al. 2005; Taylor 2005). Thus, protecting and restoring large carnivores and their habitats provides a protective "umbrella" for various other species (Duke et al. 2001; see Chapter 12). Foremost among the tasks to ensure cougar longevity is the protection and restoration of migration corridors to facilitate dispersal and gene flow among populations (Beier 1993; Maehr et al. 2002; Dickson et al. 2005; McRae et al. 2005; Anderson 2006; Thorne et al. 2006). Although science is necessary, it is not sufficient for the challenges ahead. Scientific and technological expansion, uncoupled from public values, have been major contributors to the current ecological dilemma (Yankelovich 1991, 1998).

Every collective human action, consciously or unconsciously, proceeds from three sequential questions—can we, should we, and will we? *Can we*, meaning do we have the intellectual capital or knowledge to do what we want? *Should we*, meaning do we have the moral impulse and social capital to do what we want? *Will we*, meaning do we have the political capital or support to do what we want?

As a remedy, environmentalists and academics are promoting policy-making experiments that link science to wellsprings of public values and experience (Fischer 1995, 2000; Shutkin 2000; Nie 2004b; Jacobson and Decker 2006). Environmentalists are insisting that procedural justice and grounding the new environmentalism in grassroots democratic values are necessary for successful environmental policy making (Lawrence et al. 1997; Young 2000; Smith and McDonough 2001; Parkins and Mitchell 2005). In addition, by linking democratic policy making with participatory science, environmentalists are rejuvenating both science and politics (Shindler and Creek 1999; Irwin and Freeman 2002).

Participatory democracy has yet to penetrate deeply into the politics of cougar management (Teel et al. 2002). Thus far state wildlife agencies with regulatory responsibility and state legislatures with policy-making authority have resisted overtures to share political power, but economics and demographics no longer favor politics of exclusion (Clark et al. 1996; McLaughlin et al. 2005; Jacobson and Decker 2006). More and more, though, these questions are being resolved in democratic forums that invite broad public participation, and with democracy come political support and stability (Light 2000; Vining et al. 2000). As the questions *can we*, *should we*, and *will we* continue to reverberate, the bond between the human and natural communities strengthens, and with it comes the deepening realization that they were always one and the same.

Already scientists know enough about cougar ecology to recognize that protecting cougar populations will not suffice to save them over the long haul. Corridor restoration, maintenance of genetic diversity, and habitat conservation and restoration must be part of any long-term management agenda. The science of cougar conservation is well ahead of the politics of cougar conservation, but the day seems to be approaching when both will be on a more equal footing (Clark and Munno 2005; McLaughlin et al. 2005; Logan et al. 2005).

Another major challenge to the renewal of environmentalism is religion. Traditional western religions until recently have been tepid in their support of an environmental ethic. Now that too seems to be changing. There has always been a spiritual dimension to the environmental movement. Thinkers like Emerson, Thoreau, and Muir often used spiritual metaphors to describe the benefits of wildlands and invoked ethical arguments in favor of wilderness preservation (Kline 1997; Gottlieb 2004; Dunlap 2006). Aldo Leopold's famed land ethic is fundamentally spiritual (Leopold 1949). But as science came to dominate the politics of environmental protection, the spiritual fire began to dim.

In response, the deep ecology movement began to emerge in the early 1980s (Devall and Sessions 1985). The movement sought to integrate morality into ecology, stressing that humankind was morally obliged not only to protect the fruits of creation but also to recognize the rights of the created (von Hoogstraten 2001). Currently, a parallel development is occurring among the world's major religious institutions, especially within Judaism and Christianity. The emerging green evangelical movement, in contrast to the dominionistic Christianity of the past, commands believers to protect, preserve, and perpetuate all of God's creation (Sittler 2000; Oelschlaeger 1994; Taylor 2004).

After a long, perilous journey, environmentalism seems to be coming full circle back to its aboriginal roots. The new environmentalism recognizes and values both physical and spiritual connections with nature. It acknowledges an enduring kinship with the earth and its nonhuman inhabitants. And it accepts moral responsibility for the welfare of all. The new environmental movement aims to renew itself with a future philosophy that seeks to combine the democratic values of truth, justice, and community with the spiritual values of stewardship and compassion. Organizations founded to preserve cougars are proliferating. Groups such as the Mountain Lion Foundation and Florida Panther Society acknowledge not only the ecological value of cougar preservation but the moral obligation as well. As these values begin to fuse, perhaps it is not overly optimistic to anticipate the rediscovery of genuine *soul among lions* (Shaw 2000).

The title of this chapter is adapted from Harper Lee's gripping novel, *To Kill a Mockingbird* (Lee 1960). It is a story of the relationships among truth, justice, community, and compassion. If we hope to save the cougar and other large predators from future extinction, surely we will need an abundance of all four. It is both enigmatic and true that if we are wise enough to save the mountain lion (cougar), ultimately, the mountain lion may save us. I can hear Lame Deer, Black Elk, and others chuckling in the background. This is the fundamental wisdom they tried to teach us centuries ago (Neihardt 1961; Erdoes 1976).

Chapter 2 The Emerging Cougar Chronicle

Harley Shaw

THIS CHAPTER IS a history of cougar research, addressing the question of where we have been. Extensive field studies of the cougar over the past fifty years have given us a better understanding of the big cat's natality, mortality, genetics, dispersal, diet, home range, social structure, and behavior (see Young and Goldman 1946a; Barnes 1960; Anderson 1983; Tinsley 1987; Hansen 1992). New research tools such as satellite monitoring (global positioning systems, GPS), camera traps, and DNA analysis have opened new possibilities to investigators. To explore each of these areas in greater detail, please refer to Table 2.1, which also summarizes findings on physical characteristics such as size and color. Later chapters elucidate our current knowledge on topics now becoming technical specialties in their own right and outline additional research needs.

Regarding early knowledge of the cougar, we differentiate somewhat arbitrarily between lore and science. Lore is knowledge derived from accumulated experience and tradition, transmitted verbally or in writing until it has become accepted as fact. It can encompass many subjects, including hunting methodology, regularity of cougar movements, vocalizations, methods of marking territories, frequency of killing, and danger to humans.

Science-based knowledge is derived from systematically recorded observation, using preplanned methods designed to test hypotheses. The majority of current research emphasizes cougar populations and cougar-human interactions rather than behavior or physical characteristics of individuals. Among other subjects, contemporary work may attempt to evaluate factors affecting long-term trends in cougar numbers, effects of cougars on prey densities and of prey densities on cougar numbers, seasonality and size of litters, causes of mortality, effects of habitat modification or loss, and how human infrastructure is affecting lion movements—information needed to guide cougar management. The lines between lore and science are not distinct. Lore has provided numerous untested beliefs that were ultimately used to develop testable research hypotheses. Cougar hunting and trapping lore has been essential in helping biologists capture cougars for research (Hornocker 1970; Logan and Sweanor 2001). And in the long run, scientific research may accomplish little unless it, too, has relevance to the conventional knowledge—the lore—of the stakeholders who affect political decisions regarding the cougar.

The first published fact about the cougar in biological literature was probably acknowledgment of its existence by European scholars shortly after Columbus reached the New World (Young and Goldman 1946a; Bolgiano 1995). The next steps involved description, classification, and assessment of its taxonomic relationship with other known felids. According to Edward Goldman (Young and Goldman 1946a), the cougar entered the scientific literature with Marcgrave's 1648 essay on Brazilian natural history. British naturalist John Ray (1693) described the species in 1647, and Barrere (1741) provided it with its first binomial, *Tigris fulvus*, translating roughly as "red tiger." In 1771, Linnaeus assigned its commonly accepted binomial, *Felis concolor*. This has recently been changed to *Puma concolor*, a name first applied by Jardine in 1834 and revived by modern molecular taxonomy, which defines precisely the genetic position of cougars in the evolutionary array of felids (Wozencraft 1993; Nowell and Jackson 1996). An evaluation of current nomenclature of

Table 2.1 Cougar characteristics.

Trait	Current Knowledge	Comments
Adult pelage	Grizzled gray or dark brown to shades of buff, cinnamon, tawny, cinnamon-rufous, or ferruginous. Most intense along mid-dorsal line from top of head to base of tail. Shoulders and flanks lighter; underparts dull whitish, overlaid with buff across abdomen; sides of muzzle black; chin and throat white; ears black externally with grayish median patches on some. Tail above like back; lighter below; distal 2–3 in black.	Anderson 1983; color can vary with region, individual, and season (Young and Goldman 1946).
Kitten color	Spotted with black on a buffy ground color.	Faint overall spotting and darker stripes inside front legs may carry over into young adults, stripes sometimes persisting to old age.
Adult weight, kg (lb)	Male: 50–105 (110–232) Female: 36–60 (79–132)	Records of heavier animals (e.g., Young and Goldman 1946) include Roosevelt's (1901) 232-lb Colorado cougar, the heaviest documented; a 276-lb eviscerated cougar (Musgrave 1926), based on hearsay; Hibben's 260-lb Utah cat (Robinette et al. 1961), probably an estimate; and a 375-lb animal (Tinsley 1987), beyond credence.
Adult total length, m (ft)	Male: 1.8–2.9 (6.0–9.5) Female: 1.6–2.2 (5.2–7.2)	Anderson 1983; tip of nose to tip of tail, animal lying straight.
Adult tail length, cm (in)	Male: 69–97 (27–38) Female: 63–79 (25–31)	Anderson 1983.
Height at shoulder, cm (in)	Male: 56–79 (22–31) Female: 53–76 (21–30)	Anderson 1983. Not a standard measurement, unclear how taken, depends on how leg is extended, should be taken on live animal; likely biased high. Few cougars stand 2.5 ft at the shoulder.
Paw length, fore to aft, cm (in)	*Front* Male: 8.8–10.4 (3.5–4.1) Female: 7.6–8.8 (3.0–3.5) *Hind* Male: 8.8–10.2 (3.5–4.0) Female: 7.9–8.6 (3.1–3.4)	Anderson 1983. Paw length is a variable measurement, depending upon substrate, posture, and rate of movement.
Paw width, laterally across track, cm (in)	*Front* Male: 8.2–8.8 (3.2–3.5) Female: 7.0–7.6 (2.8–3.0) *Hind* Male: 7.4–7.9 (2.9–3.1) Female: 6.0–7.4 (2.4–2.9)	Anderson 1983. Hind paws are smaller and narrower than front paws. On level ground, hind and front feet register in the same tracks and only hind paw track is visible.
Heel (interdigital) pad length, fore to aft, cm (in)	*Front* Male: 5.4–6.0 (2.1–2.4) Female: 4.2–5.7 (1.7–2.2) *Hind* Male: 4.4–5.7 (1.7–2.2) Female: 3.4–6.3 cm (1.3–2.5)	Anderson 1983. Pad measurements tend to be more consistent than total paw measurements. Substrate produces some variation. Fjelline and Mansfield (1989) recommend measuring the flat bottom of the track.
Heel pad width, laterally across track, cm (in)	*Front* Male: 5.6–6.3 (2.2–2.5) Female: 5.1–5.7 (2.0–2.2) *Hind* Male: 5.4–6.3 (2.1–2.5) Female: 3.8–5.1 (1.5–2.2)	Anderson 1983. Probably the most useful single measurement of tracks; there seems to be less overlap between males and females for this measurement.

Longevity	Maximum 19.5 years in captivity. Normal longevity in wild <12 years.	
Minimum breeding age	Female: 20–25 months Male: probably 36 months or older.	Allen 1983.
Number of kittens at birth	3–4.	
Season of birth	Can be born year-round; peak in late summer, early fall.	Allen 1983.
Kitten survival	Mean litter size for kittens >50-lb: 2.8.	Allen 1983. Kitten mortality undoubtedly varies with prey availability for adult.
Sounds	Throaty growl, low hiss, spit, gurgle, purr, bleat, birdlike whistle, mew, main call (intense form of the mew), and wah-wah. Sounds can be combined to form more complex vocalizations, such as the caterwaul the female emits during estrus.	Peters 1978; Logan and Sweanor 2001. The "scream" reported in popular accounts may be the caterwaul.
Prey	Most common prey in N. America are deer, elk, moose, bighorn and domestic cattle, horses, and sheep also taken. Smaller prey include rabbits, hares, porcupine, skunks, coatimundi, dogs, cats. In S. America, native camelids.	
Rate of killing	Varies with prey size, season, and presence of other carnivores that can usurp kills. For females, it varies depending on litter size and age of kittens. Equivalent of one deer-sized prey animal every 8–10 days seems normal. Females accompanied by large young kill more frequently.	Ackerman 1982, Murphy et al. 1998, Logan and Sweanor 2000.
Home area size	Varies with prey density, habitat type, and terrain. With high prey densities in good habitat, males may use 75–150 sq mi; females 25–50; with prey scarce and scattered, males may range over 700 sq mi or more.	
Dispersal	Young males (2–3 years of age) may disperse widely seeking available open territories. Movements of several hundred miles from area of birth have been recorded. Young females tend to settle nearer to their area of birth and may reside next to their mother's home area.	Logan and Sweanor 2001.
Territorial marks	Parallel "scrape" marks made by hind feet; normally by male but occasionally by female.	
Historic distribution	Virtually throughout North and South America south of the boreal forests.	
Current distribution	In N. America, mostly west of the 100th meridian; recent records suggest reoccupation of North and South Dakota, Nebraska, and western Kansas.	
Habitat	Normally rough terrain with moderately dense low vegetation.	
Killing behavior	Stalks prey, killing with sudden quick dash. Usually bites through nape of neck or throat, severing vertebrae or choking prey.	
Food habits	Opportunistically takes advantage of the most abundant and vulnerable prey. In tropical regions tends to prey on relatively small and diverse prey. Importance of larger ungulates increases with latitude and a corresponding increase in their distribution and abundance. In temperate North America, prey commonly includes ungulates that, as adults, weigh well over 15 kg. Smaller prey, including lagomorphs, rodents, and other carnivores are eaten opportunistically. Also preys on livestock, usually sheep and cattle, in localized areas. Pets (e.g., domestic dogs and cats) are occasionally eaten where residential areas overlap cougar habitat.	See Chapter 9 for citations and percent occurrence of prey items in cougar diets for Canada and northern United States (Table 9.1), southwestern United States (Table 9.2), and Central and South America (Table 9.3).

the species and the rationale for use of the genus *Puma* are given in Chapter 3.

These first citations tell us little about what the early explorers of the Americas thought about the cougar. This heretofore unknown felid was only one of many New World discoveries. Simply acknowledging its existence was a major accomplishment, and their failure to impart more information is understandable, given the animal's cryptic nature; its secretiveness no doubt contributed to unfounded speculations surrounding the species.

After the publication of Darwin's *On the Origin of Species by Means of Natural Selection* (1859), naturalists of the day focused on documenting subtle differences in wild species and assigning new binomials. Through the 1920s, field biologists were largely interested in naming new species and, if that failed, identifying subspecies within existing species. By 1946, Edward Goldman had listed thirty-two subspecies of cougar in the Americas (Young and Goldman 1946a). A recent revision by Melanie Culver and colleagues (2000a) has reduced this to six (see Chapter 3).

Concurrent with the efforts to classify the species came efforts by the public, often aided by governments, to eradicate it (Young and Goldman 1946a). To some extent, science and eradication supplemented each other. Many of the specimens examined by workers at the Smithsonian and other museums were provided by early government hunters and trappers. In addition to providing specimens for classification, the government hunters also began to document the killing habits of the species. In their reports, they discovered and recorded animals that had been killed by cougars and at times identified the stomach contents of cougars they

had killed (Young and Goldman 1946a, 127). These reports may have been biased toward livestock, given the mission of the hunters. Nonetheless, they represented a step toward recording population and biological data. Government and private hunters also developed basic knowledge of the behavior of the species through observation of sign; Ben Lilly's journals and records are an example (see Lilly 1998). While their motive may have been extirpation of cougars, they nonetheless developed skills that were later needed in scientific studies.

One of the first direct studies of cougar natural history was a master's thesis project carried out by Frank C. Hibben at the University of New Mexico (1937). Using a grant provided by the Southwestern Conservation League, Hibben spent a year accompanying professional houndsmen in New Mexico and Arizona (Figure 2.1). He interviewed ranchers and others reputed to be knowledgeable about cougars and investigated thirty-two reports of livestock depredation. While hunting, he documented cougar prey and collected scats for food habits analysis. Hibben's sample of kills and scats was small; his greatest contribution was, perhaps, the approach he developed—gathering field data himself rather than relying upon hunter or trapper reports. Hibben later left wildlife biology and became an accomplished, and at times controversial, archaeologist, but he retained his interest in cougars and cougar hunting throughout his life (Hibben 1948).

For nearly thirty years after Hibben published his work, little field research on cougars occurred, but two significant syntheses of cougar information were written. The first of these, *The Puma—Mysterious American Cat*, by Stanley P. Young

Figure 2.1 Frank C. Hibben relied on professional houndsmen while gathering data on cougars for his 1937 master's thesis. Photographed about 1936 are (left to right) Giles Goswick, his son George Goswick, Dorsey Buckley, Bill Goswick, and Frank Hibben. George Goswick became the principal hunter in 1972 for Arizona Game and Fish Department's Spider Ranch cougar research project. Photo courtesy of the Maxwell Museum of Anthropology, New Mexico.

and Edward A. Goldman (1946a), brought together historical information, tabulations from the files of the U.S. Biological Survey, and assessment of taxonomic literature. Until at least 1970, it remained the bible of cougar management.

Following Young and Goldman's book, the next synthesis was written by Utah attorney and amateur naturalist Claude Barnes. In *The Cougar or Mountain Lion* (1960), Barnes summarized lore that he had gathered for thirty-five years from experienced cougar hunters and naturalists in the western United States. In his introduction, Barnes noted that:

... treeing a cougar with dogs and shooting it is almost a meaningless commonplace when it comes to the study of the animal's life history; seeing it in the wildwoods when it is unpursued is a rare occurrence; hence to get the complete story one must rely on a host of observers, each of whom might be able to supply a new fact from his experience, and thus aid the ensemble. A naturalist seeks only the truth, and in ascertaining it he is sometimes grateful for a knowledge of the laws of evidence (1960, 7).

Barnes's book remains the best summary of cougar lore for North America. It was published a decade before Maurice Hornocker's (1970) groundbreaking field study and represents a transitional stage, combining lore and available scientific knowledge. Unfortunately, Barnes's book was published in a very limited quantity and rapidly went out of print. It is extremely difficult to find today.

The next, and perhaps most detailed, synthesis of cougar science was written by Allen Anderson (1983) of the Colorado Division of Wildlife. It was the first effort to bring together available scientific information on the cougar, including laboratory, zoo, and field studies. Anderson's book remains the best source for basic measurements and physiological data on the cougar (see Table 2.1). It was followed by a more popular work by Tinsley (1987), which repeats much of the lore presented earlier by Young and Goldman and Barnes but also includes a useful discussion of Native American knowledge and beliefs regarding cougars. The most concise synthesis of cougar biology was written by Kevin Hansen in *Cougar: The American Lion* (1992). In addition to these general syntheses, Ken Logan and Linda Sweanor summarized and updated cougar research in *Desert Puma* (2001), which also reports the results of their ten years of fieldwork in New Mexico.

The syntheses mentioned reflect generally the trends in cougar knowledge. Modern population studies of cougar were initiated by Hornocker (1970) in the Idaho Primitive Area now Frank Church—River of No Return Wilderness Area (Figure 2.2). Improving upon Hibben's earlier effort, Hornocker employed a skilled houndsman to tree cougars but, rather than killing the cats, he tranquilized them and fitted them with visible, numbered collars (see Plates 9, 10, 11; Figure 2.3).

Figure 2.2 In 1964, Maurice Hornocker started his groundbreaking research on cougars in the Idaho Primitive Area in central Idaho. The initial work involved intensive capture, marking, and recapture techniques to quantify the cougar population. Photo by Maurice Hornocker.

Figure 2.3 Immobilizing drugs used to handle cougars safely are often administered via a dart fired from a rifle or pistol into the heavy rump or thigh muscles of a treed or snared animal. Other methods of delivery include blow pipe and pole syringe. Photo © Linda L. Sweanor.

Figure 2.4 John Seidensticker was the first researcher to monitor cougar movements by fitting the animals with radio collars. Photo by Mark Lotz.

Figure 2.5 A receiver and antenna are used to pinpoint a cougar's location based on the signal emitted from the cougar's radio collar. Telemetry has provided valuable information on cougar movements, prey use, and social interactions. Photo by Mark McKinstry.

Figure 2.6 Harley Shaw and canine assistants are perched at the rim of the Grand Canyon, 1979. Hounds were indispensable to researchers until the advent of GPS, which allowed much more intensive tracking of cougar movements. Photo by N. L. Dodd.

Data acquisition required continuous trailing of cougars and treeing both marked and unmarked animals. It also involved documenting kills found during the process. For the first time, cougar home ranges were delineated and strides were made in documenting size and composition of a cougar population. In addition to this, killing behavior was related to the availability of prey, and efforts began to quantify the actual effect of cougars on prey populations (see Chapter 9). Hornocker was followed in the same area by John Seidensticker, who first used radiotelemetry on cougars (Figures 2.4, 2.5), thereby relocating marked animals more frequently, assigning kills to known individuals, and monitoring cougar social interactions (Seidensticker et al. 1973).

The methodology Hornocker and Seidensticker developed was quickly adopted by other western states, and by the mid-1970s cougar studies were active in Arizona (Shaw 1977; see Figure 2.6), Nevada (Ashman et al. 1983), New Mexico (Evans 1976), Utah (Hemker et al. 1984), and California (Sitton and Wallen 1976). This was all occurring at a time when the various states and provinces were assuming management responsibilities for cougars and needing increased information. Long-term cougar research was in progress in Alberta, Canada (Ross and Jalkotzy 1992). As a result, a need for a forum for sharing information developed (Bolgiano 1995; Logan and Sweanor 2001). In 1976, Nevada hosted the first Mountain Lion Workshop (Christensen and Fischer 1976), bringing together research and management personnel from thirteen states and two Canadian provinces. While some study data were presented, the workshop was carried out in a relatively informal setting, with state management reports being emphasized.

During the 1970s and into the 1980s, cougar studies developed in Wyoming (Logan et al. 1986), Montana (Murphy 1983), Texas (McBride 1976), and Colorado (Anderson et al. 1992). Again more or less emulating the work of Hornocker and Seidensticker, T. E. Smith and collaborators (1986) made initial efforts to develop noninvasive monitoring in the Guadalupe Mountains of New Mexico. In addition to this, major research and recovery efforts began in Florida aimed at saving the endangered population there. Field studies have now been carried out in virtually every state and Canadian province known to have cougar populations, and eight mountain lion workshops had been convened by the time of this writing. Attendance at these workshops now consistently exceeds 150 participants from as many as twenty-five U.S., four Canadian provinces, Latin America, and Europe and representing agencies, agriculturists, and wildlife-oriented non-governmental organizations. The workshops have grown to include students and academics as well.

A rough classification of papers presented in the workshops provides some insight into trends in research (Table 2.2). During early workshops, most reports covered baseline studies that generally replicated in other locales the kind of work initially done by Hornocker and his students in the Idaho Primitive Area (Hornocker 1970; Seidensticker et al. 1973). As might be expected, the number of papers has increased over time, and they show greater specialization and diversification of subjects, including urban conflicts, habitat fragmentation and loss, cougar genetics, and conservation concerns.

With the development of live-capture and radio-tracking technology, the accumulation of knowledge on the species has accelerated rapidly. While houndsmen clearly understood how cougars moved through the country and what they killed, the lives of animals pursued usually ended at the tree. This precluded continued study to determine movements, killing behavior, or population characteristics, and it meant prolonged observation of cougar social interactions was impossible. With radio tracking, each cougar caught became a source of ongoing information, and in some projects individual cougars have been located 200–300 times over several years. A study in southern Utah lasted fourteen years (Hemker et al. 1984) and monitored the population response after a known number of cougars had been removed. This study was perhaps the first to evaluate experimentally the effects of hunting on cougar numbers and the subsequent effects on prey. Logan and Sweanor's (2001) intensive study in New Mexico accumulated 13,947 radio-locations involving some 126 individual cats over a ten-year period. An additional 115 cougars were handled but not radio-marked during that study. Logan and Sweanor experimentally moved cougars away from the

area after establishing the initial density of animals. They then monitored the re-sorting of dominance and territories of remaining animals and the rate of replacement of those removed. They also documented the fate of the cougars taken from the initial study area and transplanted elsewhere (Ruth et al. 1998). Results of these studies are presented in Chapter 8.

A third long-term study of cougars was carried out by Laundré (2005) and colleagues (with Hernandez et al. 2002; with Hernandez, 2000b, 2003a) in southern Idaho, where the population inhabited a series of small Great Basin mountain ranges separated by grazed or farmed valleys. Laundré's work has generated several theoretical papers regarding cougar-prey-habitat interactions and effects of hunting on cougars.

The most exciting recent gains in research technology involve use of genetic markers in DNA to evaluate presence of cougars in new areas, to detect connectivity between populations, and to assess population numbers (Anderson et al. 2003; see Chapter 3 for fuller discussion of genetic markers). This methodology also provides hope of developing a noninvasive and relatively inexpensive method for cougar census within limited areas (Beausoleil et al. 2005). Use of GPS radio-tracking methodology is allowing continuous monitoring of radio-collared cougars independent of time of day or weather, providing less biased relocation records and numbers of relocations far in excess of conventional methods, and allowing new insights into nocturnal movements and social interactions (various poster sessions, seventh and eighth Mountain Lion Workshops). While complete enumeration through intensive capture-recapture and radiotelemetry monitoring is the most accurate method for estimating cougar numbers, noninvasive frameworks applying mark-recapture theory may eventually provide accurate and precise estimates of cougar numbers, but this needs to be tested rigorously on a reference cougar population. Track counts have been shown to be a useful method of indexing cougar abundance (Choate et al. 2006). At least two studies are attempting to measure detection rates for trail camera arrays established within known home areas of GPS-fitted cougars. Camera traps may lead to improved, noninvasive monitoring techniques (see Figure 2.7; Haynes et al. 2003). Along with DNA analysis, tissue samples from scats, and hair snares (see Figure 2.8; for scat studies, see Chapter 3), camera traps may also allow documentation of cougars outside their previously acknowledged ranges or tracing of re-expansion into historic ranges where the species was extirpated.

This growing body of knowledge is leading to new understanding of the habitat needs of the species, and its ability to exist in the presence of humans. It is also providing better understanding of the diversity of prey taken and helping

Table 2.2 Subjects covered in eight Mountain Lion Workshops, 1976–2005, arranged by number of studies in category.

Subject	1976	1984	1988	1991	1996	2000	2003	2005	Total	Comments
Baseline studies, general ecology	1	6	13	1	7	12	2	4	46	Population, movements, prey, social behavior combined.
Human interaction	0	0	1	14	4	5	4	3	30	1991 workshop focused on human interactions.
Habitat/corridors	0	0	0	2	2	7	8	9	28	
Census/monitoring	1	1	2	2	4	3	6	2	21	
Genetics	0	0	1	0	2	5	3	4	15	
Depredation/control	1	1	0	3	2	2	2	4	15	
Standardization of terms and methods	1	0	2	1	4	2	3	2	15	
Politics/philosophy	0	0	1	3	1	0	2	5	12	
Harvest/regulation	1	2	2	0	0	2	3	1	11	Does not include annual state reports.
Range expansion/eastern populations	0	0	0	0	0	4	2	4	10	Two eastern cougar workshops have also occurred.
Effects on natural prey	1	1	1	0	1	3	2	1	10	
Interactions with other carnivores	0	1	0	0	1	0	4	4	10	
Reintroduction/captive populations	0	1	1	1	2	2	0	0	7	
Bighorn depredation	0	0	0	0	2	1	1	2	6	
Population modeling	1	0	1	0	0	1	1	1	5	
Questionnaire/opinion	0	1	1	0	1	0	1	1	5	
Adaptive management	0	0	0	0	1	0	3	1	5	
Disease	0	0	0	0	1	1	1	1	4	
Other cat species	0	0	1	0	0	2	1	0	4	
Aging techniques	1	0	1	0	0	1	0	0	3	
Capture/handling	0	0	0	1	0	0	0	1	2	
Behavior/vocalizations	0	0	0	0	1	0	1	0	2	
Education/public relations	0	0	0	0	0	0	1	1	2	
Law enforcement	0	0	1	0	0	0	0	0	1	
Taxonomy	0	0	0	0	1	0	0	0	1	
State reports	0	14	11	0	14	8	14	13	NA	Varied with workshop format.
Total subjects	6	7	13	9	17	16	20	19	25	State reports not included in total.
Total papers	8	14	29	28	37	53	51	51	271	State reports not included in total.

SOURCES: Christensen and Fischer 1976, Roberson and Lindzey 1984, Smith 1988, Braun 1991, Padley 1996, Harveson et al. 2003, Becker et al. 2003, Beausoleil et al. 2005.

Figure 2.7 The use of camera traps either as an index to cougar abundance or for estimating cougar numbers needs to be tested on a reference population (i.e., a marked population with known numbers). This camera in Chiquibul Forest Reserve and National Park in Belize was knocked askew by an ocelot a few days prior to the cougar's visit. Photo by Marcella Kelly.

Figure 2.8 Hair snags provide another inexpensive and noninvasive method for evaluating cougar presence, population trends, and habitat connectivity between populations. Designs vary; this snag consists of two carpet pads soaked with beaver castor and catnip (to entice the cat to rub) and fitted with protruding, barbed tacks (to catch the cougar's hair). Photo courtesy of the Wildlife Conservation Society.

to refine estimates of birth peaks. One of the most difficult ongoing tasks is synthesizing this accumulating information, formulating management hypotheses based upon the information, and incorporating it into adaptive management programs by agencies. In an effort to facilitate such implementation, a group of thirteen cougar biologists and wildlife managers developed the *Cougar Management Guidelines* (Cougar Management Guidelines Working Group 2005),

bringing together the most current research, methodologies, policies, and approaches to assist wildlife managers in the United States, Canada, and Mexico. One of the intended goals of these guidelines was to establish a process for applying research results more rapidly, and incorporating feedback from agencies into their periodic revision.

Gaining new knowledge and conveying it to wildlife managers is relatively simple compared to helping policy makers understand its significance. Even more difficult is bringing new, at times highly technical, information to the concerned public—the people who ultimately determine the values driving cougar management priorities. While those who work directly with the animal may develop increasingly objective (as well as intuitive) understanding of cougars, public values remain diverse, are generally self-interested, and fluctuate with political tides. They often derive from misinformation in the popular media. For people who seek simple management solutions, only two options exist: eliminate cougars or leave them entirely alone. Given human dedication to the task, the first of these is undoubtedly possible, but most state management agencies are charged with assuring that species are present for future generations. With increasing human numbers occupying larger expanses of cougar habitat, the option of no management is not feasible. Human activities will inevitably affect cougar numbers. To assure the continued existence of the cougar, those of us who understand its ecological limits and how human activities affect those limits must find ways to convey this understanding to the larger public.

Conveying such information is among the purposes of this book. But simply listing facts, old or new, is not enough. Absorbing the increasing mass of information is difficult, even for biologists working directly with the cougar, and our realizations of the complexity of natural systems may complicate rather than simplify management decisions.

While science can help us understand the ecological limits of cougars, and hence can aid in predicting the effects of our management strategies, facts gathered through research will not necessarily modify stakeholder values. Knowledge of a species can, in fact, be used to eliminate it as well as to sustain it, and thus public values—something over which scientists have little control—will determine the cougar's ultimate fate. We face an ongoing need to condense and simplify the growing profusion of information into a lucid description of the cougar's ecological niche in terms that everyone can understand. This is not easy to do when the species and the system within which it resides are complex, dynamic, and often unpredictable.

Sweanor (1990) and Hansen (1992) suggested that the cougar and other large carnivores are "umbrella" species— wide-ranging species with large area requirements that automatically conserve a host of other species (for definitions, see

Groom et al. 2006; NOAA 2007). Cougars thus not only indicate the health of an ecosystem but can also serve to define the extent of the ecosystem required to sustain a whole suite of species needing less space (see Chapter 12). Logan and Sweanor (2001, 21) have characterized the cougar as

. . . a highly adaptive asocial predator suited for environments with rugged terrain and closed vegetative cover where prey is dispersed and clumped. The puma preys on mammals ranging in size from hares (Iriarte et al. 1990) to adult elk, *Cervus elaphus* (Hornocker 1970). Of the large felids from leopards to tigers, pumas routinely kill the largest prey relative to their own mass (Packer 1986). Selection pressures designed the puma as an ambush and stalking predator that attacks its prey with a quick powerful rush, overwhelms the victim with staggering strength, and delivers a killing bite in seconds.

The *Cougar Management Guidelines* (Cougar Management Guidelines Working Group 2005, 2) state:

Cougars are presently the only large, obligate carnivore thriving in self-sustaining populations across western North America . . . [as they have] for at least 10,000 years (Logan and Sweanor 2001). Ecologically, cougars strongly influence energy flow in ecosystems, are a potent selective force on prey animals, modulate prey population dynamics, indirectly affect herbivory in plant communities, influence competitive interactions between herbivores, and compete with other carnivores for prey (Logan and Sweanor 2001). Moreover, because self-sustaining cougar populations require expansive, interconnected wild land, conservation strategies to benefit cougars also benefit an array of other wildlife (Beier 1993; Logan and Sweanor 2001).

People can certainly exist on this planet in the absence of cougars and other large carnivores, hence the value of a human choice to sustain cougars may be more symbolic than practical: if we can learn to tolerate controversial creatures such as the cougar, we may also learn to preserve the quality of our own environment.

Chapter 3 Lessons and Insights from Evolution, Taxonomy, and Conservation Genetics

Melanie Culver

BREAKTHROUGHS IN SCIENCE are rare and always to be treasured. Breakthroughs in conservation are likewise thin on the ground. The application of genetic techniques to the field of conservation biology is intimidating for its complexity but also rewarding for its potential, and it is beginning to provide insights not available by any other route. The variety of topics that can be addressed through genetic research ranges all the way from relatedness of known individual animals to species-wide assessments of several kinds. Genetics allows the gathering of information relevant to cougar evolution and taxonomy, and also on a series of population characteristics, such as inbreeding, mutation rates, migration rates, and population size. Incorporating such information into management can be a critical tool for maintaining functioning populations with natural levels of gene flow. The purpose of this chapter is to provide an overview of the cougar's taxonomic status and evolution in light of recent genetics studies, and an account of what research in cougar conservation genetics is revealing, presented through case studies at the species, subspecies, population, and individual levels.

Historically, cougars (*Puma concolor*) have shared their distribution with a diverse array of carnivores, including extinct felids such as the America lion, *Panthera leo atrox*; the North American cheetah, *Acinonyx trumani*; and the saber-toothed cat, *Smilidon fatalis*. A wave of extinctions occurred in the late Pleistocene era in North America 9,000–13,000 years ago, among them thirty-five to forty species of large mammals (Pielou 1991). Of those, six to seven were carnivores, which included three to four felid species. One or two felids went extinct in South America during the same period (Martin and Klein 1995).

Currently, cougars are distinguished by the largest latitudinal distribution of any native extant terrestrial mammal: from the Canadian Yukon to the Chilean Patagonia (Heilprin 1974; Walker 1975). As recently as one hundred years ago, the cougar occupied every biogeographic zone in the New World except arctic tundra (Figure 3.1). Cougars ranged from desert environments to tropical rain forest and from sea level to 4,500 meters in elevation (Heilprin 1974; Anderson 1983; Currier 1983; Seidensticker 1991). Throughout the Americas, recent human encroachment and habitat destruction have caused increased fragmentation of the existing populations (Anderson 1983). In temperate North America the cougar has been displaced from two-thirds of its historic range (Hansen 1992). Cougar populations in the western United States and Canada have been increasing since the turn of the century as each state and province ceased paying cougar bounties (Beier 1991). As a result, the status of cougar populations is variable across the species' distribution, ranging from unknown to stable, decreasing, endangered, or extinct.

Evolution

Cougars are one of twelve New World felids, which evolved from four different felid lineages (Figure 3.2). Seven of the other eleven New World felids compose the ocelot lineage (ocelot, *Leopardus pardalis*; margay, *Leopardus wiedii*; tigrina, *Leopardus tigrina*; Geoffroy's cat, *Oncifelis geoffroyi*; kodkod, *Oncifelis guigna*; Andean mountain cat, *Oreailurus jacobita*; and pampas cat, *Lynchailurus colocolo*). Ocelot

Figure 3.1 Geographic ranges of the thirty-two previously recognized cougar subspecies (modified from Culver 1999, redrawn by John Norton). Question marks are areas of unknown subspecies affinity. The three-letter codes reflect the following subspecies: ACR = *acrocodia*, ANT = *anthonyi*, ARA = *araucanus*, AZT = *azteca*, BAN = *bangsi*, BOR = *borbensis*, BRO = *browni*, CAB = *cabrerae*, CAL = *californica*, CAP = *capricornensis*, CON = *concolor*, COR = *coryi*, COS = *costaricensis*, COU = *couguar*, GRE = *greeni*, HIP = *hippolestes*, HUD = *hudsoni*, IMP = *improcera*, INC = *incarum*, KAI = *kaibabensis*, MAY = *mayensis*, MIS = *missoulensis*, OLY = *olympus*, ORE = *oregonensis*, OSG = *osgoodi*, PAT = *patagonica*, PEA = *pearsoni*, PUM = *puma*, SHO = *shorgeri*, SOD = *soderstromi*, STA = *stanleyana*, VAN = *vancouverensis*.

the puma lineage (Johnson and O'Brien 1997; Pecon-Slattery and O'Brien 1998).

The puma lineage is an old, deeply divergent lineage within the cat family Felidae, and lineage members likely originated from a North American ancestor. The cheetah was the earliest to diverge, about 5–8 MYA (Wayne et al. 1989; Janczewski et al. 1995; Johnson and O'Brien 1997; Pecon-Slattery and O'Brien 1998; Johnson et al. 2006), after which it presumably dispersed through Asia to its current distribution in Africa and the Middle East (Turner 1997). Fossil evidence in North America, from 0.6–3.2 MYA, suggests that a cheetahlike Plio-Pleistocene felid, (*Miracinonyx inexpectatus*) may link the cougar and cheetah (Orr 1969; Kurtén 1976; Adams 1979; Van Valkenburgh et al. 1990). Early cougars (of the Rancholabrean time period) are intermediate in morphology between *M. inexpectatus* and modern cougars (Kurtén 1976). About 4–5 MYA, the cougar diverged from a common ancestor with the jaguarundi (Janczewski et al. 1995; Johnson and O'Brien 1997, Johnson et al. 2006). Cougar fossils have been found in the southern half of the United States and Mexico dating to between 0.01 and 0.2 MYA (Rancholabrean period; Kurtén 1976; Kurtén and Anderson 1980), and in South America to between 0.01 and 0.3 MYA (Lujanian and Ensenadan periods; Savage and Russell 1983; Werdlin 1989).

The cougar, or its ancestor, probably arrived in South America approximately 2–4 MYA during the Great American Interchange (Patterson and Pascual 1972; Webb 1976, 1978; Webb and Marshall 1981; Stehli and Webb 1985; Webb and Rancy 1996), at which time placental carnivore species first migrated to South America with the formation of the Panamanian land bridge. Since the general flow of the interchange went from north to south (Marshall et al. 1982), and because *M. inexpectatus* is found only in North America (Orr 1969; Kurtén 1976; Adams 1979; Van Valkenburgh et al. 1990), it is likely that the cougar, or an ancestor to the cougar, evolved in North America and migrated south.

Taxonomy

The cougar was originally described as *Felis concolor* by Linnaeus (described in 1758, published in 1771) and later recognized as *Felis (Puma) concolor* by Jardine (1834), with *Puma* recognized as a subgenus of *Felis*. More recently, *Puma* was recognized as a separate genus by Ewer (1973). Molecular genetic evidence indicates that the cougar is a member of the puma lineage, which, as noted, includes cheetah and jaguarundi and is not allied with the small cat genus *Felis* (Janczewski et al. 1995; Johnson and O'Brien 1997; Pecon-Slattery and O'Brien 1998).

lineage species are restricted to Central and South America (Collier and O'Brien 1985; Johnson et al. 1998) and likely diversified after migrating into South America, a dispersal facilitated 2–4 million years ago (MYA) by the closing of the Panamanian isthmus (Patterson and Pascual 1972; Webb 1976, 1978; Webb and Marshall 1981; Stehli and Webb 1985; Webb and Rancy 1996). The jaguar (*Panthera onca*), is the single member of the genus *Panthera* still found in the New World (Johnson et al. 1996; Johnson and O'Brien 1997). The bobcat (*Lynx rufus*) and Canadian lynx (*L. canadensis*) are the North American members of the genus *Lynx*. Two American cat species, cougar and jaguarundi (*Puma yaguarondi*), plus the African cheetah (*Acinonyx jubatus*), compose

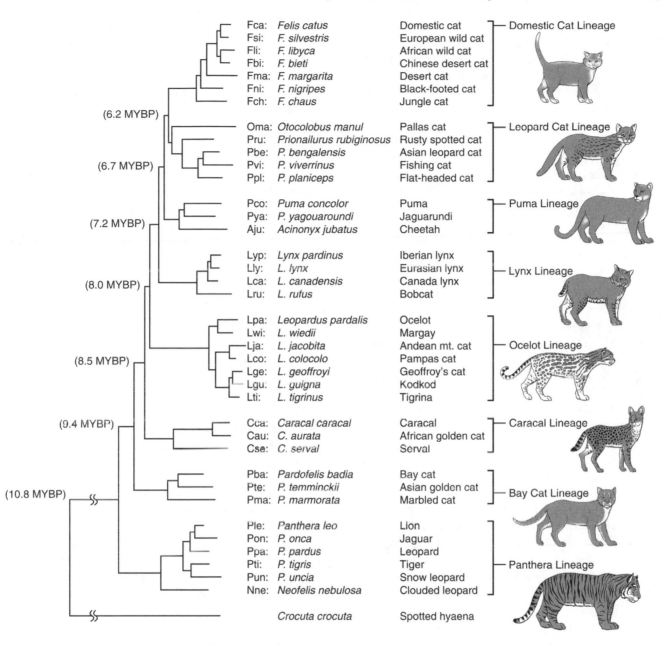

Fca:	*Felis catus*	Domestic cat	Domestic Cat Lineage
Fsi:	*F. silvestris*	European wild cat	
Fli:	*F. libyca*	African wild cat	
Fbi:	*F. bieti*	Chinese desert cat	
Fma:	*F. margarita*	Desert cat	
Fni:	*F. nigripes*	Black-footed cat	
Fch:	*F. chaus*	Jungle cat	
Oma:	*Otocolobus manul*	Pallas cat	Leopard Cat Lineage
Pru:	*Prionailurus rubiginosus*	Rusty spotted cat	
Pbe:	*P. bengalensis*	Asian leopard cat	
Pvi:	*P. viverrinus*	Fishing cat	
Ppl:	*P. planiceps*	Flat-headed cat	
Pco:	*Puma concolor*	Puma	Puma Lineage
Pya:	*P. yagouaroundi*	Jaguarundi	
Aju:	*Acinonyx jubatus*	Cheetah	
Lyp:	*Lynx pardinus*	Iberian lynx	Lynx Lineage
Lly:	*L. lynx*	Eurasian lynx	
Lca:	*L. canadensis*	Canada lynx	
Lru:	*L. rufus*	Bobcat	
Lpa:	*Leopardus pardalis*	Ocelot	Ocelot Lineage
Lwi:	*L. wiedii*	Margay	
Lja:	*L. jacobita*	Andean mt. cat	
Lco:	*L. colocolo*	Pampas cat	
Lge:	*L. geoffroyi*	Geoffroy's cat	
Lgu:	*L. guigna*	Kodkod	
Lti:	*L. tigrinus*	Tigrina	
Cca:	*Caracal caracal*	Caracal	Caracal Lineage
Cau:	*C. aurata*	African golden cat	
Cse:	*C. serval*	Serval	
Pba:	*Pardofelis badia*	Bay cat	Bay Cat Lineage
Pte:	*P. temminckii*	Asian golden cat	
Pma:	*P. marmorata*	Marbled cat	
Ple:	*Panthera leo*	Lion	Panthera Lineage
Pon:	*P. onca*	Jaguar	
Ppa:	*P. pardus*	Leopard	
Pti:	*P. tigris*	Tiger	
Pun:	*P. uncia*	Snow leopard	
Nne:	*Neofelis nebulosa*	Clouded leopard	
	Crocuta crocuta	Spotted hyaena	

Branch-point ages: (6.2 MYBP), (6.7 MYBP), (7.2 MYBP), (8.0 MYBP), (8.5 MYBP), (9.4 MYBP), (10.8 MYBP)

Figure 3.2 Phylogenetic relationships among the thirty-eight extant felid species, which form the eight felid lineages, and outgroup (spotted hyena) from a maximum-likelihood tree using 18,853 bp of nuclear DNA (modified from Johnson et al. 2006, redrawn by John Norton). One representative of each felid lineage is depicted in these drawings. Cat illustrations reprinted courtesy of John Norton.

From the mid-1700s to the early 1900s, the cougar was described as having thirty-two distinct subspecies, fairly evenly distributed throughout North and South America (Figure 3.1; Young and Goldman 1946a). Two additional subspecies were described later in the twentieth century (Jackson 1955; Cabrera 1963). These classifications were based on morphological characteristics (skull and skeletal measurements, coat color, size), habitat, and geographic location (Young and Goldman 1946a; Jackson 1955; Cabrera 1963). However, there may be inconsistencies among the subspecies descriptions based on morphology alone; often the subspecies were described based on few individuals, or sometimes the specimens compared were a mixture of juveniles and adults.

A morphological study in which cranial and mandibular measurements of 1,700 cougar skulls were subjected to geographical analysis found that cougars were grouped by similarity of distance from the equator rather than by geographical clines (Gay and Best 1995). These results support Bergman's rule of size variation (Bergman 1847) but did not conform to previous subspecies groups (Figure 3.1). Considering these results together with the inconsistencies of the early morphological analyses, the original thirty-two cougar subspecies descriptions may not reflect historic subdivision and may not be the optimal units for conservation and management of cougars.

Five cougar subspecies are currently listed as endangered or category 2 by the U.S. Fish and Wildlife Service (Hansen 1992) as a result of recent population reduction. The Florida panther, *P. c. coryi* (endangered), declined to less than fifty individuals (Belden 1986). The eastern North American cougar, *P. c. couguar* (endangered), and the Wisconsin cougar, *P. c. schorgeri* (category 2), are probably extinct (Anon. 1989). The lower Colorado River cougar or Yuma puma, *P. c. browni* (category 2), of Arizona has also declined in numbers (Duke et al. 1987), as has the southernmost population in Central America, *P. c. costaricensis* (endangered; Hansen 1992). However, McIvor and colleagues (1995) reviewed morphometric data from the literature and museums and concluded that the subspecific status of the Yuma puma is probably not warranted. Throughout the rest of the Americas, including Latin America, human depredation and habitat destruction have increasingly fragmented the remaining populations (Anderson 1983), but the status of many populations is unknown due to a lack of data.

Relevance of Genetics to Cougar Conservation

As populations decline in numbers due to a variety of causes, but most commonly from habitat destruction, a loss of genetic diversity results and some level of inbreeding is inevitable. If the inbreeding level is high, loss of heterozygosity and possibly a loss in fitness from inbreeding depression will result. Eventually, these genetic processes cause an increased risk of extinction, which is avoidable if populations can be maintained at moderately high numbers.

An understanding of genetic diversity—what it is, why it is important, and how it is maintained—is needed if we are to recognize how genetic characteristics are relevant to wildlife conservation and management. Genetic diversity can be defined simply as the genetic variation (i.e., all the different alleles) found within a population or species. Genetic diversity is important when we think of natural selection, which is the differential reproduction and fitness of genotypes (the genetic constitution of an organism). There can be no differential reproduction if there is no variation. With no variation, all individuals would respond to their environment the same way and would be equally successful. Thus, it is important for species, populations, and individuals to maintain some level of genetic variation.

Several forces interact to determine the amount of genetic variation maintained: mutation, natural selection, genetic drift, inbreeding or other mating system (polygyny, polyandry, etc.), and migration (migration in a genetic sense refers to any immigration, emigration, dispersal, or migration and successful reproduction of an individual in a population other than its natal population). Mutation, migration, selection, and drift are forces of evolution in that they cause allele frequencies to change; inbreeding causes a reduction in heterozygosity but is not an evolutionary force since it does not alter allele frequencies.

Mutation is a very slow process but acts to increase the total amount of genetic variation by introducing new variants into the population. Migration can be a weak or strong force to increase variation in a population, depending on the number of migrants and the size of the recipient population (a single migrant into a large population has a small effect; a large number of migrants into a small population can have a large effect). Selection acts to decrease the overall variation by favoring a single more fit allele at the expense of all other alleles in the population. Genetic drift acts to decrease genetic variation by chance and is basically a sampling effect (a small number of individuals are reproducing, so there is a high probability that not all existing alleles will be represented in the next generation); thus, alleles are lost at random. Because of the random nature of the alleles lost, a certain proportion will be alleles that would have been beneficial to the population either now or in the future; thus, fitness is reduced. In large populations, the alleles that are lost due to natural selection are those that are detrimental; thus, fitness is increased.

Most threatened and endangered populations have one thing in common—they are small. Some may always have existed in small numbers, but the majority are declining in numbers. In small populations, chance (genetic drift) has a greater effect than do selection and mutation; migration has potential to be a significant factor, but threatened and endangered populations are often cut off from potential migrants due to habitat loss and fragmentation (see Chapter 12). Moreover, population size reductions usually result in some level of inbreeding, which reduces genetic variation (by lowering heterozygosity) and fitness (through inbreeding depression). The level of inbreeding is directly related to the effective population size (meaning the number of individuals in the population that actually breed). Inbreeding decreases fitness by raising the probability that detrimental alleles will be found in the homozygous state and will, therefore, become expressed.

Fragmented populations undergo all the negative effects of a small population if they are truly cut off from gene flow from other populations. This means that even when a species is still present in high numbers it may nevertheless be suffering from loss of genetic diversity if all the populations have become small and isolated. Fragmented populations have all the same risk factors as small populations.

Beyond an understanding of genetic variation, one of the greatest contributions of genetics to wildlife conservation is in resolving taxonomic uncertainties. "Cryptic" species are those that have no obvious morphological differences despite having diverged from each other significantly; molecular-level studies can reveal cryptic divergence. Without molecular studies, the fact that they are separate species will remain unknown, and species may go extinct without ever being recognized. The reverse also applies: species may have significant morphological variation and yet not have the genetic divergence that would warrant their being considered separate species. In this case, resources may be invested into protection of a species that is actually a morphological variant rather than a true species.

Genetic studies on a wide variety of species have added to our knowledge of taxonomy at many levels. With respect to cougars, much of what has been determined regarding evolution and taxonomy has relied on molecular genetic data as well as the fossil record and morphometrics. The methods in molecular genetics research relevant to cougar case studies are intricate and are described in Appendix 1. The remainder of this chapter describes how those techniques have been applied to questions about cougars, ranging from matters of population genetics to relatedness, forensics, and disease (Table 3.1). Species-level questions involving the systematic relationships of the cougar to other felids are discussed first, followed by subspecies-level questions including subspecific taxonomy, then population-level studies, and concluding with individual-level work.

Species-Level Case Studies

For felids, molecular genetic studies indicate that the thirty-seven extant species are divided into eight principal lineages (Figure 3.2), with the puma lineage being one of the main ones (Johnson and O'Brien 1997; Johnson et al. 2006). It has been determined that all felids shared a common ancestor as recently as 11 MYA and that all thirty-seven species have diverged since that time, with twenty-one species having diverged in the last one million years. These studies have included combined data from mtDNA genes, nuclear DNA genes, and X- and Y-chromosome linked genes for a high-resolution molecular phylogeny (see Appendix 1).

Looking at the puma lineage, based on molecular evidence, there is no close relative to the cougar. The closest relative is the jaguarundi, which diverged from the cougar

Table 3.1 Molecular genetic studies of cougars.

Study	Geographic Region	No. of Individuals (Populations)	Genetic Marker(s)	Objectives
O'Brien et al. 1990	Florida and misc. populations	94 (9)	Allozymes mtDNA RFLP	Florida panther introgression
Roelke et al. 1993	Florida and misc. populations	81 (9)	Allozymes DNA fingerprint mtDNA RFLP	Genetic variability in Florida panthers
Carpenter et al. 1996	North, Central, South America	434 (26)	FIV pol gene	Natural history of FIV in cougars
Culver et al. 2000a	North, Central, South America	315 (32)	mtDNA genes Microsatellites	Phylogeography and demographic history
Ernest et al. 2000	Yosemite, California	62 (1)	Microsatellites	Genetic mark-recapture
Walker et al. 2000	Texas	35 (1)	Microsatellites	Population differentiation
Sinclair et al. 2001	Utah	50 (1)	Microsatellites	Population differentiation
Ernest et al. 2003	California	431 (1–2)	Microsatellites	Population differentiation
Loxterman 2001	Idaho	200 (1)	Microsatellites	Population differentiation
Driscoll et al. 2002	Idaho, Florida, South America	30 (3)	Microsatellites	Demographic history
Anderson 2003	Wyoming	200 (1)	Microsatellites	Population differentiation
McRae et al. 2005	Utah, Arizona, Colorado, New Mexico	540 (2–4)	Microsatellites	Population differentiation
Biek et al. 2006b	British Columbia, Idaho, Montana, Wyoming	352 (2–4)	Microsatellites FIV genes	Population differentiation and demographic history
Biek et al. 2006a	Yellowstone, Wyoming	128 (1)	Microsatellites	Sex-biased dispersal

FIV=feline immunodeficiency virus

about 4–5 MYA. The next closest relative to the cougar and jaguarundi is the cheetah, divergence of which was about 5–8 MYA. The cheetah originated in the puma lineage and then migrated to Asia and Africa. These three species make up the puma lineage. The lynx lineage is more closely aligned with the puma lineage than with any other felids.

Subspecies-Level Case Studies

The term *subspecies* was introduced by Esper to designate geographical varieties. In 1942, Mayr defined a subspecies as "a geographically defined aggregate of local populations which differ taxonomically from other subdivisions of the species" (Mayr 1982a). However, Wilson and Brown (1953) dismissed the subspecies trinomial as arbitrary, too subjective to be useful, a taxonomic designation that is often based on insignificant groupings or clines and is widely exploited.

Findings from Mitochondrial DNA and Feline Microsatellites

In 1981, with the emergence of conservation biology as a science and the legal incorporation of subspecies into the ESA, debate over subspecies was revived at an ornithological forum (Barrowclough 1982; Gill 1982, Mayr 1982a). Later, Avise and Ball (1990) suggested that subspecies designations should be made based on "concordant distributions of multiple independent [genetic] traits" (p. 54) and that this would reduce the subjective and arbitrary nature of trinomials (subspecies designations). Subsequent to Avise's and Ball's definition, O'Brien and Mayr (1991) proposed that members of a subspecies would share (a) a unique geographic range, (b) a set of "phylogenetically concordant phenotypic characters," (p. 1188) (c) a suite of molecular genetic similarities, and (d) derived adaptations relative to other subspecies. Furthermore, O'Brien and Mayr suggested that a subspecies is a dynamic entity and can (a) go extinct, (b) become a new subspecies by hybridization with another subspecies, (c) become a new subspecies by changing its genetic character through stochastic processes like genetic drift, (d) become a new species by becoming reproductively isolated—that is, by developing intrinsic barriers, or (e) remain the same.

Molecular genetic tools were used to examine the accuracy of the previously described thirty-two "subspecies" subdivisions for cougars. DNA sequence from three mitochondrial DNA (mtDNA) genes and ten feline microsatellites, obtained from domestic cats (Menotti-Raymond and O'Brien 1995; Menotti-Raymond et al. 1997), were used in this study. Together, these markers provided eleven independent lines

of evidence to examine subspecies designations in more than three hundred cougar samples from throughout North and South America (Culver et al. 2000a).

These molecular data indicated that South American cougars had high levels of genetic diversity for both mitochondrial and microsatellite DNA markers, whereas cougars throughout Central and North America north of Nicaragua had no mitochondrial variation (except one mutation observed in the Olympic Peninsula) and had only moderate levels of microsatellite variation (Table 3.2). Phylogenetic analyses of individuals and subspecies (Figure 3.3) indicated six groups of cougars throughout their range. These results suggested the existence of one cougar subdivision north of Nicaragua and five cougar subdivisions south of Nicaragua. Furthermore, the presumed boundaries for these six groups of cougars incorporated major geographical features (Figure 3.4). The northernmost boundary occurs in Nicaragua, potentially the "lake region," which was the original (but unrealized) location chosen for a canal to connect the Atlantic and Pacific—what is now the Panama Canal. In South America, the boundaries incorporate several major rivers (Amazon River, Río Paraná, Río Negro, and Paraguay River). This suggests that, given a choice, cougars may prefer not to cross large bodies of water.

Using the clocklike nature of neutral genetic markers (i.e., markers having no effect on physical characteristics and thus not under selection pressure), and knowing the mutation rate of those markers to calibrate the "clock," we can determine that North America is the most recently founded population; the population inhabiting Brazil and Paraguay (specifically, the Brazilian Highlands) is the oldest; and cougars as a species are only 390,000 years old (Culver et al. 2000a). The Genetics Techniques Primer (Appendix 1) gives more detail about neutral markers.

A more traditional estimator of timing for species existence is the fossil record. In the case of cougars, the fossil record is approximately 300,000 years old (Kurtén 1976; Kurtén and Anderson 1980; Savage and Russell 1983; Werdlin 1989); there is a striking agreement between the fossil and the molecular timing for the most recent common ancestor to all cougar lineages. However, the fossil record does not indicate that North American cougars are a more recent lineage; cougar fossils on both continents are equally old.

The disagreement between the fossil record, indicating that North and South American cougars lineages are equally old, and the molecular data, indicating that North American lineages are much younger than the South American lineages, leads to the conclusion that cougars were extirpated in North America, and extant North American cougars recolonized there much more recently than 300,000 years ago. Due to this proposed extirpation event, extant cougar

Table 3.2 Measures of genetic variation across ten microsatellite loci and mtDNA gene segments, in pumas grouped by phylogeographic groups.

		mtDNA		Microsatellites						
Continent or group	Number of individuals, mtDNA/ μsat	Haplotype	π (x10²)	Avg. H₀(SE)ᵃ %	Total no. alleles	No. of unique alleles	Avg. no. alleles per locus (SE)	Avg. var.	Avg. range	Max. range
Puma concolor	286/277	A-N	0.32	52	121	—	12.1	6.9	12.5	21
NA	186/78	M N	0.02	42 (0.16)ᵇ	64	5	6.4ᶜ	4.8ᶜ	9.8	18
CA	13/13	M F C	0.40	63 (0.11)	54	12	5.4	11.1	8.4	20
SA	87/86	A B D-L	0.30	71 (0.33)	110	41	11.1	9.3	10.8	21
ESA	23/22	A F H G I K L	0.22	71 (0.09)	86	5	8.6	8.0	9.3	19
NSA	25/25	B F E L	0.04	75 (0.52)	91	5	9.1	8.2	10.0	19
CSA	17/17	F J	0.10	75 (0.46)	67	1	6.7	9.6	9.2	21
SSA	22/22	F D J	0.19	64 (1.16)	60	1	6.0	8.3	8.6	20

NA=North America; CA=Central America; ESA=eastern South America; NSA=northern South America; CSA=central South America; SSA=southern South America.
ᵃAverage observed percent heterozygosity across all loci and standard error.
ᵇSignificantly different from South American subspecies (*P* < .05).
ᶜSignificantly different from South American subspecies (*P* < .01)

populations in North America are probably the result of a relatively recent recolonization event.

The timing of the proposed extirpation event that reduced overall genetic variation in North American cougars was further investigated by Driscoll and colleagues (2002). In this study, a large number of microsatellite DNA markers were able to provide a reliable measure of cougar natural history. A set of eighty-four microsatellite DNA loci were examined and compared among cougars from Florida and Idaho (representing North America), and South America. The analysis method is based on detecting the rate of regeneration of microsatellite alleles by new mutation, combined with the variance in allele sizes. This set of markers was able to detect a historic North American cougar population reduction; rate calibration of regenerated alleles indicated the North American demographic contraction event took place 10,000–17,000 years ago. This young age for North American cougars is directly related to the lack of genetic diversity and differentiation observed in current North American cougars.

This genetic work dovetails with another study (Barone et al. 1994) that provides indirect evidence of a bottleneck in North American cougars—(a bottleneck means a once large population that later persists as a small population). A trait that has been correlated to lack of genetic diversity, in populations that have experienced a bottleneck, is sperm morphology defects. Barone and colleagues examined reproductive characteristics of male cougars throughout their range. Their results showed that Florida panthers, having experienced a documented population bottleneck, had an abundance of structural abnormalities, including only 7 percent normal sperm. In contrast, abnormalities were found in western North American cougars and Latin American pumas at a much lower frequency. Western North American cougars had 14–17 percent normal sperm, and Latin American pumas had 39 percent normal sperm. This is consistent with a severe bottleneck in Florida panthers, a moderate bottleneck in western North American cougars, and constant large populations for Latin American pumas.

Subspecific taxonomy of modern cougars can be inferred through combining the species-wide genetic survey of cougars just noted and a species-wide skull and morphology study (mentioned in the taxonomy section) with what is known from the fossil record. Taxonomic revisions have been suggested for cougar subspecies, and the proposed revision includes six subspecies instead of thirty-two. Unique genetic characters can be used to define each subspecies, and the name for each subspecies corresponds to the earliest description of a subspecies in each geographic locale. The characters unique to each of the six subspecies are shown in Table 3.3; as is typical for subspecies-level classification, the unique characters are not fixed in

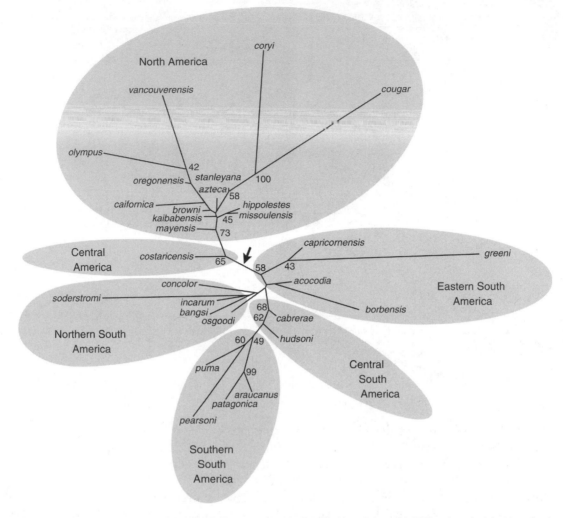

Figure 3.3 Phylogenetic relationships among twenty-nine described cougar subspecies (modified from Culver et al. 2000, redrawn by John Norton), using microsatellite genotypes for 277 individuals. The neighbor-joining tree uses a kinship coefficient distance estimator, and bootstrap values are indicated at the relevant node. The arrow indicates the position of the midpoint root for the tree which is the midpoint of the longest pathway between two taxa.

each of the six groups, but represent characters that exist in only one of the six cougar groups. There is not complete agreement among biologists and managers on whether to lump the Florida panther with the rest of North American cougars. This action could affect the protected status of the Florida population and clearly Florida panthers need to retain protection. Arguments for Florida panthers to remain their own subspecies cite unique morphological traits in the Florida cats (e.g., cowlick, kinked tail, and Harris lines), and the potential for unobserved adaptive traits to exist due to their isolation and unique habitat niche. In Florida panthers, the increased frequency of the morphological traits mentioned is likely a result of inbreeding rather than of unique adaptation. However, there is always the possibility that Florida panthers have evolved unique adaptation in the short time (~100 years) that they have been isolated from other cougars.

Findings from FIV Viral Sequences

An additional species-wide survey of cougar DNA sequences was conducted by Carpenter et al. (1996) and used sequences from the cougar feline immunodeficiency virus, known as the puma FIV virus. FIV has not been demonstrated to be pathogenic in any felid species other than the domestic cat. Puma FIV polymerase gene sequences were isolated from cougars that were infected with the FIV virus, and sequences from throughout the geographic range of cougars were compared. The FIV virus is present in most parts of the species range. It occurs at a lower frequency in South America; therefore, only Brazil was represented. A phylogenetic analysis of FIV sequences indicated two separate groups, one made up of Florida and California puma FIV sequences (Clade A), and the other containing the rest of western North and South America (Clade B). The unlikely relationship in Clade A between virus sequences in Florida

and California could indicate that these sequences are ancestral, and that a second wave of FIV infection spread widely throughout North and Central America but did not reach Florida or California.

Western North American cougars from Clade B were further subdivided into two subclades: north of Wyoming and south of Wyoming—with Wyoming cougars included in both subclades (Figure 3.5). The diversity of the Wyoming samples was striking considering that all Wyoming samples were from Yellowstone, yet they appeared in both subgroups. Cougars were eliminated from Yellowstone in the early 1900s, and the high diversity of Yellowstone sequences could indicate that cougars recolonizing Yellowstone came from multiple areas (particularly from north and south of Yellowstone). Florida contained two divergent lineages; the first was the lineage shared with California and the second lineage is likely derived from a Latin American cougar known to have been released into the Everglades in the 1950s. In this study, the FIV viral sequences achieved a slightly higher level of resolution than what was found from mtDNA and microsatellite markers, exemplifying how the natural history of puma FIV viral sequences reflects the natural history of the cougars.

Population-Level Case Studies

Another advantage of molecular markers for resolving subdivisions within a species is that the range of resolving power in genetic tools can take us from the subspecies

Figure 3.4 Boundaries of the six genetically defined cougar subspecies (modified from Culver et al. 2000, redrawn by John Norton) based on phylogeographic groups for composite mtDNA haplotypes and microsatellite genotypes. Major geographical barriers (mountain ranges and rivers) are included.

Table 3.3 Trinomial subspecies designations and diagnostic characters for the six phylogeographic groups.

Group	Subspecies[a]	mtDNA sites[b]	mtDNA haplotypes[c]	Microsatellite alleles	Citation
NA	P. c. couguar	12908	N	FCA043-104; FCA082-252; FCA090-95, -122; FCA117-152	Kerr 1792
CA	P. c. costaricensis	8630	C	FCA082-236; FCA166-223	Merriam 1901
ESA	P. c. capricornensis	3063, 12723, 12819	A, H, G, I, K	FCA008-140, -146, -150, -164; FCA043-120	Merriam 1901
NSA	P. c. concolor	12834, 12840	B, E	FCA008-162; FCA035-124; FCA043-142; FCA166-221; FCA249-251	Linneaus 1771
CSA	P. c. cabrerae	None	None	FCA008-176	Pocock 1904
SSA	P. c. puma	12809	D	FCA249-239	Molina 1782

NA=North America; CA=Central America; ESA=eastern South America; NSA=northern South America; CSA=central South America; SSA=southern South America
[a]Subspecies designations are the earliest description in the geographic locale.
[b]Sites listed relative to domestic cat mtDNA sequence (Lopez et al. 1996).
[c]mtDNA haplotypes as defined in Culver et al. 2000.

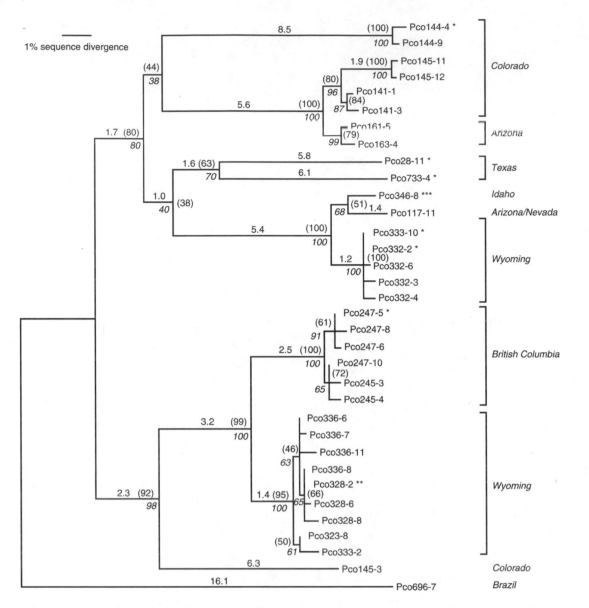

Figure 3.5 Phylogenetic tree of cougars using puma FIV sequences (modified from Carpenter et al. 1996, redrawn by John Norton) and neighbor-joining tree of sequences from western North American cougars. Branch lengths represent percentage of sequence divergence, calculated by using Kimura's two-parameter model. Bootstrap values from neighbor-joining analysis are shown in italics at the nodes, and bootstrap values from maximum parsimony analysis are given in parentheses at the nodes. Asterisks represent additional identical sequences of clones isolated from the same animal.

level to the population level. The differentiation described in cougars in this chapter has so far been at the subspecies level; at this point, discussion shifts to cougar populations. Populations of interest from a conservation perspective are those that have experienced bottlenecks and suffered a reduction in genetic diversity. In addition to the continent-wide cougar bottleneck in North America, a significant bottleneck was detected in Florida, which has been well documented in the literature. Worthy of note are two other mild but detectable bottlenecks observed—one on the Olympic Peninsula and one on Vancouver Island. While

Florida panthers showed no genetic variation at 8/10 microsatellite markers, those in the Olympic Peninsula and on Vancouver Island showed no genetic variation at 5/10 microsatellite markers.

Florida Population

The Florida panther living in Florida's Big Cypress Swamp and the Everglades is a relict population that once was distributed in the United States from Florida and Georgia to Arkansas and Louisiana (Young and Goldman 1946a).

The bottleneck population size was estimated to be six individuals based on genetic data from mitochondrial and microsatellite DNA in contemporary versus historic museum samples (Culver et al. 2008). The reduction in population size and distribution has been accompanied by both morphological and genetic characteristics consistent with severe inbreeding (Roelke et al. 1993). Morphologically, individuals from this population have a high incidence of kinked tails, cowlicks (a distinct whorl in the fur between the shoulder blades), cryptorchidism (failure of testes to descend), heart defects, and a high prevalence of infectious agents usually found in hosts with compromised immune defenses.

Genetic diversity levels were also reduced, as has been observed in other bottlenecked species (Bonnell and Selander 1974; O'Brien et al. 1986; Packer et al. 1991; O'Brien 1994). Florida panthers display reduced levels of minisatellite DNA fingerprint variation and reduced allozyme diversity relative to six other western U.S. cougar populations (Roelke et al. 1993; see DNA fingerprint in Appendix 1). The Florida population was found to be more closely related (phylogenetically) to the six North American populations than to three South American populations, based on mtDNA RFLP patterns. Genetic analyses also revealed the presence of individuals in Florida with genotypes likely derived from the release of a captive cougar from Latin America in the 1950s (O'Brien et al. 1990). Because of the severe demographic reduction and the ensuing genetic threats revealed by data collected on Florida panthers, a consensus recommendation to supplement Florida panthers with eight female individuals from Texas (*P. c. stanleyana*) was made and implemented in 1996.

The objective of the genetic restoration was to overcome the genetic problems of human-caused isolation of Florida panthers and to reinstate gene flow between the Florida and Texas populations that occurred prior to 1900 (Belden and McCowan 1996). The research of Culver and colleagues (2000) showed that Florida and Texas represent distinct populations of the same subspecies (although this might not be widely accepted), so Texas was an appropriate source population. Since the specific goal in genetic restoration was to maintain not more than 20 percent Texas alleles in Florida individuals, each Texas female was removed from the population after she contributed a set number of offspring to the Florida population. The achievement of this goal requires continued monitoring of the Florida panther population. After the introduction of Texas cougars into Florida, cryptorchidism and atrial septal defects were no longer observed in Florida panthers. In addition, the disease load was reduced and heterozygosity for microsatellite markers increased by 28 percent (see Chapter 12 for further discussion of the Florida panther).

Western North American Populations

Within a subspecies, molecular genetic data can be used at a higher resolution to identify populations, where each subspecies would comprise many populations. Utilizing higher resolution genetic data we can infer how North American cougars are subdivided at the higher resolution population level; for this, a higher density of samples and a greater number of microsatellite markers are needed (see DNA fingerprint and microsatellite DNA in Appendix 1 Primer). Seven recent population-level studies, using 9–18 microsatellite markers and 50–300 samples, have done this for many regions of western North America (Arizona, California, Colorado, New Mexico, Idaho, Montana, South Dakota, Texas, Utah, Wyoming, Alberta, and British Columbia).

The results of these higher resolution genetic studies have shown large regions with little or no population structure, such as Colorado and Utah (McRae 2004; Sinclair et al. 2001); eastern Wyoming and South Dakota (Anderson 2003); and northwest Wyoming, Montana, Alberta, and British Columbia (Biek et al. 2006b). Also found were areas with significant population-level subdivision within a single state, such as Arizona and New Mexico (McRae et al. 2005), California (Ernest et al. 2003), Idaho (Loxterman 2001), and Texas (Walker et al. 2000). When population-level subdivision was found, the population boundaries corresponded to potential barriers to gene flow. In California, the Central Valley, San Francisco Bay and Delta, Los Angeles Basin, Mojave Desert, and Sonoran Desert were potential barriers (Ernest et al. 2003). In Arizona and New Mexico, features such as low desert, semidesert grasslands, urban development, and possibly the Grand Canyon provided barriers (McRae 2004). In Idaho, the Snake River Plain and associated human development were the potential barriers (Loxterman 2001). The cougar populations in the south versus the west showed evidence of geographic subdivision indicating the possible existence of barriers (Walker et al. 2000). In addition, the California and Arizona–New Mexico studies showed that higher levels of genetic variation were found farther south in the study areas. This could be evidence of the post-Pleistocene range expansion because, as already suggested, this expansion went from south (where populations remained stable) to north (where populations had been extirpated).

One of the seven higher resolution population studies also used puma FIV polymerase and envelope gene sequences. The FIV envelope gene codes for a protein the virus needs in order to infect its host. Biek and colleagues found evidence of some population structure for cougars in the northern Rocky Mountains (Biek et al. 2006b). A significant division between the more northern areas (British Columbia, Alberta, northern Montana) and southern areas (southern Montana and northwest Wyoming) was observed in phylogenetic analysis of sequences from two puma FIV genes. In addition, spatial structure of viral sequences was much more pronounced, with some sequences restricted to small geographic regions within the study area. Based on the estimated mutation rate for these viral sequences, this structure likely represents movements of the past twenty to eighty years.

This time interval corresponds to the period of low cougar numbers due to human persecution, so the pattern observed probably reflects lineages that survived and expanded following the end of the war on predators, with the widespread lineage being one of the first lineages to recover following the years of persecution. Because large-scale cougar movements occur, we can expect that this pattern of viral sequences will be visible for a brief window of time, and that over time the viral lineages will become mixed and uninformative for small-scale resolution, as observed with the microsatellite data for this same study area.

Individual-Level Case Studies

The power of genetics is its capacity to identify all individuals as unique, to distinguish paternal/maternal relations, and to estimate kinship among individuals in a population.

Kinship

Kinship levels were investigated in male cougars compared to females by measuring the band-sharing across eleven microsatellite loci (Biek et al. 2006a); higher levels of band-sharing indicate greater kinship relation, lower levels of band-sharing indicate lower kinship or even unrelatedness. Because males are the long-distance dispersers in cougars, it would be expected that males in an area would have lower levels of kinship, and that females of an area (representing a group of related individuals) would exhibit higher levels of kinship. Biek and colleagues found that while male cougars had lower than expected band-sharing, and females did have higher band-sharing than males, females did not exhibit higher than expected band-sharing levels. These results indicate that male dispersal likely contributes to inbreeding avoidance—males share fewer alleles with each other and with previous female residents. However, the role of female philopatry—the dispersal behavior where females stay near their place of birth and males migrate away from their birth area, or natal range—is unclear because it did not result in higher than expected levels of band-sharing. Biek and co-workers suggest that female short-distance dispersal may be enough to break up spatially related female groupings, or female lineages may have high turnover.

Overall kinship levels for four cougar populations were examined in a separate study (Culver et al., in prep.), using three populations in western North America (Yellowstone, southern Idaho, and southwestern New Mexico) and one in South America (Chile). The random set of ten microsatellite markers gave heterozygosities of 0.42–0.52 in the North American populations and 0.63 in Chile. These ten markers could distinguish relatedness classes in the Chile populations with 85 percent accuracy (using calibration from Blouin et al. 1996) but, due to the lower heterozygosity levels in the North American populations, this data set could not distinguish full-siblings from half-siblings, or unrelated individuals from cousins. The implication for North American cougar populations is that given this set of microsatellites, classification of an unknown individual as related or unrelated to the population would be inaccurate. Thus for kinship studies of North American cougar populations, instead of using general cougar microsatellite loci, a separate set of loci that have high heterozygosity in each population of study should be selected.

Paternity

Paternity studies were performed for populations in Yellowstone and southern Idaho, using ten microsatellite markers (Culver et al. 2008; Loxterman et al., in prep.), where females with known litters were tested for paternity against all adult males known in the population. In Yellowstone, of twenty-four potential fathers known from telemetry, only nine were proven fathers from genetic testing. However, in Idaho, of seven potential fathers known from telemetry, all seven were proven fathers from genetic testing. This resulted in a ratio of effective population size to census population size (N_e/N) of 0.31 in Yellowstone and 0.41 in southern Idaho.

A potential reason for this difference in N_e/N ratio may be that the Yellowstone population is protected from hunting, whereas in the Idaho population the dominant male may be removed each year through hunting. Alternatively, this may be an artifact of the Yellowstone population having been extirpated in the early 1900s and still being an expanding population. This could be resolved by future examination of the N_e/N ratio in other disturbed and undisturbed cougar populations.

Forensics

The use of forensic genetic techniques in cougar conservation is relevant when noninvasive samples, such as scat, are used for genetic analyses. Ernest and colleagues (2000) laid the groundwork to show that these techniques are useful for identification of individual cougars as well as to determine the minimum number of unique individuals occupying an area of study. In this study, twelve microsatellite loci were used to genotype thirty-two scat samples collected in the Yosemite Valley of California. Individual identification was able to "capture" the same individual more than once, indicating the potential of this technique to provide an alternate method for conducting mark-recapture study of cougars.

Since population size estimates are still lacking for most cougar populations, this technique can be an important conservation genetics tool that can aid managers. This type of noninvasive monitoring is currently being conducted for cougars in Arizona and other parts of the western United States, and has been performed in other carnivore species.

Importance for Conservation

Based on the findings of these various cougar genetic studies, several conclusions can be made regarding cougar subdivision and population genetics:

1. Cougars originated in the Brazilian Highlands approximately 300,000 years ago.
2. An extirpation and recolonization occurred in North America between 10,000 and 17,000 years ago, causing North American cougars to appear bottlenecked relative to South America.
3. Molecular data do not support the traditional thirty-two subspecies divisions for cougars, but instead six subspecies are recommended.
4. Within these six groups cougars are fairly panmictic; that is, freely interbreeding because of having no barriers to movement.
5. Some genetic structure and barriers to gene flow exist within the broadly defined North American group.

Implications

These conclusions have several implications for cougar conservation. A key one is that managers should strive to maintain habitat connectivity within the six large groups defined by this data set. Management for the two North American threatened or endangered cougar populations (Florida panther and eastern cougar) should take into account the revised subspecies designations suggested by this data set.

Florida panther. In Florida, managers have indeed taken genetic findings into account with the introduction of neighboring Texas cougars to ameliorate the effects of inbreeding. Based on the genetic results, this action in Florida resulted in a mixing of two populations of the same subspecies rather than a mixing of subspecies, as originally thought. The endangered Florida panther is listed as an endangered subspecies, but this population closely fits the definition of a distinct population segment (DPS); the Florida population clearly occupies a unique geographic location relative to other North American cougars. A DPS is afforded protection under the ESA, thus a revised subspecific taxonomy for cougars need not alter the protected status of Florida panthers.

Eastern cougar. If it is determined by the general public, non-government organizations, and agency managers that a naturally breeding free-ranging population of cougars in the East is a desirable objective, a similar introduction of cougars from neighboring regions to the west should be explored. The population-level studies clearly demonstrate that broad-scale maps of cougar habitat can be used to predict where significant barriers to movement and gene flow are likely to occur and could be applied to areas in the eastern United States where no intensive gene flow studies have been performed.

Future of Cougar Conservation Genetics

There are many unanswered questions to examine for which genetic analysis would be the tool of choice. Two obvious examples are additional study of the unexpected isolation of cougars on the Olympic Peninsula (illustrated by 5/10 monomorphic microsatellite loci and one unique mutation that was fixed on the peninsula and not seen anywhere else) and possibly a further examination of the bottleneck of Vancouver Island cougars (5/10 monomorphic microsatellite loci); see Chapter 12 for more about these populations. Further exploration of barriers to gene flow and current population boundaries for cougars—both natural and human imposed—is another promising line to follow. Also, population size estimates are still difficult to obtain for large carnivores such as cougars; surveys using noninvasive (forensic-style) genetic techniques have allowed some progress, but additional data are needed in this area. Related to population size would be further study of N_e and N_e/N ratio for cougar populations undergoing different levels of disturbance to assess effective (breeding) population size.

Much of the genetic diversity measured in wildlife studies and discussed in this chapter is neutral diversity, not undergoing strong selection. As noted, the advantages of this type of genetic variation are that it can indicate the time elapsed since populations were founded or became isolated and ceased interbreeding, and it can also indicate the genetic relationships among individuals. As sequenced genomes are established for more species, direct inferences (at the single-locus DNA level) regarding detrimental and adaptive variation will be possible (Kohn et al. 2006). The study of adaptation and fitness traits is an area in which genetics can make an important contribution. For this, we would need to utilize data from sequenced genomes to find genes that have known location, known sequence, and known functions. In particular, since the domestic cat genome project—including the full sequence for the domestic cat genome—is expected to be complete in the near future, looking for adaptive genes in felids such as the cougar will be possible.

Conclusion

The intricacy of genetics studies can be intimidating, but the information generated has increased our knowledge of the evolution, taxonomy, management, and conservation of cougars. Findings can aid conservation of cougars by providing managers with a significantly clearer understanding of populations to be managed. Genetics work can illuminate the level of inbreeding, inbreeding depression, heterozygosity, allele frequencies, migration rates, mutation rates, population size (effective and census), mating system, and sex ratio.

Cougar populations should be managed to maintain genetic health so that populations retain enough genetic variation to retain reproductive fitness and evolutionary potential in both the short and long term. An additional management priority should be to maintain functioning groups of populations that are able to retain natural levels of gene flow. The uses of microsatellite DNA and FIV viral DNA may be the most useful tools to assess barriers to gene flow and thus aid the achievement of this goal. Management that incorporates genetic data along with ecological and demographic data will ultimately be more robust and successful.

Chapter 4 Cougar Management in North America

United States: Charles R. Anderson Jr. and Frederick Lindzey

Canada: Kyle H. Knopff, Martin G. Jalkotzy, and Mark S. Boyce

FROM THE MOMENT Europeans established the original colonies that would become the United States, predators were viewed as a threat, not only to livestock but also to the settlers themselves and to the other wild animals settlers relied on for food. Since then three phases of cougar (*Puma concolor*) management have evolved: attempted eradication, followed by agency management to sustain sport hunting and address depredation concerns, and more recently an effort to sustain viable cougar populations as part of the ecological community.

The first phase of cougar management emerged as an agricultural ethic that focused on eradicating "undesirable species" that potentially threatened livestock, game animals, or the settlers themselves. One of these undesirables was the cougar. This attitude of eradication dominated until the middle of the twentieth century, when states and provinces began to assume management authority for the species, initiating the second phase of cougar management.

Along with this new authority to manage cougars came the responsibility to sustain cougar populations. This phase is characterized by the hunting of cougars to provide recreational opportunity, while continuing to address livestock concerns. Complexity of cougar management accelerated rapidly from this point as a more biologically based ethic began to develop within agencies and the public, and new stakeholders entered the management arena. This increasing complexity has forced some agencies into a third phase in which they have begun to examine management approaches more critically. State and provincial management plans illustrate these evaluations by addressing cougar ecology and sociopolitical aspects of cougar management. They also incorporate stakeholder input and recently acquired knowledge in order to develop management programs more acceptable to the public at large. In the first two sections of this chapter, Charles Anderson Jr. and Frederick Lindzey discuss the phases of cougar management in North America, primarily in the United States. They describe the evolution from eradication to supporting a sport harvest, addressing cougar predation on livestock, and sustaining viable cougar populations for ecological and recreational purposes. In the third and final section, Kyle Knopff, Martin Jalkotzy, and Mark Boyce identify how management in Canada differs from that in the United States.

Earliest bounties paid by settlements, colonial governments, and later fledgling states were directed primarily at controlling wolves (*Canis lupus*). As wolf numbers declined and density of human settlement increased, bounties were also paid for other predators, such as cougars (Young and Goldman 1944, 1946a). From colonial times to the 1960s, the goal was to eradicate predators, a philosophy that moved westward across North America with European settlement. The result of eradication, along with habitat losses associated with human development, was that by the twentieth century cougars were extirpated from North America east of the Rocky Mountains, except for a remnant population in Florida (Nowak 1976; see also Range Map, p. vii).

Although wolves were initially the focus of predator control efforts in the West as they had been in the East, other predators were taken incidentally during wolf control campaigns. Later these species, including the cougar, became prime targets as well. Livestock associations often hired hunters and trappers to kill cougars in areas with depredation problems. Bounties were commonly used to reward such hunters and direct their efforts to selected areas, a practice that continued in the western states until the mid-twentieth century. Young and Goldman (1946a, 166) found nine western states offering cougar bounties in 1937, with payments ranging from $50 per animal in Colorado to $2 in Nebraska. Poisons were widely used for predator eradication and likely killed cougars as well as their canid targets. As noted by Young and Goldman (1946, 167), young cougars may have been most vulnerable to poisoning from consuming poisoned carrion baits set for wolves and coyotes, *Canis latrans*. But poisoning the carcasses of animals killed by cougars can be effective in targeting the cats, because they commonly return to feed on these carcasses (Anderson 1983).

Federal involvement began in 1907 with U.S. Forest Service wolf reductions on national forests. The federal government formally entered the predator control business in earnest in 1914 with passage of legislation providing money for the Department of Agriculture to fund "experiments and demonstrations in destroying wolves, prairie dogs and other animals injurious to agriculture and animal husbandry" (Young and Goldman 1946a, 383). The animal damage control program within the Department of Agriculture's Bureau of Biological Survey was later moved to the Department of Interior, and then eventually returned to Department of Agriculture (see Chapter 1 for fuller discussion of predator control).

Bounties for cougars were generally phased out by the early 1960s (Cougar Management Guidelines Working Group 2005). Indiscriminate use of poisons on public lands was terminated by presidential proclamation in 1972, and most western states and provinces assumed management authority for cougars between 1965 and 1973. Current Environmental Protection Agency regulations prohibit the use of poisons on public lands unless a memorandum of understanding is developed with the federal land management agency. In these cases, only target-specific methods (M-44s or 10-80 collars) can be employed. Colorado and Nevada reclassified the cougar as a game animal in 1965; Washington in 1966; Oregon and Utah in 1967; California in 1969; Arizona, Montana, and New Mexico followed suit in 1971; Idaho in 1972; Wyoming in 1973 (Roberson and Lindzey 1984, Smith 1989, Dawn 2002); North Dakota in 1991 (with closed season until 2005); and South Dakota in 2003. Except for Florida, the remaining states and provinces have not implemented active cougar management programs because viable cougar populations are not evident.

State and provincial agencies now faced the task of managing an animal that had been persecuted for two centuries to protect livestock and wild prey. Responsibility for cougars as a species accompanied management authority. For example, Wyoming statutes charge the Wyoming Game and Fish Commission and Department with providing an adequate and flexible system for the control, propagation, management, protection, and regulation of all Wyoming wildlife. The department is the only entity of Wyoming state government charged with managing wildlife and conserving it for future generations (Wyoming Game and Fish Department 2006b, Department and Program-Level Strategic Plans). Real or perceived livestock depredation and wildlife predation concerns did not go away simply because the cougar was decreed a game animal. At last, however, the attitude of some members of the public and of many professional biologists supported the notion of a balanced approach in which "positive" as well as "negative" aspects of the species would be considered in management planning.

Legal protection of the cougar as a game species signaled the entry of new stakeholders attempting to influence cougar management (Chapter 14, 15). Agencies faced the unenviable task of trying to achieve management decisions that would serve to maintain biological integrity while also balancing the demands of conflicting interests. Approaches and techniques learned from decades of managing other game species were often of little use when dealing with an obligate carnivore that occurred in comparatively low densities and traveled over large areas. The general history accounts (i.e., Young and Goldman 1946a) and the few more specific reports available in the early 1960s on breeding, food habits, and natural history (e.g., Connolly 1949; Gashwiler and Robinette 1957; Robinette et al. 1959, 1961) provided limited support for, or help in, developing management plans in what was certain to be a hostile management environment.

Regulation Begins

Typically, the first years of wildlife agency responsibility for cougars saw little more than the setting of bag limits. Hunting regulations then became progressively more restrictive as season lengths were shortened and timing shifted. Management areas were delineated to control distribution of the hunting harvest. In some areas, quotas were set to limit total kill and/or the harvest of male and female cougars. Over time, regulations protecting spotted juveniles and females with kittens have also been adopted by most management agencies (Table 4.1; management status reports in Becker et al. 2003). At the same time, most states and provinces continued to include liberal provisions for livestock owners to respond to cougar depredation problems (Roberson and Lindzey 1984, 207).

Agencies have used a combination of approaches to manage cougars, often with only a limited understanding of the effectiveness of their management prescriptions. Working with imperfect knowledge is nothing new to biologists, but it is a politically risky affair when the focus is an animal over which people are dramatically separated in their views. When faced with limited information, wildlife managers tend to be conservative in setting harvest objectives. Because knowledge of the effectiveness of various actions is slow to accumulate, state- or provincewide management strategies usually also change slowly. Political pressures, however, sometimes cause sudden and drastic shifts in cougar management programs (discussed in Chapter 15).

Data Analysis Units

Although most states and provinces began cougar management on a province- or statewide basis, many have subsequently delineated their cougar habitat based on vegetation and topography, and presumably similar cougar densities, to refine management efforts (i.e., cougar management units). These areas are typically large enough to support population-level analyses, although it is generally recognized that they rarely contain isolated cougar populations. Management units, in turn, are frequently broken into smaller contiguous areas or hunt areas that share "social" characteristics (Wyoming Game and Fish Department 2006b), such as cougar depredation history, land ownership, public access, and hunting history, where unique management strategies can be applied to address localized issues (e.g., historic depredation incidents, cougar-human interactions). This division allows agencies to direct management actions to address the sociopolitical aspects of cougar management. Traditionally, social issues revolved around cougar depredation of livestock or potential predation impacts on game species (e.g., deer and elk). More recently, human-cougar interactions have become of increasing concern as recreational activities have increased in, or human development has encroached on, cougar habitat.

Seasons

Earliest regulations typically provided year-round hunting, but as management evolved, seasons were often shortened and timed to reflect specific objectives within management or hunt areas. For example, year-round hunting has been used to direct hunting effort into areas experiencing cougar depredation problems. Seasons are often set to begin after hunting seasons for other game species are over to avoid conflicts among different kinds of hunters, particularly in areas where cougar hunting involves dogs (Roberson and

Lindzey 1984, 97). Seasons typically end in the spring, before ungulate peak birthing periods, in order to reduce the potential of disturbing ungulate neonates (Table 4.1). In northern states, timing of seasons in this manner often did little to reduce cougar hunting opportunities because hound hunting was normally limited to periods of optimal opportunity, when snow was on the ground. Additionally, most hound hunters chose to avoid situations where their dogs might be shot (e.g., during big game seasons). Season timing has also been used to reduce the likelihood of kittens being killed by dogs (e.g., Utah, Colorado). Because cougars give birth year-round, however, timing of seasons does not protect all kittens; patterns of cougar litter production are given in Chapter 5 (see Figures 5.2–5.6). When harvest quotas are used, seasons end when the quota is met.

Regulating Hunter Numbers and Distribution

Regulation of sport hunting for cougars typically follows one of three harvest strategies: general seasons, limited entry, and harvest quota systems (see Table 4.1; Cougar Management Guidelines Working Group 2005). General seasons allow unlimited hunting of cougars of either sex and have the least control over harvest levels, the only restrictions being the number of licenses issued per hunter (typically one per season) and the timing and length of the hunting season. General seasons provide the most hunting opportunity and have the least control over harvest levels. Also they may result in uneven hunting pressure—accessible areas are hunted more heavily than inaccessible areas, limiting control over the harvest level, composition, and distribution.

Limited entry programs may restrict the number of hunters per management unit through limited license allocation, using either a first-come, first-served basis or lottery license sales. This approach may be the most limiting in terms of hunter opportunity but can be useful to disperse hunting pressure and control harvest levels, and it may increase the opportunity for hunters to be selective (increasing the male harvest) in areas where hunting pressure is low. For hunters willing to travel, limited entry may continue to allow similar hunting opportunities if the number of permits allocated is high.

Harvest quota management requires setting a limit on the total harvest and/or the number of female or male cougars harvested from an area. Quotas are not goals but allowable harvest limits set to achieve specific population-level objectives. The hunting season is closed for the area once the quota has been met. Advantages of the quota approach are that hunting opportunity remains high and harvest distribution and level can be regulated. The quota system requires an agency to develop a method that allows hunters to monitor

Table 4.1 Cougar population status[a] and characteristics of management programs in the western United States and Canadian provinces in 2008.[b]

State or Province	Population Size/Trend	Legal Status[c]	Season Dates (Bag Limit)[d]	Season Structure[e]	Dogs Allowed	Female & Cub laws	Pursuit Seasons	Mandatory Inspection	Depredation Compensation
Alberta	800–1200/I	Big game	12/1–2/28(1)	FQ, MQ	Yes	Yes	No	Yes	Yes
Arizona	1,500–2,500/Unk	Big game	9/1–5/31(1±)	Gen	Yes	Yes	No	Yes	No
British Columbia	4,000–6,000/S	Big game	9/8–6/30(2)[f]	Gen, FQ	Yes	Yes	Yes	Yes	No
California	4,000–6,000/S	Protected	NA	NA	No	NA	No	NA	No
Colorado	3,000–3,600/Unk	Big game	11/19–3/31(1)	TQ	Yes	Yes	No	Yes	Yes
Idaho	2,000/D	Big game	8/30–3/31(1–2)[f]	Gen, FQ	Yes	Yes	Yes	Yes	Yes
Montana	Unk/Unk	Big game	10/21–4/14(1)	LE, FQ, MQ	Yes	Yes	Yes	Yes	No
Nevada	2,500–3,000/S	Big game	YR(2)	TQ	Yes	Yes	No	Yes	No
New Mexico	2,000–3,000/Unk	Big game	10/1–3/31(1–2)[g]	TQ	Yes	Yes	No	Yes	No
North Dakota	27–101 adults/I	Furbearer	9/1–3/11(1)	TQ	Yes	Yes	No	Yes	No
Oregon	5,700/I	Big game	8/1–5/31(1–2)[g]	Gen, TQ	No	Yes	No	Yes	No
South Dakota	200–225/I	Big game	11/1–12/31(1)	TQ, FQ	No	Yes	No	Yes	No
Texas	Unk/S	Non-game	YR/unlimited	Gen	Yes	No	No	No	No
Utah	2,528–3,936/Unk	Big game	11/21–6/1(1)[g]	LE, TQ	Yes	Yes	Yes	Yes	Yes
Washington	1,000–2,500/D	Big game				Yes			No
	21 counties		9/1–3/15(2)	Gen	No		No	No	
	6 counties		12/1–3/31(2)	LE, TQ, FQ	Yes		Yes	Yes	
Wyoming	Unk/S	Trophy game	9/1–3/31(1)[g]	TQ, FQ	Yes	Yes	No	Yes	Yes

[a]Population size and trend based on subjective information such as harvest data, sightings, nuisance incidents, extrapolation of localized field research, and/or literature-based density estimates extrapolated to suitable cougar habitat. Trend: I = increase; S = stable; Unk = unknown; D = decrease. Population size and trend information reported from most recent management summaries (Becker et al. 2003 or Martorello and Beausoleil 2005) if available, information accessible from agency websites, or Beausoleil et al. 2008.
[b]Information accessed from management agency Web sites.
[c]Legal status change from predator to game animal: Colorado and Nevada in 1965; British Columbia and Washington in 1966; Oregon and Utah in 1967; California in 1969; Alberta, Arizona, Montana, and New Mexico in 1971; Idaho in 1972; and Wyoming in 1973. Legal status in California changed from game animal to specially protected mammal in 1990, and from protected to game animal in South Dakota in 2003 and North Dakota in 1991 (with a closed season until 2005).
[d]Bag limit = maximum number of cougars harvested/hunter/year except in Arizona where some management areas allow for 1 cougar harvested/hunter/day. YR = cougar hunting seasons open year-round.
[e]Season structure: Gen = general; LE = limited entry; TQ = total quota; FQ = female quota or female subquota when used in combination with TQ; and MQ = male quota.
[f]Season dates vary among management areas within interval reported.
[g]Some management areas are open to cougar hunting year-round.

the harvest so that they know when the quota has been met. This is accomplished using a toll-free hotline that is continually updated as cougars are harvested; hunters are expected to monitor the harvest by accessing the hotline. Occasionally, quotas are exceeded because there is often a lag in the reporting of kills and their entry on the hotline. This should be recognized and adjusted for in the development of harvest quotas. Female subquotas can be used to support a management objective of sustaining harvest levels by limiting female harvest levels and reducing impact on the cougar population. Potential disadvantages of harvest quotas are that the number of hunters per management unit is unlimited until

quotas are filled, and quotas may be exceeded if several cougars are taken toward the end of the season but before the harvest is recorded on the quota hotline.

All human-caused cougar deaths (including depredation control removals and known accidental deaths, such as from vehicle collisions) may or may not be counted against the quota; Wyoming, for example, recently moved to include all such deaths in its quota for fuller accountability. Counting all human-caused mortalities toward management quotas is a desirable management strategy because mortality factors other than hunting likely contribute to cougar population dynamics (Laundré et al. 2007).

Hunting Methods

Methods of hunting cougars include opportunistic spot-and-stalk hunting, calling cougars using predator calls, and hound hunting—tracking and baying cougars using trained hunting dogs. Most western states and provinces allow hound hunting, which has traditionally been the most common and effective method for hunting cougars. However, some stakeholders dislike the idea of pursing wildlife with dogs. As a result, Oregon and Washington have banned hound hunting. Where hound hunting is not allowed, predator calling and opportunistic cougar hunting during big game seasons appear to be comparably successful, based on harvest levels observed in Washington (Beausoleil et al. 2005) and South Dakota Department of Game, Fish and Parks. "South Dakota Mountain Lion Hunting Season," http://www.sdgfp.info/Wildlife/MountainLions/MtLionhuntingseason.htm (accessed 2007). This may be a function of the increased number of cougar hunting permits issued and longer seasons in place of the more effective hunting method of using hounds.

Results from Washington (Martorello and Beausoleil 2003) suggest that opportunistic cougar hunting is less selective than hound hunting. This has made female cougars more vulnerable: relative female harvest levels increased from 42 percent to 59 percent after hound hunting was banned in Washington. Harvest data from western states (management status reports in Becker et al. 2003, Beausoleil and Martorello 2005) suggest that hound hunting results in the higher harvest of males than females. Presumably, hound hunters have better opportunity than other hunters to identify females because they can often distinguish size differences when they encounter a track. They also can spend time observing the animal once it is treed. Additionally, they are more likely to encounter males while tracking, because males travel distances that average more than twice the average distances for females (Anderson 2003). On the other hand, opportunistic hunters who are not tracking cougars are more likely to encounter the more abundant sex (females; Logan and Sweanor 2001, Laundré et al. 2007).

Bag Limits/Permits. Agencies have offered "sportsman packages" that included a cougar permit with the purchase of other game tags, but most states and provinces typically require the purchase of a separate cougar permit. Bag limits are most commonly set at one per season, but larger bag limits have been used to raise harvest levels in specific areas (e.g., where depredation incidents are high). Sportsman packages are more common where hound hunting is not allowed, in an attempt to increase the number of licensed hunters afield and thus maintain similar cougar harvest with less effective methods.

Achieving Desired Cougar Sex and Age Classes of Harvest. It is illegal in most jurisdictions (with the exception of Texas; see Table 4.1) to kill spotted juveniles or females with young at their side (prohibitions known as cub laws; management status reports in Becker et al. 2003, Beausoleil and Martorello 2005). Although there are birth pulses, typically May–October (Figures 5.2–5.6; Cougar Management Guidelines Working Group 2005, 53), cougars may have young at any time of the year. Consequently, because kittens do not always accompany their mother, particularly when very young (Barnhurst and Lindzey 1989), hunters may unknowingly kill females that have young. Based on the proportion of reproductive-age females killed by hunters each year, evaluations in Wyoming (Wyoming Game and Fish Department 2006b) suggest that juvenile loss resulting when females with young are killed average about twenty-two per year. Although undesirable, this loss should have limited effects on the statewide population. Female subquotas, cub laws, and the statutory timing of seasons to exclude summers are intended to offer juveniles and females with young some protection. Aiming in part to help hunters identify females and look for signs that they may have young, Colorado and Washington have recently introduced mandatory cougar hunter education programs, and New Mexico, Utah, and Wyoming have similar voluntary programs.

Pursuit Seasons. Currently, four states and one Canadian province provide special seasons during which cougars can be pursued by dogs and treed but not killed (Table 4.1). Some jurisdictions prohibit pursuit during ungulate seasons (Roberson and Lindzey 1984, 87). Pursuit seasons allow hound hunters increased opportunity to work their dogs by baying and releasing cougars during the pursuit season. Although cougar populations are ostensibly not directly impacted by this practice, unintentional kitten loss, potential for stress-related mortality (Harlow et al. 1992), and increased illegal take may result.

Dealing with Depredation. States and provinces have used various approaches to deal with livestock depredation. Generally, livestock owners are given a fair amount of latitude in dealing with problem cougars. Typically, they can kill the offending cougar without first obtaining a permit from the state. They are required to report any cougars killed and to justify their actions. Agencies have also responded to depredation problems by employing professional hunters, structuring hunting regulations to increase sport harvest in problem areas, reimbursing livestock owners for animals killed by cougars (four states and one Canadian province, Table 4.1), and contributing monies to federal animal damage control programs (Roberson and Lindzey 1984, 102, 204).

Refining the Procedures

Documenting the harvest level and composition is a required component of management. Hunter questionnaires have been used to estimate total harvest, location of kill, sex and age of the cougar taken, and effort expended. More commonly, agencies require that harvested cougars be checked by agency personnel, when biological data and hunter information are recorded. Cougars taken under depredation provisions are subject to similar reporting requirements, except in Texas, where cougar take is unregulated and mortality reports are voluntary (Young 2003).

Management is the attempt to achieve desired objectives. Success is measured by how closely the results of the management prescription match the desired outcome. This implies that results can be measured, but cougar densities are rarely known or measured due to the cryptic nature of this solitary large carnivore and the rugged terrain it occupies. Hunting is the most controversial component of cougar management. Besides providing recreational opportunity, hunting is also used to alter populations in an effort to achieve specific objectives, such as reducing livestock depredation or predation on other animals, both of which are assumed to be related to cougar density. Although the relationship between cougar density and ungulate predation or depredation level is assumed to be linear—that is, a given percentage reduction of the cougar population will result in a proportionate reduction in predation or depredation—this has not been demonstrated (Cougar Management Guidelines Working Group 2005). Specifically, we do not know what percentage of the cougar population must be removed, if any, or what seasonal conditions, prey number regimes, and or husbandry practices result in a given level of predation or depredation reduction.

Population Assessments

Direct measures of cougar population characteristics (density, sex, and age composition) are needed to evaluate success of management programs, but estimating cougar numbers and documenting composition are difficult. The best density estimates have come from long-term studies of relatively small areas where most cougars were captured and dynamics of the populations were monitored (e.g., Hemker et al. 1984; Ross and Jalkotzy 1992; Logan and Sweanor 2001, Laundré et al. 2007). Such studies are costly, and they often yield data from central habitats where circumstances may differ from those in more peripheral parts of the species' range. Nevertheless, these density estimates provide a starting point for management programs.

Track surveys (Smallwood and Fitzhugh 1995) can provide estimates of relative cougar abundance in areas where tracking substrates are suitable, abundant, and sufficiently spaced to support a proper sampling design. Line intersect probability sampling, proposed by Van Sickle and Lindzey (1991), and later evaluated and expanded on by C. R. Anderson (2003), can yield precise density estimates and can detect 15–30 percent changes in population size with intensive sampling of moderate to high density populations that provide reasonable sample sizes (at least ten cougar tracks per survey). This approach, however, is expensive and geographically limited in application, because it involves helicopters and requires specific snow conditions.

Indexes or indicators of population trend are valuable tools for agencies, but if population size is unknown, the actual relationship between the index (e.g., cougars harvested, depredation events, sightings) and population size is also unknown. Thus, where rigorous documentation of population status is still needed, indexes have limited use alone or even in combination. Anderson and Lindzey (2005) used the relative vulnerability of cougar sex and age classes to hunting, as described by Barnhurst (1986), and monitored changes in composition of a hunted population as it declined and recovered, in order to document compositional shifts that could be used to index population status. Various methods that use genetic identification with sight-resight analyses have been proposed (e.g., Ernest et al. 2002) and may provide useful population estimators in the future. Regardless of the monitoring methods available, multiple indicators should be used if the best evaluation of cougar management programs is to be achieved.

Cougar Management Plans

Most western states (except for Arizona, Montana, and Texas) periodically develop plans that, in effect, set the policy for cougar management and attempt to balance biological and social aspects. These plans typically include syntheses of available knowledge about ecology of the cougar and about management practices, provide for information and education efforts on how to avoid conflicts, and are intended to reflect stakeholder values incorporated after often lengthy public input sessions. Broad parameters are set for how objectives will be determined for hunt areas and at times management areas, how hunts are structured to accomplish objectives, and how accomplishment of objectives will be measured.

Development of cougar management plans took hold primarily in the 1990s in response to several factors, including an accumulation of management and biological information from the Mountain Lion Workshops (see Chapter 2, Table 2.2), new agency personnel having a broader ecological background, the reported increase in cougar populations regionwide (management status reports in Becker et al. 2003), increasing cougar-human interactions (e.g., Fitzhugh et al. 2003), and growing pressure from more diverse stakeholders challenging traditional cougar management programs. For the first time, cougar management began to include the human factor more explicitly. Approaches were developed to educate people living in and using cougar habitat on how to avoid conflicts (see Beausoleil et al. 2008).

In addition, the management planning process began to acknowledge the value of cougars beyond recreational hunting, recognize potential threats to cougar populations from factors other than hunting (e.g., habitat loss), and accept the need for justification of hunting as a management tool and for transparency in the management planning process to ensure that all stakeholders are included. These are ostensibly major shifts, but results are variable; some of the newer participants contend that even where broader public input is effectively gathered, it is nevertheless often ignored (Chapter 14).

Evaluation of Management

Experience using various combinations of seasons, permits, and sex and age restrictions to achieve management goals has provided managers insight into the effectiveness of the different approaches. Dawn (2002) conducted the first broad-scale evaluation of cougar management strategies and, despite the apparent trend of progressively greater restrictions on hunters, results showed that the number of cougars taken by sport hunting increased (see Figure 4.1; Appendix 2). She noted that increases in harvests after states and provinces assumed management authority did not necessarily (but could) mean that cougar numbers had grown since the presumed lows in the mid-to late 1960s (Nowak 1976). Other factors might also have bearing, such as number and effectiveness of hunters and liberalization of allowable harvest limits. Further, Dawn's analyses suggested that, among the season types, general seasons yielded the lowest harvest rates, a result she noted possibly reflected the fact that general seasons were most common during the early agency management period when cougar numbers were presumably low. The lowest percentages of females in the harvest occurred when female subquotas were imposed.

Ross and colleagues (1996) also found that the harvest increased under the quota system in Alberta. However, they noted that the increase may have been due in part to concurrent increases in season length. The quota system with a female subquota resulted in a reduction of the proportion of females taken by sport hunters from 43 to 29 percent, with the increased harvest being composed primarily of males. Comparisons among various harvest methods over time can be tenuous because cougar numbers and hunter effectiveness may change. For example, less restrictive management strategies—that is, general seasons—are more common where cougar hunting is less effective, namely, in

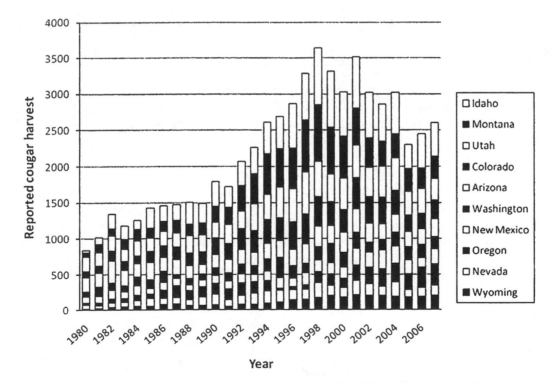

Figure 4.1 Reported number of cougars killed by sport hunters in the western United States, 1980–2007. Calendar year reported from Montana, Arizona and Oregon, and harvest year (fall of the previous year to spring of reported year) reported from Idaho, Utah, Colorado, Washington, New Mexico, Nevada, and Wyoming. The overall reported harvest increased threefold from the early 1980s to the late 1990s.

areas lacking snowfall for tracking or areas where hound hunting is not allowed. Moreover, most states have shifted their management strategies in similar ways, broadly from general seasons to setting quotas (Dawn 2002, 25), during the period when cougar populations were considered to be increasing (management status reports in Becker et al. 2003 and Beausoleil and Martorello 2005). Further evaluations of the effectiveness of type and timing of seasons relative to cougar densities will increase the confidence of managers in their application. State and provincial agencies are uniquely positioned to test the prescriptions in their management plans, if they so choose, using designs built into an adaptive framework (see also Adaptive Management on page 49).

Trends in harvest level should also be interpreted cautiously and in context. Harvest levels in western North America show remarkably similar trends, increasing during the early to mid-1990s, then leveling during the late 1990s and early 2000s, and exhibiting recent declines in some states (see Figure 4.1; Appendix 2). Although annual fluctuations in cougar harvests are in part related to changing harvest quotas/management strategies, some observers interpreted increasing harvests as endangering cougar populations, while others viewed the numbers as an indication that populations were growing at least sufficiently to support this increased harvest. States and provinces generally felt they were dealing with increasing populations (management status reports in Becker et al. 2003; Beausoleil and Martorello 2005) during the 1990s, even though evidence to support this conclusion was largely anecdotal, based on crude indices or a few long-term, localized studies. Four research efforts occurring at different but overlapping time spans and diverse locations (from New Mexico to Alberta) documented increasing cougar densities beginning in the mid-1980s and continuing through the mid-1990s (Ross and Jalkotzy 1992; Murphy 1998; Logan and Sweanor 2001; Laundré et al. 2007). Whether these localized observations reflected regionwide cougar population trends, however, is unknown. Three factors suggest that cougar populations were stationary, if not increasing, during the first three decades of agency management: (1) there was no consistent trend toward increasing adult females in the harvests, which would indicate that populations were being affected (Anderson and Lindzey 2005), (2) cougar populations were reestablishing along the species' eastern range (i.e., South Dakota, North Dakota; see Range Map, p. vii) and (3) cougar observations have come from even farther east and in Canada (Chapter 12; see also Canadian section below).

The recent downturn in harvests from some western states, however, may warrant attention. We do not know whether this trend reflects a reduction in cougar numbers and, if so, whether the reduction results from habitat conditions,

prey densities (e.g., Laundré et al. 2007), increasing harvest levels (Dawn 2002), hunter participation, or a combination of these factors. Management summaries from the 2005 Mountain Lion Workshop Proceedings (Beausoleil and Martorello 2005) are conflicting relative to recent trends, suggesting a decline in cougar populations in British Columbia, Idaho, and Washington; increasing populations in Oregon and South Dakota; and stable populations in California and Nevada. If western cougar populations have indeed increased over the past thirty years and are reaching equilibrium, expansion eastward should be expected to continue and to present new management challenges in areas being recolonized.

These population questions are controversial, and participants in the cougar debates tend to have very specific objectives. Livestock interests want management to lower depredation levels, and it has generally been assumed that the level of depredation is directly related to cougar density. Yet, the actual form of the relationship between cougar density and level of livestock depredation is not known. Addressing depredation simply by increasing the cougar harvest may find favor among livestock interests and deer and elk hunters who perceive a benefit from reduced cougar densities, but it garners less enthusiasm from the broader public and some cougar hunters, whose main interest is to ensure healthy/abundant populations. Thus, the challenge to management agencies is not merely to devise a biology-based approach that balances different interests but also to sell that approach to everyone and keep learning as we go along. While it may be relatively easy to document changes in livestock depredation or hunter opportunity, it is much more difficult, but equally necessary, to document concurrent changes in the cougar population.

Among the population characteristics that have bearing here is the fact that cougar populations tend to be genetically and physically connected over large areas (see Chapter 3; Culver et al. 2000a; Sinclair et al. 2001; Anderson et al. 2004), with segments of each population's overall area varying in suitability as cougar habitat and in terms of hunter access. The source-sink thesis described in Chapter 5 suggests that areas with low cougar survival (whether because of hunting and or other deaths) are supported by immigration from adjacent source areas, where survival and reproduction are higher. Logan and Sweanor (2001), and later Laundré and Clark (2003), used the source-sink concept to suggest a zone management and a metapopulation approach, respectively, whereby refuge areas are formally delineated within a state or province to guarantee sources. Abundance within these refuges of higher survival, and dispersal from them, should then ensure a supply of immigrants for exploited areas and assure continuation of viable populations.

Managers have long been aware that some areas of suitable cougar habitat are not hunted, or are seldom hunted; hunt areas were often initially delineated based on access, relative cougar density, and/or prevalence of depredation problems. Review of harvest records providing mortality density (number of human-caused deaths/area of cougar habitat) and sex and age of harvested cougars (Wyoming Game and Fish Department 2006b) would help confirm initial assignment of hunt areas to source or sink status. Sex and age composition of harvests are best interpreted on the basis of whether the harvest is from a source or sink area. Harvests composed primarily of young and male cougars might be expected from both source and sink areas. For a source area, a preponderance of young male animals reflects harvest of the most vulnerable age class, with reproductive-age females remaining relatively secure. For a sink area, such a harvest may simply reflect the main category of cougars present (Anderson and Lindzey 2005). Past cougar mortality records providing changes in harvest sex-age composition, human-caused cougar mortality densities, and age estimates showing changes in age structure can assist in determining initial source or sink status of cougar populations. Monitoring changes in demographics of harvested cougars over time should allow detection of declines or increases in that population (Wyoming Game and Fish Department 2006b).

Adaptive Management

State and provincial agencies do have the framework to design and conduct experiments to answer questions about the effectiveness of approaches such as using hunting to reduce livestock depredation, reduce predation on other desired wildlife, or reduce human-cougar interactions. Whether and how agencies use that framework depends on levels of interest, budget, and public pressure. *The Cougar Management Guidelines* (2005, 9) suggest that "adaptive management is characterized by the continual monitoring of indicators that measure progress toward the achievement of management goals and objectives, changing of management practices when new information indicates that better alternatives are available, monitoring relevant stakeholder values and interests, and the monitoring of natural environmental changes that may affect cougar management results." While this statement seems to fit the cougar management approach of most agencies, learning from experiments requires that careful thought be devoted to their design and that agency and stakeholder support be secured.

Management actions should be advanced as questions to be asked or hypotheses to be tested, with specific predictions made as to the outcome. For example, if an agency wished to test the question of whether sport hunting of cougars could reduce depredation on domestic sheep, the design would need to include means to measure level of cougar removal by sport hunting and level of sheep losses to cougars over the time of the experiment. Obviously, many other variables could act to influence the results. Sheep numbers and husbandry practices would ideally be held constant during the experiment, as would level of cougar harvest (this is where agency and stakeholder buy-in becomes essential). Weather patterns and general trends in cougar prey availability should be monitored as well.

The results of such an experiment conducted in an area of contiguous cougar habitat might well differ from results in a more isolated area, indicating the need to replicate the experiment in other locales in order to predict better when and where sport hunting may be an effective tool to reduce sheep depredation problems. Results of properly designed experiments could be of value to managers throughout the species' range, and if they were replicated with similar methods by a number of states and provinces, the cost and effort of gaining information could be shared.

Recent Changes in Management Status

Most western states and provinces have management authority over cougars, with the exception of Texas, where the species is not classified as a game animal and harvest is unregulated (Haverson et al. 1997; see also government mandates and jurisdictions discussion in Chapter 15). Management flexibility has been curtailed in some states and has more recently been relaxed in others. In 1990, California voters approved Proposition 117, which prohibited sport hunting of cougars. In 1994 a citizen ballot initiative called Measure 18 passed in Oregon, prohibiting hound hunting for cougars, and in 1996 Washington voters approved Initiative 655, likewise prohibiting the hunting of cougars with hounds (see Appendix 5 for more detail). Bills passed in 2004 in Washington (Beausoleil et al. 2005) and in 2007 in Oregon relaxed these prohibitions somewhat, allowing hound hunting for cougars in five Washington counties (increased to six counties in 2008) to address livestock depredation concerns and allowing the use of hounds for specific agency management actions in Oregon (e.g., depredation incidents). Cougars were reclassified in North Dakota in 1991, from state-threatened species to furbearer. South Dakota reclassified cougars in 2003, from state-threatened species to big game animal.

Cougars have long been extirpated from most areas east of the Rocky Mountains except for the Florida panther, which the state classified in 1958 as endangered, with the federal government following suit in 1967 (Lotz 2005). Management emphasis went from documenting cougar presence to protection and recovery efforts (see Chapters 3, 12).

Presence of cougar populations in other areas east of the Dakotas is currently unconfirmed, but sightings and occasional confirmed reports of dead cougars have increased in the past fifteen years (see http://www.cougarnet.org/network.html for examples). Confirmed deaths and reliable reports have been documented in several midwestern states (e.g., Iowa, Missouri, Oklahoma). Whether these cougars are of captive or wild origin is unconfirmed in many cases, but known dispersal of radio-collared cougars from South Dakota into Oklahoma and Minnesota (Daniel Thompson, South Dakota State University, pers. comm., 2008) supports the notion that many of these records, at least west of the Mississippi River, are true wild dispersers.

Additional factors also suggest potential expansion of western cougar populations eastward: the prevalence of males of dispersal age among the confirmed deaths; reemergence of cougar populations in eastern areas adjacent to the Rocky Mountain West (North Dakota, South Dakota, and possibly western Nebraska); and a recent increase in reliable cougar reports. As noted, observers in several western states considered cougar populations to be increasing during the 1990s, and this is supported by localized research documenting increasing cougar populations from New Mexico to Alberta (Ross and Jalkotzy 1992; Murphy 1998; Logan and Sweanor 2001; Laundré et al. 2007; on Canada, see below), and by genetic evidence suggesting recent demographic increase and expansion (Biek et al. 2006b). Whether establishment of cougar populations eastward succeeds will depend largely upon social acceptance and the proactive involvement of local wildlife agencies. Of the states bordering the known cougar distribution in the west, cougars are designated as protected in Louisiana and Wisconsin; game species with a zero harvest quota in Nebraska and Oklahoma; nongame in Arkansas, Kansas, Missouri, and Minnesota; and as unprotected in Iowa. Modification of management designations and implementation of management planning may be necessary to accommodate and/or prepare for cougar recolonization eastward.

Special Circumstances in Canada

Although most cougar research in North America has been conducted in the United States, findings presented in this book on cougar population dynamics, morphology, behavior, and interactions with prey are just as relevant to cougar north of the 49th parallel. Similarly, many of the general management issues described above also apply to Canada. Several aspects of the status and management of cougar in Canada are sufficiently distinct, however, that they deserve special attention: population distribution and abundance; level of protection, harvest, and control; and potential for range expansion and its implications for the northernmost cougars.

Distribution and Population Status

Prior to European settlement, cougar were present throughout the southern portions of Canada. Historical population size is unknown, but by the early 1900s a reduction of ungulates through market hunting and simultaneous persecution of predators probably restricted cougar at low densities to portions of British Columbia and western Alberta. As ungulate populations recovered by the mid-1900s cougar numbers are also thought to have increased, but as in the United States, bounty hunting and widespread poisoning for predator control may have limited recovery (Jalkotzy et al. 1992). Currently, breeding populations of cougar are known to be present in both Alberta and British Columbia (see Range Map, p. vii).

With the possible exception of the far north, breeding populations of cougar are found across the entire British Columbia mainland and have also found their way onto the coastal islands, including Vancouver Island, where densities have reached levels among the highest ever recorded (Wilson et al. 2004). The estimated number of cougar in British Columbia is between 4,000 and 6,000 individuals (Austin 2005). By contrast, in Alberta, breeding populations have traditionally been relegated only in a relatively small region along the western edge of the province, where the most recent published population estimate is 640 cougars on provincial lands—that is, excluding Banff and Jasper national parks (Jalkotzy et al. 1992). Our ongoing research data suggest that either cougar populations have increased north of the Bow River, where they had not previously been directly studied, or the original estimates were too low in this large portion of Alberta's cougar range. In either case, current cougar numbers in Alberta likely exceed the estimate provided in the early 1990s.

Breeding populations of cougar have recently become established outside the previously well-defined eastern boundary for Canadian cougar along the foothills of the Rocky Mountains in Alberta (see Range Map, p. vii). In the Cypress Hills Interprovincial Park, which straddles the Alberta-Saskatchewan border, a large number of cougar sightings, accidental trapping of an entire family group in snares set for coyote in December 2006, and high-quality photographic evidence of a second family group in August 2007 (Figure 4.2) are strong evidence that a breeding population exists in the park. Elsewhere in Saskatchewan, large numbers of cougar have been accidentally snared, shot, photographed, or confirmed by wildlife personnel since 2000. This, combined with numerous other sightings and reports across the province, has recently prompted the government

Figure 4.2 A camera trap captured this cougar family in Cypress Hills Interprovincial Park in Alberta, Canada. This is the first photographic evidence demonstrating that breeding populations of cougar have moved east to the Alberta-Saskatchewan border. Photo by M. M. Bacon, Cypress Hills Interprovincial Park Cougar Study.

to provide a provincewide estimate of approximately 300 cougars (Saskatchewan Department of the Environment 2007). Other than in the Cypress Hills Interprovincial Park, it is not clear where breeding populations might be found within Saskatchewan.

Cougar sightings also are common in Manitoba, where a female cougar was shot and a male was trapped near the Duck and Riding mountains in 2004, suggesting the possibility of a small resident cougar population (Watkins 2005). In Ontario, large numbers of unconfirmed sightings in the Great Lakes region (e.g., Ontario Puma Foundation 2007) have led the provincial government to officially accept the presence of a breeding cougar population (Ontario Ministry of Natural Resources 2007). Sightings are often unreliable indicators of cougar presence, however (Beier and Barrett 1993), and confirmation of breeding populations will require further evidence. If populations are present east of Saskatchewan, it remains unclear whether these are the remnants of original eastern cougar populations or derive from recent expansion out of the west or from the Dakotas to the south (see Range Map, p. vii).

Confirmed presence of individual cougars has been reported in many other locations across the country. In Quebec and the Maritime provinces, for example, dedicated researchers have recently obtained genetic evidence of cougars occurrence (Gauthier et al. 2005). Cougars also have been confirmed as far north as the southern portion of the Yukon Territory (Jung and Merchant 2005). Dispersing cougars can cover incredible distances (Thompson and Jenks 2005), and many isolated occurrences and even repeated sightings may represent dispersing animals (see Chapters 5, 8, and 12). Alternatively, some of the eastern sightings and confirmed occurrences could be animals that have been released from captivity. Where confirmed incidents are particularly concentrated outside the known distribution of breeding populations, they may indicate resident populations occurring at low densities.

Management and Conservation

As in most western states, cougars were managed as a bountied predator in Canada until the mid-twentieth century. They were extirpated from eastern Canada, and it was not until 1966 in British Columbia, and 1971 in Alberta, that the cougar achieved the status of big game animal. In Alberta, a comprehensive management plan is now in place (Jalkotzy et al. 1992), but it has not been updated since it was adopted in the early 1990s. At the time of this writing, there is still no official management plan for cougar in British Columbia, despite the fact that it is a species of important management concern in that province (Robinson et al. 2002).

This is quite different from the way that cougars are managed in the United States, where official management plans are now the norm and are updated regularly in many states where cougar occur (e.g., Apker 2005; Barber 2005). Lack of a plan in British Columbia and of any updates to the Alberta plan mean that official activity does not incorporate recent developments in our understanding of predator-prey ecology, such as information on the ecosystem benefits derived from the presence of large carnivores, including cougars (Ripple and Beschta 2006). Stakeholder opinions, moreover, are not officially acknowledged. In the United States, stakeholder opinion is at least designated as a component of most cougar management plans, and public action has influenced management through ballot initiatives in California, Oregon, and Washington. In Alberta and British Columbia, by contrast, public opinion and new ideas may receive consideration by provincial agency staff in their day-to-day duties and may have some bearing on cougar management, but they are not reflected in official plans.

Despite the lack of an official plan in British Columbia or regular updates in Alberta, cougar management remains an important issue. Cougar hunting is permitted and closely monitored in both provinces. In British Columbia, with the exception of some northern management units, cougar hunting is permitted throughout the province. The hunt is unlimited entry, with many units maintaining a bag limit of two cougar per hunter per season. Some management districts, however, have a female subquota limiting the number of females that can be taken each year. Cougar harvest in Alberta is regulated by a strict quota system for both males and females, with sex-specific harvest limits for each of several cougar management areas in the western part of the province (Ross et al. 1996). Both provinces permit hunting with hounds and forbid the harvest of spotted kittens or of females traveling with spotted kittens, although this regulatory constraint is a very recent development in British Columbia (Austin 2005). There is no season for cougar in the wildlife management units of eastern and northern

Alberta, nor may cougar be hunted in any of the other Canadian provinces or territories.

An important management issue for western Canadian cougar revolves around accidental snaring of the big cats. Snaring of wolves near carrion bait, often road-killed ungulates, is a common recreational and economic activity in western and northern Canada. Scavenging behavior has been demonstrated in cougars (Bauer et al. 2005), and some individuals scavenge frequently from animal remains (K. Knopff, unpublished data). Cougars are, therefore, susceptible to becoming by-catch in wolf and coyote snares. Because snaring for wolves is not currently permitted where cougars exist in the United States, cougars by-catch at wolf bait stations is a distinctly Canadian concern. Similar concerns exist, however, in states where snaring for coyotes is permitted and cougars also are present (Wyoming Game and Fish 2006b).

By law, all human-caused cougar mortalities must be reported to provincial authorities. We analyzed the data for all reported mortalities in Alberta between January 2000 and March 2006. Over that six-year period, 837 human-caused cougar mortalities were reported. The vast majority of these were cougars harvested by licensed hunters (77 percent); the second most important source of human-caused mortality was accidental snaring (9 percent). Less frequent causes of mortality were removal of problem wildlife (4.7 percent), road kills (4.7 percent), and self-defense (2.3 percent). Trappers in Alberta cannot sell cougar hides and must forfeit all cougars trapped or snared to the province, so there are neither economic nor trophy incentives for cougar snaring. The removal of so many cougars through snaring is undesirable because it neither improves the ability of managers to preserve self-sustaining cougar populations nor assists in maximizing the benefit to Albertans through optimal allocation of the resource, two key management objectives outlined in Alberta's cougar management plan (Jalkotzy et al. 1992). This issue has not previously received a great deal of attention, but our analyses show it to be an important source of human-caused cougar mortality in Canada, and it may become increasingly important in the United States if gray wolves are delisted and wolf management practices in Montana, Idaho, and Wyoming follow the Canadian model. Data currently available for cougar and wolf habitat selection, prey-site selection, and movement patterns (e.g., Alexander et al. 2006; Kortello et al. 2007; Atwood et al. 2007) might be usefully applied to help reduce the probability of unwanted by-catch of cougar at wolf bait stations, but the required analyses have yet to be conducted.

As is true wherever cougars and people share the same space, an important component of cougar management in Canada involves managing cougar interactions with livestock, pets, and people. Between 1890 and 2004, British Columbia had thirty-nine cougar attacks on people, the highest number of any jurisdiction in North America, with the majority of the incidents occurring on Vancouver Island. Alberta has only experienced a single lethal cougar attack on a human (in 2001), but complaints involving cougars are frequent. As noted, nearly 5 percent of all human-caused cougar mortalities involve removal of problem animals by wildlife officers, and a further 2.3 percent are the result of self-defense. In addition, approximately as many problem animals were relocated (thirty-six) as were killed (thirty-nine) by provincial wildlife agencies between January 2000 and March 2006. The high number of complaints by residents has led Alberta to amend its laws to allow cougars to be shot on sight on private land (Alberta Sustainable Resource Development 2007). Animals so taken may not be kept and must be turned over to provincial authorities. Compensation for livestock loss is available in Alberta from the Alberta Conservation Association but is unavailable in British Columbia (Austin 2005). As in the United States, high levels of negative interactions between cougars and people in Canada are likely a result of larger numbers of people living in and using cougar habitat, combined with a general increase in cougar numbers in recent decades.

Cougar conservation and management are becoming increasingly important topics in the provinces east of Alberta, where cougars have previously been considered extirpated but are reappearing (Watkins 2005). No official management, conservation, or recovery plans have been in place for cougars in any of these provinces. Saskatchewan, however, is now in the position of developing a strategy for managing cougar populations and addressing conflict with people in regions where cougars are becoming established. It and the other provinces east of Alberta will be responsible for determining whether and to what extent cougars are able to repopulate historic range. The Committee on the Status of Endangered Wildlife in Canada (COSEWIC 2007) lists the cougar in eastern Canada as data deficient, meaning that sufficient information is not available to assess the status of the eastern subspecies or assign it an extinction risk rating (i.e., extirpated or endangered). Therefore, if small populations of eastern cougar exist, they are not currently protected under Canada's Species at Risk Act (SARA). In 2007, the Committee on the Status of Species at Risk in Ontario independently designated the eastern cougar as endangered, stating that "there have been hundreds of sightings of cougars in Ontario over the years, and their presence here is generally acknowledged" (Ontario Ministry of Natural Resources 2007).

Consequently, the cougar now falls under the regulatory arm of the province's Endangered Species Act, affording cougar in Ontario the highest level of protection of any cougar in

Canada. Revisions to the act, which took effect on June 30, 2008, require that no person kill, harm, or harass cougars and that both industry and the public must refrain from damaging cougar habitat. In addition, the act requires that the Province of Ontario develop and implement a recovery plan to bring cougar populations within the province up to a level where they are no longer threatened with extinction (Endangered Species Act, 2007, S.O., c. C-6). The Ontario cougar is only the second subpopulation in North America, after the Florida panther, to be classified as an endangered species.

Range Expansion

Population estimates and harvest information for the states and provinces with known breeding populations of cougars (Beausoleil and Martorello 2005) suggest that Canada probably supports less than one quarter of North America's total cougar population. Relatively low human population densities, an abundance of suitable forested habitats, and large populations of ungulate prey, however, suggest potential for future population growth and range expansion in Canada. Indeed, this is already occurring. Eastward expansion of cougar may be occurring at present because it has taken western cougar populations time to rebuild from the days of bounties and general persecution. Cougar harvests and population estimates reached all-time highs in most western states and provinces around the turn of the new century (Beausoleil and Martorello 2005). Detailed genetic analysis (Biek et al. 2006b) lends additional support to the idea that North American cougars recently underwent substantial population expansion. With higher population densities, greater numbers of dispersers can be expected, facilitating expansion (see Chapters 4, 5, 12). Even so, range expansion is likely to be a relatively slow process because colonizers of both sexes must be simultaneously present and successfully produce offspring in the new habitat, and female cougars do not disperse long distances as often as males do (Sweanor et al. 2000). Anthropogenic features may also slow recolonization. For instance, while cougars do not generally avoid roads (Dickson and Beier 2002), major highways can present a barrier to dispersal if appropriate corridors or crossings are unavailable (see Chapter 12; Beier 1995). Highway 2, which runs north–south in Alberta and connects the major population centers of Edmonton, Calgary, and Lethbridge, may act as such a barrier, slowing expansion. Provided sufficient numbers of dispersing individuals are available, continued eastward expansion of breeding populations into suitable habitat where ungulate populations are high can reasonably be expected.

Expansion north into boreal habitats that are not part of historic cougar range is also possible. Continued industrial development and global warming may play a role in the potential for future cougar expansion into higher latitudes in Canada. White-tailed deer (*Odocoileus virginianus*) have increased in abundance in many parts of North America, in some cases doing so well that they become a pest species (Augustine and DeCalesta 2003). Additional forage created by industrial deforestation combined with reduced snow depth and cover in winter (Rikiishi et al. 2004) may create conditions that are favorable for their expansion northward. Indeed, in Alberta, white-tailed deer have increased markedly (Latham, pers. comm., 2009) in some northern boreal forests that were originally the domain of moose (*Alces alces*) and woodland caribou (*Rangifer tarandus caribou*) and were home to very few deer (Stelfox 1993). With increased prey densities, we speculate that cougar may successfully colonize areas north of their traditional breeding distribution (both present and historic), especially in riparian areas where prey is more plentiful. Between 2000 and 2004 in Alberta's boreal region, over two hundred cougar occurrence reports were filed with the provincial government. These reports range across northern Alberta and include sightings, livestock depredation, and road-killed cougars (e.g., districts of Athabasca, Grand Prairie, Fort McMurray, and High Level). The reports do not confirm the presence of breeding populations, but, at a minimum, they provide evidence of northern dispersal movements by individual cougars.

Expansion back into original cougar range in eastern Canada will serve to increase the resiliency of Canadian cougar populations and could serve to restore some ecosystem function in places where cougar have been absent for decades or centuries. Wolf reintroduction into the northwestern United States has had important top-down effects on ecosystems through trophic cascades (Beyer et al. 2007). Cougars have been linked to similar kinds of trophic cascades (Ripple and Beschta 2006), and effects on entire ecosystems might reasonably be expected as a result of recolonization (Chapter 10). As the Yellowstone wolf recovery program also demonstrates, however, predator repopulation can be controversial and can result in discontent among farmers, ranchers, and hunters. It is unclear how cougar expansion eastward will be received by Canadians, although there are indications that the response may be positive in some jurisdictions (Watkins 2005).

Expansion northward has similar potential to increase the resiliency of Canadian cougar populations, and because of lower human population densities in the north, it is less likely to create controversy among residents. Ecosystem effects of colonization are still likely, and may not always be "positive." Colonization of the boreal forest by cougar and white-tailed deer, for instance, may have important implications for other ungulates. Woodland caribou are a species at risk in some portions of the Canadian boreal forest (Edmonds 1991; Dzus 2001). The capacity for cougars to impact caribou populations negatively through apparent competition (Holt 1977) has been implied in other parts of Canada where cougar traditionally occur (Kinley and Apps 2001). Careful monitoring of cougar colonization (both north and east) by the provincial wildlife agencies would facilitate the identification and effective management of both the ecological and human conflict issues that surround cougar range expansion.

Conclusion

In Canada, the population distribution and size, level of protection, and management of cougar are entering a period of uncertainty and change. In the west, where cougar populations are well established, increasing human populations and development of rural areas will likely increase interactions between cougars and people. At the same time, there is great potential in Canada for cougar range expansion to both the north and east. Breeding populations of cougars have likely already become established in Saskatchewan and may also be present in Manitoba and Ontario. All this may serve to increase the profile of cougar in Canada and will likely result in the need for regularly updated management plans that account for stakeholder opinion, set protection levels and harvest objectives, and provide response guidelines for human-cougar interactions. If range expansion occurs at large scales, it will also increase the importance of Canada as a stronghold for cougar populations and will likely have important ecological consequences, both predictable and novel, for Canadian ecosystems.

In the United States, wildlife managers with responsibility for cougars will continue to face the same issues they already know. Traditional stakeholders—livestock interests and hunters—have grown to expect hunting to play a major role in cougar management, whereas the larger public does not share this expectation and calls for reducing a suite of threats to cougar populations. The often diametrically opposing views of participants in decisions make it likely that middle ground will be hard to find; each group will dislike some management decisions and insist that these reflect pressure from other stakeholders. It is easy to describe, and perhaps even to implement, the steps agencies should take in their management of cougars: well-developed proposals with clearly stated objectives and scientific support as available; opportunity for stakeholder input; and transparency in a decision-making process, addressing as many stakeholder comments as possible. But even this approach will not result in all parties being happy with management decisions.

General growth of the human population and the trend toward rural housing developments will increase contacts between cougars and humans. Agencies need to hone protocols for dealing with people and cougars in populated areas and for handling incidents in which people are harmed or threatened (see Cougar Management Guidelines Working Group 2005, chap. 7). While most western states and Canadian provinces contain vast areas of contiguous, suitable cougar habitat, and cougars show remarkable flexibility in habitats that suffice as movement corridors, managers should be aware that massive land use changes and human structures can fragment habitats and compromise dispersal corridors (Chapter 12).

In places, new issues are arising. Long before Europeans settled North America, cougars commonly moved through the plains states, and the frequency of recent confirmed reports suggests that this pattern may now be redeveloping. If so, agencies in these states are acquiring a new responsibility at a time of new complexity in public perceptions, and they will need to be both responsive and proactive about what to do in the novel situation of independent recolonization efforts by a large carnivore.

Agencies and stakeholders will face many changes in the future and need to consider their actions in the broader context of how these will affect conservation of the species. Currently, hunting is the single most controversial aspect of cougar management programs. It is, after all, premeditated killing of cougars for sport or to address depredation, predation, or human safety concerns. Success is easily demonstrated only for recreational hunting. If hunting is removed from the equation, recreation is all that will be lost. Cougars will continue to be killed to protect livestock, to protect wildlife at risk (e.g., isolated bighorn sheep populations), and to address human safety concerns.

Of potential threats to the species, sport hunting is the most visible and easily fixed by simply banning it, but it may be the least important in the long term. Alteration and fragmentation of habitats for cougars and the ungulate prey supporting them are ongoing and insidious and much more difficult to control. Loss of cougars because of habitat alteration will never be as obvious or as easily documented as cougars killed by hunters, making it much more difficult to develop support for necessary management actions. Decisions to protect cougar habitat in place of human development will be as controversial as decisions about hunting, or more so, and much more difficult to implement. The next phase of cougar management should see authority remain with state and provincial management agencies, and managers and stakeholders should recognize that habitat management is the crux of the long-term survival of the species.

Part II

Populations

Tracking for a Living

Kerry Murphy

O NE OF MY most memorable experiences in the field occurred while I was studying a cougar population in the back country of Yellowstone National Park and the surrounding ecosystem. The experience brought home to me the range of scent information that is opaque to us but is discernible daily to large carnivores.

The principal investigator of the study, Maurice Hornocker, had conducted prior work documenting the return of cougars to Yellowstone after an absence of fifty years. Now, assisted by other field biologists, I was responsible for capturing and radio-collaring cougars, monitoring their movements, predation, survival, and sources of death. This work, performed from 1987 to 1996, provided a baseline "pre-wolf" data set for comparison with similar information collected from 1998 to 2006, the period of gray wolf restoration in the park.

Every day, our assignment was to hike in close to a radio-collared cougar by "homing" on the signal, approaching closely enough to be confident of which patch of trees, brush, or rocks held the cat, and to leave without disturbing it. The second task was to return to locations obtained the previous day and record what we found— a kill, bed site, or footprint, a fresh scat or cougar latrine, or a scrape indicating communication with other cougars. We followed snow trails in the winter and used tracking hounds in the summer, and watched for subtle clues like the behavior of scavenging ravens. Once the cougar had left a kill, we examined the prey animal for sex, age, physical condition, the pattern of cougar feeding, and whether other carnivores used the kill. We did this work yearlong on foot, often in pairs but sometimes alone. There was as yet no global positioning system (GPS) technology available that we could rely on to provide frequent and accurate animal locations. New Mexico cougar researcher Kenny Logan dubbed our technique the "beat-the-bushes" method.

Hazards in the Rocky Mountains are many, ranging from avalanches to rattlesnakes, and working in grizzly country kept us watchful, especially around food sources that might attract bears. Both black bears and grizzlies readily detect carrion at a distance using their powerful sense of smell, and may walk boldly to a cougar kill looking for a free meal—unnerving for a biologist already at the site. My colleagues and I had numerous encounters with bears, often at kills. Early in our study, veterinarian Dr. John Murnane, later of the Interagency Grizzly Bear Study team, predicted we would eventually discover that grizzly bears were scent-trailing cougars for long distances to find cougar kills. Although we documented many cases of bears displacing cougars from kills, we never actually documented what Dr. Murnane suggested. My most memorable encounter with a cougar was scary but illuminating.

In June 1994, I was trying to keep up with a large male cougar while working alone. Originally captured inside the park, he eventually shifted his range northward to the adjoining wilderness. After getting a close location on him, I hiked to where I had circled him the previous day, a park-like opening in a mature stand of lodgepole pine with little understory. There, I found the remains of a mature cow elk. The elk's shoulder had been fed upon and there was drag trail, but the carcass was not covered with sticks or pine needles. I proceeded clockwise around the elk and then walked away in a broad arc up a little swale and sat down to enjoy my lunch.

I was working on my third pretzel when I noticed a dark form approaching. At first I thought it was a moose, but as it came through the trees I realized it was a grizzly male, dark brown, silver-tipped, and large. His arrival caught me by surprise, as grizzly sightings in the area were fairly rare in those years. I suppose I should have hollered immediately to signal him that I was there, but I was struck motionless by his unexpected appearance. My instinct was to remain silent and hope he would detect me and move off, as bears often do.

When the bear reached the elk, instead of feeding as I expected, he put his head down like a hound and began following my circle around the elk. When he got to where I had left the carcass, he turned and continued along my arcing path, all the time with his nose to the ground. By the time it dawned on me that he was following my scent, he was about ten meters away. At this point, it was obvious that there was not going to be an easy way out for me. Running might trigger his instinct to pursue, and he would easily overtake me. Some confrontation was likely, and I needed to prepare. I reached down and pulled the flap on the holster containing my canister of pepper spray, a recommended deterrent for bears. The velcro that secured the canister sounded its characteristic *chrrrrrt* in the quiet forest, startling the bear. He instantly started up on his rear legs, wheeled around toward me, but then completed a U-turn and galloped off, the pads of his big front feet and long claws flashing backward at me. After catching my breath, I hastily packed up my gear, made a quick inspection of the elk, and got out of there, lest he might return.

On the long downhill walk back to the truck, I reflected on my good fortune. The outcome for me could have been much different had he reacted aggressively. Perhaps his large size (and presumed old age) stemmed from a pattern of tolerance during encounters or even from outright avoidance of people. Undoubtedly, he had some experience with humans, as he lived in an area where there were livestock growers year-round and many elk hunters in the fall. He had ample opportunity for conflict with people carrying weapons.

I took from this an appreciation of the tracking ability and intelligence of grizzlies. His behavior reinforced what I have seen in other carnivores—wolves trailing wolves, and cougars trailing other cougars, both on bare ground and in snow and apparently for a variety of reasons. Training tracking hounds for work and pleasure has given me a special appreciation for the sense of smell among carnivores and the importance of scent trailing for territorial defense and acquisition of food. Undoubtedly, there is much to learn about the role of scent in the world of predators. But I hope the next time a master tracker like a grizzly teaches me something about scent trailing, I will just get to watch from a distance.

Chapter 5 Cougar Population Dynamics

Howard Quigley and Maurice Hornocker

I N LATE OCTOBER 2006, a female cougar lay down for the last time and died under a small conifer, high on a west-facing slope in the southern Yellowstone ecosystem. Some time later, probably a day or two, her kitten also died after wedging herself under her mother, perhaps in a last pursuit of comfort and warmth. Researchers reached the site shortly afterward, alerted by the signal in the mother cougar's radio-collar. They examined the area, photographed the setting, and then packed up the carcasses to carry them out of the backcountry. Four days later, the bodies were examined in a state laboratory. A definitive cause of death could not be determined, but both mother and offspring tested positive for plague. The deceased cougar mother was carrying a GPS (global positioning system) collar, storing her location coordinates six times per day. These locations provided no evidence that she had made any kills in the previous ten days, although she was in reasonable physical condition; the deceased kitten, four months of age, was emaciated.

This incident illustrates the difficulty of obtaining a clear picture of life and death in cougar (*Puma concolor*) populations. Following individual animals and documenting their fate is the most effective method for determining the ebb and flow of populations, yet natural deaths such as that described are rare events. From intensive field work and tracking comes the understanding of populations required to answer pressing questions about cougar management, the effects of sport hunting, and the role of male territoriality, for example.

Our goals in this chapter are to provide an update of the information available, interpret the current data for ecological and management implications, and suggest some

directions for future efforts to understand populations as an aspect of cougar ecology. Emphasis is on those aspects of research that can provide new understanding of cougars. Important challenges lie ahead for our understanding of cougar population ecology. New knowledge will enhance our general understanding of cougar populations, specifically addressing questions about regulation and limitation of populations, and will ultimately provide the best strategies for cougar conservation and the maintenance of this great cat's ecological role.

Components of a Population

A wildlife population, like a motorized vehicle or a watch, has many moving parts that make it function. Farner (1955) described wildlife populations using a lake analogy: a lake has an input stream and an output stream; the lake rises or falls depending on the balance between the two. In wildlife populations, quantification can be simplified to positive or negative, growth or decline, depending on the balance between new animals added versus animals lost from the population. But to understand truly the "lake level" and predict it, we must know all the parts that go into determining the flow into and out of the population and how these components function. We must also know all the components and influences not just for a year or two but over many years, under varying ecological conditions.

First, we must characterize the component parts of our population, which can be defined as a "group of organisms of the same species occupying a particular space at a particular time, with the potential to breed with each other" (Williams et al. 2002, 3). The formula can be simple: births balanced with deaths to raise or lower the overall level, as in the lake analogy, or it can incorporate a more complex approach, involving such details as age-specific birth and death rates, condition of animals as they age, and when they die. As Caughley describes it:

A simple approach is to treat individuals as if they were identical, to express the numbers in the population as an average over several years, and to investigate why the average has this value. At a more advanced level, the study might be aimed at expressing the rate of change in numbers as a difference between birth rate and death rate, the difference being related to environmental influences. More detailed again is the study which discards the simplifying but unrealistic assumption that the animals in the population are identical . . . [and] that the environment does not act directly on numbers as such but indirectly through its influence on fecundity at each age and survival over each interval of age. [And] at the final stage of this progression each individual is recognized as unique. (Caughley 1977, 3)

In laboratory studies, researchers can mark individuals and follow in an enclosed area to see when they breed, how many young they produce, and when they die, as done with mice and rats (Bixler and Tang-Martinez 2006) or insects. In the wild, such detailed studies are virtually impossible to undertake. The closest we can come in most wild situations is to count the total number of animals in a particular area at a certain time of year. Wildlife agency personnel count waterfowl, for example, and count or estimate the number of elk (Smith et al. 2006). At best, these usually provide indices to population trends—that is, whether the entire population is increasing, decreasing, or staying the same. In some cases, we survey sex and age ratios to get an idea of how the composition of the population has changed, which gives us a better idea of the "why" of population change. For example, we calculate doe:fawn ratios or cow:bull ratios to obtain herd composition in species easily sexed and aged from a distance (see Smith and McDonald 2002).

Another key need is to avoid misinformation, especially when manipulation of a population is the objective. With improper or imprecise information, management actions may produce unwanted results, such as diminishing or increasing the target population more than desired. Management actions in such a situation, given the lack of information, could produce detrimental results. The characteristics of the target species determine both the difficulty of obtaining good estimates for the population and the intensity with which population parameters must be pursued in order to obtain reliable estimates.

Characteristics that increase the challenge for wildlife managers are low reproductive rates, low density on the landscape, and low potential for observing the target species through sign or direct sightings. These aspects have been more clearly understood in the modern study of wildlife than they once were. For instance, there once was widespread comfort with the use of the harvested sample as a reasonable index to the wild cougar population. As we have recently seen in many species, it can take a suite of measurements to obtain an index to numbers, and even these indices may not be reliable (Garshelis and Hristienko 2006).

As Caughley points out (1977), the level at which one studies populations determines the complexity of information required. For perspective, what do the best data sets for other populations of large mammals look like? How does our understanding of cougar populations compare? For example, work by Clutton-Brock and colleagues (1985) with red deer (Cervus elaphus) sets an impressive standard for ungulate studies. They displayed the changes in populations over time and during growth and decline, and they documented factors correlated to individual fitness, mating preferences, and reproductive success. This and similar work took the study of large mammals to a different level, as did the enclosure studies that McCullough (1979) performed. In the world of carnivores, the level of detail with which Packer and co-workers (1991b) performed their analyses of African lion populations was considered a pinnacle of population tracking at the time and still persists as a target for high-level population study. These examples fit the highest level of the Caughley format (1977) by following individuals through their lives and adding each one to the picture of the population.

More broadly, one can examine several different populations under different ecological circumstances to obtain a picture of how a *species* behaves from the population perspective. The relative abundance of wolf studies is noteworthy: more than thirty studies are currently contributing to our understanding of wolf (Canis lupus) populations (Fuller et al. 2003); for black bears (Ursus americanus) and grizzly bears (Ursus arctos horribilis) the numbers are even higher. This abundance of information allows researchers and managers to view populations of these carnivores at the highest levels of Caughley's description (1977). Each study may not have followed each individual through life, documenting its ultimate fate, but they all made attempts (of varying completeness) at following individuals within populations. Taken together, these archives of information provide a reasonably clear picture of how the species' populations function under particular ecological and human-impacted circumstances. Research on other carnivores provides targets for excellence in science and shows that in cougar population studies, we have a significant challenge ahead. The crucial aspect of this is recognition of the need.

Cougars, as solitary carnivores and felids, have some commonly ascribed or "standard" features. Yet, some of the most important questions of science, and those that broaden the thinking, intrigue, and exploration of science, are those that do not fit the standard (Lott 1991). Wolves form packs and African lions form prides but for cougars the units of study are individuals. These individuals, or component parts, of a cougar population can be categorized as adult males, adult females, dependent young, and subadults or transients (see Chapter 8). Figure 5.1 presents these four sex and age groups as they exist in general proportion in a cougar population at any one time (based on Hansen 1993).

To understand the dynamics of the population, we require critical features of each component: survival, reproduction, and numbers. The numbers of each sex and age group are the basic foundation; Table 5.1 gives the relative numbers of each group from field investigations. The relative proportion of adult males to adult females can vary from population to population (Seidensticker et al. 1973; Logan and Sweanor 2001), but the ratio is most commonly 1:2 to 1:3, male to female adults. Due at least in part to the fact that cougars are promiscuous and polygynous, and male territories typically overlap the home ranges of more than one female, there is no indication that lower male numbers limit reproduction in the wild. Dependent young are defined as those young cougars between the ages of birth and approximately twelve to eighteen months of age that are with their mother at least some of the time and are likely obtaining part, if not all, of their sustenance either through milk or from kills made by the mother (see Chapter 8).

The number of dependent young in a population at any one time is primarily determined by the number of breeding females. Because females breed every other year, starting as early as age seventeen months (Lindzey et al. 1994; Logan and Sweanor 2001), half of the breeding-age females normally produce young in a given year. With an average litter size of between two and three young (Anderson 1983), the number of dependent kittens produced in any one year should thus be approximately equal to the number of adult females in the population. Subadults, or transients, are those individuals—normally twelve to eighteen months of age—that have separated from their mother and are traveling independently and killing on their own. However, these "independent" young may continue to associate with their mother and/or with siblings for weeks or even months after they become completely independent. The number of transients in a population can vary widely as a percentage of the total population (Seidensticker et al. 1973; Hemker et al. 1984; Lindzey et al. 1994) and can be difficult to quantify (see Logan and Sweanor 2001, 73). From a population standpoint, the key is the role transients may play in reproduction, in sustaining populations through filling home range/territorial matrices, and in recolonization. They may also be a tool for wildlife managers and ecologists in assessing populations (see discussion below) and indexing immigration and emigration as a measure of potential recruitment.

Numbers, Reproduction, and Mortality

Reiterating the standard definition, cougar populations are potentially breeding individuals in a defined area at a defined time. The boundaries of a cougar population are sometimes easily demarcated by landscape features and habitats, such as drainages, mountain ranges, rivers, or deserts. Some of these boundaries are barriers to interpopulation movement and are therefore relevant for defining the population; others may derive simply from map lines, designated study areas, or political units that have no biological significance but are convenient. Cougars do not occur consistently over the landscape but as several groups of interacting individuals separated by areas that rarely support resident cougars, across which dispersing cougars travel and link the various groups. Their heterogeneous distribution makes it imperative that we accept a geographic component for cougars across a given region, a situation best described as a metapopulation. Thus, as we build our description of cougar populations, we provide some important insights into cougar populations as part of larger metapopulations.

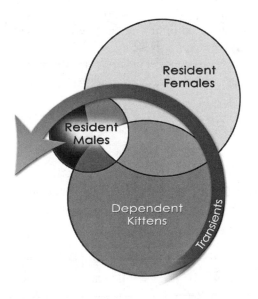

Figure 5.1 Graphic representation of an idealized cougar population. Proportions of each sex and age group are represented by the area of their symbol on the figure.

Reproduction

Reproduction in cougars has been described in great detail for captive animals (Eaton and Velander 1977), and the last thirty years of field research have supplied important details about reproduction in the wild. Cougars are considered induced ovulators (Bonney et al. 1981), which require multiple copulations (Eaton 1976) during the period of one to sixteen days when the male remains with the female for courtship and breeding (Seidensticker et al. 1973; Mehrer 1975; Logan and Sweanor 2001). Litter sizes reported from the wild range from a mean of 3.1 (Spreadbury et al. 1996) to 1.9 and 2.2 (Wilson et al. 2004; Ross and Jalkotzy 1992). In the largest sample from one population, fifty-three litters aged 9–49 days, Logan and Sweanor (2001) reported a mean litter size of 3.0; in Yellowstone National Park in 1998–2004, nineteen litters had a mean of 2.6 kittens (Ruth 2004a), and fifteen litters documented in 1986–1994 had a mean size of 2.9 kittens (Murphy 1998); and in Florida, Maehr and Caddick (1995) documented an average litter size of 2.3. Although some attention has been given to the litter size differences in the first litters produced by females compared to later litters, this was not detected by Logan and Sweanor (2001). Kittens are born after a gestation period of approximately 92 days (Anderson 1983; Logan and Sweanor 2001). Wilson and colleagues (2004) documented the lowest mean litter size (1.9, n = 14) and suggested that the small litters were the result of limited food resources. Such an observation is supported by the results of Stoner and colleagues (2006), in which low cougar fecundity rates were correlated with a 50 percent reduction in elk numbers.

The timing of reproduction in cougars is of both ecological and evolutionary significance in the natural history of this cat. As with most cat species that occur in temperate and tropical regions, cougars can breed and produce young at any time of the year (Sunquist and Sunquist 2002). However, the influence of nutrition, photoperiod, and even competitors likely all come into play to produce this flexible reproductive schedule, backed by the capability of females to re-cycle in a polyestrus pattern if they do not become pregnant during one estrus or if they lose a litter (Seidensticker et al. 1973). Despite evidence that cougars can produce young in any month of the year, as shown in Figures 5.2, 5.3, and 5.4, there is a preponderance of births during the summer months of July and August in northern latitudes (Cougar Management Guidelines Working Group 2005), such as in Yellowstone National Park (Ruth 2004a) and southern Ontario (Ross and Jalkotzy 1992). This general pattern seems supported by results in New Mexico (Figure 5.5; Logan and Sweanor 2001, 88), a more southern, desert site, but shifted earlier in Florida (Figure 5.6; Lotz et al. 2005, 18), the southern most of the studies examined here, and the most tropical in nature. Such shifts indicate that reproductive timing is driven by multiple factors, the relative influence of which can change with the ecological setting where cougars live.

Documentation of cougar dens is a relatively recent activity in field investigations. Intensive tracking through traditional very high frequency (VHF) telemetry and more recent global positioning system (GPS) technology has allowed documentation of dens and early detection of litters (see Figure 5.7; Logan and Sweanor 2001, 47). Such new knowledge is providing fresh insights into population characteristics of cougars and can potentially reveal faulty conclusions from previous work. For instance, by all indications, the sex ratio at birth for cougars is 1:1 (Logan and Sweanor 2000); however, until recently, this assertion was supported by little field data (Robinette et al. 1961). Logan and Sweanor (2001) were able to examine all cubs nine to forty-nine days old in fifty-three different dens and found a 1:1 sex ratio, supporting the assumption that cougars are born in equal sex proportions. However, in the same

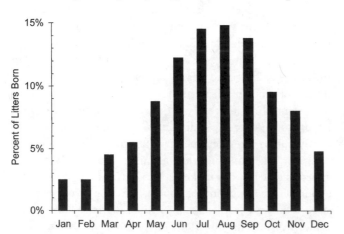

Figure 5.2 Combined annual pattern of litter production from multiple field projects (Cougar Management Guidelines Working Group 2005).

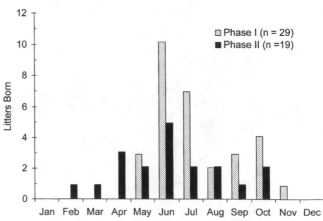

Figure 5.3 Annual timing of litters in Yellowstone National Park (Ruth 2004a).

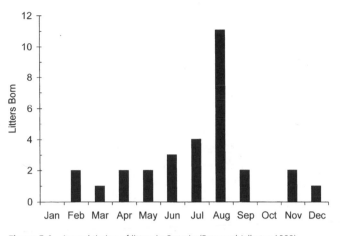

Figure 5.4 Annual timing of litters in Ontario (Ross and Jalkotzy 1992).

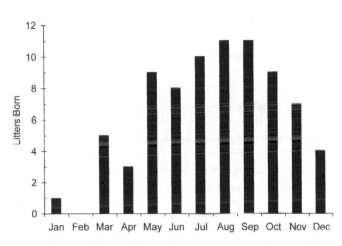

Figure 5.5 Annual timing of litters in New Mexico (Logan and Sweanor 2001).

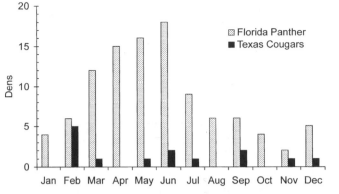

Figure 5.6 Annual timing of litters (dens) in Florida (Lotz et al. 2005).

Figure 5.7 Using radiotelemetry and GPS technology, researchers are now able to discover dens and document den characteristics and cougar demographic parameters from early life stages. This was rarely possible before such equipment came into use. Photo courtesy of Chuck Anderson.

study, researchers examined fifteen weaned, dependent litters with a ratio of 1:1.4 (male:female), indicating lower male survival during the early months of life. Such observations can provide important information for long-term monitoring and modeling of populations. Lack of sufficient sample sizes at certain points in development can have critical implications for recovery of populations (Beier et al. 2006; Maehr and Caddick 1995). Den and litter examinations are providing additional information on demographics, survival, sex ratios, and individual mating patterns (see Murphy 1998; Ruth 2004a), potentially supplying insights into the role of competitors (such as wolves) or the effects of human hunting on cougar mating patterns.

Survival and Mortality

In the most intensive form of population research, as described by Caughley (1977), all individuals in the populations would be monitored constantly, as would new individuals entering the population. More typically, researchers may simply know

the number of individuals in an area and the changes in the general numbers, up or down. Adult survival is one of the most important factors to determine in cougar society. In a population where stability is sought, low adult survival would create a need for greater immigration and recruitment to maintain the population, which would likely be characterized by a younger age structure. Higher adult survival would create a need for higher dispersal of subadults out of the population and would be characterized by an older age structure. These characteristics are important for understanding a population and its current status in relation to prey, management objectives, and responses to a variety of potential natural perturbations of local and regional populations. As part of our understanding of cougar populations as metapopulations, with source-sink dynamics, survival is also a critical metric for a complete assessment (see discussion of source-sink population, p. 69). Again, it starts with following the fate of individuals over time. As Murray and Patterson (2006, 1499) note, "Although researchers commonly consider survival to be a population parameter, it is an attribute of individual animals."

Survival of cougar kittens has been documented through sightings of litters at various times during their development. The accuracy of these data depends on the ability to observe litters (both directly and through sign), the assurance that all the offspring (of the sampled litters) have been observed by the field investigator, and accomplishing counts of kittens very soon after birth. Logan and Sweanor (2001) followed 157 cubs in New Mexico from nursling stage to independence and found a survival rate of 0.64. This compares to estimated survival rates of 0.45–0.52 in southern California (Beier and Barrett 1993), 0.50 in Yellowstone National Park (Ruth 2004a), 0.42 in Montana and southern Idaho (DeSimone and Semmens 2005; López-González 1999), 0.21 in Yellowstone (Murphy 1998), 0.98 in Alberta (Ross and Jalkotzy 1992), and 0.84 in Florida (Maehr and Caddick 1995). These last three extremes are likely due to the age of first detection of the kittens for Alberta and Florida, and the long survival calculation period (to twenty-four months) for Yellowstone. Late detection of litters can provide only a partial view of survival statistics due to the higher death rate of kittens after three months of age, as found by Logan and Sweanor (2001, 119) and shown in Figure 5.8 (Logan and Sweanor 2000). This lower survival rate post-weaning and pre-independence can be accompanied by a sex-biased survival rate in weaned cubs (López-González 1999). Murphy (1998) also found progressively smaller mean litter sizes with age of detection.

Subadult survival is an especially difficult piece of the cougar population puzzle. As noted, subadults become semi-independent from their mother and siblings over a period of months. Although data are scant, during this time they are thought to be hunting on their own, and killing prey as their skills are honed, though they may be partially dependent on kills made by their mother. They separate further from their family and finally become fully independent. Once independent, they may remain within their natal range or leave it completely to establish their own home ranges separate from or near the natal range (see Chapter 8). After independence from their mother, siblings can continue to associate for weeks or months (Ross and Jalkotzy 1992). The unpredictability of subadult movements makes these animals difficult to follow; thus, this portion of the cougar population picture is not well documented. As Sunquist and Sunquist (2002, 266) note, "Dispersal appears to be a hazardous time in a puma's life, but it is very difficult to collect mortality or survival data on this phase of life. In many cases, young pumas simply disappear, their fates unknown" (see also Hemker 1984). After ten years of study in New Mexico, Logan and Sweanor (2001, 126) stated, "Quantitative data on survival rates of independent subadults in other puma populations is practically nonexistent."

With their higher exposure to unknown areas and unknown dangers, dispersing individuals should and do display lower survival rates. Because almost all subadult males disperse, whereas some female subadults remain near or in their natal range (known as philopatry), female philopatry should also impart a sex-biased survival advantage. Subadult survival rates in New Mexico were 0.56 for males (n = 9) and 0.88 for females (n = 16; Logan and Sweanor 2001). This compares to a 0.22 survival rate for nine males in California (Beier 1995). Ross and Jalkotzy (1992) documented the fate of twelve subadult cougars, eight of which reached adulthood (a survival rate of 0.67). In Colorado, Anderson and colleagues (1992) calculated an annual survival rate of 0.64 for forty-two cougars, males and females, 12 to 24 months old; however, it is unclear what proportion of these individuals was independent (see Logan and Sweanor 2001). Only two of twelve dispersers (0.17) in the same study are known to have established themselves on independent home ranges. Of eight subadult dispersers— five males, two females, one unknown—tracked by Lindzey and colleagues (1994), seven died before establishing residency elsewhere (DeSimone and Semmens 2005).

Annual adult survival rates vary widely in cougars, from approximately 0.50 to full survival of 1.0. In the largest sample to date, Logan and Sweanor (2001) monitored 9–20 males and 7–24 females per year for eight years to arrive at mean survival rates of 0.91 and 0.82 for males and females, respectively. This compares to lower survival rates from a series of studies with smaller sample sizes: a pooled male-female adult survival rate of 0.75 in California (Beier and Barrett 1993); mean adult female survival of 0.71 in Utah (Lindzey et al. 1988); annual rates of 0.69, 0.92, and 0.80 for pooled male-female samples in three adult age classes in Colorado (A. E. Anderson et al. 1992); annual adult survival of 0.29 for males

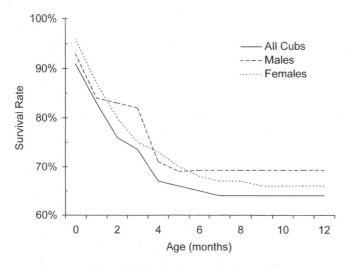

Figure 5.8 Monthly survival rate of kittens (n = 157) in New Mexico over a ten-year period. The category "all cubs" includes kittens that were documented but for which sex was undetermined (Logan and Sweanor 2001:118–119).

and 0.65 for females in northwest Montana (Ruth 2004b); and annual adult survival of 0.12–1.0 for males and 0.0–0.81 females in Arizona (Cunningham et al. 1995).

Under Caughley's plea (1977) for the upper level of understanding, the ecological interpretation of survival data is best facilitated by a full documentation of cause of death. Thus, the information provided by the disappearance of individuals from a population is greatly enhanced by determination of the cause for disappearance. For cougars, for example, a high mortality rate from road collisions may point managers to adjustments in highway design. And cause-specific mortality rates may be the most important factor in determining the relationship of all the enumerated components of a cougar population.

Mortality Factors

Much emphasis has been placed on the responses of cougar populations to mortality, especially through sport hunting (see Stoner et al. 2006). Perhaps the greatest insight can be gained through comparison of natural and human-induced mortality, examining both the response of a population to losses and the potential fitness-related aspects of the population in question (that is, those characteristics directly related to individual survival and reproduction). This may involve not only the vital rates of populations (i.e., fecundity rates, mortality rates, etc.) but also individual lifetime contributions to populations. Mortality factors—and how they affect such ecological relationships and the vitality of populations—are not easily obtained. However, the effort of obtaining such information is rewarded by greater understanding of ecological and evolutionary underpinnings. Some of the most important keys to understanding wildlife populations generally, and cougar populations specifically, can come from long-term monitoring of population responses to ecological changes, not from short-term artificial removals, which inevitably offer more limited insights.

Establishing cause of death requires that researchers examine animals soon after death. Cougars were intensively monitored over a ten-year period in New Mexico to acquire the fates of thirty adult cougars. Of these, 50 percent died from interactions with other cougars; the second and third most common causes of death for adults in this population were disease (i.e., plague and unidentified pathogens; Figure 5.9) and accidents (i.e., females injured during prey capture attempts). Beier and Barrett (1993) determined that vehicle accidents caused more than half of adult deaths, followed by disease, interspecific strife, and killing by wildlife authorities due to depredation by cougars. In Florida (Lotz et al. 2004), this pattern also applied for panthers, with intraspecific aggression, vehicle accidents (Figure 5.10), and disease/infections accounting for the top three causes

Figure 5.9 A female cougar was found dead, along with her four-month-old kitten, in the mountains of the southern Yellowstone ecosystem. The female was found to be in good physical condition. The kitten was emaciated. Both cats tested positive for plague. Photo courtesy of Dan McCarthy, Craighead Beringia South.

of death. All three of these investigations were in cougar populations in which hunting was not permitted, except in the case of depredation.

Younger age classes might be expected to be vulnerable to other forms of stresses and causes of death. In the same three unhunted situations, subadults appear to continue the same pattern as adults; internal social strife was the main cause of death in New Mexico (Logan and Sweanor 2001); subadults succumbed to intraspecific strife and/or vehicles in California (Beier and Barrett 1993). More than 80 percent of kittens (n = 27) in the New Mexico study were either killed by male cougars or starved to death, in almost equal proportions. Hemker and colleagues (1984) described a situation in Utah where nonhunting mortality was the primary cause of loss of cougars in the core study area, including six kittens.

In hunted populations, the primary cause of death is direct killing through sport hunting (Hornocker 1970; Murphy 1983; Logan et al. 1986; Lambert et al. 2007). Cunningham et al. (1995) found the same human-caused mortality in a population under intensive depredation control. Importantly, as indicated by Logan and Sweanor (2000), even in populations protected from hunting for research purposes, human-caused mortality (through vehicle collisions and depredation removal) can be the primary cause of death (Shaw 1977; Spreadbury et al. 1996; A. E. Anderson et al. 1992; DeSimone and Semmens 2006).

Two important extremes are evident in mortality factors in cougar populations, one human caused and one driven by internal strife (supplemented by other factors, such as disease and old age). Few field investigations have documented an even distribution of these two extremes, and few have

Figure 5.10 Collisions with vehicles are a leading cause of cougar mortality in urban areas, such as south Florida and southern California, and a major hazard for cougars recolonizing the midwestern and eastern United States. Places with high traffic volume on a dense road network will likely be sink habitats maintained by immigration from more secure areas. Photo courtesy of Darrel Land, Florida Fish and Wildlife Conservation Commission.

Dispersal, Immigration, Emigration, and Recruitment

Because of the great distances attained by some dispersing individuals, dispersal is viewed and presented as the most dramatic phenomenon in cougar populations. From both wildlife professionals and the general public, it garners more attention than other aspects of population dynamics. In part, this may be a residual effect of the earlier mystery that surrounded the disappearance of young individuals from study populations (see Seidensticker et al. 1973). With improved telemetry-tracking technology, documenting the movements of dispersing individuals has become more effective, and more data are becoming available. Even with telemetry technology from the 1990s and earlier, dispersals of over 100 km are not uncommon (see Logan et al. 1986; Ross and Jalkotzy 1992; Sweanor et al. 2000). GPS technology offers new insights into the dynamics of dispersal, beginning with simply allowing researchers to track more dispersing cougars without losing contact but also allowing better documentation of the routes of dispersal chosen by individuals (Figure 5.11).

Such distances—impressive as they are and important as they may be for landscape approaches to cougar conservation through metapopulation analysis and corridor planning—are secondary to the mandatory and initial examination of the basic components of population function. From a population standpoint, the more fundamental requirement is the characterization, basic description, and identification of new breeders in a population, some of which are supplied through dispersal while others are not. Without diminishing the importance of dispersal movements to metapopulations, recolonization, and maintenance of isolated populations, we must begin with the building blocks of populations: input and production.

When a cougar kitten has matured to independence, it either remains to recruit into the local population as a breeder or die, or leaves to die or become recruited into other populations. Individuals that leave their local population are labeled "dispersers." Logan and Sweanor (2001, 145) defined dispersal as beginning when "a subadult made its first movement outside its natal home range and did not return." Although some individuals remain near their mother's home range (the offspring's natal home range), more than 50 percent of offspring that survive until independence leave the immediate area of their natal home range. This "immediate area" definition is important. What of the *degree* to which cougars leave their natal home range? Logan and Sweanor (2001) defined degrees of dispersal or philopatry for cougars. A philopatric individual was one that overlapped its natal home range by 5 percent or more; dispersers were defined as those individuals that established an adult home range overlapping less than 5 percent of their

documented a preponderance of natural deaths; exceptions are the work of Logan and Sweanor (2001) and Hemker and colleagues (1984), which involved isolated, unhunted populations where human access was limited, and of Beier and Barrett (1993) in California, closed to cougar hunting. These extremes allow for additional comparisons to be drawn between mortality causes and for researchers to examine in more depth the effect of human-induced mortality. They also point up the possibility that the inverse relationship between human-caused and natural mortality suggests a compensatory mechanism in which human-caused mortality is replacing natural mortality in human-impacted ecosystems. Thus, a significant question is: what are the differences between a naturally driven cougar population and a human-driven population? And what is the evolutionary significance of such a difference for questions of fitness?

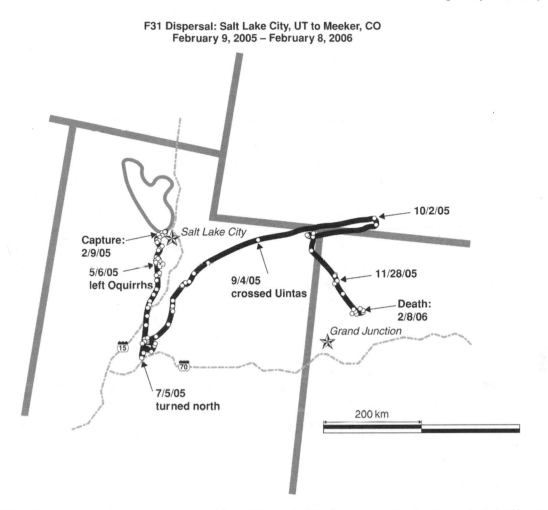

**F31 Dispersal: Salt Lake City, UT to Meeker, CO
February 9, 2005 – February 8, 2006**

Figure 5.11 Dispersal movements of a female cougar documented through the use of a GPS collar recovered when the cat was taken in legal harvest (Stoner et al. 2007).

natal home range (see Logan and Sweanor 2001). For the purposes of the present discussion, we define a dispersing individual as a cougar that leaves a defined study area boundary, an area defined by the researcher (for further discussion about dispersal see Chapter 8). Although this is somewhat arbitrary biologically, most study areas are small enough to be populated by interbreeding individuals, with little isolation except for distance between home ranges or territories. And study areas for cougars are normally large enough to encompass ten to twenty adult cougars (Table 5.1).

For describing contiguous cougar populations, as opposed to metapopulations, the foundation of understanding is the potential of new individuals in a population. The fact that dispersal happens is important; it is one phenomenon whereby new individuals are supplied to a population. But if young animals remain in the study area—and thus in the local population—we define them as nondispersers, or philopatric individuals; this is another phenomenon whereby individuals are supplied to a local population. Essentially, we

need to know if an individual becomes a breeder, and if so, whether it came to be in its local area through philopatry or dispersal. These new breeders are the individuals that make up recruitment numbers for a population.

Recruitment is the final product of birth and survival statistics, as well as the currency by which populations persist. These new individuals replace the individuals lost through mortality and allow a population to continue on a trajectory of stability or growth (see Hemker et al. 1984, 1278); if attrition losses are not replaced with recruitment, the population decreases. Logan and Sweanor (2001) concurrently observed cougar population dynamics in adjacent manipulated and unmanipulated subpopulations of equal size and found an equal number of recruits from philopatric individuals as from immigrating cougars (six and six, respectively) after removal of residents in the manipulated area. All six of the philopatric recruits were females; five of the six immigrants were males. In addition, overall, during the entire ten-year research effort, pre- and post-manipulation, Sweanor and colleagues (2000)

Table 5.1 Estimates of cougar population composition in North America from field studies employing intensive capture-recapture, ground tracking, and radiotelemetry techniques.

Location and Reference	Resident Males	Resident Females	Transients or Subadults	Dependent Offspring
Alberta[a]	4–5	8–12	3–10	6–18
British Columbia[a]	1–2	4	2–4	11
Idaho[a]	3	2–6	0–5	1–7
New Mexico:[b] TA	4–6	3–9	0–5	3–15
RA	7–11	6–16	1–5	6–23
Utah	0–4	5–8	0–10	3–1
Montana	0–2	6–10	1–3	9–14
Wyoming	3	7–9	1–7	13–17
Yellowstone Nat'l Park[a]	3–5	6–11	0–6	5–17
Yellowstone Nat'l Park[b]	3–5	9–14	NA	NA
Southern Yellowstone Ecosystem[a]	2	6	0–2	6

[a]Numbers estimated when pumas and prey were on winter range.
[b]Study area was divided into two parts. In the reference area (RA) puma numbers were protected over a ten-year period. In the treatment area (TA), pumas were protected for five years, then the numbers of adult and subadult pumas were experimentally reduced by 53 percent and 100 percent, respectively. After the one-time reduction, the TA was again protected and monitored for the remainder of the ten-year period.
SOURCES: Alberta, Ross and Jalkotzy 1992; British Columbia, Spreadbury 1989; California, Hopkins et al. 1986; Idaho, Seidensticker et al. 1973; New Mexico, Logan et al. 1996; Utah—1900 km², Lindzey et al. 1994; Utah—Oquirrh, Stoner et al. 2006; Utah—Monroe, Choate et al. 2006; Montana, DeSimone and Semmens 2005; Wyoming, Logan et al. 1986; Yellowstone—1500 km², Murphy 1998; Yellowstone—1700 km², Ruth 2004a; southern Yellowstone ecosystem, Quigley and Craighead, 2004.

reported approximately equal numbers of philopatric and immigrant recruits into the overall study population (twenty-one versus twenty-two, respectively).

Lindzey and colleagues (1992) also performed an experimental removal on a cougar population in Utah and found that the male portion of the adult population recovered completely with the immigration of males from outside the population. The female portion was somewhat slower to recover, but two recruits were identified, one philopatric and one an immigrant. Of fourteen progeny born on the Utah study area, seven remained on the study area and seven dispersed. However, none was known to be recruited into local or distant populations. In a later enumeration of results, Lindzey and colleagues (1994) documented one female disperser that survived to age 12.5 and was presumed to have been recruited as a breeder outside the study area in which she was born. Ruth (2004a) documented at least one disperser old enough to have bred in an area outside the study population (3.5 years old, sex unknown).

Siedensticker and colleagues (1973) documented no philopatric kittens remaining in their central Idaho study area (n = 12, 6 males and 6 females); two subadult females with undetermined natal areas successfully bred and produced their first litters in the study area. Likewise, Stoner and co-workers (2006) could document no kittens recruited into an unhunted population over a seven-year period; during that same period, of five documented dispersers (three males and two females) from the study area, one female survived to adult age and was assumed to have produced a litter in an adjacent mountain range before she was legally killed during the hunting season. In Montana, over a similar period (1998–2004), DeSimone and Semmens (2005) documented two philopatric females recruited into their study population and two dispersing females that lived long enough to breed in other populations; male offspring born in the study area were all dispersers and none lived beyond its third year. Hemker and colleagues (1984) also recorded the survival of one disperser, a female, from their study area that survived long enough (3.5 years) to have bred and produced young prior to being shot; two females immigrated into the study area and became residents. In all these research efforts, it is not uncommon to find reports of philopatric or dispersed subadult cougars, male and female, killed in legal sport harvest just prior to the age at which potential home range establishment and breeding would occur (see Logan et al. 1986; Lindzey et al. 1994).

Dispersal provides genetic integration of populations and subpopulations, and it provides for the revitalization of populations heavily impacted, or even eliminated, by natural and/or human-caused mortality factors. The capability for dispersal is one of the greatest conservation assets of large carnivores (Noss et al. 1996). Dispersal is a key driver of population stability and vitality. Again, however, dispersal data, like recruitment data, are obtained slowly, and full enumeration requires following individuals over time to determine if and where the dispersing individual contributes, and (in the case of recruitment) whether young become breeding contributors to the population.

Male-biased dispersal in cougars is well documented (see Logan and Sweanor 2001, 240–43), as is female-biased philopatry. In the scientific record, even when males did not leave a defined study area, such as in New Mexico (Logan and Sweanor 2001), they moved far enough from their natal ranges that they might be considered dispersed in other studies with smaller areas of research focus. The evolutionary significance of male-biased dispersal has been covered extensively by other authors, both in carnivore social organization (Pusey and Packer 1987a; Waser et al. 2001) and in vertebrate social organization in general (Lidicker 1975; Pusey and Packer 1987b; Logan and Sweanor 2001, 240–43). From the genetic standpoint, this can be a critically important characteristic that promotes the vibrancy of populations and metapopulations (see Chapter 3; also Hedrick 1996); for colonization or recolonization, dispersal and its sex-biased characteristic can also be an important element of the geographic expansion of cougars (Chapter 12); from a population standpoint, however, the main issue is the characterization and quantification of the sex-biased dispersal and philopatry and their role in the functioning of cougar populations. These characteristics, overall, are only vaguely understood in cougar populations. Although scores of dispersals and philopatric young have now been documented from field research on wild cougars, the number of recruits documented in contrasting environmental situations is relatively small. This information, derived under different influences, could provide valuable insights into the effects of human exploitation, changes in prey abundance, and the effects of other disruptions such as disease.

Density

Comparing cougar densities in different field projects is complicated by the lack of a format that makes cross-project comparisons valid. The main reasons for this involve the methodology for calculating densities and the extensive areas used by such large mammals. Issues are exacerbated by their low density compared to that of most other vertebrates, and the difficulty of consistently tracking all individuals in a population. Yet providing densities—and monitoring densities—is essential to describing the overall influence and impact of ecological conditions and human impacts on cougar populations.

Table 5.2 provides an overview of densities for cougars. Despite limited application to our overall understanding of cougar populations, these figures at least supply a framework for comparison with other species and between cougar field projects. As Hemker and colleagues (1984, 1275) point out, ". . . differences in density might be explainable by varying environmental conditions," and the identification of the specific influences of various environmental factors could provide some of the most insightful information for understanding cougar population dynamics. However, until a reasonable approach to standardization of densities is developed and samples can be compared directly, only the most general comparisons can be performed on these data.

Regulation, Metapopulations, and the Source-Sink Concept

Our understanding of cougar population dynamics remains superficial in comparison to that for many other mammals, especially the game animals of North America and other large carnivores. This is unfortunate in view of the threats and issues—and potential—surrounding cougars (see Figure 15.1). As the most successful land mammal of the Western Hemisphere, the cougar doubtless holds some important keys to ecological discoveries and answers to long-term questions about the plants and other animals with which it lives. Thus we must illuminate those aspects of cougar populations that will advance science and conservation of the species as rapidly as possible.

What are the factors that most influence the number of individuals in a population? Some of the most pressing questions can be distilled to those of population regulation and limitation. Complementary and integral to that distillation are questions about density-dependent versus density-independent population changes. One approach to viewing population regulation and limitation is to categorize influences as biotic and abiotic as they act on the four factors influencing population growth: birth, death, immigration, and emigration. Because cougars are not first-order consumers, like elk and deer, they are not directly dependent on sunlight and water for primary productivity. As such, we can say that cougar population levels are likely, and most directly, dependent on biotic factors. As B. K. Williams and colleagues (2002, 5) point out, however, biotic factors "can

Table 5.2 Estimates of cougar population density in North America from field studies employing intensive capture-recapture and radiotelemetry techniques.

Location and Reference	Study Area Size (km^2)	Density (Cougars/100km^2)	
		Resident Adults	Total
Alberta	780	1.7–2.1[a]	2.7–4.7
British Columbia	540	0.93–1.1[a]	3.5–3.7
California	600	1.2–2.3	
Idaho	520	0.96–1.7[a]	1.7–3.5
New Mexico:[b] TA	703	0.84–2.1[c]	2.0–4.3
RA	1,356	0.94–2.0[c]	1.7–3.9
Utah	1,900	0.32–0.63[a]	0.58–1.4
Utah Oquirrh	1,300	2.8[e]	
Monroe	950	2.0[e]	
Montana	2,500[f]	0.32–0.44	0.83–1.08
Washington, NW	1,500	1.0	
Wyoming	741	1.4–1.5[a]	3.5–4.6
Yellowstone Nat'l Park	1500	0.6–1.1[a]	1.2–2.6[a]
Yellowstone Nat'l Park	1700	0.76–1.1[d]	NA
Southern Yellowstone Ecosyst.	600	1.3[a]	2.3–2.6
Vancouver Island N	700	2.6–7.3	
S	1,000	1.4–2.0	

[a]Density was estimated when cougars and prey were on winter range.
[b]Study area was divided into two parts. In the reference area (RA) cougar numbers were protected over a ten-year period. In the treatment area (TA), cougars were protected for five years, then the numbers of adult and subadult cougars were experimentally reduced by 53 percent and 100 percent, respectively. After the one-time reduction, the TA was again protected and monitored for the remainder of the ten-year period.
[c]Density was estimated during January of each year in an environment where neither cougars nor prey migrated.
[d]Study area defined by a 95 percent fixed kernel estimator using all winter locations of radio-collared cougars.
[e]Average winter density estimate adjusted for slope variation.
[f]Area of entire study area; capture and marking took place on a smaller, core area.
SOURCES: Alberta, Ross and Jalkotzy 1992; British Columbia, Spreadbury 1989; California, Hopkins et al. 1986; Idaho, Seidensticker et al. 1973; New Mexico, Logan et al. 1996; Utah—1900 km^2, Lindzey et al. 1994; Utah—Oquirrh, Stoner et al. 2006; Utah—Monroe, Choate et al. 2006; Montana, DeSimone and Semmens 2005; Washington NW, G. Koehler, preliminary data, pers. comm. 2006; Wyoming, Logan et al. 1986; Yellowstone—1500 km^2, Murphy 1998; Ruth 2004a; Quigley and Craighead, 2004; Vancouver Island, Wilson et al. 2004.

affect more than one of the primary population processes," and there can be multiple indirect effects. So, oversimplification can divert us from the real population picture. This has been the case with much of the dialogue on cougar population regulation and limitation: we seek to determine primary population drivers for cougars. But often many factors interact.

The history of perspectives on cougar population regulation is easily defined in three stages. The first stage was based on little scientifically derived information during a period of heavy exploitation (Nowak 1976; Riley et al. 2004; Cougar Management Guidelines Working Group 2005). Early accounts of hunters and trappers in North America likely led to the general impression that cougars occurred at high densities in some areas. These sometimes overstated accounts provided fodder for cougar population control. Cougars were able to persist in places even in the face of intensive persecution, and it was just such control actions and overstatements that led to the impression of high density. For example, Young and Goldman (1946, 15) describe the exploits of "a private puma hunter" who was "reported to have killed more than 600 pumas on the Kaibab Plateau between 1907 and 1919." At the same time, the authors reported several accounts of very low abundance in states such as Minnesota and Michigan. These accounts provided no useful information about density and abundance of cougars on the landscape or the factors that might regulate their numbers.

The second stage of understanding cougar populations started with scientific investigation. In the first attempt to document scientifically the dynamics of a cougar population, Hornocker (1969, 1970) studied cougars in central Idaho. From this work, and the extended study of the population using radiotelemetry (Seidensticker et al. 1973), came the conclusion that cougars were regulating their own numbers through a system, a matrix, of individual use areas to which each adult cougar established rights, and by which the population established its level. These individual areas were inviolate as long as the adult animal was alive and advertised its presence. They concluded that individual cougars interacted with one another through scent marking and sometimes through direct contact, which "manifested through territoriality, [and] acted to limit numbers and maintain population stability" (Hornocker 1970, 33). Seidensticker and colleagues (1973, 59) concluded that "the land tenure system maintains the density of breeding adults below a level set by food supply in terms of absolute numbers of mule deer and elk."

These statements endured for more than twenty years, with little additional information to support or refute the hypothesis that cougars maintain their population numbers through social organization and territoriality. Lindzey and colleagues (1994), citing the results of nine years of work in Utah, concluded that the cougar population had not responded to an increase in prey during this time, which "appeared to support" the land tenure hypothesis. However, they also hedged; the magnitude of the prey increase could not be quantified and may not have been sufficient to test the influence of prey abundance. Still, it was their conclusion that "cougar populations are not controlled by prey abundance alone" (Lindzey et al. 1994, 619). In additional field work on the same central Idaho study area mentioned earlier (Hornocker 1969; Seidensticker et al. 1973), Quigley and colleagues (1989) documented an increase in cougar numbers during an increase in elk. The increase was detected only in the female portion of the adult population, going from between four and six females in the 1960s and 1970s to ten females in the 1980s. The adult male numbers remained at three during the two periods. Hemker and colleagues (1984), in a three-year study of cougars in Utah, characterized proximal and ultimate cougar regulation factors as social behavior and prey, respectively, but they also stated that the Idaho and Utah cougar populations in the two comparative studies (i.e., Seidensticker et al. 1973; Hemker et al. 1984) might be controlled by different factors. Thus, this second period of cougar population understanding, from approximately 1970 to 1995, can be characterized by the land tenure "paradigm for understanding how populations of mountain lions are regulated" (Pierce et al. 2000b, 1538), modified by evidence that prey numbers may also play a key role in cougar population regulation. The influence of prey numbers, although probed during this period, was still not quantified or tested in such a way as to index its influence adequately against social tolerance and avoidance.

In the late 1990s, final analysis of two field studies was completed; with the publication of the results, a third phase of cougar population understanding began. In California, Pierce and colleagues (2000b) examined the land tenure hypothesis utilizing an analytic design to test the movement and kill site locations of cougars in relation to one another and to the distribution of their primary prey, mule deer, *Odocoileus hemionus*. These tests were specifically designed to measure the influence of social tolerance and prey distribution. In most cases, the authors rejected land tenure as a regulator of cougar numbers and concluded that they "observed no indication of a land-tenure system that would lead to regulation of the population" (Pierce et al. 2000b, 1542). The authors called for additional research to document more thoroughly the role of intraspecific interactions they were not able to measure adequately. In addition, they suggested that more intensive monitoring of movements might provide for more robust conclusions in this case and supply better information about this strongly seasonally migratory predator-prey system.

However, working in a New Mexico system that was seasonally stable compared to the California study area, Logan and Sweanor (2001) came to the same conclusion as Pierce and colleagues (2000b). They stated that "puma social organization apparently did not function to limit the population below the level set by the prey" and that "the puma population is ultimately limited by food" (Logan and Sweanor 2001, 298, 339). The foundation for these conclusions was placed squarely on evidence for territoriality in male cougars, nonterritorial behavior in female cougars, and a lack of response by cougars to a decline in their primary prey, mule deer. Even with the substantial length and intensive methods of both studies, Pierce and colleagues (2000b) expressed ambiguity about their conclusions, and the Logan and Sweanor (2001) data raised additional questions about the relative roles of prey numbers and cougar social tolerance in cougar population regulation. For instance, what role does male territoriality play in population regulation? What is the response time for cougar populations faced with changing prey numbers? Until additional information is added to the cougar population regulation questions, prey numbers will be seen as the primary factor in cougar population regulation (see Chapter 8) and the role of social tolerance will be less than fully understood.

Density dependence is less in question in cougar populations than the question of population regulation. Although population regulation is often framed in the question of density dependence, for cougar populations there is a general acceptance that density dependence is at work. That is, population growth and population decrease are inversely related to the density of individuals. However, as with population regulation, the foundation for such acceptance lacks definitive and specific evidence of how it functions. Again as with the population regulation argument, it is reasonable to assume that biotic factors are working directly to influence population increases or decreases. The population regulation debate is centered on the relative influence of intraspecific versus interspecific biotic influences. But both are considered to be operating—perhaps in combination—in a density-dependent manner (see Cougar Management Guidelines Working Group 2005). When one examines the evidence, there is little supporting a direct and definable relationship between the density-dependent population behaviors and the specific vital rates. That is, with high population density or low population density, which factors are moving up or moving down? Litter size? Survival of kittens? Survival of subadults? Or even age at first reproduction?

Logan and Sweanor (2001) documented density-dependent population growth during their ten-year study of cougars in New Mexico. The growth of the population was facilitated by female recruitment and male immigration. During high cougar density, female cougars used smaller home ranges, but this was confused by an increase in the deer population (which would reduce the area required to sustain a cougar); male territories increased during increasing cougar density. However, these relationships were weak, "probably because of the dynamic nature of the [cougar] society" (Logan and Sweanor 2001, 211). A mixed set of results was produced for the relationship between vital rates and density. Survival rates did not differ between their control area and the area in which cougars were experimentally removed; subadult survival rates were also ambiguous, although female philopatry was at least partially density dependent, and such philopatry is associated with higher survival; fecundity did appear density dependent, as it was highest in the area where more than 50 percent of the adult population was removed; and litter size did vary inversely with density, although the authors cautioned against broad conclusions due to the limited sample size (Logan and Sweanor 2001, 85). Thus, even in field projects of extensive length, sample sizes limit our capability to answer completely the larger questions related to density dependence in cougar populations. "Unfortunately . . . research has not established which vital rates of cougars are density-dependent, much less what equation describes the relationship between a vital rate and cougar density" (Cougar Management Guidelines Working Group 2005, p. 65).

Cougar populations, because of their distribution on the landscape (both natural and human influenced), may be well suited to the metapopulation description (Gilpin 1996). Although cougars occur in a wide variety of habitats throughout the Western Hemisphere, they are not ubiquitous on the landscape and can occur only in places they can colonize, having food sources they can capture, and offering habitat suitable for their protection, physiological capabilities, and behaviors. Some of these habitats are marginal for their survival, some are those to which they

are moderately adapted, and some are areas of highly suitable habitat where they become relatively abundant. Unfortunately, there are not enough field results from which to draw sweeping conclusions such as those of Fuller and colleagues (2003) on population density and prey biomass, nor to draw sweeping conclusions about cougar metapopulations and how they function.

Cougar metapopulation dynamics may never be adequately studied to provide the information pertinent for our understanding of landscapes used by cougars and of the long-term conservation needs of cougars in a metapopulation framework. We may have to be satisfied with short-term studies and the derived models. Even long-term studies by Beier (1996) and Sweanor et al. (2000), both of which supplied substantial information on cougars in metapopulations, were hampered by loss of study animals, trapping bias, difficulty tracking dispersers, and lack of knowledge about the surrounding populations to which individuals dispersed and from which individuals originated. Information on metapopulations is important, both for ecological understanding and from a landscape population standpoint, and it will become increasingly important with expanding human development and resulting fragmentation of wild habitat.

As more information becomes available from additional field studies, more accurate portrayals of cougar metapopulations will be possible. And, with time, due to the fragmentation of habitats, the descriptions of these metapopulations will become more important as human activities separate populations into subpopulations. The dispersal capabilities of cougars, and of large carnivores in general (Noss et al. 1996), make them good candidates for persistence, even in highly fragmented, widely dispersed subpopulations. Beier (1993) determined that one to four immigrants into small populations greatly increase the probability of persistence of the small population in a hundred-year projection. Thus, movement corridors become crucial components of landscapes fragmented by human development or characterized naturally by an abundance of zones that are marginal or uninhabitable for cougars. As more data become available on cougar subpopulations, dispersal, and metapopulation dynamics (see Sweanor et al. 2000), one subset and characteristic of metapopulations that may provide insights and immediate applicability is the concept of source-sink population dynamics.

Like metapopulations, source populations and sink populations occur naturally on the landscape (Pulliam 1988). Source populations are those where productivity exceeds mortality; they sustain themselves and they supply surplus individuals to other populations. Sink populations are those where mortality exceeds productivity; they are not self-sustaining and rely on immigration for persistence. The source-sink characterization is not only useful for describing the functioning of populations and subpopulations (Pulliam 1988; Dias 1996); it can be very important for the development of functional conservation strategies for single species and multiple species (McCoy et al. 1999). Cougar populations have long been recognized for their source-sink characteristics, although the terminology was different. Young and Goldman (1946, 15) noted that "infiltration" from Mexico allowed the continued occurrence of cougars in Arizona, despite efforts to reduce their numbers on the Arizona side of the border. And as Lindzey and colleagues (1988, 666) point out, "Mountain lion populations in national parks or other areas where they are protected from hunting" contribute to adjacent populations. Offspring that emigrate may help sustain adjacent populations that are reduced by sport hunting or predator control programs. In essence, these protected areas become "biological savings accounts" (Logan and Sweanor 2000, 368) that serve to maintain or repopulate subpopulations cut off from other populations, reduced through some natural or human-caused perturbation, or unable to sustain themselves in marginal habitat. Although we know little about the functioning of cougar metapopulations and source-sink dynamics, as our knowledge of sub population functioning continues to improve (see Sweanor et al. 2000; Stoner et al. 2006), we can begin to formulate keys to subpopulation survival and frames of reference for evaluation. For example, through independent analyses in different locations, researchers estimated the areas required to provide for these biological savings accounts, or source populations, in California (approximately 2,000 km²; Beier 1993) and New Mexico (3,000 km²; Logan et al. 1996; Logan and Sweanor 1998). Monitoring such populations over time will provide validation for these claims, or provide adjustments to this size, and will allow for the advancement of our knowledge of cougar populations. In addition, regional and state management of cougars has been proposed as best viewed on a source-sink basis (Cougar Management Guidelines Working Group 2005).

C. R. Anderson (2003, 66) described cougar management as "traditionally conservative," with temporary extremes in cougar removal to respond to a "problem" of depredation or predation. As he pointed out, much of the issue in this type of management is simply the ability of the manager or managing agency to assess the status of cougar populations or subpopulations. There is no "silver bullet" by which we can make such assessments. As habitats become more fragmented, subpopulations become more isolated, human-cougar interactions become more likely, and the role of cougars in ecosystems becomes clearer and more valued, managers will require more tools to assess cougar populations. Anderson and Lindzey (2005) suggested that harvest composition can play a role in assessing the status of populations. That is, source and sink populations can be characterized by the sex and age composition of the harvest. Such new tools improve upon previous management approaches and will become increasingly important.

At the same time as examining affected populations, it is equally important that we examine naturally functioning populations (see Logan and Sweanor 2001; Ruth 2004b; Stoner et al. 2006) to add to the base of knowledge about how they function under minimal human impact. In doing so, we may develop additional tools for evaluating populations and their status. For example, subadult recruitment is central to understanding almost any terrestrial vertebrate population (see Lidicker 1975). The composition of subadults in the cougar population—male or female, within or not within natal home range—can be highly variable from year to year and between populations (see Table 5.1); their presence and abundance may be useful indicators of stability and turnover (Logan and Sweanor 2001; Anderson 2003). Their detection may thus provide insights into population status.

New Developments and Additional Needs

The features of a cougar population—birth rates, litter sizes, age at first reproduction, survival, and causes of mortality—are more easily defined than are their functions. The fact that we have yet to characterize the features fully is a good indication of how far we have to go in describing cougar populations. Our understanding of the species is miniscule compared to the state of population science for many other large vertebrates. The study of wolves, for example, has produced more than thirty reference populations, whereas only a few intensive studies have supplied enough information to contribute to our knowledge of cougar populations. Why might wolves gain more attention? Part of the answer lies simply in the characteristics of the species. In contrast to wolves, cougars lead largely solitary lives, create fewer conflicts, and are mostly undetected by people, resulting in less public interest and fewer research projects. And it should be noted that some investigations were undertaken by independent researchers rather than the agencies with authority over the species (e.g., Murphy 1998). The focus now should be on continued research in the areas of greatest need for the cougar. Such understanding will be important for the species in the framework of wildlife management and conservation in the United States and may be even more important in setting the stage for the species in Latin America (see Chapters 6, 7). New technologies and methods that are making the study of cougar populations more productive include use of GPS (Anderson and Lindzey 2003) and new approaches to the application of technology (Murray 2006). Figure 5.12, for example, displays a field download of locations of a female cougar showing the location of her den and two confirmed kills made during the denning period.

As we apply these new approaches, we must be aware of the persisting gaps in our knowledge about cougar population dynamics. Some of these are overarching questions, such as the relative importance of social tolerance and prey abundance in cougar population regulation. Specific target answers must be the focus upon which we design investigations. For instance, we can speculate about the role of transient males and their function as competitors for mates. This small piece of the population puzzle could have large implications in the evolutionary picture of the vitality of cougar populations. What happens when dispersal is disrupted? Can we quantify the role of transients as breeders? How often do transient males breed? What percentage contribution do these transients make in stable, increasing, and decreasing populations? What effect does hunting have on the percentage of transients breeding in a population?

Genetic techniques can now provide some of these answers. What might the answers mean to the vibrancy, adaptability, and resilience of a population or to its evolutionary potential? And how many times in how many situations will we have to find the answers in order to understand how this phenomenon operates and affects populations under different environmental conditions and at different population levels? Murphy (1998) found that resident males fathered all the offspring in the northern range study area of Yellowstone National Park, a population relatively unaffected by human influences. How does that compare with populations that are heavily affected by human influences? Scores of other questions can be derived from the incomplete picture we have of cougar population dynamics.

One of the overarching questions for many wildlife populations is the effect of humans, via sport hunting, fragmentation of habitat, and changing habitat availability. Logan and Sweanor (2001) and Beier (1993) call for large, contiguous cougar populations to provide source areas supplying dispersers to other areas. They also promote these "reserves" to provide areas in which ecological processes can be acting with little human influence. However, what is that influence? As earlier noted, in human-impacted populations, intraspecific strife appears to be replaced by human-caused mortality (i.e., sport hunting, vehicle collisions); that is, very little natural death deriving from intraspecific strife is documented in populations that are heavily hunted or that live in areas of high human density. Is this replacement of natural deaths by human-caused deaths exactly compensatory in ecological and evolutionary function and impact? We are also aware of the biases and potential biases related to the different movement patterns and rates of cougar sex and age groups (Barnhurst 1986; Anderson 2003). Movement rates and male and female movement patterns can translate into differential vulnerability to hunting. Can we continue to document other instances of such differences between human-influenced and "wild" populations? And what is the best design through which to find the answers? Will it be experimental removals (Lindzey et al. 1992; Logan and Sweanor 2001) or paired comparisons (Stoner et al. 2006)? Whatever the conditions developed to seek these answers, there is a pressing need to find answers. Holt and Talbot (1978) speculated about possible changes within populations due to human exploitation. Their concern focused on changes in age structure as a result of selective harvest of larger, older individuals. They cautioned about the influences this might have on social

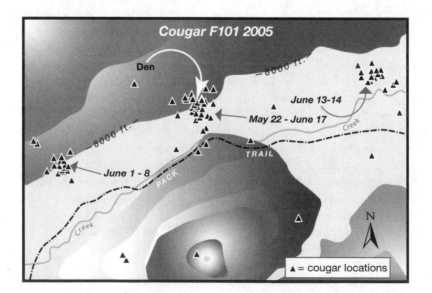

Figure 5.12 Locations of an adult female cougar obtained by a collar using GPS technology to determine a den site versus sites where the cougar made kills. The central cluster of locations indicates the den; two separate kills (both elk) are indicated by the east and west clusters of locations. Courtesy of Craighead Beringia South.

organization and behavior, two driving forces in evolutionary change.

The stability of the social organization of cougars can have bearing on both male and female reproductive success. Such was the case in New Mexico (Logan and Sweanor 2001) and in northern Yellowstone National Park (Murphy 1998). In both cases, adult males and females that lived the longest and maintained their home ranges and territories were the most productive individuals. These individuals displayed a high level of fitness to assume an adult home range or territory. With this level of fitness comes a higher production of litters than among their intraspecific competitors. In addition, these stable female home ranges tend to form neighboring home ranges occupied by their offspring, building matrilineal groupings (Logan and Sweanor 2001, 99). What would be the effect of disruption of these matrilines? Stoner and colleagues (2006) noted that as estimated density dropped during a time of increased hunting, so did mean age of females. They concluded that "survival and fecundity were all negatively associated with sustained high per-capita hunting pressure" (Stoner et al. 2006, 1596).

Thus this search for human impacts can drive some of the questioning from which investigations can be derived. In the meantime, we do not know the long-term effects of human impacts. A conservative approach to cougar management—and to wildlife management generally—would be to establish areas of populations not impacted by humans. Some of these may be de facto, and some may require some type of regulatory designation. But their role in allowing natural regulation of populations may be an important one if human impacts disrupt the "traditional patterns of natural selection in pumas" (Logan and Sweanor 2000, 368).

A significant change is taking place in the West with the reintroduction of wolves (Smith and Ferguson 2006). Wolves can affect cougar populations directly by killing cougars (Boyd and Neale 1992; Ruth 2004b) and indirectly through the alteration of cougar habitat use and the distribution of prey (Chapter 11; Ruth 2004a; Kortello et al. 2007). These interactions have high potential for broad effects on cougar populations and their management and conservation in the West. In addition, knowledge of this interaction can offer insights into the ecology of carnivore guilds and their effects on prey and use of different habitats (Ruth 2004).

Conclusion

Returning to our opening incident, the death of a cougar mother and kitten illustrates the data collection challenges in the study of cougar populations. Lack of specific and robust information on particular aspects of cougar populations means that much of our knowledge still derives from population modeling. Modeling has limitations, and it can only be as good as the data provided to the model (Law and Kelton 2000). Although, as we have seen, information on vital rates in cougar populations is extremely limited, models can extend our understanding of the relative importance of vital rates in the functioning of cougar populations. For example, Beier (1993) provides some parameters for testing the long-term survival of populations. When framed in a metapopulation approach (Beier 1996), these can offer valuable insights into populations across landscapes and the interactions between populations. Additional data on vital rates will allow further honing of the models to provide realistic predictors for conservation and management of the species. However, even the best models cannot compensate for inadequacy of data. The basic refrain in cougar population studies persists, "We need more data and better data" (Beier 1996, 319). Nevertheless, the last ten years have provided important insights into the function of cougar populations. We now have some very revealing windows of knowledge into the lives of this secretive, enduring, and captivating species. No doubt, with finely honed field research and monitoring, and the new techniques of modern science, these windows will open wide in the coming years, adding to our knowledge of cougar populations and enhancing our abilities in cougar conservation.

Chapter 6 What We Know about Pumas in Latin America

John W. Laundré and Lucina Hernández

P UMA CONCOLOR IS the most widely distributed large predator in the Americas and most of its original range lies south of the U.S. border. Yet, information about cougars in South America is striking for its absence from most discussion in this book, and many people in North America are scarcely even aware of the southern populations. This chapter presents the current state of information on the species in South America, commonly called the puma. This vast geographical range is referred to as Latin America because the official languages spoken there have Latin roots (Spanish, Portuguese, and French), except in Belize, Suriname, and Guyana. For simplicity, we include those three countries as part of Latin America in our discussion of pumas.

The reason for the awareness gap is that, although Latin America includes most of the range of pumas, the greater part of the scientific information we have on this species comes from the United States and Canada. Pacheco and Salazar (1996) reported only eight published studies on Latin American pumas up to 1994. Studies have multiplied recently, but the amount and diversity of information on pumas in Latin America remains low. Much of what we know about puma biology in the north may be applicable to their southern cousins. However, there are major environmental and, perhaps more important, sociological considerations in Latin America that caution against too broad an application of northern knowledge to southern populations. Across Latin America, pumas are found in a higher diversity of habitats than in the north, ranging from desert to tropical rain forest. Except in the extreme south (see Chapter 7 on Argentina and Chile), Latin American pumas live in a snow-free environment. In contrast, most northern pumas occupy areas where snow in winter is a major environmental element. Only in the studies of pumas in Arizona, southern California, southern Florida, and New Mexico (Shaw 1977; Beier et al. 1995; Maehr and Cox 1995; Logan and Sweanor 2001) do we have data on northern pumas in a snow-free world. However, these studies are primarily from arid areas. This obviously leaves a large gap in information applicable to pumas living in the many diverse habitats of Latin America.

Another major consideration, especially relative to the management and conservation of this species, involves the differences between the north and Latin America in political and social environments. In most Latin American countries, the structure of political and social systems is influenced by their Spanish heritage, departing in many respects from the English-influenced heritage in the north. Because of these historical differences, there are some striking dissimilarities between the two regions, which can have profound effects on management and conservation of wildlife. Not the least of these is the rate of human population growth in Latin America, where the overall total more than tripled between 1950 and 2000, from some 167 million to 520 million people (Brea 2003). Thus, much of what we know about living with pumas in the United States and Canada may not be directly applicable to southern populations.

All this is not to say that we know nothing about pumas in Latin America. Researchers have conducted various studies and some of the resulting information has been published in better-known journals, but much of it is found in theses or regional journals and is written in Spanish or Portuguese. Unfortunately, these works get little attention in the scientific circles where English predominates. Among our goals is to spotlight these works and the dedicated puma biologists working in Latin America, thereby providing a state-of-the-art synthesis of current knowledge. From such a synthesis, we can evaluate the differences and similarities between Latin American pumas and their northern cousins and can provide a scientific basis for further work necessary on pumas in the south. Because pumas in Argentina and Chile are covered in Chapter 7, we focus primarily on those found north of these countries and south of the U.S. border.

We begin with an overview of the taxonomic status of pumas in Latin America and then try to provide an analysis of their current status and abundance. Most of the discussion is devoted to the state of our knowledge on the ecology and behavior of pumas, their role as predators, and their interaction with other species, including domestic livestock. Last, we describe what is known about the conservation efforts for this species and what the future holds for pumas in Latin America.

Taxonomic Status of Pumas in Latin America

Originally, taxonomists recognized approximately twenty subspecies of pumas in Latin America (Currier 1983; Culver et al. 2000a). However, recent mtDNA analyses have reduced the total number of subspecies to six (see Figure 3.4). All six subspecies are represented in Latin America, with Mexico sharing the subspecies *Puma concolor cougar* with the United States and Canada (Culver et al. 2000a). Of importance for Latin American pumas is the conclusion by Culver and colleagues (2000a, 2007) that the most studied subspecies in North America is actually a founder population, recolonized from South America approximately 10,000–12,000 years before the present, after the Pleistocene extinctions. If this interpretation is correct, the cradle of modern pumas is South America (see Chapter 3). However, it is in this cradle that we know the least about pumas. It is also in this region where pumas face the greatest threat from increasing human populations and activity. Because the diversity of the South American subspecies correlates with geographic barriers (Culver et al. 2000a), there is the real danger that we could lose vital information on the ancestral populations of pumas. All this makes it imperative that we expand our knowledge base about pumas in Latin America, starting with the synthesis of existing information.

Current Distribution and Abundance

Historic distribution maps for pumas in Latin America depict their occurrence in all parts except the Caribbean island countries. However, as with eastern populations of pumas in North America, human activity and population

expansion have affected puma distribution. Large sections of the tropical forest regions of Latin America are highly fragmented by resource exploitation (Chiarello 1999). In Brazil, Chiarello (1999) found that pumas were absent from the small fragments (<300 ha) and from highly disturbed moderate-sized fragments (>3,000 ha). Even in less disturbed areas, such as the Chihuahuan Desert of northern Mexico (Figure 6.1), puma distribution is affected by natural habitat fragmentation and proximity to urban areas (Guerra-Benítez et al. 2007, in review).

Current distribution maps still tend to show pumas occurring throughout the Latin American countries (e.g., Reid 1997; Chávez Tovar 2005); an exception is the map in Nowell and Jackson (1996), which depicts reduced populations of pumas in eastern Brazil, Paraguay, and northeastern Argentina. Given the rapidly increasing habitat fragmentation and human population growth in most of Latin America, we have attempted to refine the picture. Human activity can impact wildlife abundance in immediately surrounding areas (Escamilla et al. 2000, Guerra-Benítez et al. 2007, in review). We therefore used the 2000 population figures, locations of urban centers, and road densities to construct a more realistic distribution map for the southern pumas

Figure 6.1 This young adult male was trapped in dense scrub in the Mapimi Biosphere Reserve in northern Mexico. Although the puma was fitted with a radio collar, he disappeared shortly thereafter. This is typical for young males in this region because of the large territories pumas have in the mountain island habitat of the Chihuahuan Desert. Photo by John W. Laundré.

(Figure 6.2). Based on our assessment, we estimate that approximately 40 percent of the original habitat for pumas in Latin America is either lost or threatened by urban expansion. As we discuss later, the remaining 60 percent is either desert or under threat of future urban expansion. Thus, we can expect the actual range of pumas in Latin American to continue to decrease.

Given its poorly described distribution, it is not surprising that the current status of pumas in most Latin American countries is relatively unknown, being based largely on old distribution maps and "common knowledge." Pumas are legally protected in seventeen of the twenty-one Latin American countries (Figure 6.3; Nowell and Jackson 1996). However, in many of these countries pumas are still killed because of their depredation on livestock (Pacheco and Salazar 1996; Costa et al. 2005; Michalski et al. 2006). This loss plus the impacts of fragmentation and other human disturbances probably greatly impact the status of pumas. Pacheco and Salazar (1996) report that there are insufficient data to determine the status of pumas in Bolivia. However,

Anderson (1997) reports the Convention on International Trade in Endangered Species (CITES) status of pumas in Bolivia as Appendix II; that is, it is not necessarily threatened with extinction but may become so without protection. In Costa Rica, the puma is considered endangered (Carrillo et al. 2000). In Brazil, pumas in northeast and southeast are considered vulnerable (Costa et al. 2005), but they are reported as common in the Pantanal (Trolle 2003). In Colombia, the status of pumas is listed as vulnerable (FVSN 2006). In Mexico, the puma is listed as a game species but is classified as a species requiring special protection (Chávez Tovar 2005), and its status varies considerably from state to state. In the Mexican state of Chiapas, the puma is considered threatened even within the large protected areas (Medellín 1994). In the states immediately around Mexico City, pumas are rare (Sánchez et al. 2002) and their status is considered critical (Chávez Tovar 2005). In all the Mexican examples, the status of pumas is the impression of those who have studied them rather than any official designation. For the remaining countries in Latin

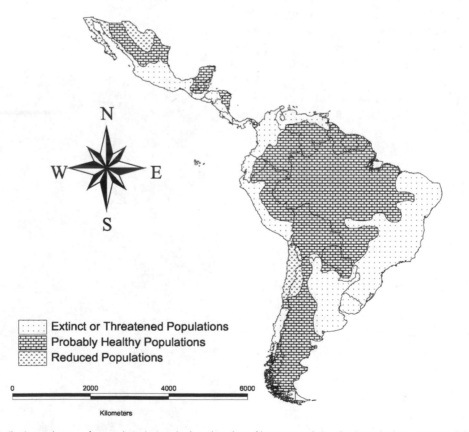

Figure 6.2 Estimated distribution and status of pumas in Latin America based on data of human population density and urban centers and data on abundance of pumas in hot deserts. In zones of high urban concentration (light stippled areas), pumas are probably extinct or exist in extremely reduced numbers. The small areas of Mexico and Chile with darker stippling represent the major desert regions of Latin America in which pumas probably occur in reduced numbers (Guerra-Benítez et al. 2007). Only in the areas of open blocks are pumas probably present in healthy numbers. However, even in most of these areas, lack of data prevents any actual assessment of their status. For clarity, country names are given in Figure 6.3.

Figure 6.3 Legal status of pumas in Latin American countries based on Nowell and Jackson (1996). Countries: MEX = Mexico, BLZ = Belize, HND = Honduras, GTM = Guatemala, NIC = Nicaragua, SLV = El Salvador, CRI = Costa Rica, PAN = Panama, ECU = Ecuador, PER = Peru, CHL = Chile, COL = Colombia, VEN = Venezuela, GUY = Guyana, SUR = Suriname, GUF = French Guyana, BRA = Brazil, BOL = Bolivia, PRY = Paraguay, ARG = Argentina, URY = Uruguay.

America, excluding Chile and Argentina (see Chapter 7), the status of pumas is unknown.

In the areas where pumas are still found in Latin America, their densities are quite variable. In the Pantanal region of Brazil, densities of independent animals ranged from 3 pumas/100 km² (Trolle et al. 2007) to 4.4 pumas/100 km² (Crawshaw and Quigley, unpublished data, cited in Nowell and Jackson 1996). Franklin and colleagues (1999) estimated densities of 2.5 to 3.5 resident pumas/100 km² in Patagonia. Based on camera trap data, researchers have reported total densities of 0.5–0.81 pumas/100 km² for Argentina, 2.4–4.9 pumas/100 km² for Belize, and 12.4–19.4 pumas/100 km² for Bolivia (Noss et al. 2006; Kelly et al. 2007). Polisar and colleagues (2003) reported 12 adult pumas in their 632 km² study area in grass-shrubland of Venezuela, or a density of 1.9 pumas/100 km². Mazzolli (2000) reported densities of 6–13 pumas/100 km² for the Atlantic rain forest of southern Brazil. These reported density estimates come from semi-deciduous forest/shrubland (Brazil, Venezuela), dry forest (Bolivia), subtropical forest (Argentina), tropical rain forest (Belize, Figures 6.4, 6.5), Atlantic coastal forest (Brazil), and grassland steppe (Chile) (Franklin et al. 1999; Trolle et al. 2007; Kelly et al. 2007).

Figure 6.4 This animal was photographed by camera trap on a mountain trail in the Chiquibul Forest Reserve and National Park in Belize. The bumps on its body arise from botfly larvae living underneath the skin. The larvae eventually hatch out, leaving open sores that heal into dark scars, which fade over time. These scars are useful to researchers for individual identification of pumas. In Belize, pumas are more likely than jaguars to be infested with botfly larvae. Photo courtesy of Marcella Kelly.

Most of these density estimates are comparable to the range of 0.8–2.2 pumas/100 km² reported for northern pumas (Logan and Sweanor 2001; Laundré et al. 2007).

Figure 6.5 This female puma, bearing a large and distinctive healed scar, has lived for many years in the Chiquibul Forest Reserve and National Park in Belize. The scar may be a result of intra- or interspecific aggression or struggle with a prey animal. She was photographed by a camera trap on a small footpath near Las Cuevas Research Station. Photo courtesy of Marcella Kelly.

The unusually high density estimates in Bolivia might have been influenced by the small size of the survey area and the close spacing of the cameras (Kelly et al. 2007). Although additional data are needed from tropical rain forest habitats, it does appear that puma densities in these habitats tend to be higher than in other areas. The differences may reflect prey abundance differences, but little comparative data exist on the prey base of rain forest versus other habitat. Polisar and colleagues (2003) estimated vertebrate biomass in the grass-shrubland of Venezuela to be around 750 kg/km^2 with 241 kg/km^2 being mammal species. Peres (2000) reported vertebrate biomass of 240–720 kg/km^2 for twenty-five Amazonian sites. Although these data indicate little difference between the two biome types, more data are obviously needed.

Still unknown are the densities of pumas from southern deserts and the Amazon lowlands. Logan and Sweanor (2001) reported resident densities of 0.8–2.1 pumas/100 km^2 for the northern Chihuahuan Desert in the United States. However, this was in an area of high prey abundance (mule deer, *Odocoileus hemionus*) and included just the mountain ranges. In the Mexican portion of the Chihuahuan Desert, mule deer densities are substantially lower (Laundré and Hernández, personal observation). Consequently, puma densities would probably be comparatively lower in Mexican and possibly South American desert environments. Because of the density estimates from Belize, we might assume that puma densities in the Amazon lowlands might also be high. However, Mares (1992) cautioned against making broad statements regarding mammalian richness in the Amazon lowlands. Obviously, additional information on puma density in all Latin American habitats is needed.

Ecology, Behavior, and Habitat Use

A tremendous amount of work has been conducted in the United States and Canada on the ecology, behavior, and habitat use of pumas in northern habitats. Much of the resulting information is presented in this book and provides a considerable data base for the management and conservation of pumas in the north (Cougar Management Guidelines Working Group 2005). Unfortunately, little of this type of information exists for pumas in Latin America. Considering that southern pumas survive in many more diverse habitat types than in the north, this lack of information is indeed a critical obstacle to sound conservation efforts. We present what is known and indicate where future research is most needed.

One of the first and most basic aspects of puma ecology is home range size. In the dry forest habitat of Venezuela, home range sizes of three female pumas in the dry season averaged 33 km^2 (Farrell and Sunquist 1999; Scognamillo et al. 2003). Home ranges of two males averaged 60 km^2 but ranged from 17 to 104 km^2. During the wet season, home ranges for a female and a male were 23 and 83 km^2, respectively (Scognamillo et al. 2003). In the tropical dry forest of western Mexico, Núñez and colleagues (2002) reported home ranges in the dry season of around 25 km^2 for two females and 60 km^2 for two males. Home ranges for all animals increased during the wet season to around 60 km^2 for females and 90 km^2 for males (Núñez et al. 2002). Mazzolli (2000) reported a home range size of 75.5 km^2 for an adult female in the Atlantic coastal forest of southern Brazil. In the Pantanal/Cerrado region of Brazil, Silveira (2004) reported that home ranges of three males averaged 222 km^2 and those of three females averaged 89 km^2.

Based on this limited data set, it appears that home range sizes of pumas in northern Latin America are similar to those reported for southern Chile (Franklin et al. 1999). However, their home ranges in most cases are considerably smaller than most reported in the United States and Canada (Logan and Sweanor 2001; Laundré and Loxterman 2007). Because of the limited number of studies and animals, conclusions and even speculation as to whether home ranges of pumas in Latin America are smaller than those of northern animals and, if so, why, are premature. Consequently, this is another area where significantly more data need to be collected.

Relative to habitat use within the home ranges, female pumas in Venezuela seem to select open dry pastures more than expected, while males did not demonstrate any selection (Farrell and Sunquist 1999; Scognamillo et al. 2003). However, pumas were mostly (81 percent of the time) found within 0–500 m of the edges of forest patches larger than 300 ha. In western Mexico, pumas seemed to use the arroyos or river bed areas most during the dry season (Núñez et al. 2002). In southern Brazil, Mazzolli (2000) found approximately equal proportions of native Atlantic coastal forest and eucalyptus and *Pinus* spp. plantation habitat within the core area of use by an adult female. In the Cerrado region of Brazil, pumas used the forest habitat most (Silveira 2004). Because of the limited data, not much is actually known about habitat use by pumas, and this represents an exciting area for future research.

Because suitable habitat for pumas in Latin America is becoming fragmented, an important aspect for the conservation of this species is how pumas react to fragmentation. In the Brazilian Amazon, sizes of forest patches used by pumas averaged 1,372 ± 2,962 ha, while patches not used by pumas averaged only 177 ± 287 ha (Michalski and Peres 2005). Pumas in the region around São Paulo, Brazil, used patches of dense native savanna and exotic eucalyptus plantations (Lyra-Jorge et al. 2008). Mazzolli (1993) reported for the state of Santa Catarina just to the south of São Paulo that pumas were limited to remnants of native vegetation more than 800 m above sea level. In the Atlantic forest of southeastern Brazil, pumas were absent from small (<300 ha) reserves (Figure 6.6). Thus, from the limited data available, it seems that patches of <300 ha are not readily used by pumas. These results are similar to our findings in Idaho and Utah, where pumas rarely used forest patches smaller than 100 ha (Laundré 2009 in review). Because of the conservation implications of patch use, this is a topic that warrants further study.

Concerning daily activity patterns of pumas in Latin America, again we have limited data. In three studies

Figure 6.6 A camera trap in a forested area in southern Brazil photographed this female puma in 1999. Pumas in Brazil are protected, but killing them to protect livestock is nevertheless permitted (Crawshaw 2003). In the eastern and southern parts of Brazil, the puma is considered vulnerable (Costa et al. 2005), although it is reported as common in the Pantanal. Photo courtesy of Marcelo Mazzolli.

(Venezuela, Brazil, and Mexico), pumas could be active at all times of the day but typically had a more crepuscular-nocturnal activity pattern (Scognamillo et al. 2003, Núñez et al. 2002, Silveira 2004). This crepuscular pattern of activity is mirrored in data from southern California (Beier et al. 1995) and southern Idaho/northern Utah (Laundré 2008). However, additional data are needed.

Predator-Prey Interactions

Understanding predator-prey interactions is a key to the ecology of the puma, a top carnivore in all ecosystems where it is found. The puma's impact on prey populations in Latin America (top-down effects) and, conversely, the impacts of prey populations on pumas (bottom-up effects) are important in understanding the dynamics of puma distribution, habitat use, and abundance. The first aspect to be considered is: what do pumas eat? Once we know something about puma diets, we can start to determine the possible impacts on their native prey and vice versa; predation on livestock is addressed separately.

Puma Diets

Fortunately, puma diets have been studied relatively often and in a variety of habitats (Table 6.1). In the United States and Canada, pumas have a diet of 90–100 percent mammalian prey, with the majority being large species (>15 kg; Iriarte et al. 1990). Although the puma's diet in Latin America seems to include birds and reptiles, diet composition from twenty-one different studies still averaged 90 percent mammalian species (Table 6.1). Thus, Latin American pumas in general are similar to their northern cousins, even though there seems to be a higher availability of reptilian biomass in the more tropical areas (Polisar et al. 2003).

There does seem to be a major habitat effect regarding the pumas' reliance on large (>15 kg), medium (1–15 kg), and small (<1 kg) species (Table 6.1). In eight of the ten studies of puma diets in tropical rain forest habitat in six different countries, use of large mammals by pumas was less than 20 percent occurrence, with 0–8 percent being found in five studies. In contrast, some of the highest use of small mammals by pumas was in the tropical rain forests. The highest uses of large mammalian prey by pumas (>50 percent) are in the northern desert of Mexico (McBride 1976) and the semi-arid habitats of Venezuela, Brazil, and Mexico (Scognamillo et al. 2003; Núñez et al. 2000; Oliveira 2002). The large native prey most commonly taken were white-tailed deer (*Odocoileus virginianus*), brocket deer (*Mazama* spp.), and mule deer (Oliveira 2002). In eight of the twenty-one studies, pumas preyed on livestock. The highest livestock predation level was in the grassland/dry forest of Brazil (51 percent occurrence; Crawshaw and Quigley 2002).

Medium-sized mammalian prey were consistently used (20–76 percent) in all habitats. The most commonly used species were large rodents (pacas, *Agouti paca;* agoutis, *Dasyprocta punctata*), and armadillos (*Dasypus novemcinctus;* Novack et al. 2005; Oliveira 2002, Leite and Galvão 2002). In the tropical rain forests use of medium and small mammals was often higher than in other habitats (Table 6.1). Though not used extensively, nonmammalian prey species—birds and reptiles—do show up in the diet of pumas (0.2–17 percent). The most commonly used reptiles were caimans (*Caiman crocodiles*) in Venezuela (10 percent occurrence; Scognamillo et al. 2003) and iguanas (*Ctenosaura pectinata*) in Mexico (12.7 percent occurrence: Núñez et al. 2000).

The difference in the diets of pumas in the tropical rain forest ecosystems is probably a reflection of lower large mammal availability in most rain forest habitats. For example, in the rain forest of Guatemala, Novack and colleagues (2005) reported white-tailed and brocket deer densities of 1.2 and 2.4/km², respectively. In contrast, the collective density of medium-sized mammals was around 120 individuals/km². On a biomass basis, medium mammals represented approximately 65 percent of available biomass (our calculations). In the Amazonian area of Peru, Kuroiwa and Ascorra (2002) also reported low abundance of deer and peccaries and a higher abundance of medium-sized mammals. In one tropical rain forest site in Mexico, puma diets did consist of a high percentage of large mammals (60 percent), mostly deer (50 percent; see Table 6.1; Aranda and Sanchez-Cordero 1996). Escamilla and colleagues (2000) recorded white-tailed and brocket deer and collared peccary (*Pecari tajacu*) as the most abundant of large and medium mammal species in this area. In the dry woodlands of Venezuela white-tailed deer biomass represented 51.6 percent of total estimated native mammalian biomass (Polisar et al. 2003). This is also the area where, after livestock, deer were the most common prey item.

Impact on Native Prey Populations

The impact of northern pumas on their prey populations is unclear and controversial. Some consider that pumas can have a controlling or top-down effect on deer populations (Logan and Sweanor 2001), while others have found little evidence that pumas can affect ungulate populations (see Chapter 10; Hornocker 1970; Murphy 1998). Ballard and colleagues (2001) summarized the results of various studies on predation effects on mule deer and concluded that depending on which studies are chosen, one can show either

that predators did or that they did not have an effect on deer populations. Recently, Laundré and co-workers (2006, 2007) demonstrated that in some instances, pumas do not impact deer populations but, in contrast, cycles in deer abundance can influence puma abundance, a bottom-up effect.

In Latin America, even less is known about potential impacts of pumas on prey or vice versa. In Chile, the puma predation rate on guanacos (*Lama guanicoe*) during a severe winter was only 2.2 percent of the population (Bank et al. 2002). This corresponded to predation rates on mule deer estimated in Idaho (Laundré et al. 2006). In Argentina, Branch and colleagues (1996) found that pumas relied on the plains vizcacha (*Lagostomus maximus*), and that the use of this species was reduced when vizcacha populations declined. However, it is unknown if there was a corresponding decline in puma abundance. Apart from this and other studies in southern South America (covered more extensively in Chapter 7), we did not find any studies on predator-prey interactions from the rest of Latin America. If puma abundance is impacted by prey abundance, reductions in the prey base could result in reduced puma populations (Leite and Galvão 2002). Consequently, this is obviously an area of interest in the conservation of pumas in Latin America. It is also of importance in conserving the integrity of ecosystems since loss of predators can cause major impacts (Terborgh et al. 2001). Future fieldwork is needed in this area and should provide interesting and useful information.

Relations with Other Predators

The major interaction pumas have with medium to small predators across their entire range is that pumas eat them (Iriarte et al. 1990; Oliveira 2002; Laundré, personal observation). Pumas in northern areas have had to compete with wolves (*Canis lupus*) and grizzly bears (*Urus arctos*), and usually do not fare well (see Chapter 11; Murphy et al. 1998; Akenson et al. 2005; Buotte et al. 2005; Kortello and Murray 2005). In much of Latin America, the only large predator competing with the puma is the jaguar (*Panthera onca*). Obviously, the major area of potential competition between these two large felids is in their diet. In a summary of nineteen studies of puma and jaguar diets, Oliveira (2002) found that jaguars and pumas consumed large and medium prey at about the same level. Consumption levels can be affected by availability, which could mask any difference when comparing different study sites for pumas and jaguars.

However, for eight studies, Oliveira (2002) had data on both pumas and jaguars. Again, there were no significant differences in use of medium prey by the two species. These

Table 6.1 Food habits of Latin American pumas. All values are in relative percentage frequency of occurrence. Where there were estimates of percentage biomass, we indicated these in parentheses. Studies are arranged relative to general habitat type as described in the individual studies.

Habitat	Large Mammals	Medium Mammals	Small Mammals	Total Mammals	Reference
Desert Mexico	68.8	14.4	0	83.2	McBride 1977 As reported in Iriarte et al. 1990
Hot desert Mexico	53.5 (88.0)	37.5 (19.3)	0	90.0	Rosas Rosas 2003
Hot desert Mexico	63.4	34.6	0	90.6	Bueno-Cabrera 2001
Dry woodland Paraguay	18.2	72.6	4.5	100.0	As reported in Iriarte et al. 1990
Dry woodland Paraguay	23.6 (51.1)	25.4 (34.3)	43.7 (12.8)	92.7 (98.2)	Taber 1997
Dry woodland Venezuela	52.2 (71)	21.7 (18)	17.1 (1.0)	91.0 (90)	Scognamillo et al. 2003
Tropical dry forest Mexico	44.3 (75.3)	17.5 (16.9)	18.6 (1.6)	80.4 (93.8)	Nuñez et al. 2000
Grass/dry forest Brazil	51.6	35.4	3.2	96.8	As reported in Iriarte et al. 1990
Grass/dry forest Brazil	67.7	32.2	0.0	100.0	Crawshaw and Quigley 2002 As reported in Oliveira 2002
Grassland/forest Bolivia	50.4 (61.3)	49.0 (38.7)	0	99.4	Pacheco et al. 2004
Subtropical forest Brazil	29.7	54.1	5.4	89.2	Crawshaw 1995 As reported in Oliveira 2002
Rain forest Belize	0	33.3	66.7	100.0	Rabinowitz and Nottingham 1986 As reported in Iriarte et al. 1990
Rain forest Peru	0.0	58.3	16.7	75.0	Nuñez et al. 2000 Emmons 1987
Rain forest Guatemala	38.4 (44.5)	50.1 (49.5)	0	88.5 (94.0)	Novack et al. 2005
Rain forest Costa Rica	8.3	49.9	25.1	83.3	Chinchilla 1997 As reported in Oliveria 2002
Rain forest Mexico	60.0	40.0	0	100.0	Aranda and Sanchez 1996 As reported in Oliveira 2002
Tropical forest Mexico	19	79	0	98	Amin 2004
Rain forest Peru	0.0	33.3	66.7	100.0	Romo 1995 As reported in Oliveira 2002
Atlantic rain forest Brazil	17.9	76.3	0.0	94.2	Brito et al. 1998 As reported in Oliveira 2002
Atlantic rain forest Brazil	5.0	40.5	18.3	63.8	Leite and Galvão 2002 As reported in Oliveira 2002
Atlantic rain forest Brazil	38.7 (42.8)	54.9 (52.0)	2.0	96.6	Mazzolli 2000

NOTE: Large native animals were primarily deer (*Odocoileus* spp. and *Mazama* spp.); medium mammals included pacas (*Agouti paca*), agoutis (*Dasyprocata punctata*), and armadillos (*Dasypus novemcinctus*). Small mammals were usually referred to as unidentified rodents.

results agree with those of three later studies (Scognamillo et al. 2003; Amin 2004; Novack et al. 2005). In contrast, jaguars seem to rely slightly more on large prey (mean 42.2 percent, our calculations) than do pumas (mean 31.6 percent, our calculations). These results agree with Scognamillo and colleagues (2003) and Silveira (2004), who found a large difference between jaguars and pumas in use of large prey (81.0 vs. 36–50.0 percent occurrence). Amin (2004) found no difference in use of large prey, and Novack and colleagues (2005) found pumas using large prey more than jaguars (44.5 percent vs. 27.1 percent occurrence), but the majority of the results indicate that where pumas and jaguars are sympatric, jaguars rely on larger prey slightly more than pumas do.

Within the large prey categories, based on the above twelve studies of sympatric pumas and jaguars in all tropical habitats, we calculated that pumas rely more on deer (pumas 19.7 percent, jaguars 10.5 percent) and jaguars rely more on peccaries (pumas 6.3 percent, jaguars 23.9 percent). In the montane temperate forest of northern Mexico, both species were using white-tailed deer, but no comparative use levels were available (López-Gonzalez and

Carrillo-Percastegui 2003). For medium-sized prey, one of the more commonly used species was the armadillo (Oliveira 2002). In eleven studies, jaguars relied slightly more on this species (11.7 percent occurrence) than did pumas (8.7 percent occurrence), but our calculations indicate that the difference was not significant. Thus, there is some diet niche separation regarding the largest prey used by pumas and jaguars, but deer and peccaries represent only about 30 percent of both species' total diet. It is difficult to say if this difference is sufficient to explain the coexistence of these two species, considering that for the majority of their diet, they overlapped significantly (Oliveira 2002).

Less is known about habitat use. Few studies exist on habitat use of pumas and jaguars in Latin America and fewer still of sympatric populations. Where there were diet differences between sympatric pumas and jaguars, some habitat differences have been assumed (Novack et al. 2005). Based on a radiotelemetry study in the llanos of west-central Venezuela, jaguars seem to use the open and forested habitats proportional to availability (Scognamillo et al. 2003). Pumas showed a slight preference to the more open habitats but did not avoid forested habitats (Farrell and Sunquist 2000; Scognamillo et al. 2003). Additionally, there did seem to be some small-scale separation, with only about 37 percent of the puma radio locations being inside the home ranges of jaguars. Only about 15 percent of the puma locations were within 200 meters of jaguar locations (Scognamillo et al. 2003). In the Cerrado and Pantanal region of Brazil, jaguars used the closed forest more than pumas did, while pumas used agricultural and pasture lands more (Silveira 2004). However, again, there was substantial use of the forest areas by both species. In the dry forest of western Mexico, telemetry data indicated that pumas and jaguars both used dry arroyos and in general the same habitats (Núñez et al. 2002). In montane temperate forests of northwestern Mexico, data from camera traps indicated pumas using mountain habitats according to availability, while jaguars seemed to be using oak woodlands more than expected (López-Gonzalez and Carrillo-Percastegui 2003). But again, no mutual avoidance was noted.

In conclusion, there is good evidence that diets overlap extensively in pumas and jaguars. Though the data are more limited, there is some evidence for small-scale spatial and temporal separation, but overall, pumas and jaguars can generally be found in the same areas at the same times. This coexistence seems to be relatively peaceful; none of the diet studies of jaguars and pumas that we reviewed listed remains of either species in the other's diet. How these two species of similar size with overlapping diets and habits coexist is an area of scientific and conservation interest that needs to be pursued with further study. Considering that puma conservation is often tied to the conservation of jaguars (see later discussion), it is important to know more about their level of coexistence. Additional studies using the new global positioning systems (GPS) technology should provide the type of information needed to assess habitat and activity overlap of these two species.

Puma-Human Conflicts

As detailed in Chapter 1, the pre-colonial attitude toward pumas in North America was one of tolerance and even reverence; they were viewed as a totem of grace and silent power. In Central and South America, little information exists on pre-colonial attitudes because the Spanish routinely destroyed the historical records of the native people (Hemming 1970). The puma is occasionally found in ancient carvings and paintings, but in most areas it is overshadowed by the jaguar in the symbolism of Central and South American cultures.

For European cultures, the conflict between predators and humans is an age-long struggle between two competing predatory species. As mentioned earlier, many view predators as impacting prey species we value. However, the ownership of wild prey and the impact of predation are debatable (Laundré et al. 2006). The greatest arena of conflict is when predators begin to kill domestic livestock. Here we have a predator killing prey that specifically belongs to people. Every one they kill is one less for our use and represents an economic loss. Not surprisingly, this competition has led to large-scale campaigns to eliminate predators, with success exemplified by the extermination of wolves and grizzly bears in most of the United States and all of Mexico (Boitani 2003).

Livestock Losses

Do livestock constitute a large proportion of puma diets? If the answer is yes, then pumas could be considered a major threat to livestock. There are two ways to assess the impact of pumas on livestock. The first is through an analysis of their diet. Livestock remains were found in scats from eight of the twenty-one studies in Table 6.1. The percentage occurrence ranged from 0.9 percent in the Atlantic rain forest of Brazil (Brito et al. 1998) to 51 percent in grass/dry forest habitat (Oliveira 2002), and averaged 26.1 percent occurrence for these eight studies. Based on all twenty-one studies, average percentage occurrence would be 9.9 percent. Hence, based on this analysis, only 38.1 percent of the studies found evidence of puma predation on livestock. However, in these eight studies, domestic livestock represented the majority of the large mammals in puma diets. Consequently, puma predation on livestock based on scat analysis indicates that pumas are not generally an important predator but can be in certain areas. The disadvantage of using scat analysis to assess puma predation on livestock is that without information on livestock numbers, it is difficult to relate this directly to predation rates and economic consequences.

In six Latin American studies, the rate of loss of livestock due to puma predation was estimated (Hoogesteijn et al. 1993; Mazzolli et al. 2002; Conforti and Azevedo 2003; Polisar et al. 2003; Bueno-Cabrera 2004; Michalski et al. 2006). Polisar and colleagues (2003) reported a 13 percent loss of cattle to large cats (pumas and jaguars) over ten years at a ranch in north-central Venezuela. Similar to some U.S. areas (Shaw 1977), most losses were calves, with an average of twenty individuals killed per year. Polisar and colleagues reported that approximately 3,200 calves were born per year, which according to our calculations would represent an annual predation loss of calves to big cats of 0.5 percent. In the llanos of Venezuela, Hoogesteijn and co-workers (1993) reported that 6 percent of all losses of cattle on one ranch and 31 percent of calf losses were from big cats.

Forty-five of sixty-two ranches (72.5 percent) in central Brazil reported cattle losses to large cats of 0.4 to 1.7 percent of the herds, with losses at medium-sized ranches (500–1,500 head) being the highest (Michalski et al. 2006). Economic loss per year was estimated at US $290–$1,770. In southern Brazil, pumas killed cattle at four of twelve ranches (33.3 percent), representing 0.1–2.5 percent of the herds on three ranches and 16 percent on the fourth (Mazzolli et al. 2002). For all ranches combined, predation losses to pumas were $1,890 per year or 0.27 percent of total livestock value (Mazzolli et al. 2002). Losses to other causes accounted for 3.3 percent of total losses and represented $23,520 lost per year (Mazzolli et al. 2002). In a similar region in southern Brazil, Conforti and Azevedo (2003) reported that big cats on forty-two rural properties killed about 2.6 animals/year/property and pumas represented only one out of thirty-nine attacks, with the rest from jaguars.

In northern Mexico, puma predation was reported for sixteen of ninety-three ranchers interviewed (17.2 percent; Bueno-Cabrera 2004; Bueno-Cabrera et al. 2005). Between 2000 and 2002, approximately 1,500 head of livestock were reported lost. Of these, pumas were responsible for killing about 100 head. Twenty-six of these were cattle, representing approximately 8 percent of total losses (Bueno-Cabrera 2004). This loss represented 1.9 percent of the average herd size and a mean economic loss of $49/ranch/year. Losses due to causes other than predation (e.g., drought, robbery, and disease) were estimated to be 771 head over three years, with an economic value of $105,666 (Bueno-Cabrera 2004). These results mirror those from Bolivia, where camelid livestock losses to disease were two to six times higher than those from predation (Zacari and Pacheco 2005).

Although predation by pumas on cattle appears to be light, predation on sheep and goats seems to be higher and often more localized. In southern Brazil, ten out of twelve ranches reported sheep and goat losses that averaged 37 percent and 38 percent of the flock, equivalent to $5,900 per year and $4,332 per year (Mazzolli et al. 2002). In northern Mexico, seventy of the hundred head of livestock killed by pumas were goats and sheep. In both studies, the losses were not uniform, with some ranches losing more sheep and goats than others. In Brazil, the losses seemed related to poor husbandry practices. On four of the twelve ranches, sheep were placed in corrals at night. Only two of these ranches reported losses, of 3 and 14 percent of the flock value, respectively. On the ranch with 14 percent loss, the sheep were penned in a corral about 60 m from the house near a deep and forested canyon (Mazzolli et al. 2002). On the other four ranches, sheep and goats were free ranging and had depredation losses of 20 to 80 percent of the flock value. In northern Mexico, no relationship was found between losses and husbandry practices, but that is probably because sheep and goats were all free ranging with limited care (Bueno-Cabrera 2004).

For all types of livestock, there does seem to be a correlation between losses and habitat. In central Brazil, presence of forest areas and distance to riparian corridors correlated with depredation losses (Michalski et al. 2006). In the forested areas of southern Brazil, proximity to forested areas and canyons tended to increase the predation risk (Mazzolli et al. 2002). In the desert areas of northern Mexico, ranches with mountainous terrain suffered higher losses than those in valley areas (Bueno-Cabrera 2004).

Although the data suggest that predation losses to pumas are often at an economically acceptable level, most rural people still have strong negative opinions of pumas (Bueno-Cabrera 2004; Michalski et al. 2006). Thus pumas, along with jaguars, are often killed illegally in an effort to reduce predation loses. During a two-year period in a 34,200 km^2 area in the Amazon, 185–240 pumas and jaguars were killed to address livestock depredation (Michalski et al. 2006). To reduce this type of loss of pumas, it is necessary to ease the puma-livestock conflict.

Public Awareness and Remedial Measures

Because puma depredation on cattle is relatively low, the issue is less a question of reducing predation levels than a matter of convincing people that pumas are not a major threat and have ecological value. One possible way of reducing the conflict is through public awareness via environmental education (Bueno-Cabrera 2004). Many rural people in Latin America have limited education, which results in a poor understanding of ecological processes, especially the role of predators (Conforti and Azevedo 2003). There seems to be a relationship between knowledge of a species and perceptions (Conforti and Azevedo 2003). Consequently, local workshops and direct interactions between local people and natural resources professionals could help change people's perceptions and thus attitudes about the role of pumas and other predators (Conforti and Azevedo 2003; Bueno-Cabrera 2004). Organizations such as the Jaguar Conservation Fund (2007) in Brazil and Pronatura (2007) in Mexico are examples of nongovernmental organizations (NGOs) working to increase environmental awareness in Latin America. The obvious limitation to this approach is the time it takes and the cost of conducting such programs on a large enough scale to be effective. In areas with parks or reserves, such programs staffed by official personnel offer potential to change attitudes in surrounding areas. This could result in more protection for pumas in conserved areas, which are often surrounded by intense human activity (Conforti and Azevedo 2003).

In contrast to the situation with cattle, puma predation on sheep and goats is higher and can account for substantial financial losses. Again, environmental education programs would probably be beneficial in increasing awareness of the role of large predators. However, the main goal for these species is to try to reduce puma predation with nonlethal methods. Fortunately, many livestock owners are open to the use of such nonlethal methods (Conforti and Azevedo 2003; Michalski et al. 2006). Because there does seem to be a relationship between husbandry practices and high predation by pumas, altering herd management offers promise for reducing predation losses. Since pumas prey most on free-ranging goats and sheep, programs that encourage closer management of flocks, such as by putting them in corrals at night, could effectively reduce predation levels (Mazzolli et al. 2002). Mazzolli and colleagues (2002) suggested that the effectiveness of these changes in management practices could be increased if they were implemented and supervised by local environmental officers. They also suggested that subsidizing fencing costs could be effective in reducing ranchers' reluctance to make these changes. Besides building corrals, other nonlethal methods suggested are the use of guard dogs or llamas, loud sounds, and aversive conditioning, such as nauseating substances in carcasses of killed animals (Mazzolli et al. 2002; Crawshaw 2003).

Habitat characteristics can also affect predation levels on sheep and goats (Bueno-Cabrera 2004; Mazzolli et al. 2002). Because of this relationship, Bueno-Cabrera (2004) suggested that local or regional evaluations of predation risk could be made to advise ranchers of their risk level. This information can be used by local environmental personnel to better inform ranchers about predation risk and to target nonlethal control efforts in potential predation hot spots. Such an effort would help concentrate often limited resources with focus on those ranches needing the most help in reducing predation losses.

Often the conversion of an area from native vegetation to ranching results in fragmentation of the habitat and a decrease in native prey species. This reduction may be the result of habitat loss or, most often, of an increase in human access and hunting (Alvard et al. 1997; Chiarello 1999; Carrillo et al. 2000; Escamilla et al. 2000; Peres 2000, 2001; Novack et al. 2005). Often the hunting preferences of humans overlap with those of pumas (Carrillo et al. 2000; Escamilla et al. 2000) and can reduce prey populations (Carrillo et al. 2000; Novack et al. 2005). The reduction of native prey and its replacement with domestic species seems to be related to increased livestock predation (Polisar et al. 2003). Proposed solutions would include enhancement of native prey populations via stricter control of illegal hunting or the reintroduction of native prey that might have been extirpated (Conforti and Azevedo 2003). The main idea behind this action is to increase native prey populations, which could lead to reduced livestock depredation. Because of the potential benefits of this action, more research is needed in this area.

Compensation

Apart from trying to reduce the incidence of puma predation on livestock, an additional method to reduce the economic loss and in turn increase tolerance of predators is economic compensation for losses (Crawshaw 2003; Naughton-Treves et al. 2003; Conforti and Azevedo 2003). These programs are used in some states in the United States (see Table 4.1; Keiter and Locke 1996; Naughton-Treves et al. 2003). They usually have the support of ranchers (Crawshaw 2003) and have been proposed for predation losses in Latin America (Crawshaw 2003; Conforti and Azevedo 2003). Crawshaw (2003) proposed a national program for Brazil that would designate a small percentage of the funds from existing fiscal taxes. This program would be administered by the Brazilian Federal Environment Agency. Because livestock losses affect a proportionally small number of producers, Crawshaw (2003) estimates that this program would be a permanent self-sustaining operation. However, compensation programs in the United States have been criticized as being inadequate, fraudulent, or cumbersome (Naughton-Treves et al. 2003). Also, there is some evidence that compensation may not change negative attitudes about predators (Naughton-Treves et al. 2003). Obviously, these are important considerations for Latin American countries where most claims would be made by small-scale owners of sheep and goats—people who might have difficulty in pursuing compensation claims. Consequently, though they seem promising, compensation programs for Latin American countries need to be investigated further.

An alternate form of compensation for losses of domestic livestock to pumas is by generating a secondary income because of their presence (Crawshaw 2003). Sport hunting in general in the United States and Canada generates millions of dollars annually, much of the money spent locally. The harvest of most populations in the north is managed by state game agencies with one objective being to maintain viable populations. Thus, the management of a legal harvest of predators to generate income can be an effective tool in their conservation (Crawshaw 2003). This method can involve controversy, but it could be especially useful with problem cats, allowing the hunter to pay the rancher a fee for the privilege of removing a particular animal, and would be a way of generating economic compensation for losses (Crawshaw 2003). Obviously, for such a system to work, Latin American countries would have to have effective game management agencies staffed by trained personnel. Unfortunately, in many countries there are few wildlife professionals in comparison to the amount of wildlife they have to conserve and manage. For example, in the northern two-thirds of Mexico, it is difficult to identify more than a dozen people with advanced wildlife training (personal observations).

Another way of generating economic support for the presence of large cats is through ecotourism. The reintroduction of wolves in Yellowstone National Park has generated millions of dollars as thousands of people come to the park specifically to see wolves (Fritts et al. 2003). Most Latin American countries have an active tourist industry, mainly centered around ocean beaches. Some countries, such as Costa Rica, have well-developed ecotourism programs and could be used as the model for such development in other countries. The development of a tourist industry around the large predators in an area might be possible and, if so, could provide income that would offset depredation losses. Obviously, there would need to be investments into infrastructure for such activities as well as training of local personnel. However, since only a small percentage of the land mass in Latin America is within protected areas (Costa et al. 2005), the development of tourism provides a tremendous opportunity for conservation on private lands, where wildlife can often be better protected (Polisar et al. 2003).

In conclusion, puma conservation in agricultural areas depends on educating the local people about the ecological role and value of large carnivores and on reducing predation losses or compensating people for them. Owners of large and small livestock constitute the segment of society that has the closest contact with pumas and often determines whether the cats remain in an area. Thus, through the methods outlined and possibly via more ingenious methods

waiting to be discovered, the goal is to build consensus in rural areas that the presence of pumas is desirable or at least tolerable. Fortunately, there seems to be hope for this consensus (Quigley and Crawshaw 1992). Crawshaw (2003) observed increasing numbers of ranchers wanting to understand the need for the conservation of large felids, as long as the cost is not too high. He further stated that the presence of significant populations of large felids will depend on this consensus because of the limited extent of protected areas.

One last kind of conflict arises when pumas attack humans (Beier 1991, 1992). Although the rate is not high, there is ample documentation of cases in which pumas have attacked and killed people in the United States and Canada (see Chapter 13; Beier 1991). Many of the recent records are associated with increased human use of puma habitat (Beier 1991). As in the north, people in Latin America are making rapid inroads into puma habitat. However, apart from one researcher in Costa Rica being attacked (Foerster 1996), and one reported being killed in Patagonia (see Chapter 7), we could not find any other records of pumas attacking or killing humans in Latin America. This absence of records possibly reflects lack of reporting of such incidents rather than any real difference between pumas in the south and the north. This is another area where more investigation is needed.

Threats and Conservation Status

In the conservation of any species it is important first to identify the threats to the species and then to evaluate how these threats can be reduced. Obviously, one of the major threats to any species is the direct killing of individuals. As mentioned, the puma is officially protected in fifteen countries, and there are regulated hunting laws in two more. Only in Ecuador, El Salvador, and Guyana is this species still unprotected (Figure 6.3). However, even in the countries where pumas are protected, illegal killing still occurs (Mazzolli 2000), much of it associated with puma predation on livestock (Mazzolli 2000; Mazzolli et al. 2002). In Brazil, where pumas are protected, it is still permitted to kill them to protect livestock (Crawshaw 2003). The direct killing of pumas is probably a major cause of the loss of individuals in many areas.

Illegal or subsistence hunting is a major problem for most wildlife species in Latin America (Robinson and Redford 1991; Alvard et al. 1997; Merriam 1997; Chiarello 1999; Escamilla et al. 2000). Because the prey favored by subsistence hunters overlaps with that of pumas (Redford 1992; Carrillo et al. 2000; Escamilla et al. 2000), a secondary impact of illegal hunting is the reduction of prey needed to support viable populations of pumas. Since the level of subsistence hunting is usually highest in areas of heavier human use, where puma populations are already low, this reduction of the prey base can have a major impact on the survival of the remaining individuals.

Besides the direct and indirect impacts of hunting pumas and their prey, the most significant factor affecting puma populations is the loss and fragmentation of habitat. Native vegetation types have been and are still being converted to human-dominated ones at an alarming rate in all of Latin America. In some areas, such as the Atlantic coastal forest of southern Brazil, 88 to 95 percent of the original forest habitat has been removed (Chiarello 1999). Approximately 20 percent of the Amazon forest has been removed, and it is estimated that in twenty-five years this figure will increase to 40 percent (Wallace 2007). What remains in these areas is usually a mosaic of native vegetation patches embedded in a growing human-dominated matrix of towns, fields, and pastures (Chiarello 1999; Michalski et al. 2006). It is within this matrix that pumas and their prey must survive. Not only is there an ever increasing loss of native habitat but the usefulness of the remaining patches decreases as greater human access exacerbates the impacts of illegal hunting (Peres 2000, 2001; Michalski and Peres 2005; Laurance et al. 2005). An additional impact is the presence of more roads between the patches, which increases direct mortality as animals move from one patch to another (Mazzolli 2000; Michalski and Peres 2005; Ceballos et al. 2005). The patches of native vegetation vary in size and interpatch distance and effectively establish a metapopulation system where smaller, more isolated patches become sinks for pumas (see Chapter 5; Sweanor et al. 2000; Laundré and Clark 2003).

Human Population Growth and Land Reform

Having identified the major threats to the conservation of pumas in Latin America, it is important to analyze why these threats exist. The first and most obvious reason for habitat loss and fragmentation is the rapid human population growth in most of the Latin American countries. In Brazil alone, where most of the Amazon basin lies, the human population grew from 54 million to 170.7 million between 1950 and 2000 and will reach 250 million by 2050 (Brea 2003). As already noted, the overall population of Latin America more than tripled from 167 million in 1950 to 520 million in 2000, and it is expected to reach 802 million by 2050 (Brea 2003). The growth is not equal in all countries but ranges from 1.5 times more people in Uruguay to 4.7 times more people in Venezuela (Brea 2003).

Much of the population growth in South America has occurred in coastal areas, especially along the northern Pacific and central Atlantic coasts, and there are increasing numbers of people in the Amazon basin, encouraged by government resettlement programs. In Mexico, the growth is primarily from Mexico City to the south; Mexico City is one of the largest cities in the world, with over twenty million inhabitants (City Population 2007). In Central America, the growth is primarily along the Pacific Coast. This dramatic increase in human population is the driving force behind the loss and fragmentation of native habitat to provide places for people to live and grow food. The increase in population has also resulted in dramatic increases in the number and size of towns and cities and expansion of the network of paved roads. Because of the anthropogenic effects associated with population centers and roads (Laurance et al. 2005), puma populations in these areas can be considered extinct or extremely threatened (Figure 6.2).

Although the human population in Latin America is growing rapidly, this in itself need not necessarily impact puma survival and abundance as much as it has done. In the western United States and Canada, healthy puma populations can be found near fairly large metropolitan centers (Shuey 2005), and there is evidence that pumas are returning to fairly heavily populated areas in northeastern North America (see Chapters 4, 5, 12; Gauthier et al. 2005). Apart from the dramatic increase and spread of the human population in Latin America, social factors exacerbate the problem. The most important factor is that a majority of wealth is concentrated in the hands of a very small percentage of the population (Lozada 2003). Most countries have only a small middle class, with the remaining vast majority of the people being poor (CEPAL 2002, Lozada 2003).

Poverty means most of the people are more concerned with sheer survival than with environmental issues (Leonard et al. 1989). When the level of education is low, an environmental ethic can scarcely flourish. People view large predators as competitors or threats. Moreover, inherent distrust of governmental authority results in disregard for regulations or laws established to protect natural resources. This is especially the case in rural areas where human impact should be the least but wild prey populations are often overexploited, which in turn affects puma abundance (Hernández and Laundré 2000; Loredo-Salazar 2003; Laundré et al., in press).

Another factor intensifying the impact of population growth is the land reform programs operated by some governments to distribute people, usually poor, into undisturbed rural areas (Leonard et al. 1989). Besides the increased wildlife exploitation that follows such settlements, agricultural activity for food crops and pasture land reduces the native vegetation. Introduction of domestic livestock creates conflict with large predators, which are soon killed, as in the previously mentioned death of 185 to 240 jaguars and pumas along the tropical deforestation frontier of Brazil (Michalski and Peres 2005; Michalski et al. 2006). The settlements bring nonlocal people with low environmental concern into areas that would represent the best possibilities for conservation of puma populations.

In the Shadow of the Jaguar

How do all these threats affect puma conservation efforts in Latin America? The first obstacle pumas face is name recognition. Their secretive behavior results in their routinely being overshadowed by other large predators (Kellert et al. 1996), in this case the jaguar. Where the two co-occur, local people are often unaware of the presence of pumas (Conforti and Azevedo 2003). This lack of recognition is double edged; without spots, pumas are saved from fur hunting pressure. However, pumas also do not generate the same public enthusiasm for their conservation as do jaguars. Thus, even though pumas are officially listed as protected in most of Latin America (Figure 6.3), we rarely find reference to conservation efforts specifically aimed at this species.

Fortunately for pumas, their diet and habitat use overlap extensively with those of jaguars, and except for fur hunting, they face the same threats. Thus conservation efforts for jaguars often benefit pumas. Conservation of jaguars generally centers around the following possible actions:

1. Reduction of habitat loss and protection of existing habitat
2. Establishment and protection of movement corridors
3. Reduction of illegal hunting
4. Reduction of impacts of depredation of livestock
5. Environmental education
6. Increased investigation

The first of these actions involves status surveys to identify priority areas for conservation (Sanderson et al. 2002; Vaughan and Temple 2002), establishment of habitat conservation units (Silveira and Jácomo 2002), and enforcement of protective laws (Leite et al. 2002). Coupled with habitat protection is the concept of identifying and establishing connecting corridors. Such corridor systems can be on the country level (Leite et al. 2002) or be multinational, as in the case of the proposed Paseo Panthera project, which is intended to provide movement corridors for jaguars from southern Mexico to southern Panama (Vaughan and Temple 2002).

Because big cats are often killed for preying on livestock, the next two kinds of actions are closely connected. Since illegal hunting of big cats and subsistence hunting of their food species are a major immediate threat, the obvious conservation action is to try to reduce this harvest through enforcement of game laws (Vaughan and Temple 2002;

Silveira and Jácomo 2002). Regarding livestock conflicts, most observers propose alternate plans, such as the improved husbandry options noted (Quigley and Crawshaw 1992; Perovic 2002; Vaughan and Temple 2002; Leite et al. 2002). However, implementing these action plans can often be difficult because of the socioeconomic factors described.

The first four types of actions can be improved by environmental education and additional investigation (Leite et al. 2002; Vaughan and Temple 2002). As noted, many people are simply unaware of the role and value of large predators and the benefits of native areas. Public awareness campaigns are needed, with emphasis on environmental education at the elementary school level to raise awareness among the young people who will inherit the natural legacy. Although we have seen an increase in the study of pumas and jaguars in recent years, much more is needed. We have tried to point out knowledge gaps so that future research can focus on material needed for education efforts to support conservation.

The Future of Pumas in Latin America

What is the future for pumas in Latin America? Will conservation efforts for jaguars be sufficient to save puma populations also? What will the picture look like a hundred years from now? What can we learn by comparing what we know about Latin American pumas with fuller information from northern populations? Although to some observers the future of northern pumas may seem secure, others are sounding alarms even in the north.

Despite the benefits to pumas from conservation efforts designed for jaguars, the future is not a bright one for many populations of either species. In the face of expanding human settlement and dramatic habitat changes, the cougars of eastern and central North America are gone. The original great eastern forests of the United States have been cut down. Much formerly wooded country joined the Great Plains in becoming the great corn field, approximating the size of the Amazon basin (Wallace 2007). The central valley of California—the U.S. version of the Pantanal—is now the largest vegetable garden in the world. Puma populations in many parts of Latin America are similarly being edged toward extinction. Current deforestation in the Amazon, the Atlantic coastal rain forest of Brazil, and many other sections of Latin America closely mirrors what happened to the eastern and central United States after European settlement began.

As in the north, the repercussions of massive habitat destruction in Latin America will be immense, especially for large carnivores. We lose plant and animal species at an accelerating rate. Some people fight to reverse the trend, but as in the United States a hundred years ago, most lack the concern and the political will to prevent a similar environmental disaster. Large areas of native vegetation will be lost, protected zones will continue to become isolated islands—ships too small to ensure the survival of their passengers in the stormy sea of surrounding humanity. Like their northern cousins, pumas will eventually remain only in the extreme environments of Latin America: high mountain ranges, hot deserts, and humid "wastelands." In these retreats, the pumas in Latin America are likely to survive, as they have in the north—a remnant of their past distribution and abundance, but surviving nonetheless. Unfortunately, we could still lose at least two of the six puma subspecies in Latin America (see Figure 3.3). This would represent a significant loss in genetic diversity and would impoverish the cradle of puma evolution.

Although we have painted a rather dim picture for the puma in Latin America, there is some hope on the horizon. Human birth rates across Latin America have begun to decline and a stable population is projected for around 2050 (Brea 2003). Environmental education is making inroads and changing attitudes, and increased study of pumas aids in this endeavor. The economic status of people in some countries is rising, notably in mega-diverse Mexico (Calva-Mercado 2007), which could reduce human impacts on wildlife (Carrillo et al. 2000). If the people of Latin America are eventually able to afford it, they may recognize and address the environmental consequences of their actions, as in the eastern United States, where new awareness is beginning to foster the return of some native habitats—and perhaps of the puma.

Conclusion

We have tried to include all relevant information available about pumas in Latin America north of Chile and Argentina to supply a sense of what is and is not known about pumas in this vast region. Recent work by diligent puma researchers has significantly increased our knowledge base for this region. Because of the extensive geographic reach of this chapter, we have doubtless missed some work, and we apologize for these omissions.

Comparing what is known about Latin American pumas to the more extensive information base for their northern cousins, we find both differences and similarities. Because the most data exist for diets, it is now obvious that southern pumas rely on a more diverse spectrum of prey species, albeit still primarily mammals. As in the United States and Canada, southern pumas are often blamed for livestock depredation and impacts on native ungulate populations. However, again as in the north, this blame is misplaced and is more an irrational human response to a fellow predator than a real assessment of all the interacting forces. What emerges from the science is a picture of a highly adaptive species that makes fine-tuned adjustments in its ecology and behavior to survive in the diverse habitats found in Latin America. Because of this adaptability, the puma presents both challenges and opportunities. Its ubiquitous distribution results in conflicts with humans and population losses in areas most desirable for human occupation. But it also means that remaining populations will have refuge in those areas less desirable for people. It is in these corners or *rincones* of Latin America that the puma will survive. Though impoverished, the cradle of pumas will endure.

The current state of knowledge of pumas is a valuable start. However, if we are to aid pumas in Latin America, a tremendous amount of work still needs to be done to fill the gaps in understanding and geography. For several Latin American countries, we lack even basic information on pumas. What is needed is a massive international effort to raise the state of knowledge about this species. With such an effort, perhaps one day a book like this can be devoted solely to Latin American pumas.

Chapter 7 The World's Southernmost Pumas in Patagonia and the Southern Andes

Susan Walker and Andrés Novaro

PUMAS OF THE far south are treated separately from those elsewhere in South America because conditions in the south differ markedly from those across most of the rest of the continent. In some ways, the story of these pumas more closely resembles that of pumas in North America. In Patagonia and the southern Andes, pumas suffered drastic declines during the twentieth century, and then launched a population rebound when the rural human population decreased in the region. They have shown the classic adaptability of their species, changing their diet as prey conditions shifted and, later, dispersing to reoccupy lands where they were once persecuted. Research on these pumas began in the late 1980s, but even now studies are few. In this chapter, we present conservation, research, and management needs for pumas and what is known about their recent history.

Patagonia is the southernmost region of Argentina and Chile, comprising over 700,000 km² and spanning several provinces in the two countries. Except for a band of humid forest along the Andes, it is an arid area of steppe and scrub habitats. The southern Andean steppe extends northward from Patagonia along the Andes. The puma is the largest carnivore of Patagonia and the southern Andes. The larger-bodied jaguar (*Panthera onca*), was once found in the northernmost part of Patagonia but was extirpated before Europeans settled the region (Parera 2002). Prior to the arrival of Europeans, the puma was distributed throughout virtually every habitat type in this large region—including forests, scrub, and grasslands—except for the island of Tierra del Fuego. A large cat species occupied Tierra del Fuego 12,000 years ago, but whether this was the puma has

yet to be confirmed (Massone Mezzano 2001; G. Lheureux, pers. comm., 2007).

In the temperate habitats of southern South America, the puma (*Puma concolor*) coexists with five small cat species. It shares the southern Andean steppe with the Andean cat (*Oreailurus jacobita*), colocolo (*Lynchailurus colocolo*), and Geoffroy's cat (*Oncifelis geoffroyi*); in the southern Andean forests, it occurs with Geoffroy's cat and the kodkod (*Oncifelis guigna*). It shares the scrub and steppe with Geoffroy's cat and colocolo, and, in the northernmost parts of those ecosystems, the jaguarundi (*Herpailurus yaguarondi*). The second largest carnivore, however, is a canid of 10–12 kg called the culpeo (*Pseudalopex culpaeus*), which is found everywhere that pumas are found (Figure 7.1). The puma outweighs the culpeo by 50 kg and can prey upon adults of that species (Novaro et al. 2005).

The guanaco (*Lama guanicoe*) is the main native prey of the puma throughout this region, except in the southern forests (Iriarte et al. 1991; Franklin et al. 1999; Bank et al. 2002). The guanaco is a 120-kg camelid, the dominant herbivore of the Patagonian and southern Andean steppes (Figure 7.2). In the northern part of the southern Andean steppe, a smaller, 50-kg camelid known as the vicuña (*Vicugna vicugna*) is also a major prey species, although the guanaco appears to be preferred when these two camelids are sympatric (Cajal and Lopez 1987; Figures 7.3, 7.4, 7.5). Other important prey species in the arid and semiarid regions include rheas (*Pterocnema pennata*), meter-high, flightless birds; maras (*Dolichotis patagonum*), deerlike rodents of 8 kg; and, in the northern scrub, the plains vizcacha (*Lagostomus maximus*), a large, burrowing rodent

Figure 7.1 In the far south, the second largest carnivore is the 10–12 kg culpeo, seen here in San Guillermo National Park, San Juan, Argentina. Pumas provide carrion for scavenging culpeos but also prey on them. Poisoning of these foxes is widespread, and some pumas undoubtedly succumb as well. Photo by Andrés Novaro.

with males weighing up to 7 kg. In the forests, the principal native prey are the southern Andean deer or huemul (*Hippocamelus bisulcus*) and the pudu (*Pudu pudu*).

Large prey populations and frequent puma attacks on horses and people suggest that pumas were abundant in

the region until the late 1800s (Moreno 1997). In southern Chile, guanaco numbers increased greatly after the Torres del Paine National Park was expanded in 1975. Puma sightings were rare in the park during the 1970s, but by the early 1990s, the cats were seen frequently, with nine sightings recorded in a single day. This park is the only place in all of Patagonia and the southern Andes where pumas have been radio-tracked. In mixed grassland-forest habitat, minimum home ranges for females were 27–102 km^2 and for males 24–100 km^2 (Franklin et al. 1999). Based on radiotelemetry data, the estimated density of adult pumas in one area of the park was more than 3/km^2, which is similar to the highest density reported for North America (Cougar Management Guidelines Working Group 2005). In one part of the park, puma density was reported to be 30/100 km^2, although how this estimate was obtained is not entirely clear (Franklin et al. 1999).

Pumas may have followed large migratory herds of guanacos, as did the Tehuelche people, who depended on the guanaco for food and other resources (Bank et al. 2002; Musters 1964). Guanaco migrations persist only in a few isolated areas. Some pumas specialize on patchily distributed, locally abundant prey, as provided by plains vizcacha colonies (Branch et al. 1996). When the vizcacha population in Lihue Calel National Park in La Pampa province began to decline, vizcachas were still selected by local pumas, and when that prey population went completely

Figure 7.2 Guanacos, in the foreground, are the principal native prey of pumas in Patagonia; in the background are exotic red deer. Photo by Kerry Lock.

Figure 7.3 A mother vicuña and her young in San Guillermo National Park, San Juan, Argentina. Vicuñas are the main native prey of pumas in this part of the southern Andean steppe. Photo by Emiliano Donadio.

Figures 7.4, 7.5 Incessant Andean winds ruffle a puma's coat in San Guillermo National Park, San Juan, Argentina. The rocky cliffs and grassy meadows typically found in the high Andes provide poor stalking cover; thus, a puma must crouch low and move carefully to remain undetected by its prey. After successfully killing a vicuña, this puma covered its kill with the only available material—sand and pebbles. Photo by Andrés Novaro.

extinct, the pumas broadened their diet, consuming a wider variety of prey (Pessino et al. 2001). As the largest predator in the region, the puma provides carrion for culpeos and the much smaller chilla fox (*Pseudalopex chilla*), and for a broad assemblage of avian scavengers, including the Andean condor (*Vultur gryphus*), black vulture (*Coragyps atratus*), turkey vulture (*Cathartes aura*), caracaras (*Polyborus plancus* and *P. megalopterus*), and chimango (*Milvago chimango*).

Effects of European Colonization on Pumas and Their Prey

After the indigenous Tehuelche and Mapuche people were conquered and driven from their lands in the late 1800s, most of Patagonia was occupied by Argentine and European sheep ranchers, with the number of sheep in Patagonia peaking at 22 million in the 1950s (INDEC 2002). Guanacos and other native wildlife were heavily persecuted and their numbers were drastically reduced. Guanacos declined to 2–9 percent of their pre-conquest numbers and were pushed to the more arid and isolated areas (Baldi et al. 2006). During the 1900s, European immigrants introduced their favorite game species. Wild pigs (*Sus scrofa*) and European hares (*Lepus europaeus*) spread throughout the Patagonian steppe and forest and much of the southern Andean steppe to the north. Red deer (*Cervus elaphus*) were introduced in the ecotone between steppe and forest in the northwest and in the scrub of northernmost Patagonia. In the northwest, at first the deer became well established only in the forest, but since the late 1980s they have expanded well out into the steppe. Along with the eradication of its native prey, the puma was extirpated throughout most of its range in Patagonia. By the 1980s, pumas were found only in isolated refuges in the mountain forests of the west and in some areas of scrub in eastern and northern Patagonia (Bellati and von Thungen 1990).

During the last half of the twentieth century, the number of sheep in Patagonia began to decline, largely due to decreased carrying capacity caused by overgrazing (Golluscio et al. 1998). A further decrease came in the 1990s, because of low international prices for wool and the relatively high cost of production in Argentina, where the local currency was pegged one-to-one to the dollar. In the southernmost province of mainland Argentina, many sheep ranches, which were already suffering reduced profits, were abandoned altogether after a 1991 volcano eruption left them covered with ashes. Ranchers turned to cattle in other areas where the range was suitable, or began to look for other alternatives. With the decline in sheep numbers, and the concomitant decline in the rural human population, guanacos and other wildlife began to recover in some areas, and exotic wildlife, such as the red deer, expanded their distributions.

The puma also began to recover its previous distribution. In 1985 it did not occupy any of the thirty-two ranches in southern Neuquén province where ranchers were surveyed, but by 1995 it occupied 91 percent of those same ranches (Novaro et al. 1999; Novaro and Walker 2005). Interestingly, local people in one of the areas most recently recolonized, northeastern Neuquén province, had lost all memory of the previous occurrence of the species, and when pumas began to reappear they were considered an exotic species. The puma has recovered to a much greater extent than have its native prey species. Pumas have been expanding from refugia in the eastern scrub and western mountains; in northern Patagonia these two fronts of expansion have met. In the south, pumas have now reached the Atlantic coast. Today, they once again inhabit almost every corner of Patagonia and the southern Andes.

Current Interaction between Pumas and Prey

The role of the puma in Patagonian and southern Andean ecosystems has changed over the last hundred years due to human-induced changes in the prey base (Novaro and Walker 2005). Over 95 percent of arid Patagonia is private land, mostly dedicated to sheep husbandry or other livestock management. The distribution of guanacos is patchy, and the species is therefore absent from many areas inhabited by pumas. Pumas in most parts of Patagonia and the southern Andes today prey largely on livestock and exotic wildlife rather than on guanacos and other native species. European red deer and hares represent 90 percent of the biomass in the diet of pumas in southern Neuquén and La Pampa provinces (Novaro et al. 2000). In the forests of southern Chile, the European hare has replaced the pudu as the primary prey of pumas (Rau and Jiménez 2002). Throughout much of Patagonia, native species are ecologically extinct as prey for pumas and other carnivores, meaning that although they are still present their numbers are too low to carry out their functions in the ecosystem (Estes et al. 1989; Novaro et al. 2000).

Since the removal or reduction of sheep, guanacos have increased dramatically on some of the large ranches of southern Neuquén province, while on others they have not recovered or have continued to decline (Novaro and Walker 2005). The ranches where guanacos have recovered started with initial population densities of more than 10/km², as opposed to those where they have not recovered, where densities of fewer than 6 guanacos/km² exist. These ranches have high densities of wild boars and exotic red deer, and puma densities are high at ranches where guanacos have and have not

recovered. Therefore, we hypothesize that puma populations are supplemented by wild boar, red deer, hares, and sheep, and that predation by pumas may limit population recovery of guanacos when they are below a threshold density.

In another area, it appears that puma recolonization may have stabilized the growth of a large population of guanacos. In the early 1980s, after the creation of the Payunia provincial reserve in southern Mendoza province, guanacos increased at a rate of 4 percent per year (Puig 1986). At that time, pumas had not yet recolonized the area. After 1990, when recolonization by pumas occurred, the rate of increase of guanacos in Payunia appears to have declined to near zero (Puig et al. 2003; Guichon and Novaro, unpublished data, 2007).

Conflicts over Livestock

Since Europeans and Argentines arrived with sheep, the puma has been considered a major pest in Patagonia because of its predation on sheep and other livestock. So effectively were pumas eliminated from large areas during the twentieth century that for many decades conflict was limited to a few areas near the Andes and in the northeastern scrub, where pumas remained. With the recovery of pumas throughout the region, however, have come growing and more widespread conflicts. Of 165 livestock producers interviewed in the province of Santa Cruz, 50 percent reported puma predation on their stock (Travaini et al. 2000). This sample of producers was distributed throughout the province, the largest in Patagonia. Puma predation on livestock is now considered a regionwide problem. Although depredation is intense only in limited areas, wildlife managers in Patagonia have come under increasing pressure from ranchers during the last five years to provide solutions to the puma problem (M. Failla, S. Rivera, B. Alegre, and N. Soto, pers. comm., 2007).

On four large sheep ranches adjacent to Monte León and Los Glaciares national parks in southern Patagonia, ranchers reported annual sheep losses to pumas between 30 and 800, representing between 3 and 9 percent of sheep stocks (Pía and Novaro 2005). These are not confirmed puma kills and may actually include losses due to other causes. These ranchers killed up to fifteen pumas every year on their properties, but puma numbers remained high, likely due to recolonization from adjacent parks. This recolonization contributes to animosity between sheep ranchers and managers of protected areas, as in other parts of the world. Similar levels of loss were reported in Chilean Patagonia, where ranchers killed as many as fifty to seventy-five pumas per year (Franklin et al. 1999). In the scrub of northeastern Patagonia, native prey of puma are scarce, due to the same processes of poaching, habitat degradation, and competition with livestock operating throughout most of Patagonia. In this corner of Patagonia, cattle are the main livestock. Attacks by pumas on calves and horses are widespread. In addition to large-scale ranches, there are many goat and sheep herders with small flocks in northern Patagonia and the southern Andean steppe. Surplus killing of sheep and goats by pumas can be economically devastating to these small producers.

A newer conflict has arisen between pumas and ranchers who make use of red deer. With the decline in profitability of sheep ranching in recent decades, some northern Patagonia ranchers have turned to red deer sport-hunting operations or farming the red deer found on their properties by fencing them into large paddocks. Red deer are currently the major prey of pumas in northwestern Patagonia (Novaro et al. 2000). Producers see the pumas both as competitors with sport hunters and as predators of farmed and free-ranging red deer. In a survey of six farms with red deer and other exotic ungulates, five reported losses to pumas totaled $28,700 annually (Novaro et al. 1999).

Puma Management in Patagonia and the Southern Andes

In Argentina, wildlife management is regulated by each province, whereas in Chile it is managed at the national level. Wildlife in Argentina is the property of no one, but provincial governments have the responsibility to manage and conserve it. Nevertheless, provincial wildlife agencies have little capacity to monitor compliance with the laws and, in general, ranchers do as they please with the wildlife

on their property. On many ranches there is indiscriminate killing of pumas. One rancher in Neuquén, for example, reported killing ten pumas in 2006. They were preying on the abundant but declining red deer on his ranch, and he killed the cats to prevent possible future killing of his cattle when deer became scarce.

There are three major legal management strategies in the region: total protection, bounty hunting, and sport hunting. In La Rioja, San Juan, and Mendoza provinces of Argentina, as well as in Chilean Patagonia, no killing of pumas is legally allowed. This prohibition is partly in response to the historic scarcity of pumas, and legislation has not been updated to fit the new situation.

Since 1995, three provinces—Río Negro, Chubut, and Santa Cruz—have reinstated the bounty system (Table 7.1). These bounties are considerable by Argentine economic standards, especially for low-paid ranch workers. Close to 2,000 puma bounties are paid per year in these three provinces, which have a total area of approximately 670,000 km². This number of pumas killed is extremely high if we compare it to the 1999 total of 3,200 cougars removed from the western United States and Canada (Cougar Wildlife Management Guidelines Working Group 2005); the western U.S. states alone cover an area of more than 3 million km². In spite of the high number of pumas removed in the Argentine provinces with bounties, complaints about puma predation on livestock continue and have increased in recent years (M. Failla, director, Province of Río Negro Department of Wildlife, pers. comm., 2007).

Hunting pumas for control has not been proven to be an effective method of reducing depredation by pumas on livestock (Cougar Management Guidelines Working Group 2005). Although most skins are turned in to stations in

Table 7.1 Management strategies and numbers of pumas legally reported to be removed from provinces in southern Argentina, and amounts paid to those who removed them.

Province	Area in km²	Main Management Strategy	Number of Pumas Hunted per Year	$ per Bounty
Santa Cruz	242,633	Bounties	1129±785	$70 (total paid in 2004–2006: $118,333)
Chubut	224,187	Bounties	Approx. 400	$115
Neuquén	93,556	Control and sport hunting	No numbers for control 5±4 (sport)	No bounties
Río Negro	202,846	Bounties and sport hunting	457±176 (bounties) 5–15 (sport)	$70
La Pampa	142,632	Sport hunting	40–60 legal trophies	$2000–3300 per trophy paid to hunting operator

areas of high conflict, there is no way the provinces can be guaranteed that the skins they pay for actually represent problem animals from problem areas; they could just as well be skins brought in from neighboring provinces. Thus, the bounty system does not appear to be either effective or efficient as a method for reducing livestock loss to pumas. Moreover, until now the bounty system in Patagonia has not been accompanied by periodic assessment of the level of puma attacks on livestock or the level of rancher complaints, making it impossible to determine whether bounties have ever played any role in reducing the conflict. Most managers are fully aware that the system is ineffective for reducing livestock depredation but see it as a way to "show that they are doing something," a desperate attempt to keep ranchers from becoming too unhappy.

Sport hunting of pumas is legal in La Pampa, Río Negro, and Neuquén provinces, although the number of pumas killed through sport hunting is probably significant only in La Pampa (Table 7.1). This type of hunting, however, is far from becoming an effective management tool. In La Pampa, 40–60 puma trophies are reported every year, on over ninety ranches that are registered as puma hunting concessions. The true number hunted is probably much greater—perhaps in the thousands—because many ranchers and hunting operators do not follow the legal restrictions.

But the pumas hunted do not all come from within the province. Every year between the late 1990s and early 2000s, up to 200 hunters, primarily from Europe, paid large sums to hunt pumas in La Pampa (Romero 2007). The properties average 10,000 ha in size and cannot produce enough animals for the flow of client hunters. Hence, ranchers pay trappers about $300 per animal to bring in pumas captured illegally on other ranches and even in neighboring provinces. Reports in local newspapers (El Diario 2007a, b, c; La Arena 2007a, c) reveal the level of irregularities in many of these concessions, where inspections have turned up 114 skulls at one site, and between six and thirty live pumas in cages at several others. Because the animals are captured with legholds, snares, and dogs, many die in this black market trade—perhaps several for each one that is eventually delivered to a hunting concession. How many pumas are affected is unknown, but based on the number of live pumas found during inspections, it is likely that hundreds of pumas are involved (Romero 2007). Responding to public criticism by a conservation group, the province has cracked down on this illegal trade in pumas and suspended the 2007 puma hunting season (La Arena, 2007b).

Sport hunting began only in the early 2000s in the other three provinces, where it is restricted to a handful of ranches, with only a few pumas hunted every year. Thus, in Patagonia sport hunting is neither sufficiently widespread nor well regulated enough to be an effective management tool. It could perhaps become so in the region if properly managed and if ranchers comply with hunting regulations. Sport hunting of pumas could help ranchers offset the costs of livestock losses through hunting revenue and could also help build a constituency of hunters or hunting operators who favor the recovery of puma populations, as has occurred in some parts of the western United States (K. Alt, Montana Department of Fish, Game, and Parks, pers. comm., 2005).

Many methods of reducing predation on livestock by pumas and other carnivores have been put to use in this region. Poisons are used widely among Patagonian ranchers against predators (Travaini et al. 2000). Poison is administered in carcasses, with the main target being the culpeo, although undoubtedly some pumas succumb. To kill livestock predators more selectively, some sheep ranchers have experimented with toxic collars imported from the United States (J. Bellati, pers. comm., 1994). These collars are placed on sheep and kill a predator that bites into the collar. Tests were limited and not monitored closely, except on one ranch in Neuquén province, where the collars did not appear to be very effective against pumas.

The selective killing of pumas that have attacked livestock was approved in Neuquén province in 2003. However, in practice, the province has no way of monitoring whether pumas killed by ranchers are actually problem animals, and most pumas that are killed still go unreported. Selective killing of problem pumas in Monte Leon National Park in southern Patagonia and the adjacent sheep ranches is under consideration and may be approved if a system using authorized professional hunters, with government oversight, can be implemented and properly monitored. Only pumas tracked from livestock kills would be killed under this system. To date, ranchers have killed pumas in this area indiscriminately, which has been ineffective for reducing predation on sheep, and rancher animosity toward the park is high. Indiscriminate killing of pumas may, in fact, be exacerbating the problem because killing territorial pumas that do not prey on sheep but rather on the abundant guanacos on some of the ranches may open up territories for transient pumas, which may be more likely to prey on sheep (Pierce et al. 2000b). Thus, *selective* killing of offending pumas may help reduce attacks on livestock and local animosity toward pumas and the park (Cougar Management Guidelines Working Group 2005).

The use of special breeds of guard dogs to protect sheep from puma predation has been tried on some Patagonian ranches with limited success. On the large ranches, sheep are too numerous and the areas they travel are too great to be dealt with by dogs. On the other hand, small ranches that produce goats in an area adjacent to a provincial reserve with abundant guanacos and pumas in northern Patagonia have had some success with guard dogs, using local dogs crossed with border collies and raised by nanny goats. Although there are no quantitative data to demonstrate the success of these guard dogs, six producers report reduced losses over a period of five years and are satisfied with their results (J. Fernandez, pers. comm., 2005). The Wildlife Conservation Society (WCS) team we lead has replicated this method with goat herds in another area in northern Patagonia. In addition, a few sheep ranchers in northern Patagonia are experimenting with the use of donkeys to protect their flocks.

Increasing Contact with People

Although pumas have increased in numbers and recolonized most of their former range in Patagonia, the probability of conflicts between pumas and people differs between the open rangelands of the steppe and scrub and the mountain forests. Despite the fact that the rural human population has declined since the 1950s, the frequency of puma sightings in rural areas of the Patagonian rangelands has grown greatly. In most parts of the rangeland, however, pumas are still persecuted to some extent and remain shy and wary of humans. Attacks on people in the rangelands have not been reported.

In the Patagonian forests, on the other hand, the increase in urbanization and tourist visitation to national parks is putting thousands more people within close proximity to pumas every year. While human population in the rural rangelands has declined in recent decades, that in towns and cities in the Patagonian forests has boomed. The largest of these towns are San Carlos de Bariloche, San Martín de los Andes, and Villa la Angostura, all in the northern part of the forest. The people who have moved into these urban areas include people who formerly lived in rural areas nearby, as well as people moving from Buenos Aires and other large cities in the eastern part of the country in search of a more peaceful existence. In the national parks and urban areas of the forest, pumas are not hunted. So far, only one attack on a human has been reported, in the Torres del Paine National Park in Chilean Patagonia (Franklin et al. 1999). These pumas are becoming bolder and less fearful of humans, frequenting campground dumps in parks and backyards of residences in newly urbanized areas. The rising level of contact, the lack of fear by pumas, and the naïveté of park visitors and residents of forest towns regarding a large predator combine to greatly increase the likelihood of attacks on humans in the future.

Research Needs

Very little research has been carried out on pumas in Patagonia and the southern Andes. Most published research is on diets and predator-prey relations (Wilson 1984; Yañez et al. 1986; Cajal and Lopez 1987; Iriarte et al. 1991; Rau et al. 1991; Branch et al. 1996; Novaro et al. 2000; Pessino et al. 2001; Bank et al. 2002; Rau and Jiménez 2002; Novaro and Walker 2005). There is only one published radiotelemetry study on spatial ecology, habitat use, and predator-prey interactions (Franklin et al. 1999), and one attempt at estimating relative abundance among sites (Muñoz-Pedreros et al. 1995). One published study, although oriented mostly toward the culpeo (see Figure 7.4), reports human attitudes toward pumas and human perceptions of puma damage to livestock (Travaini et al. 2000). Much of this published research is from a single protected area, Torres del Paine, in southern Chile (Wilson 1984; Yañez et al. 1986; Iriarte et al. 1991; Franklin et al. 1999; Bank et al. 2002), with three studies from the Patagonian forests of Chile (Rau et al. 1991; Muñoz-Pedreros et al. 1995; Rau and Jiménez 2002), two from a single protected area in the scrub (Branch et al. 1996; Pessino et al. 2001), one from the southern Andean steppe (Cajal and Lopez 1987), and one from ranches in the steppe (Novaro et al. 2000).

The ecology of the puma in the far south is perhaps more similar to that of the cougar in western North America than in the rest of Latin America, which is mostly tropical. However, in Patagonia and the southern Andes, the puma is the only large carnivore, compared to the north, where cougars coexist with bears (*Ursus arctos* and *U. americanus*) and wolves (*Canis lupus*), and tropical Latin America, where they coexist with jaguars. Also, native prey are far more widespread and abundant in North America than in Patagonia, where pumas prey mostly on livestock and introduced wildlife. While North American cougars have recovered from previous population reductions following recovery of native prey populations, the recovery of Patagonian pumas began in a different way. Their rebound resulted from a decrease in the rural human population and the presence of abundant populations of alternative, exotic prey. Although much can be learned from research and experiences in North America, research to address the specific situation in Patagonia and the southern Andes is necessary for the conservation and management of pumas and their prey in this region.

We believe there are three pressing areas for puma research in southern South America. First, in relation to the conservation of native prey, it is vital to determine the role of abundant pumas in the recovery of threatened or reduced populations of guanacos, rheas, and other species that occur in low numbers in many areas. Is there a threshold density above which these species escape from control by puma predation? How is this threshold affected by density of different species of exotic prey? Can native prey escape limitation by pumas without the need for artificially (and temporarily) reducing puma numbers? What are the potential positive impacts of puma predation on Patagonian biodiversity?

Second, research in methods for reducing livestock losses to pumas is urgently needed. Controlled studies comparing the efficacy of different methods currently in use, or that could potentially be used, are critical for the conservation of the puma in this region. To address the growing problem of livestock predation in Patagonia, effective solutions compatible with puma conservation are needed.

Finally, it is essential to develop and test appropriate methods for monitoring puma population trends in the steppe, scrub, and forest habitats of Patagonia and the southern Andes. In a southern Chilean forest agroecosystem, scent stations were used to estimate relative abundance of pumas. Scent stations were set systematically and consisted of an olfactory attractant placed on a smooth surface. An animal that was attracted to the scent would leave identifiable tracks. The average visitation rate by pumas was 44 percent (Muñoz-Pedreroz et al. 1995). However, years of scent station surveys in the Patagonian steppe of Argentina and two Argentine national parks in the southern Andean forests have rarely registered visits by pumas, despite apparently high puma densities (Funes et al. 2006). The only site where visitation rates were high was on one ranch in the scrub of La Pampa province (M. Pessino, pers. comm., 1995). Counts of puma signs (feces and tracks) along dry river beds and rocky cliffs were used successfully at a steppe site in southern Patagonia (Beier and Cunningham 1996; Pía and Novaro 2005), but this method is not practical for all areas.

The current research of our team working for WCS and collaborators from various agencies in this region includes identifying the effects of puma predation on the recovery of guanacos, the efficacy of locally bred guard dogs in preventing livestock predation by pumas and culpeos, and patterns of puma recolonization in northern Patagonia. We are studying puma predation on guanacos by comparing predation rates and prey selection patterns among sites with different densities of guanacos and alternative prey (both native and exotic). We are assessing puma recolonization by studying genetic structure. A. Marino and R. Baldi are studying behavioral responses of guanacos to cope with puma predation pressure in different habitats.

Colleagues from other institutions in the region are addressing other key topics. A. Travaini and S. Zapata (Zapata 2005; Zapata et al. 2007, 2008), J. Zanon (Zanon, 2006), and M. Pessino (Pessino et al., 2001) are studying predator-prey interactions and testing methods to assess relative densities. S. Montanelli and S. Rivera of the Chubut wildlife agency in Argentina (S. Montanelli, pers. comm., 2006) and N. Soto of the agriculture and wildlife agency for the Magallanes region of southern Chile are leading efforts to map conflict areas, assess the magnitude of livestock losses to pumas, and test the efficacy of guard animals and other mitigation methods (N. Soto, pers. comm., 2007).

Conservation and Management Issues

The puma is now widespread in Patagonia and the southern Andes and abundant in portions of the area. It provokes conflict by preying on livestock and other species of economic importance. Yet its ecological role as top predator for the prey species native to this region remains threatened. Today, the sites where pumas predominantly hunt native prey are very few—mostly large protected areas that still harbor substantial populations of guanacos and other native prey species. These sites include Argentina's Monte León (Pía and Novaro, 2005), Payunia (Berg 2007), San Guillermo reserves (Donadío et al., 2008), and Torres del Paine Park in Chile (Franklin et al. 1999). We propose the reestablishment and preservation of the interaction between pumas and their native prey as a key conservation goal for Patagonia and the southern Andean steppe—preserving what distinguishes the world's southernmost pumas from all others. Achieving this goal would require increasing populations of native prey in some areas, controlling exotic prey, and reducing livestock predation.

Protected areas are vital to achieve the goal of conserving puma interactions with native prey. These are the places where livestock are absent or low in number and where populations of native prey are sufficient to support a puma population. Serious attempts at control of exotic species are most feasible in parks and reserves. Toleration of abundant puma populations is also more likely in protected areas. Nevertheless, control of puma-livestock conflict in the vicinity of reserves is crucial to the success of the protected areas.

Pumas in Patagonia and the southern Andes are not currently threatened, but history has shown us that they can be extirpated through concerted human effort, even in a place as sparsely populated as Patagonia. Although this does not currently seem likely, conditions could change, exposing pumas to greater threats. For this reason, it is important that pumas in the region be managed and monitored responsibly and that human tolerance of them be increased by implementing effective methods to reduce livestock depredation and prevent attacks on humans.

Gross misunderstandings and a general lack of information about pumas prevail among the public in the region. Many rural residents view pumas as a recently arrived, invasive plague, while many urban residents believe them to be an endangered species or do not comprehend the potential danger they pose. For example, a recent scientific publication claimed, with no supporting data, that pumas are scarce or extinct throughout Patagonia (Bortolus and Schwindt 2006). We received a phone call from a family with five small children, telling us about the puma that was frequenting the back porch of their new house in the mountains; they were blissfully unaware that the cat posed any danger to children, adults, or pets. Public education about the history and status of the puma, advice on handling encounters, and information about predation on livestock and the efficacy of different methods for reducing stock losses are essential to ensure that the people and the great cats in Patagonia and the southern Andes can coexist peacefully for years to come.

Conclusion

The world's southernmost pumas once preyed principally upon guanacos, vicuñas, rheas, vizcachas, and maras in the arid regions, and upon pudu and huemul deer in the forests. In the last hundred years, however, the puma's prey base has been significantly altered because of the introduction of millions of heads of livestock and the successful establishment of exotic wildlife. Thus, the role of the puma as top predator in these ecosystems has clearly been altered by human activities. Pumas in most areas prey on exotic species, and in some areas where exotic prey are more abundant than native prey, pumas may be limiting the recovery of threatened native prey populations because exotic prey support larger puma populations.

After being wiped out of most of the region until the late 1980s, the puma is back throughout most of Patagonia and the southern Andes; it is probably here to stay. Livestock producers, tourists, rural residents, and residents of towns in the southern forests are coming to terms with pumas in new ways, as in the western United States. Effective methods are desperately needed for reducing predation on livestock as well as ways to effectively educate the public about the nature of pumas, their important role in the landscape, and how to respond during encounters.

Research and management of pumas are in their infancy in the region. Research has been largely restricted to dietary studies, and management practices are generally neither based on data nor adequately evaluated. Provincial wildlife departments are underfunded, and existing laws cannot be enforced due to lack of resources. There are parallels with the situation in the western United States but also many differences. Management lessons and research results from the U.S. experience can often be applied but must be tested under the specific conditions of the region. On the other hand, Patagonia provides opportunities to test hypotheses and management practices that may be applicable in the north.

Finally, we believe that conservation goals for the puma in this vast region should address preserving what makes these southernmost pumas unique—their predation on the region's particular suite of ungulates, birds, and large-bodied rodents. Pumas that eat sheep, red deer, and hares imported from Europe are not playing the unique role of Patagonian pumas. Therefore, we propose the restoration and preservation of the ecological interactions between pumas and their native prey as a main goal for conservation in this region.

Part III
Cougars and Their Prey

Notes from the Field

Linda L. Sweanor

December 29, 1991

IT WAS my Sunday ritual. I drove sixty miles east across the Tularosa Basin to the tiny Alamogordo airport, where I met Bob Pavelka, the pilot I entrusted to help me radio-track cougars. We lifted off a little before 7:00 A.M., just as the San Andres Mountains began to blush with the morning sun.

During the six-hour flight, we located thirty-three radio-collared cougars. One cat's behavior caught my attention. F28 was in Squaw Canyon, only 800 meters from her location of two weeks ago. Since we located F28 with two different adult male cougars (M38 and M46) in the first week of September, Ken (our field research leader and my husband) and I suspect she may now have a new litter. Poor weather conditions forced me to miss a critical flight last week, so we cannot be sure of her status. If we locate F28 in the same site next week, then she likely has newborn cubs.

January 6, 1992

More foul weather kept our airplane grounded yesterday. But we still needed to determine F28's condition. There was only one thing to do—go and find her. Ken and I left Las Cruces early, drove north along the east flank of the mountains, then followed the winding two-track road into Hembrillo Canyon. We headed for her last-known location, hoping to pick up a signal.

11:03. The radio receiver emitted a faint plucking sound as we reached our tent camp in Hembrillo Basin. The peak signal of F28's transmitter was coming from the west slope of Kaylor Mountain.

11:40. We ate lunch by the Cave of the Bloody Hands, an undercut ledge where ancient people once dipped their hands into red ochre and pressed them against the limestone. Then we ascended the ridge north to the Hembrillo-Squaw Canyon divide. Wispy trailers of descending clouds fingered the piñon and juniper-clad mountain tops high above us. As we picked our way up the slope, we wove through ocotillos, sotols, and prickly pear cactus while trying to avoid slipping on the crumbled rocks and limestone ledges. I managed to stumble anyway, landing solidly on one knee. Ken and I immediately inspected the flimsy, H-shaped antenna. Without it we would never find F28. I shook off Ken's grim look and grinned under the pain; the antenna had survived the fall intact.

12:38. Just as we came even with Apache Peak, I turned on the receiver and waved the antenna. The steady beat of F28's signal was reassuring. Because we were at a high point, I scanned for other cougars. Suddenly, M38's signal came booming in along the same bearing as F28's. This was unexpected, but it assured us of an interesting day. Unfortunately, I also picked up a weak, rapid pulse from F57's collar toward the north. A collar's signal beats twice as fast when the collar has not moved for more than six hours. This meant the female cougar was probably dead. We quietly noted what we would be doing tomorrow and moved on. As we crossed the north fork of Victorio Canyon and ascended Kaylor Mountain, we continued to monitor F28 and M38. There was little doubt the two cats were together.

14:05. Sharp, pinging sounds from the receiver cued us that we were getting close. Although we had been climbing steadily for over two hours, the last two hundred meters were the most arduous. We had to proceed very quietly if we were to remain undetected.

14:20. Three rasping growls emanated from the bluff before us and drifted up the canyon. As we maneuvered through the tangled scrub to a better vantage point, more cries reverberated off the canyon walls. The vocalizations became constant and sounded so threatening that I envisioned F28 desperately trying to protect newborn cubs, or perhaps fighting for her own life. We have seen the results of past fights. A female cougar has virtually no chance against a larger, more powerful male. I was growing apprehensive about what we might find.

14:30. Visual! We stood fifty meters upslope from the two cougars, and neither yet sensed us. This was no fight. F28 and M38 were lying down, hip to hip, facing away from each other. M38 was licking his tail while F28 emitted a hodgepodge of meows and growls in a continual stream of what sounded like verbal protest. Or was it enticement? The growls were interrupted only when she took a breath. M38 appeared to take no notice. He finished grooming and began to look around. Because we were exposed on the slope, his eyes immediately locked on us. There was power in that gaze; I felt elation at the contact, but also vulnerable—a bit like a deer in the crosshairs. With little apparent concern, M38 rose and slowly moved downslope about fifteen meters. His coat was a rich rust and he looked all of the 62 kg he had weighed at first capture, almost four years ago. He lay down again, his back partly to us. His head began to bob and his eyes closed; M38 was dozing! Apparently, he did not consider us much of a threat.

In the meantime, F28 began to roll in the grass, rubbing her head and neck as she slid along the ground. We guessed her weight at about 35 kg. Back in the summer of 1987, with the aid of our hound Spotty (and accompanied by our mentor, Maurice Hornocker), we caught F28 for the first time. She was just three months old then and living with her mother (F6) about 22 km south of here. She bore her first and only known litter in 1990. Today, her smooth white belly hair showed no evidence she had recently been suckling cubs. As F28 finished a roll, she came up facing our way, saw us and froze. So did we. Five minutes into the stand-off, F28 slowly began to pull her feet beneath her. The process was almost undetectable. Finally, loaded like a spring, she bolted, quickly vanishing behind a small rock outcrop below us. Her radio signal indicated that she simply moved out of visual range, and then stopped. M38 did not respond to her hasty departure; he was still dozing. I had the distinct impression he was tired.

14:40. We left the area so as not to further disrupt what we now thought was cougar courtship. As we retraced our steps, I glanced northward and wondered if an encounter with a male had resulted in a different fate for F57.

Afterward. Our impressions were correct. Two days later, we located F28 with another territorial male, M5—perhaps M38 really was tired!—and three months later she gave birth to three male cubs. We also found F57's remains. She had been killed and eaten by a male cougar. We suspected it was M46, one of the males we had located with F28 back in September in what we believe was an amiable encounter.

Chapter 8 Behavior and Social Organization of a Solitary Carnivore

Kenneth A. Logan and Linda L. Sweanor

OTHER THAN THE sheer physical presence of the cougar (*Puma concolor*), the most fascinating aspects of this animal are its behavior and social structure. In this chapter, we focus on cougar behavior, particularly how individuals interact with one another in a population and with their environment, which we refer to as cougar social organization (Seidensticker et al. 1973; Logan and Sweanor 2001, pt. 3). We use reliable information on cougar behaviors that has been reported in the peer-reviewed scientific literature on the species, particularly behaviors that are consistent among cougar populations from different locations in North America. We use North America as our sampling frame, examining various cougar research efforts across the continent, making sense of consistent behavioral patterns in cougars to explain why they might act the way they do. Other aspects of cougar behavior, including actions toward prey and toward people, are covered in Chapters 9 and 13, respectively.

By considering patterned behaviors among cougars from one population to another, we illuminate aspects of behavior and social organization that appear to be conserved phenotypic traits. Such traits have presumably emerged through selection of properties that maximize individual survival and reproductive success, the basic elements of population persistence. In this sense, we think of cougar social organization as a template of evolved adaptations that maximize fitness within the cougars' environments (Wrangham and Rubenstein 1986). The most reliable information on cougar behavior and social organization comes principally from long-term research of non-exploited cougar populations and populations that were experimentally manipulated. Although this chapter presents a synthesis of that information and our best explanations for cougar behavior, there is still much work to be done testing current hypotheses and speculations.

The phenotypic behavioral traits discussed in this chapter are synthesized into a single explanation, a *reproductive strategies hypothesis* (Logan and Sweanor 2001). In this perspective, we think of conserved traits in the context of Darwinian evolution, and the attendant processes for change—natural selection and sexual selection (Darwin 1859; Mayr 1982b, 1996; Freeman and Herron 1998). Both are nonrandom processes that favor individuals with greater survival and reproductive success, over generations, and that lead to a systematic shift in the average form, or phenotype (Mayr 1996; Freeman and Herron 1998; Dawkins 2004). Natural selection favors traits that enhance survival and reproductive success among individuals with certain phenotypes compared to individuals with other phenotypes. Sexual selection favors traits that enhance reproductive success among individuals of the same sex (Mayr 1996; Freeman and Heron 1998).

Thus, the reproductive strategies hypothesis was developed as an explanation for why cougars behave toward one another within the concepts of Darwinian evolution and theory in natural selection and sexual selection (Logan and Sweanor 2001). The hypothesis states that male and female cougars have evolved different strategies for maximizing individual survival and reproductive success (Logan and Sweanor 2001). If these behaviors contribute to survival and reproductive success, and thus to the adaptability of cougars, the influence of human selection pressures upon the ability of cougars to adapt to changing environments is an important consideration for wildlife managers.

Solitary Living

Cougars, like most other wild felids, live primarily solitary adult lives (Sunquist and Sunquist 2002). Solitary means that individuals do not cooperate with one another to forage, rear young, achieve matings, or defend against predators (Sandell 1989). Solitary living in wild cats seems to be associated with structural complexity in the habitat formed by terrain and vegetation, and with relatively low density and patchy distribution of prey (Logan and Sweanor 2001). In contrast, sociality in the African lion and male cheetahs is associated with relatively open habitats and large concentrations of large-bodied prey (Packer et al. 1990; Caro 1994). Although properties of felid habitats certainly played a key role in the evolution of behaviors, other behaviors evolved that helped individuals to maximize survival and reproductive success as they interacted with one another and competitors of other species in those habitats. Because competition is expected to be greatest among individuals of the same species in the same population (Mayr 2001), competition among individual cougars has probably been the main factor affecting evolution of the cougar's social system (Logan and Sweanor 2001).

Although adult cougars are mostly solitary, events that bring them together are vital to individual survival and reproductive success, and thus to population persistence. Males live alone, except when they consort with estrus females and compete directly with other males. On very rare occasions, males associate with their mates and dependent offspring. Adult males are not involved in provisioning their offspring with food. They contribute to the defense of mates and offspring, but usually not while in immediate association with them (see the section on the adult male cougar below). Females generally avoid other cougars, except for the cohesive social unit they form with their dependent offspring, and to breed with adult males for a few days at a time (Logan and Sweanor 2001). In rare instances, female cougars have been observed together (Padley 1990), as have members of different cougar families. Females with cubs rarely associate with sires (Logan and Sweanor 2001).

Adult and subadult cougars generally avoid one another. This behavior is thought to reduce direct competition, the likelihood of death, and the cost of injury for an obligate carnivore that is dependent upon its own well-being to acquire prey (Hornocker 1969; Seidensticker et al. 1973). Mutual avoidance would be strongly selected for in cougars, particularly when death from other cougars appears to be the main cause of mortality for both sexes and all life stages in cougar populations that are not exploited by humans.

Cougars that live longer lives have more opportunities to breed and leave offspring (Logan and Sweanor 2001).

Mating System

Cougars have a polygamous and promiscuous mating system (Seidensticker et al. 1973, Anderson 1983, Logan and Sweanor 2000, 2001), which seems to be strongly influenced by sexual selection. An adult male cougar might breed with multiple adult females that reside within his territory, whereas an adult female might breed with more than one male during individual or subsequent estrus cycles (Logan and Sweanor 2000, 2001). Mating cougar pairs consort for one to sixteen days (Seidensticker et al. 1973; Anderson 1983; Logan and Sweanor 2001) and copulate frequently. High copulation rates, perhaps as many as fifty or more times per day, may have evolved as a means by which females assess male vigor and thus contribute to their own reproductive success (Eaton 1976).

During estrus, the female cougar visits scrapes (visual and olfactory markers usually made by male cougars; see section on communication below) and vocalizes, apparently to advertise her breeding condition. The extent to which pheromones may be involved in advertisement of female cougar breeding condition and location is unknown. Multiple males may be attracted to an estrus female and compete directly for the opportunity to breed. Thus, besides an ability to copulate repeatedly, adult male cougars also are rigorously tested by their ability to compete with other males for territory and access to mates. Competition sometimes leads to fights between male cougars that can result in the death or expulsion of one of the combatants (Figure 8.1). Winners enhance their opportunities to breed with estrus females (Logan and Sweanor 2001).

Figure 8.1 This adult male cougar has severe head wounds inflicted during a fight with another male cougar. Photo © Ken Logan.

Promiscuity enables an adult female cougar to assess male vigor, allows her to establish positive relationships with all the males that share or abut her home range, and might confuse paternity among male mates. Subsequent encounters between a female and a male that were in a prior breeding association would tend to be amicable. In theory, when an estrus female breeds with a successful male, she is contributing to her own reproductive success (i.e., selection for reproductive success; Mayr 2001).

Promiscuity provides an adult male cougar with a greater number of mates because he can compete for breeding opportunities with multiple females that reside within his territory. Although adult male cougars enhance their reproductive success by being aggressive and territorial, adult female cougars do not exhibit territoriality. Instead, adult females use their energy to hunt, in actively avoiding other cougars (particularly males), and in mating, pregnancy, and rearing of the young. Adult females fight in exceptional circumstances to defend their lives and the lives of their offspring, usually against aggressive male cougars.

Female cougars are polyestrous, meaning that their reproductive cycle is continuous until they become pregnant. Consequently, female cougars can bear young during any month of the year (see Chapter 5). Because they co-evolved with females, adult male cougars are physically prepared to breed during any time of the year. However, in western North America and Canada, there is a birth pulse for cougars from May through October in which over 70 percent of births occur (see Chapter 5; Cougar Management Guidelines Working Group 2005; Laundré and Hernández 2007). Given an average gestation of about ninety-two days (Anderson 1983; Logan and Sweanor 2001), the corresponding pulse in cougar breeding activity would occur during February through July. Hypothetically, the birth pulse is influenced by a peak in number and vulnerability of ungulate prey (i.e., births of wild ungulates occur during spring and summer), which would tend to satisfy the energetic demands of cougar mothers better, improve maternal care of cubs, and thus enhance the survival of cubs born at that time (Logan and Sweanor 2001; see Laundré and Hernández 2007). Thus, the polyestrous trait enables female cougars to exploit favorable environmental conditions for reproduction (Logan and Sweanor 2001, 90).

Sexual selection, the process that results in variation in mating success principally for the male (Darwin 1859; Freeman and Herron 1998), appears to operate in the cougar breeding system in two ways, intrasexual selection and intersexual selection. *Intrasexual selection* is selection for traits that confer breeding advantages for certain individuals over others of the same sex and results from competition among members of the same sex (Freeman and Herron 1998; Mayr 2001). An example of the result of this process

could be sexual dimorphism in cougars, with adult males weighing on average about 1.4 times more than adult females (Anderson 1983). Theoretically, selection for larger mass in male cougars is the result of larger and better conditioned males being more successful in competing directly with other males for access to estrus females and territories that contain prospective mates and the male's dependent offspring (Logan and Sweanor 2001). *Intersexual selection* is a type of selection that results from female mate choice (Mayr 2001). Again, using male physical traits as an example, large body size and superior condition in males may attract female attention and demonstrate male fitness. In either intra- or intersexual selection, the female's offspring inherit the male's superior genes, which presumably enhance their chances of survival and reproductive success. Adult male behavioral traits that might result from sexual selection include aggressive temperament and infanticide, which are discussed later in this chapter.

Relatively high variation in male cougar reproductive success has been demonstrated in protected cougar populations, where a few males exhibited substantially higher reproductive success than other males in the same population (Murphy et al. 1998; Logan and Sweanor 2001). In cougar populations where males are exploited at relatively high rates, we expect low variation in male reproductive success. The high male turnover (i.e., low survival) should result in each male exhibiting relatively low reproductive success. Thus, in cougar populations subject to heavy annual exploitation, sexual selection would be expected to be relaxed because female mate choice and male–male competition would be constrained to the current male survivors. In other words, human exploitation removes males that naturally might have been in competition for mates or subject to mate choice. Attendant phenotypic changes in the cougar, such as shifts in average morphological and life history traits, might take many generations to become manifest, if they are not already occurring.

Development of Young

Direct observation of the physical and behavioral development of cougar offspring in the wild is limited. Cougar cubs, most frequently three per litter, are born in a protected nursery located in spaces among boulders, undercut ledges, and dense lateral and overhead vegetation. They are altricial, or highly dependent—born tiny (averaging about 508 grams, Anderson 1983) and barely able to move. Their eyes and ears are closed (see Plate 1). Newborns are fully furred with black spots on reddish to gray-brown coats and with black rings on their tails. The newborn's life is dependent upon instinctual behaviors that evolved to improve its chances of survival as an offspring of a large obligate carnivore (its mother), where its mother must kill large prey in a complex environment and compete with other cougars and other big carnivores. Among those behaviors are movements in response to the sound, scent, and vocalizations of the mother to find her milk-laden nipples or to crawl for cover to avoid danger. Within about two weeks, the eyes are fully open and are blue in color (see Plate 2). The ears are open, too.

At about four weeks old, the cubs are physically able to explore their immediate surroundings and actively interact with siblings and their mother to start the process of learning to augment their instincts. Play behavior and mock attacks and battles help develop the brain, skeleton, and muscles needed to help the offspring to acquire food and interact with other pumas.

By about six to eight weeks, the mother somehow coaxes the offspring to leave the familiar security of the nursery to follow her to animals she has killed to feed the young (see Plate 6). As mother's milk becomes less of a staple in the cubs' diet, the high quality diet from meat becomes paramount. The cubs' dentition is more developed and robust enough to masticate animal soft tissues. Physically, cubs are capable of traversing rugged terrain. The greater movements to acquire food during this stage of cub development increase the family's risk of encounters with predators, including coyotes (*Canis latrans*) (Logan and Sweanor 2001), wolves (*Canis lupus*) (T. Ruth, pers. comm. 2007), and other cougars—mainly males that are not the sires of the cubs. In a nonhunted cougar population in New Mexico, the greatest cause of death in cougar cubs was infanticide and cannibalism by male cougars (Logan and Sweanor 2001).

By the time the cubs reach five months old, their eyes turn from blue to the pale brown or amber-gray color of adult cougar eyes (see Plate 5). Black spots and rings fade on a brown pelage, appearing as light brown dapples. Now, cubs can easily climb trees to avoid predators. At about one year of age, the pelage of cubs appears for the most part like the tawny brown of adults, except for light brown dapples on the shoulders, thighs, and legs (see Plate 7; Logan and Sweanor 2000). A family of a mother and two to four cubs may be mistaken for a "pack" of unrelated cougars at this stage. As a unit, the family has roughly three times the energy demands of an adult male cougar and up to six times that of a lone adult female. Still, the cubs remain highly dependent upon provisioning from their mother; she must hunt and kill a deer-sized prey roughly once every four days (Ackerman 1982). As the cubs continue to develop, learning intensifies as the offspring experience the complex environment of their natal area, deal with enemies, and develop images of prey and hone rudimentary techniques for killing them.

As young cougars reach one to two years of age and approach the mass of their mother, they become independent and enter the subadult life stage. Yet, subadult cougars still are not developed enough to breed. Independence of cougar young at this time would contribute to the lifetime reproductive success of the cougar mother by ending maternal care for offspring that no longer need it and enabling the adult female to devote her energy to raising a new litter. The large offspring are capable of provisioning themselves, even by killing large ungulate prey. As such, the self-sufficient offspring would not be direct competitors for the mother's energy required for another cycle of mating, pregnancy, and rearing offspring. Apparently this is the case, because independence of cougar offspring has been linked to the resumption of breeding behavior in mothers (Seidensticker et al. 1973; Logan and Sweanor 2001).

Dispersal

In the subadult stage, life history strategies of the female and the male cougar diverge substantially. The most obvious behavioral difference between females and males involves dispersal from natal areas (i.e., the area where the cougar was born and raised). Some female subadults, as many as one in two (Sweanor et al. 2000), are philopatric, meaning that they establish home ranges overlapping or adjacent to their natal areas. Philopatry is a strategy of female pumas, presumably enhancing fitness by taking advantage of their knowledge about resources in the area in which they were raised and of their close social ties with neighboring related females (Logan and Sweanor 2001; see matrilineal structure described later).

The rest of the independent female subadults disperse to different cougar habitats. Dispersal in female offspring has been hypothesized to be partially density dependent, and thus a strategy that allows individuals to exploit habitats with less competition and more abundant resources compared to the natal area. Female cougars do not normally disperse as far as males. Dispersal has the added advantage of allowing cougars to colonize available habitat. Of the two strategies, philopatry probably confers higher fitness to female cougars in good quality habitat because the matrilineal structure probably results in inclusive fitness (Logan and Sweanor 2001). Inclusive fitness is an individual's total fitness, the sum of its indirect fitness due to reproduction by relatives made possible by its actions, plus its direct fitness due to its own reproduction (Freeman and Herron 1998).

In contrast, practically all male subadults disperse from their natal areas (Sweanor et al. 2000). Average dispersal distances of male cougars in eight studies in the western United

States and Canada ranged from about 49 to 483 km (Logan and Sweanor 2000; Laundré and Hernández 2003), with an extreme of 1,067 km (Thompson and Jenks 2005). This common tendency of long-distance dispersal in male cougars might result in a panmictic breeding structure (i.e., interbreeding of populations from all parts of their range) where habitat is relatively continuous (Anderson et al. 2004). But where cougar habitat is not continuous, such as in the basin and range habitat configuration of the American Southwest, constraints on dispersal due to distance between cougar habitat patches, and attendant impediments to gene flow, can result in a geographically subdivided genetic population structure (McRae 2004).

Dispersal of male cougars appears to be independent of local male cougar density (Seidensticker et al. 1973; Logan and Sweanor 2001). Explanations for this almost obligate long-distance dispersal for male cougars include inbreeding avoidance and competition with other male cougars for mates (Sweanor 1990; Logan and Sweanor 2001; Laundré and Hernández 2003), both of which could contribute to fitness. The inbreeding avoidance hypothesis would be supported by male dispersal distances that would minimize the probability of a male breeding with close relatives, even with female relatives that disperse. Moreover, data on genetic relatedness of individuals in a population might reveal the frequency with which close relatives breed. Nothing is known about the extent to which cougars can recognize close relatives and whether such a property influences mate choice.

If avoidance of male competition were the strongest factor determining subadult dispersal (Laundré and Hernández 2003), then a low frequency of subadult male dispersal should occur in cougar populations where there are few adult male competitors (e.g., populations subject to sport hunting where territorial males are killed). Subadult male cougars have been documented using small transient home ranges during their dispersal moves, theoretically to acquire prey, yet also to avoid local territorial males (Beier 1995; Logan and Sweanor 2001). Presumably, they move on when they are forced to do so because of the presence of a territorial male cougar (i.e., direct encounter, or passive advertisement such as scent). As yet, there have been no published observations of male subadults remaining philopatric and subsequently establishing their adult territories in areas experiencing heavy hunting pressure (i.e., with attendant reduced male competition) and void of dispersal barriers (i.e., the habitat is expansive and connected). A preponderance of such observations would support the competition hypothesis.

Yet, these two reasons, avoidance of inbreeding and reduction in competition, could operate in tandem to cause strong selection for male dispersal. In other words,

the two effects are complementary, not mutually exclusive (Wrangham and Peterson 1996). Dispersal could have the proximate effect of competition avoidance for a young male cougar that is too small and inexperienced to be physically capable of competing directly with adult males in his natal area and other habitats through which he moves. But the ultimate effect is that long-distance dispersal minimizes the chance of a male breeding with closely related females, particularly females in matrilines (Logan and Sweanor 2001).

The term *transient* appears in the biological literature to refer to cougars with a temporary presence in a particular area (Hornocker 1970; Seidensticker et al. 1973; Laundré et al. 2007). In some of those instances the cougars were probably dispersing subadults. In other instances, the actual life stages of the cougars were not adequately known. They could have been adults with home ranges or migratory movements peripheral to study area boundaries, or individuals that died, never to be seen again (see Logan and Sweanor 2001, 244–45 for fuller discussion of transient behavior).

Both male and female cougars disperse during the subadult life stage, apparently before they would normally be reproductively active. Subadult animals generally disperse through habitat suitable for cougars. But they also traverse unsuitable areas or non-habitat relatively quickly and then use available habitat patches as "stepping stones" during their dispersal moves (Sweanor et al. 2000; Logan and Sweanor 2001). As subadults these cougars are investing in somatic growth, dispersal moves, and avoidance of other cougars but not in reproduction. Lower fitness would be expected if cougars dispersed during the reproductive stage of life when energy and time budgets in females should be allotted principally to mating, pregnancy, and rearing offspring, and in males to mating and competing for mates and territory.

The behaviors of dispersal in both males and females and philopatry in females also affect the growth rates of local cougar populations. Female rates of increase seem to be higher than for males, partly because both philopatric and immigrating females contribute to population growth. Because most male subadults emigrate, male recruitment to a local population segment is mostly dependent upon immigrants. Differential subadult survival rates (i.e., higher for females, lower for males) probably also contribute to the property that the adult female portion of a population grows at a faster rate than for males (Logan and Sweanor 2001). Dispersal contributes as well to the maintenance of cougar subpopulations that receive the immigrants and to colonization or recolonization of available habitats (Logan and Sweanor 2001). As female and male cougars become sexually mature at about two years of age and enter the adult life stage, their behaviors continue to diverge.

The Adult Female Cougar

Behavioral traits of the female cougar in North America seem consistent with strategies that contribute to her fitness. The reproductive success of the female cougar is defined as the number of offspring she has that survive to have offspring of their own. Behavior of the adult female cougar, then, should in large part be directed toward raising young successfully (Trivers 1972). Thus, the female cougar must survive as long as possible, successfully breed repeatedly in her lifetime, and successfully raise young by providing them with adequate nourishment for growth, security, and learning for their survival. Statistics on adult female cougar reproduction and behavior support this. The most successful females are those that live the longest lives and spend the majority of their adult lives bearing and raising young. The adult female can accomplish these feats by maintaining a home range in an environment that provides adequate resources to raise offspring, such as prey and nursery sites; by practicing avoidance of cougars that could pose a threat to her and her offspring; and by developing amicable relationships with territorial male cougars (Logan and Sweanor 2001).

Adult female cougars in a particular geographic area consist of some unrelated individuals that immigrated from elsewhere and groups of closely related individuals or matrilines (Sweanor et al. 2000; Logan and Sweanor 2001). Matrilines form as a result of philopatry in female offspring, where mothers, daughters, sisters, and aunts establish home ranges that overlap or are adjacent to each other (Figure 8.2).

Because adult female cougar life history strategies are geared to raising offspring, they have smaller home ranges and less extensive daily movements than adult males. In contrast, more extensive daily movements of adult males seem to be influenced by their greater territory sizes and strategy for searching for prospective mates and competitors (Sweanor 1990; Logan and Sweanor 2001; Sweanor et al. 2004; Arundel et al. 2007).

Adult female cougar annual home ranges in North America vary in area from about 55 km² to over 300 km². Size of the home range seems to depend upon several variables, including habitat quality (i.e., properties of terrain and vegetation); prey numbers, distribution, and movements; interrelationships between cougars; and the reproductive stage of the individual female (Seidensticker et al. 1973; Logan and Sweanor 2000, 2001; Pierce et al. 1999; Holmes and Laundré 2006). Cougars are obligate carnivores, which choose habitats where prey are available and terrain and vegetation features provide cougars with stalking cover and security (Logan and Irwin 1985; Laing and Lindzey 1991; Pierce et al. 2000b). Quality habitat is especially important to female cougars because they raise cubs alone. Although adult females tend to exhibit strong fidelity to such areas, females sometimes shift their

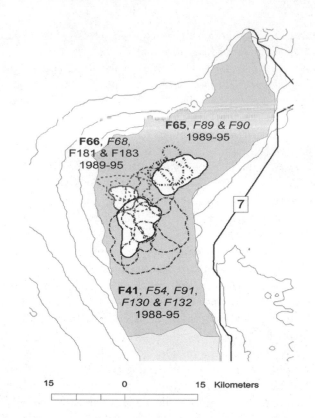

Figure 8.2 Several matrilines, formed as the result of philopatry in female offspring, in the San Andres Mountains of New Mexico. This figure depicts three matrilines forming from three P1 individuals (F65, F66, F41). Two matrilines (F65, F41) were formed from P1 and F1 generations, though with F41, the F1 cougars were daughters from two separate litters. The third matriline was formed from P1 (F66), F1 (F68), and F2 (F181, F183) generations.

activity areas within their home ranges to accommodate activity of other females, to avoid dangerous males (Logan and Sweanor 2001), and to coincide with seasonal movements of ungulates (Seidensticker et al. 1973; Pierce et al. 1999). Females living in environments with prey that migrate long distances between winter and summer habitats sometimes establish distinct seasonal home ranges (Seidensticker et al. 1973; Pierce et al. 1999). In this context, gene flow in cougars might be influenced by migration patterns in addition to dispersal (Pierce et al. 1999).

A female's home range expands and contracts in area in accordance with her reproductive status and attendant energy demands of her family. Females with nursing cubs (usually the two-month period following birth) use the smallest areas. As the offspring are weaned and begin to consume meat, the female expands her area of activity to meet the nutritional demands of the growing cubs (Ackerman 1982; Logan and Sweanor 2001). When the cubs reach independence (i.e., the subadult stage), usually between about eleven and eighteen months of age, the adult female might use an even larger area to distance herself from the newly independent offspring and to locate prospective mates. As noted, timing of independence of cubs has been linked to the subsequent breeding behavior of adult females (Seidensticker et al. 1973; Logan and Sweanor 2001). The adult female spends the bulk of her adult life in recurrent reproductive activities, which are marked by birth intervals of about seventeen to twenty-four months (Ashman et al. 1983; Maehr et al. 1991; Lindzey et al. 1994; Logan and Sweanor 2001).

Female home range overlap is a common characteristic among cougar populations that have been studied in North America (Logan and Sweanor 2001, 248). For example, the degree of overlap between female neighbors in a non-hunted cougar population in the Chihuahuan Desert was roughly 40–60 percent of annual home ranges and 10–30 percent of the annual core areas (Figure 8.3; Logan and Sweanor 2001). Overlap in female home ranges should be influenced by variations in habitat properties, particularly prey abundance, distribution, stability, and vulnerability (Brown and Orians 1970; Hixon 1980; Sandell 1989; Pierce et al. 2000b). Yet throughout the adult life of the female cougar, she probably has to exploit several habitat patches to acquire the resources she needs to survive and reproduce (i.e., food, mates, nurseries). The best strategy would be to share those areas with other females, particularly if a number of them are closely related, but to avoid direct competition for resources. To maintain a home range to the exclusion of other females would be too costly. Such behavior would require that females have large daily movements to patrol the area and defend it from other cougars. Direct competition would result in less time consorting with mates and bearing, raising, provisioning, and protecting young. It would also sometimes result in injury or death.

Adult female cougars share most of their home ranges with one or more adult male cougars. One or more of those males will sire her offspring. Adult females do not need to compete directly with one another for access to mates because there seem to be ample males for breeding in naturally functioning populations in good habitat. On the other hand, the estrus female is the limited resource to adult males, and this property results in male–male competition for mates. This is the case because the majority of adult females in a population will be pregnant or raising offspring. Therefore, the operational sex ratio favors males, even though females outnumber males in the population (Logan and Sweanor 2001). The number of males might become a limited reproductive resource in some heavily hunted or controlled populations and in highly fragmented habitats with barriers to male immigration (Padley 1990; Beier and Barret 1993).

Yet this spatial and breeding arrangement between adult females and males is risky for the females and their progeny because some males, usually not the sire of the cubs, threaten the lives of the cubs and of mothers that attempt to defend the cubs. Hence, females generally avoid male cougars, except during consort associations for the purpose of breeding, to reinforce pair bonds, and to confuse paternity. Considering the latter two properties, female cougars have been observed to consort with male mates for a few days even while the females were raising cubs, presumably with the effect of reinforcing amicable relationships with the males and confusing paternity where multiple male mates were involved. All these behaviors would contribute to the female cougar's survival, reduction of infanticide by males, and overall maximization of individual reproductive success (Logan and Sweanor 2001).

Meetings with unfamiliar males can be deadly for the female cougar, or offspring, or both (Logan and Sweanor 2001). The greatest cause of mortality in cougar cubs appears to be male-induced infanticide. Moreover, the greatest cause of mortality in nonhunted cougar populations appears to be male-induced intraspecies strife, with males killing some females that attempt to protect their offspring and in some cases their prey caches (for more on mortality factors see Chapter 5; Logan and Sweanor 2001). Clearly, avoidance of male cougars by females, especially while raising young, would be an effective survival tactic.

Although male cougars may attempt to usurp kills from females (Logan and Sweanor 2001), adult female cougars do not generally compete with one another for food in this fashion. One explanation might be that females are avoiding

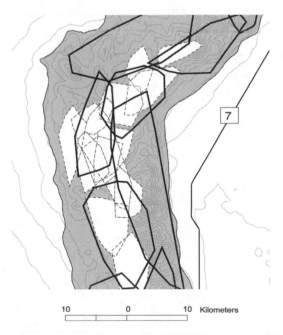

Figure 8.3 An example of home range overlap between cougars as observed in the San Andres Mountains, New Mexico. Female home ranges (white polygons with dashed outlines) overlap home ranges of other females extensively. Male territories (polygons with solid outlines), are typically much larger and overlap the ranges of several female cougars.

conflicts that might result in injury or death, which in turn could hamper, if not end, their reproductive success. Another possible explanation is that a number of females in the same area are closely related; therefore, they shun direct competition with relatives. Yet, indirectly, adult females should compete with other cougars for food where there are a number of cougars in the same geographic area using the same prey resources. Apparently, a cougar population is ultimately limited by food and, hence, prey (Pierce et al. 2000b; Logan and Sweanor 2001; Laundré et al. 2007). Competition for food should be greatest among the female segment of the cougar population because breeding females have greater total energy demands than adult males (Ackerman 1983) and they comprise the greater proportion of the adult population (Logan and Sweanor 2001). But even though adult females in an area compete for food, their inclusive fitness would be greater when females in an area are closely related (Logan and Sweanor 2001). In this sense, it is more advantageous to use a limiting resource (in this case prey) with a close relative than with a nonrelative.

The adult female cougar thus lives her life mainly in an effort to survive, successfully reproduce, and rear young, principles of the reproductive strategies hypothesis. The purpose of her male counterpart is also to survive and reproduce. But because the male's objective is to breed successfully with multiple females, this pits males against each other in direct competition.

The Adult Male Cougar

Adult male cougar behavior, as documented in North America, also seems to contribute to male fitness. The reproductive success of a male cougar is the number of offspring he sires in his lifetime that survive to reproduce themselves, similar to the definition of success for the adult female. Yet, there are major differences between adult male and female cougars in a population. Males within a local area are generally not related, while a substantial proportion of adult females are. Males are territorial, but females are not. Adult males do not invest directly in the raising of young, while females invest directly in the young through pregnancy and rearing the young to the stage of independence.

Male cougars are territorial. They aggressively attempt to establish dominance in a particular area by competing directly with other males for access to mates and space (i.e., territory). The most dramatic form of that direct competition is combat. Because territorial males are generally not related, it is advantageous for males to compete directly with other males for access to mates. Because adult males compete for access to mates, male reproductive success is

more variable than is female reproductive success (Murphy et al. 1998; Logan and Sweanor 2001).

Male cougar territories in North America are typically one and one-half to three times the size of female home ranges, much larger than what is needed to satisfy the energetic demands of adult male cougars alone (Logan and Sweanor 2001, 198). Territories vary in area from about 150 km^2 to more than 700 km^2 (Logan and Sweanor 2000, 2001). Male territories overlap several female cougar home ranges extensively, encompassing some of them completely. Neighboring male territories have been found to be relatively exclusive where less than four males have been studied (Beier and Barrett 1993). But larger samples indicate substantial territory overlap (Logan and Sweanor 2001, 248). For example, in a nonhunted cougar population in the Chihuahua Desert, roughly 50–70 percent of a male's annual territory was overlapped by other male territories; annual core areas overlapped by 15–40 percent (n = 10–12; Logan and Sweanor 2001, 247–53). But males tended to avoid using shared areas at the same time (Logan and Sweanor 2001). Boundaries of male territories are dynamic over time because of the territorial behavior of neighbors, the distribution of prospective mates, and changes in the status of male competitors. Yet, mutual avoidance on the part of males can reduce interactions that might otherwise result in costs associated with direct competitive interactions for mates, territory, or food. Because males are territorial and attendant territories are much larger than female home ranges, there are naturally fewer adult males than adult females in a cougar population (Logan and Sweanor 2000, 2001).

Adult male cougars exhibit fidelity to their territories, similar to females with their home ranges. But male fidelity has a greater tendency to decline with time. Dynamic use of area and fluctuation in territory boundaries are associated with challenges from male competitors—either neighboring territorial males or immigrant males. Direct challenges result in some males shifting their territory boundaries to include more area if they were victorious, or avoidance of an area if they were defeated. The extreme result would be the death of a male cougar and an attendant territory vacancy. Shifts in territory are also associated with the changes in distribution of potential mates caused by changes in local female cougar numbers or female reproductive status (Logan and Sweanor 2001).

Territorial male cougars perform most if not all of the breeding. Yet, as mentioned, adult male reproductive success is variable, with a few males exhibiting substantially higher reproductive success than other males (Murphy et al. 1998; Logan and Sweanor 2001). Beyond the need to search for prey, adult males tend to patrol their territories in an effort to assess the location and the breeding condition of females, and to compete with other males. In addition, adult males communicate their presence in time and space by scraping throughout their territories. Not only do the scrapes appear to function to inform breeding females of the presence of males but they likely also inform other males of the presence of potential competitors (see the next section on communication). Meetings between male cougars, particularly in the presence of an estrus female, can result in violent combat. Fights between males can result in the death of a male, shifts in territory use, or continued dispersal of subadult males. In a nonhunted cougar population in New Mexico, the major cause of death in independent males was due to mortal injuries inflicted in fights with other males (Logan and Sweanor 2001).

By not investing directly in the raising of his offspring, the adult male cougar can invest a greater proportion of his energy in competing for and breeding with estrus females. This way, the male cougar has the capacity to sire many offspring with multiple females during his lifetime. For the male, breeding opportunities at any particular time are limited because females are dispersed in low densities on the landscape, are in estrus at variable times, and are not available for breeding during the time they are raising young. The male also has to compete for estrus females with other males. Hence, to be successful, adult males must traverse huge territories to search for prospective mates and fend off male competitors. On average, adult males move more extensively on a daily basis, compared to adult females and subadults (Logan and Sweanor 2001; Dickson and Beier 2002; Arundel et al. 2007).

Although the adult male cougar is not directly involved in raising his offspring, he apparently is indirectly invested in the safety of his offspring and his mates. Through aggressive mate and territorial defense, the male cougar discourages and sometimes eliminates the activity of other male cougars that can threaten the lives of his offspring and prospective mates via infanticide or competition for prey carcasses. Other territorial neighbors or immigrant males will kill cubs they have not sired and sometimes kill mothers that attempt to defend cubs. Infanticide can accelerate breeding access to females by eight to ten months (Logan and Sweanor 2001). Thus, infanticide of cubs by unrelated male cougars might be a strategy by which males maximize their individual reproductive success. By providing a male an earlier opportunity to sire a litter, this strategy enhances the fitness of the male that commits the infanticide. Killing of cubs is termed *sexually selected infanticide* when the male is not their sire, the killing is linked to male–male competition for access to limited breeding females, loss of cubs shortens the female's birth interval, and the infanticidal male

mates with the mother of the dead cubs (Swenson 2003). But this strategy lowers the reproductive success of the sire of the cubs that are killed. Hence successful territorial males are those that directly (i.e., via confrontations and fighting) and indirectly (i.e., via visual and olfactory cues) defend prospective mates and territory (that contains mates and offspring) from male competitors. In the process, the male contributes directly to his own fitness.

In the context of the reproductive strategies hypothesis, the adult male cougar lives his life to survive and successfully reproduce. Relatively large mass, superior physical condition, and aggressive temperament should be traits that contribute to male fitness within the cougar social organization where males exhibit territoriality. Such traits should assist male survival and success in direct competition with other males for access to mates, and in indirect protection of mates and their offspring within the territory. Those traits should also confer advantages to superior males that confront other males through behavioral interactions, such as posturing, and thus settle some disputes without a fight. As the products of natural and sexual selection that have contributed to the fitness of cougars in past and current environments, these traits should theoretically figure into the ability of cougars to adapt to changing environments.

Communication

Communication among cougars appears to function to find mates, maintain family cohesion, convey social status, avoid or attract direct competition, or avoid threats from other cougars. Methods of communication between cougars can be categorized as vocal, visual and olfactory, and postural. Little is known about any of these for wild cougars. Specific messages that body postures convey between cougars are the least known, mainly because interactions between wild cougars are very rarely observed. Because there is more information on cougar vocalizations and visual and olfactory signals, this discussion focuses on those types of communications.

Vocalizations

Cougars emit a variety of vocalizations. Messages are difficult to determine because observers must be able to interpret behavior of both the sender and receiver. However, four general message systems have been outlined for carnivore vocalizations: agonistic, integrative, sexual, and neonatal (Peters and Wozencraft 1989).

Agonistic vocalizations identified in cougars seem to convey defensive and offensive threats (Logan and Sweanor 2001). These include low guttural growls, spitting, snarls, and hissing. Cougar mothers growl and hiss when their nurseries are threatened (Sweanor et al. 2005). The snarl seems to be similar to the growl (Sunquist and Sunquist 2002) and may convey the same meaning. However, snarls have also been heard between consort cougar pairs (Logan and Sweanor, direct observations). Spitting is considered to be a defensive threat, and hissing is suspected to be an offensive threat (Peters and Wozencraft 1989; Wemmer and Scow 1977).

Several cougar vocalizations seem to be integrative or neonatal and may convey satisfaction, distress, solicitation, contact, and maternal alarm (Logan and Sweanor 2001). Cubs and mothers sometimes purr when together. Purring is thought to be a simple, efficient way for individuals to convey a state of well-being or satisfaction to all family members at once or to a consort. Nursling cubs emit high-pitched, birdlike chirps and mews, which are thought to be distress or solicitation calls. Older cubs and adults emit whistles, some of which are elaborate (Logan and Sweanor 2001, 271). Chirps, mews, and whistles may also be contact calls, enabling family members to find one another when they get separated in exploratory activities around nurseries or during more extensive movements in the natal home range.

Some cougar vocalizations seem to convey sexual messages (Logan and Sweanor 2001). The most spectacular of these is the caterwaul, which is a combination of variably pitched, forceful sounds. Caterwauls seem to be emitted by female cougars as advertisement calls intended to communicate breeding readiness and location. The caterwaul, or certain audible parts of this vocalization, may be what has been popularly described as a "scream." Growls, gurgles, snarls, and mews of variable pitch and volume have also been heard from mating cougars (Logan and Sweanor, direct observations).

Reasons for other cougar vocalizations are more uncertain (Logan and Sweanor 2001). Female and male cougars yowl, the function of which might be to convey sexual identity or territory advertisement (in the case of the males). The yowl has been heard from a female apparently in estrus and from a male making scrapes. Cougars also emit an *ouch* call. One observer has interpreted this call to signify individual frustration, perhaps not associated with social interactions (Padley 1996). However, limited observations suggest that the *ouch* call might be given by both adult males and females while in communication with one another (Logan and Sweanor 2001).

Visual and Olfactory Signals

Cougars leave visual and olfactory signals in the environment that might convey identity, social status, and reproductive condition (Logan and Sweanor 2001). The most obvious markers left by cougars are scrapes (or scratches). These are made principally by adult male cougars by pushing backward with their hind feet, creating a shallow depression roughly 13–25 cm long with a small mound of loose ground material (soil, leaves, sticks, stones, snow) on one end. Adult females and subadult males apparently rarely scrape. Sometimes cougars urinate and defecate on the mound. The extent to which they use anal glands for scent communication is unknown (Seidensticker et al. 1973).

Male cougars seem to scrape throughout their territories. The scrapes are usually located along cougar travel routes, such as ridgelines, canyon rims, drainage bottoms, under large trees and ledges, and at kill caches (Hornocker 1969; Seidensticker et al. 1973; Hemker 1982; Shaw 1983; Logan and Irwin 1985; Sweanor 1990; Logan and Sweanor 2001). In areas where male territories overlap, scrape locations are often used by more than one male cougar. This property suggests that adult males communicate with one another via scrapes, perhaps to convey spatial and temporal activity and social status. Some adult males might focus activities around sites used by other males as part of the process of competition for mates and territory. Adult females visit scrape sites and sometimes vocalize in the vicinity (Logan and Sweanor, direct observations). Some females urinate on the scrapes. This suggests that females use scrapes to interpret male activity and as an area to focus advertisement of their breeding condition to attract mates. In other situations, scrapes might be used by both male and female cougars to avoid one another, such as subordinate males trying to avoid superior males, and females trying to avoid males that might be a danger to them and their offspring (Seidensticker et al. 1973; Logan and Sweanor 2001).

Other types of marking by cougars require more investigation to learn their worth as potential modes of communication. Cougars scratch on logs and trees, leaving claw gouges in the wood, flaked bark, and lamellae from the animals' claws. This may merely be an activity useful for maintaining the claws, but scent from the feet might also be left behind (Figure 8.4). Because the same trees or logs seem to be inspected and used by different cougars, such clawed sites could be used in cougar communication. In addition, captive cougars have been observed rubbing their heads and cheeks on logs and boulders. If wild cougars also do this, then those behaviors might leave scent to communicate with other cougars (Logan and Sweanor 2001). Other visual and olfactory signs that might also be used in communication between cougars are tracks, toilets (i.e., feces covered with ground debris, particularly near ungulate kills), and urine deposits and associated scents. However, their functions as such are currently unknown.

Plate 1 The cougar's gestation period is about ninety-two days. Litters most often consist of three kittens, which are born fully furred, spotted, and weighing about one pound each. These ten-day-old kittens will have to wait two to four more days before their eyes open fully. Photo by Becky Pierce.

Plate 2 Although physically able to explore their immediate surroundings, these four-week-old kittens, their eyes still blue, remain fully dependent on their mother's milk. When mothers go on hunting excursions, the kittens hide in crannies of the birth nursery for protection. Photo by Becky Pierce.

Plate 3 A cougar mother may move her cubs to an alternate nursery if she senses disturbance or danger. Kittens as young as six weeks can maneuver over steep, rocky terrain and follow their mother to the new location. The mother may carry younger cubs one by one. Photo © Susan C. Morse.

Plate 4 This kitten, with her mother, is about six to eight weeks old; typically, kittens become independent when they are eleven to eighteen months of age. Orphaned cubs younger than that have a reduced chance of survival because they lack the skills to kill large prey. Photo by Richard Badger Photography.

Plate 13 Once the cougar is immobilized, researchers obtain vital information on cougar sex, age, size, and condition. Blood, hair, and fecal samples may be collected for disease and genetic testing. This immobilized cougar has already been fitted with an identifying ear tag and radio collar. Photo by Chuck Anderson.

Plate 14 Presence of kittens is often determined by tracking the movements of a radio-collared mother. These three males are four weeks old, and because of their age can be caught by hand while the mother is away. Mothers often move the kittens to an alternative nursery after they have been handled by researchers. Photo © Ken Logan.

Plate 15 By identifying kittens, researchers can determine the timing of births and litter size. They record sex and general health and mark kittens with ear tags and tattoos to obtain future information on kitten survival. Expandable radio collars placed on kittens can provide additional information on survival, cause of mortality, time of independence, and dispersal movements. Photo by Bryan Bedrosian/Craighead Beringia South.

Plate 16 This four-week-old cub has a tattoo in one ear (note green ink) and an ear tag in the other. The ear tag may eventually be torn out, but the tattoo will remain visible throughout the cat's life. Photo © Linda L. Sweanor.

Plate 17 Maurice Hornocker's pioneering research on cougar behavior was aided by two kittens raised in captivity for four years. In 1966, Hornocker transported the growing cats from McCall, Idaho, to a facility in the Idaho Primitive Area (now the Frank Church-River of No Return Wilderness Area). Photo by Bill Dorris.

Plate 18 This female and male were among the first Texas cougars released in 1993 during the Florida Panther Reintroduction Feasibility Study. Nineteen Texas cats were released after two weeks in these pens (Belden and McCown 1996). Photo courtesy of Florida Fish and Wildlife Conservation Commission.

Plate 19 A yawn reveals the powerful dentition of a strict carnivore. The relatively small number of teeth may be the result of selection for shorter, more powerful jaws good for biting and holding large, potentially dangerous prey. Photo by Maurice Hornocker.

Plate 20 Little is known about specific messages that body postures convey because interactions between wild cougars are rarely observed. A cougar may give a defensive signal, such as lowering its ears, when threatened by pursuing dogs or people. Photo by Maurice Hornocker.

Plate 21 A cougar's tail can comprise 35–40 percent of the animal's total length (tip of nose to tip of tail). This long, muscular appendage, with its signature black tip, helps the cougar maintain balance while attacking prey and maneuvering over broken terrain. Photo by Richard Badger Photography.

Plate 22 Forelimbs are shorter and heavier than the hind limbs and support larger feet on flexible wrists—all adaptations for handling prey and for climbing. The longer hind limbs propel the cougar during pursuits, helping it to leap over obstacles. Vertical and horizontal leaps of fifteen feet and forty-five feet, respectively, have been recorded (Hansen 1992). Photo © Susan C. Morse.

Plate 23 Except for large lakes and major rivers, water is not typically a barrier to cougar movement. Cougars have been documented crossing swift-flowing rivers up to a mile wide (Seidensticker et al. 1973; A. E. Anderson 1983). Across their range, cougars prey on a broad variety of vertebrates, including mammals, birds, reptiles, and even some aquatic species. Photo by Richard Badger Photography.

Plate 24 Adult male cougars sometimes fight for territory and access to mates. In extreme cases, fights can end in the death of a combatant and a resulting territory vacancy. Photo © Ken Logan.

Plate 31   A young cougar pauses on a rocky ledge in Idaho's Frank Church-River of No Return Wilderness Area. The cougar's stalking and ambush hunting style is aided by the ample structural cover provided by vegetation and rugged, broken terrain. Photo by Maurice Hornocker.

Plate 32 The puma is the largest carnivore of the arid habitats of Patagonia and the southern Andes. This animal is at 3,500 meters above sea level in the southern Andean steppe. Photo by Andrés Novaro.

Plate 33 A mature female, marked previously in an Idaho study, is recaptured to change her radio collar. Photo by Maurice Hornocker.

Plate 34 A mature female escaped the pursuing hounds by climbing to this niche in a rocky cliff. Photo by Maurice Hornocker.

Plate 35   A three-year-old male, part of a study of captive cougars in Idaho, crosses an icy creek. Photo by Maurice Hornocker.

Plate 36   A young cougar pauses on a rocky ridge in Idaho's Frank Church-River of No Return Wilderness Area. Photo by Maurice Hornocker.

Figure 8.4 When cougars maintain their claws by scratching on logs and trees, scent from the feet may also be left behind, providing another way for these solitary animals to communicate with one another. Photo by Maurice Hornocker.

Management Implications of Cougar Behavior and Social Organization

Cougars in a population are members of a society where individuals interact and compete with other cougars, hunt prey, compete with other carnivores, and deal with predators, all in environments subject to spatial and temporal variation. In this process of life, cougars evolved life history strategies and behavioral tactics that apparently maximize individual survival and reproductive success. This aspect of cougars is one that deserves greater attention in efforts to manage and conserve the species.

Cougar management in western North America since the mid-1960s has involved activities principally focused on regulating the sport harvest and the removal of cougars involved in human safety incidents, depredation of domestic animals, and predation on species of concern (e.g., desert bighorn sheep). Regulated cougar sport hunting in particular has been conducted with a generalized goal of providing hunting opportunity while managing for healthy self-sustaining cougar populations in western North America, except for Texas, which does not afford the cougar legal protections from human exploitation. In California, cougar hunting is prohibited. Actual quantitative effects of removal (i.e., direct killing or relocation of cougars) on cougar populations have been the realm of seven intensively studied cougar populations in Alberta (Ross and Jalkotzy 1992), southern Idaho and northwestern Utah (Laundré et al. 2007), south-central Utah (Lindzey et al.

1992), central Utah (Stoner et al. 2006), New Mexico (Logan and Sweanor 2001), Wyoming (Anderson and Lindzey 2005), and a region overlapping portions of southern British Columbia, northeastern Washington, and the Idaho panhandle (Lambert et al. 2006, Robinson et al. 2008). Consequently, effects of annual population-wide removal of cougars are generally unknown to wildlife managers for all other regions not subject to such intense biological scrutiny. Moreover, even though these seven studies provide useful information for managing the effects of removal on cougar population dynamics, there is relatively little information about the consequences of cougar hunting upon processes of natural and sexual selection (see also Harris et al. 2002). In addition, loss and fragmentation of cougar habitat is occurring as a result of the ever-increasing human population and attendant developments. The presence of more people in cougar habitat can also increase the frequency of cougar-human encounters that then spawn additional cougar removals. What is known and understood about cougar behavior and social organization can be used in several ways to manage these issues and potential unintended consequences that might result from unknowns and uncertainties associated with cougar management.

Nine aspects of behavior and social organization useful to applied cougar management and science are indicated in the paragraphs that follow.

1. *Information on cougar behavior, social organization, and population dynamics suggests that cougar populations are ultimately limited by food and regulated by competition.*

Cougars are large obligate carnivores that have apparently evolved strategies that maximize individual survival and reproductive success. Thus, cougar populations are probably ultimately limited by the food supply (Pierce et al. 1999, 2000b; Logan and Sweanor 2001; Laundré et al. 2007). Population regulation mechanisms (i.e., density-dependent factors, Caughley and Sinclair 1994, 110–15) are expected to be competition for food in females, and competition for mates in males (Logan and Sweanor 2001; this chapter). Thus, healthy, self-sustaining prey populations are essential for cougar populations. On the other hand, cougar populations might need to be adaptively managed where domestic animals are an important prey source and where small ungulate populations (e.g., desert bighorn sheep) are threatened by cougar predation and conservation of those ungulates is a priority (Cougar Management Guidelines Working Group 2005).

2. *Cougar social organization and dispersal patterns indicate the need for habitat inventory and conservation.* Self-sustaining cougar populations need quality habitat and attendant prey populations that support sufficient numbers of adult cougars that interact, compete, and reproduce, and where females successfully raise offspring. Some offspring are recruited into the local population, while some long-distance dispersers are recruited into more distant cougar population segments. Inventory and mapping of cougar habitat and movement linkages would be useful for cougar management. Such information would assist managers to project potential cougar populations in management regions, contemplate effects of habitat loss and fragmentation on cougars, consider habitat conservation measures, and prepare management actions and research in urban-wildland ecotones where dangerous human-cougar encounters might increase.

3. *Female cougar maternal investment is essential to population persistence and growth.* Because adult female cougars raise offspring by themselves, protections for females in cougar population units managed for stable or increasing populations is a biologically sound management strategy. Where hunters cannot be convinced to refrain voluntarily from killing females to reach a harvest objective, then hunting regulations that limit female offtake can be imposed. Where the management objective is to reduce cougar numbers, killing adult females in addition to males should have the greatest negative impact on the cougar population (see Logan and Sweanor 2001; Lambert et al. 2006; Laundré et al. 2007). Managers can explore potential effects of changing female cougar survival in a theoretical manner by employing cougar population simulation modeling (Cougar Management Guidelines Working Group 2005).

4. *Cougar dispersal and recruitment patterns strongly influence population dynamics.* Dispersing cougars contribute numeric and genetic flow to other cougar population segments (Sweanor et al. 2000; McRae 2004; Robinson et al.

2008). Most cougar populations in western North America, with the exception of California, are hunted for sport. Some cougar population segments are reduced because of management objectives or by error (i.e., unintended chronic overkill), while other segments are stable or increasing. Natural habitat conditions can also produce variations in cougar population dynamics (Logan and Sweanor 2001, 175–79). Spatial variation in cougar population dynamics probably forms a source-sink metapopulation structure (see Chapter 5; Sweanor et al. 2000; Logan and Sweanor 2001; Cougar Management Guidelines Working Group 2005). Thus, the maintenance of metapopulation stability is dependent upon robust source populations (i.e., populations characterized by stable or positive growth, high survival, and emigration that exceeds immigration) that augment or rescue sink populations (i.e., populations characterized by negative growth or low phase due to high mortality, immigration that exceeds emigration, and young age structure). Managers can manage for source regions by imposing conservative sport harvest rates and by establishing refuges from sport hunting exploitation (Logan and Sweanor 2001).

5. *Migratory shifts in cougar numbers and distribution affect population inventory and management.* Cougars sometimes migrate with ungulate prey, consequently cougar abundance and distributions shift seasonally. Variations in cougar abundance and distributions need to be considered in efforts to sample population size (Pierce et al. 1999). In addition, migration and attendant variations in number and distribution of cougars might alter the level of risk to people in developed areas from dangerous encounters with cougars and to prey species of concern (e.g., livestock, threatened and endangered species).

6. *Natural and sexual selection processes might be influenced by extensive annual sport hunting pressure.* In instances where hunting off-take causes additive mortality, cougars that would have survived and had a chance to reproduce, are killed. The majority of the sport harvest in most western states consists of male cougars (Dawn 2002; see also state status reports in Becker at al. 2003, and Beausoleil and Martorello 2005). Yet cougars, males in particular, have high survival rates in populations protected from human exploitation (Logan and Sweanor 2001). Evidence for sexual selection in cougars includes sexual dimorphism, infanticide, aggressive male temperament, and territoriality. Thus nature seems to favor relatively long-lived, large, aggressive, highly conditioned male cougars as the most successful sires. In theory, traits and their associated genes that have survived in the past tend to be ones that will be successful in the future. But human exploitation of male cougars could reduce the population of males that participate in direct competition for access to mates and in female choice. Thus, sexual selection is potentially relaxed if human exploitation of males—whether it is directed toward

large males or more or less random (in reality it is probably a mixture of the two)—imposes different selection pressures than sexual selection, which is non-random selection for traits that apparently contribute to male reproductive success.

To the extent that people consider this to be important, cougar management could be structured to diversify selection pressures on a landscape scale (e.g., statewide) by managing different areas or zones for cougar hunting, control, and refuges. Refuges, where human interference in cougar survival would be at a minimum, presumably would help to maintain genetic diversity and population size and age structure, which is believed to enable evolutionary adaptations to changing environments (Tenhumberg et al. 2004). In this context, refuges would allow nature to choose the fittest cougars. Cougars dispersing from refuges would carry successful genotypes to other population segments, including exploited ones (Logan and Sweanor 2001; Tenhumberg et al. 2004). Zone management is an approach developed from empirical information on cougar population dynamics in a source-and-sink-metapopulation-structure that allows for cougar refuges along with zones managed for cougar population control and sport hunting (Logan and Sweanor 2001; Logan et al. 2005).

Potential effects of sport hunting on cougar breeding dynamics need investigation. Individual traits (e.g., sex, age, mass, condition, longevity) associated with reproductive success of individuals in hunted and nonhunted populations could be quantified. We would predict that the variance in male reproductive success is greater in protected cougar populations than it is in populations annually exploited by sport hunting or control, where male survival is substantially reduced. In addition, we would predict that comparatively younger and smaller males increase their reproductive success in situations where there is substantial hunting selection for large males (Law 2001; Coltman et al. 2003; Tenhumberg et al. 2004).

7. *Territorial male stability might influence adult female and cub survival.* Hunting male cougars probably cause instability in territorial male residency (Logan and Sweanor 2001). Removal of territorial males that dissuade activity of other males might increase infanticide and thus lower cub and adult female survival. Beyond the need to maintain enough adult males to breed with adult females, managers should consider that adult male removal could lower survival of adult females and offspring, and thus reduce cougar population growth (see Swenson 2003).

Comparing male territory stability and female and cub survival in protected and exploited cougar populations is another area in need of investigation. We would predict that lower female survival and reproductive success and lower cub survival would be associated with reduced male territory stability. In association with this, we would expect lower cougar population growth rates. In addition, we would expect that increased frequency of infanticide would reduce variance in male reproductive success.

8. *The cougar matrilineal structure and nearly obligate dispersal of males present opportunities for research that might enable managers to identify potential source populations.* Developing genetic methods (see Avise 1995; Kurushima et al. 2006; Biek et al. 2006) may reveal sufficient population structure or relatedness to link the origins of hunter-killed male cougars with the population regions that are relatively high producers of dispersers. Such information could reveal source populations to managers and conservationists.

9. *Long-term research of experimentally manipulated cougar populations is needed.* There is still much to be learned about cougar behavior and social organization and the attendant implications in life history strategies, variation in behavioral tactics, population dynamics, and related management and conservation strategies. Such efforts would require support for long-term research (i.e., ten years minimum) on properties of non-exploited cougar populations that are experimentally manipulated or where properties of protected populations are compared with properties of exploited populations to address specific questions and hypotheses. Again, zone management provides a structure of protected and exploited cougar populations that would be useful for science (Logan and Sweanor 2001; Logan et al. 2005).

The management relevance of some of these aspects of cougar behavior—landscape requirements, maternal investment, or dispersal, for example—seem self-evident. Other aspects involve theory and suggest cautious management and further investigation, particularly in view of growing human impacts on cougar populations. These matters include the effects of hunting on selection processes, the effects of territorial male stability on female and cub survival and individual reproductive success, and the use of genetics to infer population structure and dynamics. This view of cougar social organization as a template for adaptations that maximize fitness, with different male and female behaviors that maximize individual survival and reproductive success, throws into sharp focus the need for wildlife managers to consider the influence of human selection pressures of all kinds upon the ability of cougars to adapt to changing environments.

Chapter 9 Diet and Prey Selection of a Perfect Predator

Kerry Murphy and Toni K. Ruth

THE LONG FASCINATION of the public with predation and the parallel debate among scientists about it was aptly captured by Errington (1946, 144): "Whatever else may be said of predation, it does draw attention." Carnivory, animals eating animals, is striking—a bull elk killed and freshly covered with sticks and shaved elk hair by a cougar (*Puma concolor*) is a sight never forgotten, more than ample for extended discussion among elk hunters, gathered around a campfire. Hunters may be concerned that cougars compete directly with them for particular types of prey, such as buck deer (*Odocoileus* sp.) or bull elk (*Cervus elaphus*). Knowledge of the extent to which cougars choose particular sex and age classes of prey over others is thus important from a practical standpoint. For wildlife managers, acquiring knowledge of a carnivore's morphological, behavioral, and physiological adaptations for food acquisition and processing, its food habits and energy needs, and its predilections for particular types of food (prey selection) is the first basic step in understanding its predation ecology. These aspects of cougar predation are addressed in this chapter.

A second and more difficult step is understanding how a predator ultimately affects the composition (e.g., number of prey in particular sex, age, and condition classes) and overall numbers of its prey, and how factors such as prey condition, prey numbers relative to carrying capacity, and habitat conditions are relevant. These aspects are covered in Chapter 10. Together, Chapters 9 and 10 summarize much of what biologists currently understand about cougar predation from field studies and from predator-prey theory involving large carnivores. Chapter 11 addresses how cougars interact, both directly and indirectly, with other carnivores

that rely on similar food sources. It summarizes emerging knowledge that promises to advance our understanding of how cougars affect biological communities and, ultimately, whole ecosystems.

For scientists, predation is of universal importance because, like species interactions such as competition and parasitism, it has immediate effects on the lives of individual prey and predator as well as on the evolutionary trajectories of species. Many morphological and behavioral attributes of prey are responses to forces of natural selection created by predation pressure (Taylor 1984). Likewise, predators respond in kind with counter adaptations to keep pace with new or improved adaptations in their prey. This "evolutionary arms race," first recognized by Charles Darwin (Dawkins and Krebs 1979), accounts for the endless diversity in adaptive coloration, vigilance, fleetness, and cranial form in both prey and predator.

History of Cougar Predation Research

Prior to the 1960s, information about cougar predation was largely limited to anecdotal observations and food habit studies from explorers, naturalists, hunters, and scientists concerned with how this predator might influence human enterprise (Dixon 1925; Wright 1934; Hibben 1937; Young 1946b; Cronemiller 1948; Connolly 1949; Quigley and Hornocker 1992). As immobilization and radiotelemetry techniques improved biologists' abilities to study wildlife, more detailed and objective information about cougars emerged through key field studies in central Idaho and

Arizona (Hornocker 1969, 1970; Seidensticker et al. 1973; Shaw 1977). These efforts and other landmark investigations of large carnivores during this period (Craighead and Craighead 1965; Perry 1965; Schaller 1967, 1972; Mech 1970; Kruuk 1972) also dramatically increased public awareness of the ecological importance of predation, including its beneficial effects, and drew public attention to the precarious status of many large carnivore populations.

In the 1980s and early 1990s, the need for better information throughout the species' range led to numerous field or modeling studies of cougar predation (Ashman et al. 1983; Ackerman et al. 1984, 1986; Yáñez et al. 1986; Emmons 1987; Iriarte et al. 1990; Maehr et al. 1990; Cashman et al. 1992). Results indicated the complexity of cougar predation—that a variety of interacting prey, predator, and habitat-related variables were important in understanding how cougars might affect their prey. During the 1990s, cougar predation research progressed to tests of more explicit hypotheses (Murphy 1998; Kunkel et al. 1999; Pierce et al. 2000a; Logan and Sweanor 2001; Husseman et al. 2003) that were better grounded in predation theory than the previous work. At the same time, scientists were advancing new ideas about predator-prey systems (Messier 1994). In the early 2000s, estimation of cougar predation rates improved when collars were equipped with global positioning systems (GPS) prearranged to provide more frequent and accurate locations of cougars than VHF-based (traditional) telemetry (Anderson and Lindzey 2003; Mattson et al. 2007). In sum, while cougar research remains difficult and costly relative to studies of more visible and numerically abundant species, knowledge of cougar predation has progressed rapidly with hypothesis testing and new technology and will likely continue to do so (for more history of cougar research, see Chapter 2).

A Morphologically Specialized Predator

Felids typically are solitary hunters and feeders and, among carnivore families, are the most specialized for solo predation (Ewer 1973; Kleiman and Eisenberg 1973). They eat little vegetable matter and fish, prey principally upon mammals, and have less diverse diets than members of other carnivore families (Kruuk 1986). The various felid species tend to prey on vertebrates sized similarly to themselves (Kruuk 1986), although lone cougars can kill prey up to five times larger than they are (Ross and Jalkotzy 1996), a feat that requires considerable strength and morphological tooling.

Felids are short-range cursorial predators; that is, they accelerate rapidly and, if necessary, sustain high speed during brief pursuits of prey (Plate 27). They exhibit numer-

ous adaptations that increase speed by improving the length and rate of their stride. Limb length is increased through a digitigrade (walking on toes without heels touching the ground) foot posture and elongated bones of the lower limbs (Taylor 1989). Felid scapulas are variously modified, and their clavicles are reduced to facilitate extension of the lower limbs (Taylor 1989). In the cheetah (*Acinonyx jubatus*), and likely to some extent in the cougar, stride length is increased by flexion along the long axis of the spine, strong forward and backward reach of the limbs, and an increase in the time feet are off the ground (Hildebrand 1959). As opposed to that of horses (*Equus caballus*), which rely primarily on muscles of the limbs to propel the legs forward, the speed of cougar strides is improved by use of muscles inserted on both the limbs and the spine (Hildebrand 1959).

Living mainly in environments dominated by vegetative and topographic cover, cougars are intermediate among felids in their adaptations for speed (Gonyea 1976)—fast and agile but not adapted for extended pursuit. By necessity, the cougar's cursorial hunting strategy is balanced by the ability to hold and control large, powerful prey using well-developed muscles of the forelimbs and shoulders that detract from acceleration and body control at high speeds (Ewer 1973; Gonyea and Ashworth 1975; Gonyea 1976; Taylor 1989). The rear limbs of cougars are long relative to the length of their forelimbs, presumably an adaptation for jumping (or bounding) in variable terrain (Gonyea and Ashworth 1975). However, a small lung capacity also reduces their ability to pursue prey over long distances or to escape from runners with superior stamina, such as gray wolves (*Canis lupus*).

Cougar feet are large and equipped with claws that are highly flattened laterally and sheathed for sharpness but that protract as needed to grapple prey (prehension), for self-defense, or to climb trees for escape (Figures 9.1, 9.2; Eisenberg and Leyhausen 1972; Ewer 1973; Logan and Sweanor 2000). The claw on the opposing pollux (thumb) is well developed for predation and fighting (Logan and Sweanor 2000).

Like other carnivores, cougars require strong skulls, jaws, and teeth to kill and dismember their prey safely (Biknevicius and Van Valkenburgh 1996). Targeting large and potentially dangerous prey as a solo predator, cougars must make few but effective deep bites (stab wounds) to incapacitate their quarry quickly, a strategy that puts large and unpredictable stresses on canines. Consequently, cougar canines are robust and resistant to bending in both an anterior-posterior and mediolateral (side-to-side) plane (Figure 9.3; Van Valkenburgh and Ruff 1987). Premolars and molars are particularly bladelike for slicing soft tissues (Ewer 1973). Compared to other carnivores, blade lengths

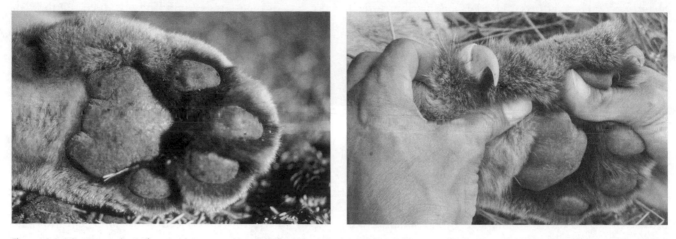

Figures 9.1, 9.2   At rest, claws of cougars are retracted into sheaths and remain sharp. When stalking prey, the cougar's soft fur and toe and heel pads help reduce noise. When attacking prey, claws are protracted (extended) for seizing and subduing prey while applying a killing bite. Although the pollex (fifth toe) is important for predation because it is large and opposes the other four toes, it is not weight bearing. Retracted claw photo by Kerry Murphy; protracted claw photo © Linda L. Sweanor.

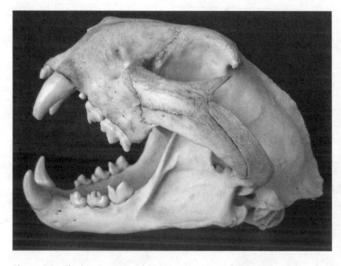

Figure 9.3   The large canines of cougars are adapted for deep penetration in prey and are resistant to bending stresses from any direction. Premolars and molars are bladelike for slicing soft tissues. Photo by Toni K. Ruth, Hornocker Wildlife Institute.

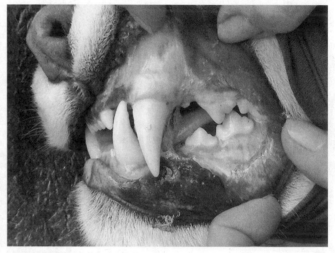

Figure 9.4   Cougar teeth are reduced in number and the rostrum (nasal area) is shortened, positioning canines near the axis of the jaw and the jaw muscles to improve biting power. Photo by Toni K. Ruth, Hornocker Wildlife Institute.

of felid carnassials—the cheek teeth specialized for shearing soft tissues (M_I and P^4)—are long relative to basal length (Van Valkenburgh 1989); but, lacking grinding surfaces on their premolars and molars, felids are not adapted for bone chewing (Biknevicius and Ruff 1992). The power of felid bites is enhanced by short jaws afforded by a reduction in numbers of teeth (thirty for cougars) and temporalis muscles that are large and mechanically advantaged through the placement of their insertions on the skull and jaws (Figure 9.4; Van Valkenburgh 1989; Biknevicius and Van Valkenburgh 1996). Felid jaws are also well buttressed to resist bending stresses (Biknevicius and Ruff 1992).

Cougars rely on their keen vision and hearing to locate prey (Kleiman and Eisenberg 1973). The importance of olfaction (sense of smell) in locating and stalking prey is not clear. Cougars commonly detect and successfully stalk bedded prey, indicating that olfaction may be important. Because cougars hunt mainly at night or during the crepuscular (dawn or dusk) hours (Seidensticker et al. 1973; Sweanor 1990; Beier et al. 1995; Anderson and Lindzey 2003; Mattson et al. 2007), they require high visual sensitivity in poor light. Eyes of cougars, the domestic cat (*Felis catus*), and other felids have several adaptations that enhance collection of and sensitivity to light: large pupils, a retina predominated by rods, nerve cells that function well under low light, and a well-developed tapetum lucidum that reflects light back into the interior chamber of the eye (Walls 1942; Hughes 1977). Rods improve vision but reduce visual acuity during the day (Walls 1942; Ewer 1973; Hansen 1992). Slit (oval) pupils controlled by interlacing muscle fibers of the iris reduce the amount of bright light entering the posterior chamber of the cat's eye (Walls 1942). Felids have a greater binocular field of vision compared to other carnivores, an adaptation that improves depth perception needed for accurate restraint of prey with their paws (Figure 9.5; Hughes 1977). Due to a highly curved cornea, they have a broad total (binocular and monocular) visual field as well (Walls 1942). Supplementing the cat's vision are whiskers on the muzzle that act as tactile organs that help direct the killing bite (Leyhausen 1979).

The hearing of cougars is not well studied, but domestic cats and jaguarundis (*Herpailurus yaguarondi*) hear well over a wider range and at a higher upper limit of frequencies than do humans and most other carnivores (Peterson et al. 1969; Ewer 1973). Large pinnae (earflaps) of felids enhance collection and location of sound (Ewer 1973).

Although highly variable across their geographic range, cougars are brownish, reddish, or tawny along the back and sides, probably related to concealment that facilitates stalking or enhances concealment for self-defense (Young 1946b; Ewer 1973). S. P. Young (1946b) suggested that similarity in color to deer, a favorite prey, may also aid cougar predation.

Felid digestive efficiencies are low compared to other carnivores; that is, they extract fewer nutrients per unit of food that passes through their gut (Kendall et al. 1982; Houston 1988). This lower efficiency is due to a shorter digestive tract achieved through reduction of the caecum and hind gut. By reducing inertia, a light gut may improve acceleration needed for short pursuit of prey and help offset the loss in digestive efficiency (Houston 1988).

Hunting Behavior

Cougars use their acute senses, primarily vision and hearing, to detect prey and use stealth when hunting. Movement patterns of individuals that are searching for prey or traveling between hunting sites are different than cougars exhibiting nonhunting behaviors such as breeding, denning, or feeding near preexisting kills. Travel bouts of hunting cougars are longer, more frequent, and less focused about a central location, and their rest periods are shorter (Beier et al. 1995). Cougars stalk or lie in wait for short periods, then travel to other locations when unsuccessful. When actively hunting, they move silently while remaining alert for prey, stand erect or crouch, and stop frequently and sit (Beier et al. 1995). When prey are seen, cougars stalk to short distances, or wait in ambush, and their hunting success undoubtedly improves if they can shorten (<2 m) their distance to prey (Figure 9.6; Beier et al. 1995). They then make a short pursuit or quick rush, bounding or galloping toward prey (Young 1946b; Robinette et al. 1959; references in Anderson 1983, app. D; P. Wilson 1984). Laundré and Hernández (2003) found that cougars usually made contact with mule deer (*O. hemionus*) within 10 m of initiating a successful pursuit. Accounts of cougars subduing prey after successful pursuit are highly variable: prey may be killed at the site of contact, or struggles may occur over distances of up to 90 m

Figure 9.6 Cougars are stalking predators that use vegetation and other landscape features for concealment. They attack prey after stalking undetected to within 1–5 m, or by sprinting after prey over longer distances. Photo by Maurice Hornocker.

Figure 9.5 The broad field of binocular vision characteristic of felids such as cougars enhances the depth perception needed during pursuit and apprehension of prey. Photo courtesy of Wildlife Conservation Society.

(Young 1946b; references in Anderson 1983). Laundré and Hernández (2003) found that deer traveled only 10–15 m after contact.

Cougars use their well-developed forelimbs and protracted claws to strike, grasp, and hold prey, typically about the shoulders, neck, or face (Eisenberg and Leyhausen 1972; Gonyea and Ashworth 1975). A killing bite is directed at the facio-cranial region, typically the throat or nape of the neck (Hibben 1939; Connolly 1949; Robinette et al. 1959; Cunningham 1971; Leyhausen 1979; P. Wilson 1984). The throat region is a common bite location for cougars that are attacking large prey, such as adult elk, bighorn sheep (*Ovis canadensis*), or moose (*Alces alces*). Heavy-boned and broad skulls, large vertebrae, antlers, or horns likely make dorsal bites infeasible and put canines at risk of breakage (Figure 9.7). In contrast, medium-sized and smaller prey, such as coyotes (*Canis latrans*) and other cougars, may be killed with lethal bites to the rear cranium or cervical vertebrae. Prey killed by cougars often have claw marks in the nose, muzzle, or behind the jaw (Connolly 1949; Cunningham 1971). Killing large prey in steep, rocky, or wooded terrain is also dangerous, and cougars may be injured or killed during encounters with prey (Figures 9.8, 9.9, 9.10; Gashwiler and Robinette 1957; Hornocker 1970; Ross et al. 1995; Logan and Sweanor 2001).

In general, the stalking approach felids use is more successful than a coursing strategy that "tests" prey vulnerability through extended pursuit, as by gray wolves (Husseman et al. 2003). However, not all cougar stalks result in attacks, and not all attacks are successful. Investigating cougar attempts to kill deer or elk, Hornocker (1970) found that eight of

Figure 9.7 The successful cougar attack on a bighorn sheep was witnessed in 2008 by boaters on Lake Minnewanka, Banff National Park, Alberta. After stalking, cougars rush their quarry. Upon contact, the cat uses its strong forearms and protracted claws to subdue ungulate prey such as this ram and enable a suffocating bite in the throat. Smaller prey such as coyotes may be killed by a bite in the skull or dorsal (spinal) portion of the neck. Photo by T. Phillips, P. Baker, K. Pranger, S. Geniole, and C. Phillips.

forty-five (17 percent, excluding aborted approaches) were unsuccessful.

Studies of habitat characteristics at sites where cougars make kills (or caches) suggest that the benefits of cover vary with vegetation and terrain type. At several study sites in the Rocky Mountains and southern Utah, cougar kills were located disproportionately more in habitats that afforded the greatest cover (Logan and Irwin 1985; Laing 1988; Williams et al. 1995; Jalkotzy et al. 2000; Husseman et al. 2003). Brushy terrain and encroachment of trees and shrubs contributed to higher predation rates of translocated bighorn sheep in central and southwestern Arizona, as compared to the northwest, northeast, or southeast of the state (McKinney et al. 2006). Similarly, Ockenfels (1994) found that rugged terrain and brush facilitated cougar predation on pronghorn (*Antilocapra americana*). But, in eastern California, researchers concluded that cover did not facilitate cougar predation; kills were more frequent in low shrub than tall shrub communities, even though mule deer foraged more frequently where cover was greatest (Pierce et al. 2004). Similarly, Katnik (2002) found in British Columbia that horizontal visibility in forest understories where cougars made kills or caches, and the percentage of cover provided by tall vegetation, rocks, and woody debris, did not differ from nearby habitat patches, although cover of low shrubs was greater at kills.

The ideal habitat structure thus may be one that provides sufficient visibility for hunting cougars to see prey and remain undetected at a distance, yet still provides adequate cover to stalk within close range without detection (Laundré and Hernández 2000). In northwest Utah and southern Idaho, Laundré and Hernández (2003) found that cougars killed wintering mule deer more often in edge habitats than in open areas, and that the cats made relatively few kills in the forests (principally Douglas fir and pinyon-juniper). Stalking cougars were probably detected by deer in open areas, and the relatively high tree densities characteristic of the forests likely obscured the predator's view (Laing 1988). Further, because the mule deer typically used the forest for resting and bedding, and were likely stationary and vigilant, they may have been more difficult to stalk successfully (Laundré and Hernández 2003), especially where dense conifers were difficult to penetrate silently (Katnik 2002). In contrast, edge areas supported deer traveling between open areas (feeding) and forest, and presumably deer were more easily detected and stalked than in the forests or open environments (Laundré and Hernández 2003). However, open habitats may confer some benefit to hunting cougars. After noting that grasslands played an important role in cougar movement, Dickson and colleagues (2005) suggested that grasslands might facilitate penetration of habitat mosaics, stalking, or pursuit of prey by cougars. Information about

Figures 9.8, 9.9, 9.10 Cougars may be injured or killed when trying to subdue and kill large prey. In this series of photos, a five-year-old male cougar and the bighorn ram it was trying to kill both died in a fall off a -foot cliff in Yellowstone National Park. Photos by Toni K. Ruth, Hornocker Wildlife Institute/Wildlife Conservation Society.

hunting behavior of cougars at small spatial scales is needed for better understanding of the terrain and vegetation attributes that facilitate predation and their interactions with other factors contributing to prey vulnerability. Interpreting past studies is difficult due to differences in the spatial scale of data analyses and the difficulty in accurately locating and characterizing exactly where prey were stalked and attacked, as opposed to where cougars cached kills. Indeed, Laundré and Hernández (2003) found that cougars cached prey at sites with greater tree densities and diameters than found at kill sites, although slopes and shrub characteristics were similar. Because cougars use a broader array of habitats when traveling and hunting than when resting (e.g., daybeds), use of diurnal locations to infer relationships among hunting cougars, cover, and prey is also problematic (Dickson et al. 2005).

Feeding Behavior and Prey Consumption

The kill accomplished, cougars often carry or drag prey a fair distance, typically up to 80 m, to vegetative cover, presumably to improve cooling and reduce carcass spoilage, to reduce visibility to scavengers, or to reach locations that apparently provide security (Robinette et al. 1959; Beier et al. 1995; Laundré and Henández 2003; Mattson et al. 2007). They may also cover carcasses of large prey with sticks, vegetation, or the prey's own hair after an initial feeding and when subsequent feedings are likely (Connolly 1949; Beier et al. 1995; Bank and Franklin 1998).

Initially, they may gorge on fresh kills (Ackerman 1982). A fifteen-month-old captive cougar consumed up to 10 kg within twenty-four hours (average 6.8 kg) after its first opportunity to feed from a fresh carcass, but intake declined to a 4.1 kg average over eleven days (Danvir and Lindzey 1981). Daily consumption reported by Ackerman (1982) for twenty-six feeding trails of one to sixteen days' duration for three cougars ranged from 1.1 to 7.2 kg. Hornocker (1970) estimated that on two occasions wild male cougars consumed approximately 9–14 kg during initial feedings. However, the rates at which cougars consume new kills probably do not reflect their average daily intake when calculated over longer periods; cougars fast while involved in other activities, such as travel in pursuit of mates (Ackerman 1982). Robinette and colleagues (1959) and Spalding and Lesowski (1971) found that 277 of 401 (69 percent) and 99 of 132 (75 percent) of the cougar stomachs they collected, respectively, were empty. Cougars typically select the nutritious thoracic organs, diaphragm, and liver for consumption first, followed by other abdominal organs, fats, the rib cage, and muscle tissues (Connolly 1949; Robinette et al. 1959; Danvir and Lindzey 1981; Ackerman 1982; Beier et al. 1995). Gastrointestinal tracts may be disconnected from the carcass and are not consumed (Hornocker 1970; Danvir and Lindzey 1981; Beier et al. 1995). Lacking a dentition adapted for bone grinding (Van Valkenburgh 1989), cougars often leave uneaten the limb bones, vertebrae, and skulls as well as some skin of large prey (Danvir and Lindzey 1981; Harrison 1990).

Estimates of the total fraction of ungulate prey consumed by cougars are variable (63 percent, Robinette et al. 1959; 70 percent, Hornocker 1970; 73 percent, Danvir and Lindzey 1981; 62 percent, Murphy 1998; 75–95 percent, Mattson et al. 2007). Consumption of small ungulates and other small mammals is nearly complete (Hornocker 1970). Among seven different variables, Mattson et al. (2007) found that the fraction of carcasses consumed by cougars varied most by prey mass, elevation of carcass, cougar sex and age, and the time elapsed between cougar feeding and the site investigation by researchers. Small carcasses (<100 kg) were consumed more thoroughly than the largest ones. Consumption also increased at higher elevations, apparently due to reductions in spoilage, in competition with arthropods, and in scent that can attract competing scavengers. Young females consumed the greatest fraction of carcasses and adult males the least.

Kills typically have strong localizing effects on cougar movements (Seidensticker et al. 1973; Shaw 1977; Beier et al. 1995). The predators bed in the vicinity (<400 m) of kills during the day, feed at night, and spend more than half of the nocturnal hours close by (Beier et al. 1995). They typically feed for two to five nights from kills, although longer associations have been reported (Thompson and Stewart 1994; Beier et al. 1995). Cougars infrequently abandon spoiled carcasses without eating large fractions, or lose portions to scavengers when avoiding anthropogenic disturbances, such

as mining activity or other sources of noise (Harrison 1990; Beier et al. 1995). Mattson et al. (2007) found that the time spent at a carcass (typically 0.5 to 2.5 days) varied with prey mass (peaking for intermediate prey sizes); decreased for mature males compared to other cougar classes; and increased with precipitation that retarded depletion by microbes and arthropods during feeding. Cougar social motivations (e.g., those of adult males in pursuit of mates) and environmental factors (e.g., carcass spoilage or competing scavengers) are likely to be highly relevant to carcass dragging, extent and rate of consumption, and burial, and this behavioral variability in prey handling and consumption after the kill may well be adaptive (Laundré and Hernández 2003; see also Mattson et al. 2007). Thus, although Sunquist and Sunquist (1989) suggested that the ideal prey is the largest item that can safely be killed, consumption, caching, and ultimately cougar reproductive strategies are also important influences on the profitability of a particular prey type (Mattson et al. 2007).

Food Habits

Food habits are a relatively well understood facet of cougar ecology (Anderson 1983), with studies conducted across a broad spectrum of environments. Studies of food habits typically reflect actual predation, but cougars also scavenge and deliberately or accidentally ingest vegetation associated with carnivory (Ackerman et al. 1984; Maehr et al. 1990; Logan and Sweanor 2000; Bauer et al. 2005). Cougars are adaptable and opportunistic predators (Logan and Sweanor 2001). Across their range, they prey on a variety of terrestrial and semiterrestrial vertebrates, including mammals, birds, and reptiles (Tables 9.1–9.3), and even some aquatic species (McClinton et al. 2000).

As predicted from models, the preferred prey size of cougars is 70 to 165 kg, or greater than 45 percent of their weight (Carbone et al. 1999). As with other predators, the range of prey sizes and prey types cougars kill may be constrained by their own energetic requirements, the time to search for and consume their prey, energy and nutrient value of their food, and the risk of injury when attacking prey (MacArthur and Pianka 1966; Rosenzweig 1966; Stephens and Krebs 1986; Sunquist and Sunquist 1989). Family groups of large felids may be obligate hunters of large prey (>15 kg, including deer fawns and elk calves) because of high collective energy requirements (Ackerman 1982; Novack et al. 2005). For cougar families, the biomass of small prey may be insufficient to offset the energy costs of foraging, whereas large prey would provide more biomass per unit of hunting, except when small prey are highly abundant and vulnerable to capture. For example, Branch

Table 9.1 Percent occurrence of prey animals in cougar diets in western Canada and the northern United States, as calculated by chapter authors.[a]

Food Item	British Columbia		Alberta		Washington			Oregon		Idaho		Montana		Montana/Wyoming		Wyoming	
Season	A	W	W	A	A	A	A	W	A	W	W	W	A	A	A	A	A
Data type	K;G[b]	G[c]	K[d]	K[e]	K[f]	S[g]	K	G[h]	K[i]	K	K[j]	K[k]	S	K[l]	S[m]	K	K[n]
Mule/Black-tailed deer	14	58	60	33	59	25	46	72	55	44	31	5	29	18	3	85	59
White-tailed deer	4		6									87	15				1
Elk	69		10		38	9	49	11	29	51	51	6	12	62	30	2	20
Bighorn sheep	4		9	58						2	5		20	1	2		
Moose		4	15								4	1		1			1
Pronghorn														T	T		8
Mountain goat					1					1							
Porcupine	2	10						17	2				2	7	7	8	7
Marmot														2	9	2	
Beaver (2 spp.)		3			1	6	1				1				3		
Other rodent						25			3				12	1	24		
Lagomorph	2	11				34							8	2	14	4	
Bird		1							8		1			1			
Raccoon																	
Coyote	2	1		9	1		1		1		4			4	6		1
Cougar	4	2					1				2			1	1		
Badger									1								
Bobcat											2						
Black bear															1		
Cattle		2															1
Sheep/goat		5					1										1
Horse		?															
Domestic cat/dog					1									2			
Sample size (# items)	51	97	334	43	175	32	72	18	89	105	85	135	27	302	170	52	75
Source	1	2	3	4	5	6	7	8	9	10	11	12	13	14	14	15	16

[a]A = annual, W = winter, K = kill, S = scats, G = gastrointestinal tracts, T = trace (<0.5%).
[b]Includes 3 gastrointenstinal samples from vehicle strikes. Excludes 1 unconsumed cougar and 1 ermine.
[c]Excludes 6 occurrences of "carrion"; 21 stomachs contained vegetation or unidentified material. Samples may be biased toward cougars associated with livestock depredation.
[d]Excludes 34 unspecified, non-ungulate prey, including ≥3 cougars.
[e]Excludes 1 ruffed grouse that was killed but not eaten.
[f]Excludes 4 unknown ungulates; data from searches of radiolocation clusters that may include instances of scavenging.
[g]Excludes 2 scats with unidentified bones and hair.
[h]Excludes 7 total stomachs that contained unidentified hair, vegetation, or cougar hair apparently from grooming.
[i]Data from solitary females and family groups only.
[j]Sample size = 85 (H. Akenson, pers. comm., 2006).
[k]North Fork Flathead River, Montana; small and medium-sized kills possibly obscured by heavy cover.
[l]Weighted average of kills detected opportunistically and through daily monitoring of focal individuals.
[m]Values adjusted for differential digestibility of cougar prey and calculated following Ackerman (1984).
[n]Data from searches of GPS location clusters which may inconsistently detect predation or scavenging on small (<15 kg) animals.

SOURCES: (1) Spreadbury 1989; (2) Spalding and Lesowski 1971; (3) Ross and Jalkotzy 1996; (4) Harrison 1990; (5) G. Koehler, pers. comm., 2006; (6) Schwartz and Mitchell 1945; (7) Spencer et al. 2001; (8) Toweill and Meslow 1977; (9) Nowak 1999; (10) Hornocker 1970; (11) Akenson, Akenson, and Quigley 2005; (12) T. Ruth, pers. comm., 2006; (13) Williams, McCarthy, and Picton 1995; (14) Murphy 1998; (15) Logan and Irwin 1985; (16) Anderson and Lindzey 2003.

Table 9.2 Percent occurrence of prey animals in cougar diets in the southwestern United States, Mexico, and Florida, as calculated by chapter authors.[a]

Food Item	Colorado		Utah		Nevada	Nevada/Utah	New Mexico	Arizona				California		Texas		Mexico			Florida		
Season	W	W	A	A	A	A	A	A	A	A	A	A	A	A	A	A	A	A	A	A	A
Data type	K	Kb	Kc	Sd	Ge	Ge,f	Kg	Sh	Ki	Sj	Kk	Sl	Sm	K	K	Sn	S	S	So	So	S
Mule/Black-tailed deer	89	91	89	66	47	68	91	63	61	33	39	64	52	21		13					
White-tailed deer							27				T			26	49	37	50	23	34	65	
Elk		4	4	T							44										
Bighorn sheep							2			6						31					
Pronghorn							1	2			T										
Collared peccary										21				37	24	11	7	10			
Wild pig												17			12				59	21	1
Oryx							1														
Porcupine	11	4		1	21	14	1	4		2	T		3								
Marmot				1		T															
Beaver				1		1							1								
Paca																		5			
Other rodent		1		11		T				7		8	8			2	19		1	3	2
Lagomorph			3	15	5	4	T	4	7	3	2	5	1			25			3	6	19
Nine-banded armadillo																9	5	6	11	3	
Opossum (2 spp.)													10				4	5			
Mole													1								
Bird				1		T	T					1	2	1		2	2	5			1
Reptile										2						5	18				1
Raccoon										2	T		1				4	23	7		
Ringtail							T														
Coati																	4	20			
Coyote			2		5		1				10		9			2					
Gray fox				T				1			T	3	1								T
Cougar			1	1	5		2				1			1	3	2				2	
Badger			1	T			T			4	1	1									
Bobcat				2	5	T				1	1	2			1	2					
Skunk (1–2 spp.)						T	1	1		3		2	1	8	1	4					T
Otter																					2
Long-tailed weasel																2					
Cattle			1			T	1	38		11		4	3	1		3					
Sheep/goat					10	10							1		8						
Horse						T										1					

(Continued)

Table 9.2 (Continued)

Food Item	Colorado		Utah		Nevada	Nevada/Utah	New Mexico		Arizona			California		Texas		Mexico			Florida		
Domestic dog/cat						T						1									
Sample size (# items)	9	75	110	293	19	281	525	193	61	187	218	53	184	89	75	79	97	20	155	119	291
Source	1	2	3	3	4	5	6	7	8	9	10	11	12	13	14	15	16	17	18	18	19

[a]A = annual, W = winter, K = kill, S = scats, G = gastrointestinal tracts, T = trace (<0.5%).
[b]"Several" porcupines estimated at 3.
[c]Excludes 2 cases of scavenging carrion.
[d]Excludes 1, 8, and 14 occurrences of carrion (cattle), unknown animal, and vegetation, respectively. Twenty-two scats contained cougar hair associated with grooming.
[e]Samples may be biased toward cougars associated with livestock depredation.
[f]Samples from winter and summer combined; 15 occurrences of vegetation excluded.
[g]Excludes 14 cougars, 4 gray foxes, 1 coyote, and 1 long-eared owl that were killed but not eaten.
[h]Excludes 3 occurrences of vegetation; data specifically from New Mexico, with those from Arizona excluded.
[i]Excludes 1 cottontail rabbit killed but not eaten.
[j]Excludes 1 unidentified canid and 2 occurrences of beetles.
[k]Data were from searches of cougar radio location clusters, and likely included scavenging.
[l]Excludes 33 occurrences of felid hair apparently from grooming; 13 of vegetation.
[m]Excludes approximately 3 unidentified canids. Number of items reported by Beier and Barrett = 193; our calculations: n = 184, plus 3 unidentified canids.
[n]Excludes 2 instances of unidentified mammals; values for mule deer may include instances of unidentified white-tailed deer.
[o]North and south study areas in southwest Florida.
SOURCES: (1) Dixon and Boyd 1967; (2) Anderson, Bowden, and Kattner 1992; (3) Ackerman, Lindzey, and Hemker 1984; (4) Ashman et al. 1983; (5) Robinette, Gashwiler, and Morris 1959; (6) Logan and Sweanor 2001; (7) Hibben 1937; (8) Shaw 1977; (9) Cashman, Pierce, and Krausman 1992; (10) Mattson et al. 2007; (11) Hopkins 1990; (12) Beier and Barrett 1993; (13) Waid 1990; (14) Harveson 1997; (15) Rosas-Rosas et al. 2003; (16) Núñez, Miller, and Lindzey 2000; (17) Aranda and Sanchez-Cordero 1996; (18) Maehr et al. 1990; (19) Dalrymple and Bass 1996.

Table 9.3 Percent occurrence of prey animals in cougar diets in Central and South America, as calculated by chapter authors.[a]

Food Item	Guatemala	Panama	Venezuela	Brazil	Peru	Paraguay		Chile		Argentina		Patagonia
Season	A	A	A	A	A	A	A	A	A	A	A	A
Data type	S[h]	S[c]	S[d]	K	S	S	S[e]	S[f]	K	S[g]	S[g]	S[h]
White-tailed deer	12		10									
Brocket deer (2+ spp.)	19	10		14		18	14					
Elk									6			2
Pudu								83				
Peccary (2 spp.)	7	11	19			14	10					
Guemal								1				
Guanaco								8			2	
Wild pig										4	18	T
Anteater (3+ spp.)		2				18	5					
Sloth (2 spp.)		16										
Armadillo (5 spp.)	2	2	2	4		32	13			1	5	1
Capybara			10	32								
Coendou	1											
Agouti	20	22	2		33							
Paca	19	11			25							
Plains vizcacha										85	52	
Opossum (2 spp.)		3					9					
Small rodents/marsupials		10	17		17	26	4			2	15	30

(Continued)

Table 9.3 (Continued)

Food Item	Guatemala	Panama	Venezuela	Brazil	Peru	Paraguay		Chile		Argentina		Patagonia
Lagomorph			7			9	11	53		6	5	61
Monkey	10	5				5						
Caiman			10									
Iguana		1										
Other reptile		2			17							2
Bird	6	1		4			8	9				4
Bat					8							
Ringtail												
Coati	1	3				5						
Kinkajou	4											
Fox (2–3 spp.)	1							5				
Hog-nosed skunk								1				
Small felid* (1–3 spp.)								4		1	2	
Cattle			24	46				T			2	T
Sheep/goat								14			2	
Horse								1				T
Domestic bird									11			
Sample size (# items)	189	115	42	28^i	12	22	96	669	18	93	67	99
Source	1	2	3	4	5	6	7	8	9	10	10	11

[a]A = annual, W = winter, K = kill, S = scats, G = gastrointestinal tracts, T = trace (<0.5%).
*May include juvenile cougar.
[b]Omits 1 "unknown" deer. Includes scats from areas with hunted and nonhunted prey.
[c]Excludes 6 total occurrences of vegetation and unidentified small mammals.
[d]Livestock included cattle and possibly several equines and water buffalo. Scats collected primarily during the dry season.
[e]Excludes approximately 30 total unidentified rodents, lagomorphs, and "other" taxa.
[f]Excludes approximately 13 unidentified mammals. Data for multiple seasons, and inside and outside Torres del Paine National Park combined.
[g]Study years 1986 (peak numbers of plains vizcacha) and 1994 (low populations).
[h]Relative frequency of prey calculated after correction for digestibility (Ackerman et al. 1984). Small rodent/marsupials include an unspecified frequency of an unidentified mustelid and a small armadillo (pichi). Reptile species were unspecified, and may include caiman or iguana.
[i]We calculated the number of cougar kills at 28, but the authors reported 31.
SOURCES: (1) Novack et al. 2005; (2) Moreno, Kays, and Samudio 2006; (3) Scognamillo et al.; (4) Crawshaw and Quigley (2002); (5) Emmons 1987; (6) Stallings, unpub. data in Jorgenson and Redford 1993; (7) Taber et al. 1997; (8) Yáñez et al. 1986; (9) Courtin, Pacheco, and Eldridge 1985; (10) Branch, Pessino, and Villarreal 1996; (11) Novaro, Funes, and Walker 2000.

and colleagues (1996) documented extensive foraging on plains vizcacha (*Lagostomus maximus*), a 2- to 5-kg Argentinean rodent, and hypothesized that this prey's predictable location in colonies and low risk of handling permitted cougars to specialize, particularly when vizcacha occurred in high numbers. The risks of injury cougars incur with predation appear to be the principal constraint on their maximum prey size (Sunquist and Sunquist 1989; Murphy 1998; Kunkel et al. 1999; Husseman et al. 2003), presumably about that of an adult male moose (410 kg). Predation on tree-dwelling or highly social species, such as some birds and primates, may be limited by the difficulty of a surprise attack (Jorgenson and Redford 1993).

Prey of tropical cougars are diverse in size and life history and include ungulates, anteaters, armadillos, marsupials, primates, rodents, and reptiles; the breadth of cougar diets here—that is, diversity of prey consumed—is greater than in the temperate zone (Chapters 6, 7; Iriarte et al. 1990). Native ungulates typically comprise less than 50 percent of food items in South and Central American cougar scats (10 of 10 studies), whereas in North America, ungulate occurrence in scats usually exceeds 50 percent (11 of 14 studies; Tables 9.1–9.3). The link between diet breadth and latitude may be related to greater abundance of vulnerable midsized and small taxa in tropical versus temperate regions (Wallace 1878; Dobzhansky 1950; Fischer 1960; Darlington 1975) as well as to the absence of a seasonal reduction in access to hibernating species characteristic of the temperate zone. The influence of closed forest structure on the distribution of large prey, or competition with sympatric jaguars

(*Panthera onca*), may also influence dietary breadth of cougars (Iriarte et al. 1990; Moreno et al. 2006). Diet breadth of panthers in southern Florida, a subtropical region, also exceeds that for cougars in western North America (Iriarte et al. 1990; Dalrymple and Bass 1996) due to greater year-round diversity of prey.

Ungulate and Carnivore Prey

Diets of North American cougars are typically dominated by large prey, particularly mule deer and white-tailed deer (*O. virginianus*) (Anderson 1983; Iriarte 1990). Diets in the northern United States and Canada also include elk, bighorn sheep, and moose; species that may exceed deer in percentage occurrence in diets or in amount of biomass provided to cougars (Ross and Jalkotzy 1996; Murphy 1998; Akenson et al. 2005). Cougars in South Dakota and northeastern Wyoming prey primarily on mule deer and white-tailed deer, but elk and bighorn sheep are taken as well (D. Thompson, pers. comm., 2006). In addition to deer, collared peccary (*Pecari tajacu*) and wild pigs (*Sus scrofa*) are major prey in the southwestern United States, Mexico, and Florida. In northern Mexico, Rosas-Rosas and co-workers (2003) found that bighorn sheep exceeded mule deer and white-tailed deer in both percentage occurrence and biomass consumed. Cougars throughout North America take pronghorn if they venture into terrain with sufficient topographic and vegetative cover (Shaw 1977; Ockenfels 1994; Murphy 1998; Logan and Sweanor 2001; Anderson and Lindzey 2003; Ruth and Buotte 2007). Mountain goats (*Oreamnos americanus*) are infrequently taken (Hornocker 1970; G. Koehler, pers. comm., 2006; D. Thompson, pers. comm., 2006; Ruth and Buotte 2007), presumably due to their affinity for highly incised terrain and the risk of injury for cougars that attack them there. American bison (*Bison bison*) appear to be the only indigenous ungulate in temperate North America that is not subject to cougar predation, possibly related to the species' use of open grasslands, its large body size (>225 kg for adults), active defense of young, and positioning of calves in the interior of the herd when approached by predators (McHugh 1958; Carbyn and Trottier 1987). The cougar's apparent historic scarcity in the Great Plains region of North America may be due to these prey characteristics and behavior, and the cougar's disadvantage in open grasslands when confronted by gray wolf packs or grizzly bears (*Ursus arctos*). However, cougars may have resided in riparian corridors associated with deer in states such as Kansas and Arkansas (Young 1946b; Nowak 1976).

Cougar prey includes about eighteen species of other carnivores (Tables 9.1–9.3). Collectively, carnivores reach 23 percent occurrence in scats and 19 percent of prey biomass

(raccoons) consumed by Florida panthers (Maehr et al. 1990). Predation on other carnivores may be motivated by interspecific competition and often occurs at sites where cougars have cached other prey (Boyd and O'Gara 1985; Murphy et al. 1998; Logan and Sweanor 2001; see Chapter 11). Carnivore prey may or may not be consumed; similarly, cougars prey on and may or may not consume other cougars, and they may scavenge conspecifics that have died from causes unrelated to predation (Young 1946b; Spalding and Lesowski 1971; McBride 1976; Ackerman et al. 1984; Boyd and O'Gara 1985; Spreadbury 1989; Harrison 1990; Waid 1990; Koehler and Hornocker 1991; Beier and Barrett 1993; Ross and Jalkotzy 1996; Harveson 1997; Murphy 1998; Núñez et al. 2000; Logan and Sweanor 2001; Akenson et al. 2005; D. Thompson, pers. comm., 2006).

Native prey typically dominate cougar diets, although wild or domestic exotic prey may be more important in areas where native prey have been reduced (see Chapters 6 and 7; Novaro et al. 2000). Occurrence of cattle (*Bos taurus*), domestic sheep (*Ovis aries*), and horses in cougar scats typically totals less than 10 percent, but higher fractions are documented in South America and Arizona (Chapter 7). Livestock depredation is conspicuously infrequent in the northern United States and Canada, no doubt related to reduced presence of livestock in cougar habitats during the late fall, winter, and early spring (Ackerman et al. 1984). Predation on domestic and feral horses is uncommon (see Young 1946b), possibly related to the large size of most equines, although foals younger than six months of age are readily taken (Turner et al. 1992). Predation on domestic dogs (*C. familiaris*) and cats is also infrequent; most incidents occur near residential areas at the periphery of wildlands (Torres et al. 1996).

Seasonal, Individual, and Other Variation

Cougars readily switch between prey, apparently related to favorable changes in prey vulnerability, availability, and distribution (Leopold and Krausman 1986; see also discussion of apparent competition and prey switching in Chapters 6, 7, 10). Researchers have suggested that cougar diets shift due to seasonal changes in access to midsized or small prey in the temperate zone during the summer (Seidensticker et al. 1973; Ackerman et al. 1984; Williams et al. 1995; Murphy 1998; Nowack 1999); to birth pulses of offspring prey, including livestock calves (Shaw 1977; Hopkins 1990; Cunningham et al. 1995); to differential habitat and elevation use by ungulates (Williams et al. 1995; Nowack 1999); and to changes in the dispersion of ungulates related to seasonal moisture or breeding (Hopkins 1990; Dalrymple and Bass 1996). However, panthers showed no appreciable difference in food habits between wet and dry seasons in subtropical southwest Florida (Maehr et al. 1990), possibly related to the area's low climatic variability. Similarly, Rosas-Rosas and colleagues (2003) found no differences in diets between cold-dry and hot-wet seasons in northern Mexico.

From year to year, cougar diets vary in response to increases in abundance of snowshoe hares (*Lepus americanus*; Spalding and Lesowski 1971), to decreases in rodents (Branch et al. 1996) and deer (Leopold and Krausman 1986), or to changes in access to prey during drought (Dalrymple and Bass 1996). Cougar food habits may also vary on longer time scales (e.g., decades) as a result of changes in the presence of other carnivores. Moreno and co-workers (2006) documented a shift in cougar and ocelot (*Leopardus pardalis*) food habits in Panama, in response to a local extinction of jaguars and apparent response to a release from competition. Cougar food habits also vary spatially across study sites due to the effects of biotic, abiotic, and anthropogenic factors on the abundance of natural and exotic prey (Spalding and Lesowski 1971; Yañez et al. 1986; Taber et al. 1997; Murphy 1998). For example, percentage occurrence of white-tailed deer (*Odocoileus virginianus*) and wild pigs in cougar scats varied from 23 to 65 percent and from 1 to 59 percent, respectively, across three study sites spanning only 500 km in southern Florida (Maehr et al. 1990; Dalrymple and Bass 1996). However, researchers in Guatemala detected no appreciable differences in food habits between a protected area and an area subject to subsistence hunting by humans, despite disparity in numbers of large prey (Novack et al. 2005). Apparently, subsistence hunting did not limit the numbers of large prey available to cougars.

Cougars of different gender, or individuals, may specialize in killing particular types or species of prey. In Hornocker's (1970) Idaho study, only one cougar out of forty-six killed two bighorn sheep over four years. Ross and colleagues (1997) concluded that predation on bighorn sheep was a learned behavior characteristic of individual cougars rather than of cougar populations as a whole. A radio-marked female cougar alone killed 9 percent of the bighorn sheep population and 26 percent of the lambs over a single winter. Predation on bighorn sheep by other cougars in the vicinity was far less frequent. Similarly, Logan and Sweanor (2001) found that a single male cougar accounted for three of four bighorn sheep kills they documented in southern New Mexico. In Alberta, Ross and Jalkotzy (1996) found that moose, primarily calves, represented 69 percent of kills and 92 percent of prey biomass in male cougar diets. Females also preyed on moose but to a lesser extent—4 percent of kills and 12 percent of biomass. However, Shaw (1977) observed little individualized specialization among cougars that preyed on cattle in Arizona: six of twelve radio-marked individuals preyed on livestock, but male cougars relied more on cattle and less on mule deer and other small prey than did females.

Prey Selection

Prey populations obviously do not consist of individuals with equal survival and reproductive value, and prey selection can accentuate or lessen the effects a uniform off-take would have on prey numbers (Rosenzweig 1978; Mills 2007). For example, selection for individuals with disproportionately low reproductive value (e.g., very old or young individuals) lessens effects on long-term numbers of prey (MacArthur 1960). Selection may focus on poor-condition prey, in which case losses to the predator may be compensated by decreases in losses to other mortality sources or by increases in birth rates of prey (Mills 2007). Because not all members of a prey population are equally "killable," selection may also alter functional responses of individual predators and, consequently, total predation losses (Mills 2007).

A predator is termed "selective" when the frequency of a prey item in its diet differs from that prey's frequency in the environment (Chesson 1978). Selective predation thus refers to hunting that deviates from an opportunistic strategy (whereby prey are attacked and killed in the proportion they are encountered, and the predator hunts equitably across its home range). Presumably, prey selection naturally involves comparisons, as in a stalking cougar that evaluates the relative vulnerabilities of different individuals in a group. Thus, the primary question in selection studies is what factors cause the predator to use a particular prey type more than would be expected based on the prey's availability (i.e., to select for it) or to underuse it relative to other prey taxa or classes (i.e., avoidance).

Typically, cougars are selective hunters, although they are also known to hunt opportunistically (Emmons 1987; Novaro et al. 2000). A central tenet of predation theory is that predators select the most vulnerable individuals from a group, vulnerability itself being mediated by the physical or behavioral characteristics of prey and by abiotic factors such as snow depth (Figure 9.11; see Errington 1946; Hornocker 1970; Mech 1970; Curio 1976; Temple 1987; Mech et al. 1995; Kunkel et al. 1999; Pierce et al. 2000a; Husseman et al. 2003). Stalking predators such as cougars, however, might take prey randomly; unlike coursing predators that scrutinize prey during extended pursuit, they would seemingly have little opportunity to identify weaknesses (Schaller 1972). Thus, the dichotomy in hunting techniques predicts stronger selection in the coursing than the stalking predator. Yet, field studies have frequently detected selective predation by cougars due to factors directly and indirectly related to the vulnerability of their prey. Ultimately, hunting technique appears to be only one of many factors that interact with prey vulnerability (Mech et al. 1995; Kunkel et al. 1999; Pierce et al. 2000a; Husseman et al. 2003).

Cougar Prey Selection

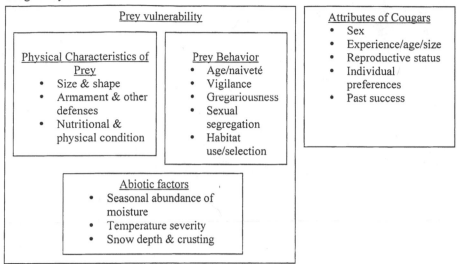

Figure 9.11 A conceptual model of cougar prey selection. Prey vulnerability and characteristics of individual cougars are overarching; within and between these categories many factors interact.

The preponderance of deer (both *Odocoileus* and *Mazama* spp.) in cougar diets suggests that they may be preferred over other prey species (Ackerman 1982). Although mule deer and white-tailed deer are clearly an important food source (Anderson 1983; Ackerman et al. 1984; Iriarte et al. 1990), they are not consistently selected over other mammals (Table 9.4). Apparently other prey attributes, such as flight capability, habitat use patterns, defenses, and naïveté, may override affinities cougars have for deer.

Cougars consistently select old-age prey, presumably because of their lower strength, vigilance, and speed as compared to prime-age prey of the same species—old age increases vulnerability (Table 9.5). How a stalking predator is able to detect weaknesses of older animals in a group is unclear. In general, traits that predispose prey to being taken by large carnivores are subtle (Mech 1996). The 83 percent success rates for stalking cougars (Hornocker 1970), and their ability to select old-age prey, attest to this felid's extraordinary ability to perceive weakness in prey (Curio 1976). Juveniles of many ungulate species are particularly vulnerable to cougars due to their naïveté and physical immaturity (Geist 1982). Among larger prey such as elk, moose, and cattle, strong selection for juveniles may also reflect a constraint on the cougar's ability to kill large prey easily (Murphy 1998).

Although they readily kill prey that are injured or in poor condition (Harrison 1990; Ross and Jalkotzy 1996; Ross et al. 1997; Logan and Sweanor 2001), cougars apparently do not consistently select this class among deer-sized and smaller prey, at least not to the extent characteristic of coursing predators (Husseman et al. 2003). Cougars apparently

kill healthy deer easily, and debilitated individuals may be rare in populations at levels well below carrying capacity (Bleich and Taylor 1998). Actively searching for weak or injured deer may be less effective than taking healthy individuals already at hand. However, weakness may enable predation on the largest prey, such as moose and elk (Ross and Jalkotzy 1996). But, perhaps most important, traditional assessments of prey condition, namely, marrow-fat content, do not necessarily detect all sources and levels of weakness that predispose individual prey to cougar predation (see section on the limitations of diet and selection studies in this chapter). That cougars consistently select for old-age prey, expected to be in poorer condition than prime-age individuals, suggests that data from past condition assessments are suspect. Thus, condition may be a more important criterion for prey selection than previously indicated. It seems reasonable that cougars would select for sick or malnourished individuals that prefer lower slopes and occur predictably in locations such as drainage bottoms due to poor health (O'Gara and Harris 1988). Miller and colleagues (2008) found that cougars were nearly four times more likely to prey on mule deer infected with prions (chronic wasting disease) than uninfected deer, even though most infected individuals showed no obvious signs of infirmity associated with the disease. Cougars apparently relied on subtle behavioral and physical cues that indicated infected individuals were more vulnerable to successful attack (Miller et al. 2008).

Numerous studies of cougar predation have addressed whether cougars select one sex of adult deer over the other; studies for elk have been fewer. Results for mule and

Table 9.4 Cougar selection for deer (*Odocoileus* sp. *or Mazama* sp.) in the presence of other large prey.

Prey Combination & Selection Pattern	Rationale	Support (+;–)
Mule deer or white-tailed deer over elk	Deer occur in smaller groups and use habitats with more stalking cover than elk.	(+) Murphy 1998 (MD bucks versus elk cows and bulls); Kunkle et al. 1999 (WTD); Anderson and Lindzey 2003 (MD, female cougars).
		(–) Hornocker 1970 (MD fawns versus elk calves); Murphy 1998 (MD versus elk calves); Nowack 1999 (MD fawns versus elk calves); Anderson and Lindzey 2003 (MD, male cougars).
Mule deer over moose	No explanation.	(–) Ross and Jalkotzy 1996 (male cougars).
Mule deer over bighorn sheep	Cougars avoid bighorn sheep because the social nature and use of dangerous terrain by bighorns thwart predation.	(+) Logan and Sweanor 2001[a].
		(–) Hornocker 1970[b]; Harrison 1990 (rams; 2 female cougars).
Cattle over mule deer	Cattle are easy to see and are noisy. Deer are cryptically colored and quiet.	(+) Shaw 1977 (calves); Cunningham et al. 1995 (calves).
White-tailed deer over collared peccary	Solitary deer present lower risk of injury to attacking predators than group-living collared peccary.	(+) Harveson 1997.
		(–) Scognamillo et al. 2003.
Mule deer over white-tailed deer	No explanation.	(+) Robinson et al. 2002; Cooley et al. 2008.

Positive (+) signs indicate support for selection of deer; negative (–) signs indicate lack of support (neutral selection or avoidance). WTD = white-tailed deer, MD = mule deer.
[a]Our visual comparison of cause-specific mortality rates (cougars) for radio-collared mule deer and desert bighorn sheep.
[b]Our likelihood ratio test preformed on counts of kills and available prey indicated no selection for mule deer over bighorn sheep.

white-tailed deer are variable—either sex may be favored. Some characteristics of adult male deer and elk that may predispose them to predation are vulnerability during and immediately after breeding periods, due to antler wounds or generally reduced body condition, and the fact that after the rut some males occur singly and lack the numerous eyes and ears that female-offspring groups have for predator detection (Geist 1981). Apparently the antlers and larger size of male deer do not dissuade cougar attacks; to the contrary, these attributes may be focal points for cougars stalking deer in groups of mixed sex (Geist 1981, 1982). A bull elk's large body size and antlers presumably present extreme risks of injury to cougars (Hornocker 1970), but selection for male elk has nevertheless been documented (Kunkel et al. 1999; Anderson and Lindzey 2003). Similar to the case for male deer, bull elk recovering from rutting activities are vulnerable to predators, and bulls stand out in mixed groups because of their large antlers, large body size, and lighter color (Geist 1982).

Sexual segregation is common among North American ungulates (Geist 1981, 1982; Main and Coblentz 1990), and this behavior may influence cougar prey selection, particularly if the segregation also includes differential use of habitats. Ungulates of both sexes attempt to optimize their reproductive fitness but use different strategies: females and their offspring select habitats that provide a combination of critical elements such as quality forage,

proximity to water, and security from predators, even at the expense of an optimal diet. In contrast, males tend to optimize body condition in preparation for breeding activities by exploiting superior, dispersed foraging opportunities using their greater mobility, but they may also expose themselves to greater predation risks than females (Main and Coblentz 1990, 1996; Bleich et al. 1997). Differences in ungulate habitat use that follow from these strategies could lead to contrasting prey selection by cougars if one sex is more vulnerable than the other. Hornocker (1970) suggested that cougars selected for male mule deer because males preferred sites at higher elevations and tended to occur alone. Similarly, Murphy (1998) hypothesized that adult female mule deer in Montana were relatively invulnerable because they often occurred in large groups with high collective vigilance and used lowland agricultural habitats that provided insufficient hunting cover and security for cougars (but see findings of Pierce et al. 2000a, discussed later in this section). Indeed, male mule deer in the Rocky Mountains and Columbian black-tailed deer (*O. h. columbianus*) in coastal western North America use higher elevations than females (Miller 1970; Robinette et al. 1977; King and Smith 1980), but this pattern may not characterize desert mule deer (*O. h. crooki*, Ordway and Krausman 1986). Also, mule deer in southern California (*O. h. fuliginatus*) may not segregate seasonally by vegetative type (Bowyer 1984).

Table 9.5 Cougar selection patterns among different classes of a prey species, and support from field studies.

Prey Type & Selection Pattern	Rationale	Support (+;−)
(+) Old-age ungulates	Old-age prey have less vigilance or less strength to escape cougar attacks than prime-age individuals.	(+) Spalding and Lesowski 1971 (MD)[a]; Ackerman, 1982 (MD); Kunkle et al. 1999 (WTD, elk)[a]; Pierce et al. 2000 (MD)[a]; Spencer et al. 2001 (BTD)[a]; Husseman et al. 2003 (MD, elk)[a]. (−) Spencer et al. 2001 (elk)[a].
(+) Native juvenile (≤12 mo.) ungulates	Compared to adults, young prey have less strength and speed than adults, and are naïve to predation. Among the largest prey such as elk and moose, juveniles better meet prey size constraints of cougars.	(+) Hornocker 1970 (MD, elk); Spreadbury 1989 (elk); Ross and Jalkotzy 1996 (MOS); Ross et al. 1997 (BS); Murphy 1998 (MD, elk); Kunkle et al. 1999 (WTD, elk)[a]; Nowak 1999 (elk; female cougars); Pierce et al. 2000 (MD); Spencer et al. 2001 (elk); Husseman et al. 2003 (elk); Mattson et al. 2007 (elk). (−) Spalding and Lesowski 1971 (MD)[a]; Shaw 1977 (MD); Hopkins 1990 (BTD); Spencer et al. 2001 (BTD); Husseman et al. 2003 (MD).
(+) Poor-condition or debilitated prey	Injured prey or those in poor condition have low vigilance or vigor relative to healthy individuals. Selection for animals in poor condition poses less risk and requires less energy for cougars.	(+) Hibben 1937 (male MD)[b]; Ross and Jalkotzy 1996 (MOS)[c]; Ross et al. 1997 (BS)[d]; Kunkle et al. 1999 (male WTD)[b]; Miller et al. 2008 (MD). (−) Hornocker 1970 (MD, elk)[e]; Spreadbury 1989 (elk)f; Nowak 1999 (MD, elk; female cougars)f; Bleich and Taylor 1998 (MD); Kunkle et al. 1999 (female WTD)[b]; Pierce et al. 2000 (MD).
(+) Male (adult) ungulates	Adult males are more solitary, more frequently use terrain favored by cougars, show less vigilance during the rut and exhibit poorer post-rut body condition than adult females.	(+) Hibben 1937 (MD)[g]; Robinette et al. 1959 (MD, winter); Hornocker 1970 (MD, elk; winter); Spalding and Lesowski 1971 (MD); Shaw 1977 (MD); Hopkins 1990 (BTD); Harrison 1990 (BS); Murphy 1998 (MD); Kunkle et al. 1999 (elk); Schaefer et al. 2000 (BS)[h]. (−) Robinette et al. 1959 (MD, summer); Spreadbury 1989 (elk); Bleich and Taylor 1998 (MD); Nowak 1999 (MD, female cougars); Pierce et al. 2000 (MD, female cougars); Logan and Sweanor 2001 (MD); Spencer et al. 2001 (BTD); Anderson and Lindzey 2003 (MD, female cougars); McKinney et al. 2006 (BS).

Positive (+) signs indicate selection for a prey class; negative (−) signs indicate lack of support (neutral or avoidance) for the pattern. MD = mule deer; WTD = white-tailed deer; BTD = black-tailed deer; BS = bighorn sheep; MOS = moose.
[a]The age structure of the prey population (i.e., prey availability) was estimated using a sample of kills by hunters or automobiles, sources that may carry biases. See Pierce et al. 2000 for consideration of age related biases associated with automobile kills.
[b]A sample of hunter kills was used to estimate the condition (physical measurements of prey or diastema lengths) of members of the prey population overall (i.e., prey availability), a source that may carry biases.
[c]Estimates of marrow-fat levels for the moose population as a whole were not available. All moose (n = 30, all calves and yearlings) killed by cougars had depleted fat reserves and >one-third had marrow-fat levels <10 percent.
[d]Estimates of condition for the bighorn sheep population as a whole were not available. More than one-third of cougar-killed bighorn sheep had apparent or possible disabilities.
[e]Our likelihood ratio test performed on counts of cougar-killed prey in poor and debilitated condition versus the same class among randomly selected prey indicated no selection.
[f]Data on condition of prey population members (i.e., prey availability) were not available.
[g]Data on the sex structure of the prey population (i.e., prey availability) were not available.
[h]Annual survival of males was significantly lower than females and cougar predation accounted for 75 percent of all deaths.

In view of the diversity of ungulate prey taken by cougars, and the range of characteristics and spatial patterning of resources that might be used differentially by ungulates of opposite sex, the differences in prey selection observed across cougar studies are not surprising. However, the common thread undoubtedly is differential vulnerability, with the caveat that sexual segregation leads members of the more vulnerable sex to use habitats that facilitate hunting success of cougars, despite the risks. Thus, integrating an understanding of behaviors such as sexual segregation and predator defenses that stem from the contrasting reproductive strategies of male and female prey may be helpful in explaining cougar prey selection.

Factors not directly related to prey vulnerability and habitat use also affect cougar prey selection, but data on them are few. Prey selection and food habits of individual cougars may reflect their hunting skills and individual preferences. Their selection may change as they mature and may ultimately align with different reproductive strategies of the two sexes. Adult males, typically weighing 1.4 times more than solitary adult females, have greater energy requirements (Ackerman et al. 1986; Lindzey 1987). Murphy (1998) documented that subadult cougars, the smallest and least experienced independent individuals in the population, took proportionately more small and midsized prey than did solitary adults. Mean weights of prey were positively correlated with weights of individual cougars. Selection patterns also varied—subadult male and female cougars favored calves over adult elk, but adult males and females selected equitably among the combined classes of elk and deer. Mattson and colleagues (2007) also found important differences among sex classes: male cougars took mainly elk and few small (defined as <30 kg) prey; females took fewer elk and far more mule deer and mesocarnivores, such as coyotes.

Pierce and co-workers (2000a) found differences among cougars in mule deer selection. Female cougars selected female deer, but adult males did not. Compared with other sex, age, and reproductive classes of cougars, females with offspring killed a disproportionately large number of young mule deer. Sexual segregation in both prey and predator apparently explained the differences. Maternal female cougars, having high protein requirements to support lactation, may have used lowland habitats that provided consistent access to female deer with vulnerable offspring. Such differences in selection among cougar social classes may occur where prey segregate into different habitats by sex, reproductive class, or even species. A proviso is that the prey be suitable in its physical and behavioral attributes: the search, pursuit, and killing must consistently occur within the energy constraints of the relevant cougar social class and must align with the reproductive strategy (see Mattson et al. 2007). Of course, terrain, vegetative cover of habitats, and how these conditions are configured on the landscape must also facilitate cougar predation and any need for security from other cougars or other predators (see Chapter 11).

Because cougar families have high collective energy requirements, mothers may use an opportunistic hunting strategy that exploits all viable predation opportunities, particularly those for small and midsized prey that can be taken incidentally during hunts for larger prey (Ackerman 1982; Ross and Jalkotzy 1996; Mattson et al. 2007). Cougar mothers may also take small prey during communal hunts to help kittens identify a range of vulnerable prey and to perfect their hunting techniques (Spreadbury 1989). Small or midsized prey may be important to newly self-sufficient cougars transitioning from dependence on their mother's hunting skills to predation on large prey by themselves (Hopkins 1990). The habit of male cougars in taking larger prey (and fewer small prey) appears consistent with a strategy that minimizes time spent on prey acquisition, processing, and feeding, thus maximizing time available for reproductive pursuits (Mattson et al. 2007).

Limitations of Diet and Prey Selection Studies

Because diet studies are often used to support conclusions regarding the effects of cougars on prey numbers, their results should accurately reflect cougar diets (Ackerman et al. 1984). However, not all items detected in scats or at kill sites result from predation because cougars also scavenge carcasses of wild and domestic animals (Young 1946b; Robinette et al. 1959; Spalding and Lesowski 1971; McBride 1976; Akerman et al. 1984; Ross and Jalkotzy 1996; Logan and Sweanor 2001; Akenson et al. 2005; Bauer et al. 2005). Cougars treat animals they scavenge similarly to prey they have killed, so scavenging can be misidentified as predation, a source of error that should be considered in estimating sources of death in prey mortality studies (Bauer et al. 2005). Similarly, cougars do not feed on all animals they kill; as noted, they may or may not eat other cougars or competing carnivores they have killed (Spreadbury 1989; Harrison 1990; Koehler and Hornocker 1991; Murphy 1998; Núñez et al. 2000; Logan and Sweanor 2001). Therefore, a sample of scats may not fully indicate the extent of predation.

Food habits data from cougar kills may underestimate the proportion of small or medium-sized prey, including ungulate newborns, which are difficult or impossible to detect because their remains are scant (Robinette et al. 1959; Shaw 1977; Anderson 1983; Ackerman et al. 1984; Spreadbury 1989; Logan and Sweanor 2001; Crawshaw and Quigley 2002). Consequently, occurrence of small prey in scats typically exceeds that estimated from kills in studies where both approaches are used (e.g., Ackerman et al. 1984; Beier and Barrett 1993; Williams et al. 1995; Scognamillo et al. 2003; but see Shaw 1977 and Núñez et al. 2000 for exceptions). This bias of underestimating kills of small prey may be particularly significant for cougar kills detected by searching telemetry location clusters, especially when there is a substantial delay (seven or more days) between data retrieval from store-on-board GPS radio collars and searches of location clusters (Anderson and Lindzey 2003). Remains of large prey such as skulls and long bones may persist as evidence longer than those of small prey such as porcupines.

Although occurrence data from scats reveal the diversity of prey in cougar diets better than a sample of kills can, scats are poor indicators of biomass consumed, that is, of the relative nutritional contribution of a prey item to the diet. Biomass consumed may be estimated from an analysis of scat samples that corrects for differences in digestibility and fecal deposition associated with consumption of small and large prey (Ackerman et al. 1984). Compared to estimates of percent occurrence, large prey typically increase and small prey decrease in the estimates of biomass contributed to cougar diets, the latter approach indicating less relative importance of small and midsized prey (Ackerman et al. 1984; Dalrymple and Bass 1996; Murphy 1998; Núñez et al. 2000; Logan and Sweanor 2001; Rosas-Rosas et al. 2003).

Data from scats are not independent of those from kills because they originate from the same source; independence is improved if only the scats collected distant from kill sites are included in analyses (Shaw 1977). But these are also less likely to be detected (Anderson 1983). Sound estimates of biomass consumed thus require addressing the matter of scat detection through modified field sampling protocols or by adjusting field data for biases.

The sampling methodology used to detect prey is also highly relevant. For example, if the data include reports by ranch workers, livestock kills may be overrepresented due to search and reporting biases (Shaw 1977; Crawshaw and Quigley 2002). Some study sites are specifically selected because some aspect of predation (e.g., suspected high levels of bighorn sheep or livestock losses) is of particular interest. In these instances, results do not necessarily represent regional trends (Harrison 1990; Cunningham et al. 1995). Similarly, contents of gastrointestinal tracts from cougars killed specifically in response to livestock depredation are predisposed to contain this prey type and often represent a biased sample (Robinette et al. 1959; Anderson 1983). Inadequate sample sizes may also constrain conclusions from studies. A. E. Anderson (1983) estimated that 90–100 scat (or gastrointestinal) samples characterized cougar food habits. Núñez and colleagues (2000) estimated 50 scats as an adequate sample size.

Biases in prey selection studies also warrant cautions. Sampling and analyzing habitat selection of animals are not straightforward (Aebischer et al. 1993); prey availability is difficult to quantify accurately; and abundance estimates must be representative of the geographic scope and duration of sampling of kills or scats (Anderson 1983; Murphy 1998). Cougar kills are often treated as independent samples, a statistical requirement, despite natural pairings of some predation events (e.g., fawns killed after and near where their mothers have been killed). This can be remedied by treating individual cougars as the sample unit (e.g., Murphy 1998), but this typically means a significant reduction in sample size. As in diet studies, the proportion of small prey items, including ungulate newborns, may be underrepresented in a sample of kills because cougar consumption of small prey is more complete. In this case, specialization on small prey may be undetected.

Selection studies often use thresholds of percent femur marrow fat as indicators of especially poor condition of prey, but this index may not reflect low vigor or other weaknesses associated with generally poor body condition, injuries, or disease that predispose predation (Mech 1970; Mech and DelGiudice 1985; Mech et al. 1995).

Finally, prey selection studies do not necessarily indicate the overall importance of a prey item, only that an item is used disproportionately more—or less—than available in the environment. For example, a common prey item such as mule deer may dominate a cougar's diet and be essential for the predator's survival and reproduction, yet not be selected.

Conclusion

A recurring theme in this chapter is that cougars are opportunistic and adaptable predators that use a foraging strategy based on evolved, specialized morphological and behavioral adaptations that enhance the species' ability to find, kill, and process prey. These activities must be accomplished efficiently, with minimal risk of injury, and with time remaining for social behaviors that more directly support reproductive success. With the exception of cougar physiology, this species' adaptations for carnivory, its food habitats, and prey selection are now reasonably well described for a diverse range of environments. Although some patterns of cougar prey selection (e.g., for young prey) are intuitive, there is still much to learn about the range of variation and causal mechanisms that explain choices of sex, age, or condition class, and different prey species.

Our understanding of cougar prey selection might advance significantly if we better considered the antipredator strategies of different sex-age classes and species of prey. Does selective predation by cougars reflect unsuccessful antipredator strategies of their prey, and are prey selection and antipredator strategies really just opposite sides of the same ecological coin? Prey often respond to predation pressure by altering their duration and timing of feeding and traveling, group sizes, vigilance, and habitat use patterns (Geist 1981, 1982; Altendorf et al. 2001; Laundré et al. 2001; Hernández and Laundré 2005). And to varying extents, these behaviors are no doubt used successfully to minimize individual risks of predation. Thus, antipredator strategies, in context with the prey's requirements for food, water, and thermal cover, may further explain cougar food habits and prey selection and help place the observed predation patterns in a broader, more meaningful ecological context.

In particular, knowledge of how prey selection interacts with habitat use by different sex-age and reproductive classes of prey, and with the prey's seasonal needs for specific kinds of forage (e.g., forbs) and for water, could be helpful. Prey vulnerability is conferred not only by physical condition, size, maturity, and armament but also by behaviors such as habitat selection and related patterns of sexual segregation. Despite the theoretical and practical difficulties, cougar-centric literature abounds with reasonable quantitative assessments of cougar prey selection, but only recently (e.g., Pierce et al. 2000a) have biologists begun to look more closely at effects related to the prey's sexual segregation. Also, because North American ungulates typically respond to predation pressure from a variety of carnivores, ungulate habitat use may collectively disfavor or favor hunting cougars, depending upon the mix of environmental conditions and predator strategies (see Chapter 11).

The importance of small prey to cougars remains a significant unanswered question. To what extent can cougars use small prey, such as porcupines (*Erethizon dorsatum*), snowshoe hares, or agoutis (*Dasyprocta* sp.), solely as an alternative to predation on large prey such as deer or capybara (*Hydrochaeris hydrochaeris*)? And if cougars can use such a strategy successfully, what is the minimum density and level of vulnerability required for these prey in order for cougars to hunt them extensively? Small prey, particularly those that are highly abundant and occur in predictable locations, may improve the survival of cougars at particular life stages, such as newly independent subadults. Small prey may also allow cougars to extend their range locally to habitats that do not support typical (deer-sized) prey due to environmental extremes, such as deep snow, or may act as temporary but critical supplemental food when the large prey decline in abundance. Examples of cougars using small prey extensively, both seasonally and annually, abound in the literature (Spalding and Lesowski 1971; Yáñez et al. 1986; Branch et al. 1996; see also Tables 9.1–9.3).

Closer study, as in Mattson and colleagues (2007), is also needed concerning the differences in food habits and feeding strategies among cougar social classes, and their interactions with environmental conditions. Each social class may have special requirements that link directly to individual survival and ultimately to reproductive success. To maximize pursuit of mates and discourage incursions of breeding competitors, resident males may take the largest prey (Murphy 1998) but minimize time feeding by gorging or consuming prey incompletely (Mattson et al. 2007). Because of their large size and dominant status in the cougar population, mature males may feel more secure at food caches than members of other social classes and when not otherwise inclined to travel.

In contrast, the ability to obtain food efficiently near natal dens and nursery areas, and with minimal risk of injury, may be critical for maternal females with newborns. Maternal females may require hunting areas inhabited by abundant "safe" prey, such as doe-fawn groups and diverse prey items that can be hunted opportunistically. Female cougars (with and without offspring) prey more extensively on young deer, small prey, and mesocarnivores than do mature males, possibly as part of a generalist feeding strategy (Pierce et al. 2000a; Mattson et al. 2007). The time spent at or in the vicinity of kills or carrion may be less constrained, and thus more variable, for cougar families than for adult males, although security for offspring may be more important. Thus, the availability of escape terrain (trees, boulder fields) may influence how completely prey are consumed and how closely food caches are attended.

The needs of subadult cougars may differ from those of adults—this group preys on small mammals and mesocarnivores more than solitary adults do (Murphy 1998) and may use these items when transitioning from reliance on their mother's support to their ability to kill dangerous, and typically less abundant, large prey. Avoiding confrontations may be critical for subadult males if they are unable to distance themselves widely from dominant males, but as with family groups, the time invested in processing and protecting food caches may be less important, given that their security needs are otherwise met. Thus, closer attention to the differences in predation and feeding patterns among cougar social classes could extend our knowledge of the adaptive significance of many cougar behaviors.

Based on the number and quality of new cougar studies completed in the last decade, the potential for applying existing and future information to benefit cougar-prey systems and for ecosystem planning has never been greater. Examples range from using knowledge of stalking cover to select translocation sites that minimize losses of prey of special concern or endangered prey such as bighorn sheep (McKinney et al. 2006) to discouraging human visitation to wildlands at locations and times when hunting cougars are typically active. Unfortunately, although good science about cougar predation can and should inform decisions that affect cougars and their prey, new information often does little to resolve conflicts and simplify decisions, a common source of frustration among biologists, policy makers, and members of the public alike (Shaw 2006a). As is the case for other predators, passionate and vocal constituencies often influence management policies for cougars, but the values and goals of those constituencies often remain unchanged despite new science (see Chapter 14; Shaw 2006a). Nonetheless, new information can ultimately serve to clarify and redefine the issues and can act as a strong catalyst for change, particularly when the correct course of action is indicated by replicated studies that address similar questions and use a range of approaches. Indeed, the cougar already offers an excellent example of how reliable knowledge can bring positive changes in the conservation of a carnivore often embroiled in contention. The elevation of the cougar's status from an unprotected mammal (varmint) to a game animal in nearly all western states during the late 1960s and 1970s was strongly motivated by the work of Hornocker (1969, 1970), Shaw (1977), and colleagues

(Seidensticker et al. 1973) in Idaho and Arizona (also see Hornocker 1971, 1996). Gray wolf and grizzly bear science and recovery work provide other good examples.

Our driving interest in cougar predation has long been that they eat other large animals, with the central question being which and how many. Our interest in whatever else they eat was largely incidental. Today our interests are broader and our questions more refined. We are also better poised for applying new science to help identify management alternatives and to monitor the results of decisions that may need frequent course corrections (Shaw 2006a). Thus, although predator-prey science does not carry promises for constituents of cougars or their prey, it offers a fascinating and evolving story for those willing to listen objectively.

Chapter 10 Cougar-Prey Relationships

Toni K. Ruth and Kerry Murphy

How do cougars (*Puma concolor*) affect their prey? Does cougar predation reduce prey numbers, or can it at times simply compensate for other sources of mortality? When does predation regulate a prey population? When does predation limit its growth? Advocates for cougars see the species as an integral component of healthy ecosystems. Some people acknowledge an intrinsic aesthetic value of the species and accept the prey requirements of cougars as prerequisites for their survival and reproduction. Big game hunters and ranchers in western North America may view cougars as competitors (Nowak 1976; Shaw 1989) and are equally interested in answers to questions about cougar predation. These varied public perceptions of cougars frequently influence the regulatory bodies that set hunting seasons and livestock depredation policies.

Knowledge of cougar-prey relationships in context with environmental and human factors is essential for formulating valid management strategies. Cougars may have direct influences on the abundance and growth of prey populations, aspects that historically have been the focus of predator-prey theory and of strong interest to management agencies. While understanding the numerical interactions between cougars and their prey is foundational to cougar-prey relationships, management that focuses on cougar removal is highly controversial. Under some conditions, reducing cougars increases prey numbers (Ballard et al. 2001). For endangered prey or other species with low population viability, cougar predation may be of particular management concern (Turner et al. 1992; Wehausen 1996; Sweitzer et al.

1997; Hayes et al. 2000; Schaefer et al. 2000; Logan and Sweanor 2001; Kamler et al. 2002). At the other end of the spectrum is the argument that predation as a baseline of mortality with positive selective pressures is too often overlooked and that fluctuations in prey numbers should be seen as part of natural ecosystem functioning (see Mills 2005).

Many morphological and behavioral attributes of prey are responses to natural selection created by predation pressure (Taylor 1984; Mills 2005). Because predation is complex, with effects that may be counterintuitive, management decisions made without considering these ecological processes can have unintended consequences (Goodrich and Buskirk 1995; Katnik 2002; Courchamp et al. 2003; Mills 2005). Removal of dominant male cougars may relax aggression toward subadults, relax territorial exclusion of subadults, and result in temporary local increases in overall cougar population density, as is documented for black bears (*Ursus americanus*; Sargeant and Ruff 2001). Predator control may also trigger increases in numbers of other predators by reducing competition (Soulé et al. 1988). Predator reduction intended to increase one prey species may not be effective unless an abundant alternative prey that also supports the predator is reduced (Gibson 2006).

Finally, predation is of interest because carnivores may play keystone roles in biotic communities. Cougar predation, therefore, may have a disproportionately strong impact on herbivores, altering prey behavior and habitat use, changing foraging patterns (Hornocker 1970; Altendorf et al. 2001), and having indirect effects on lower trophic levels in the process termed *trophic cascades* (Estes 1996; Ripple and Beschta 2006). Thus cougar predation may be an important component of top-down versus bottom-up processes that are critical to conserving biodiversity (Ray et al. 2005).

In this chapter, we discuss field studies that have addressed cougar predation on native and domestic prey. As a backdrop, we provide a review of central concepts of predator-prey theory used to understand the numerical effects of predation. The concepts of *limitation* and *regulation* are important tools for understanding predation processes.

Cougar-prey interactions are greatly affected by environmental and human variables. Which factors interact, and how, to influence the size and stability of both predator and prey populations is the basic question of population biology (Rickleffs 1990). Important variables in cougar-prey interactions include the size of the prey population relative to its food supply; how the number of cougars responds to changes in prey density (*numerical response*); and how many prey each cougar kills (*functional response*) as prey density changes (Caughley and Sinclair 1994; Mech and Peterson 2003; Mills 2007). Additional factors include presence of multiple predators, humans, and multiple prey species, which in turn influence density of cougars, density of their primary prey, and the predation rate of cougars through processes such as *apparent competition* (see discussion below) and *interspecific competition* (see Chapter 11). Environmental influences such as drought, winter severity, and diseases also affect cougar impacts on prey.

Predator-Prey Theory

The number of animals in a prey population and the population's rate of growth are a function of deaths, births, emigration and immigration, and factors that affect these (Ballard et al. 2001). Equilibrium, k, is reached when births equal deaths, assuming that immigration and emigration are negligible (different from K, ecological carrying capacity, discussed later). Does cougar predation contribute to substantial losses within the prey population so that deaths greatly exceed births, or does cougar predation merely substitute for other sources of mortality? Perhaps other sources of mortality so greatly exceed cougar predation that cougars have only a minor influence on the size of prey populations.

Limiting versus Regulating Factors

A *limiting factor* is any factor that causes mortality, affects birth rate, or affects the rate of population growth (Figure 10.1). The combination of factors that determines the upper limit a population can attain is defined as *limitation* (Caughley and Sinclair 1994). Limiting factors such as disease can respond to changes in prey density (that is, act in a *density-dependent* manner) or, like an avalanche or a drought, can affect birth and death rates randomly and independent of population size (*density-independent*). Any limiting factor, including predation, may have weak or strong effects and may act alone or in combination with other factors to set an equilibrium point for the prey population.

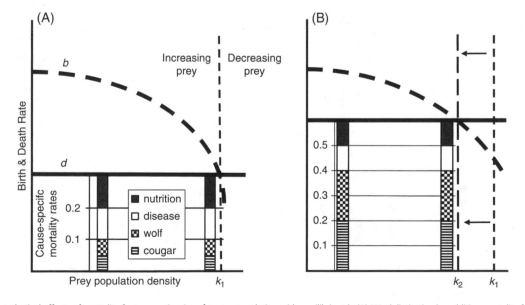

Figure 10.1 Hypothetical effects of mortality factors on the size of a prey population with equilibrium k: (A) *Weak limitation* by additive mortality factors that are density independent (i.e., constant with increasing prey density); density-dependent birth rates (*b*) regulate prey numbers. Bars indicate cause-specific mortality rates stacked to show the total death rate (*d*). (B) *Strong limiting effects* on prey due to uniformly increased mortality from cougars and gray wolves. Adapted from Caughley and Sinclair (1994).

A *regulating factor* is one that acts to keep a population increasing or declining toward its approximate equilibrium. A temporary change in a limiting factor, such as a severe winter, drought, or influx of a predator, can shift a prey population downward from its equilibrium (Caughley and Sinclair 1994). Conversely, improved forage due to precipitation or a reduction in predators can shift a prey population upward. In nature, after disturbance, the population naturally tends to return toward k, by increasing or decreasing in numbers, through the action of density-dependent factors, which are said to be regulating population size. The regulating factor is density-dependent because its influence changes with prey population size. If immigration and emigration are approximately equal, the prey population declines toward equilibrium when deaths exceed births and increases toward equilibrium when births exceed deaths.

How do we know when cougar predation strongly limits versus when it regulates a prey population? Limiting factors determine where the equilibrium density of prey occurs because they affect where births equal deaths (note the shift of k_1 to k_2 in Figure 10.1). Cougar predation is strongly limiting when it becomes a strong primary source of additive mortality (see discussion below). For cougar predation to regulate a prey population it must be density dependent; that is, cougar predation must increase as prey density increases and decrease as prey density decreases (Figure 10.2). Although a primary factor in the death of many ungulate species, cougar predation has not been demonstrated to regulate prey populations (Table 10.1). Most of the literature on cougar predation addresses limitation, not regulation, because density-dependent responses in cougar-prey systems have not been widely studied, although Bleich and Taylor (1998) discuss the possibility that cougar predation regulates migratory prey inhabiting arid and unpredictable environments.

Carrying Capacity of the Prey Population

Ecological carrying capacity, K, is the natural limit of a population set primarily by environmental resources such as food, space, or cover (Caughley and Sinclair 1994, 117; Ballard et al. 2001; Cougar Management Guidelines Working Group 2005). If cougar predation is suppressing a deer (*Odocoileus* spp.) population at low densities (i.e., deer population well below K), then experimental removal of cougars through translocations or controlled hunting should allow the prey population to increase at a greater rate than prior to cougar removal (Ballard et al. 2001, 102). Conversely, removal of cougars should do little to increase a prey population that is near K because predation mortality will be replaced (i.e., compensated; see later discussion) by other mortality factors, or prey overpopulation may harm the habitat, resulting in a population crash.

Numerical and Functional Response of Cougars

How cougar numbers respond to prey numbers will influence their effect on prey. As prey increase, more cougars potentially survive and reproduce (a *numerical response*), and collectively they of course eat more prey (Caughley and Sinclair 1994). As prey density declines, cougar numbers are likely to decline because less food is available. In general, as prey increase, predator numbers increase to an asymptote or leveling off, the level of which is determined by territorial behaviors that cause resident numbers to stabilize, through intraspecific killing and dispersal (Caughley and Sinclair 1994). In multicarnivore systems, changes in other carnivores may affect cougar density through interspecific killing, competition for prey, or both.

As prey numbers increase or decrease, the kill rate of each individual cougar could increase, decrease, or not change at all. The kill rate, or *functional response*, is the number of prey killed per cougar per unit of time. The rate is limited by search time required to locate prey; handling time to pursue, kill, and eat prey; and, generally, satiation (Messier 1995; Mills 2007). Holling's pioneering work (1959, 1965) described three basic types of predator functional responses to increasing prey density, which he termed type 1, type 2, and type 3

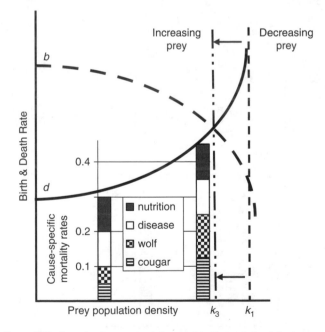

Figure 10.2 Example of a prey population regulated by predation following from Figure 10.1, where predation was a limiting factor. Here, predation from cougars and wolves is low at low prey densities (cause-specific mortality bars on left), but predation increases (bars on right) as prey density increases, a density-dependent response. The prey population declines to the equilibrium k_3 as deaths exceed births. In this example, cougar and wolf predation regulate prey numbers.

Table 10.1 Summary of studies where the limiting and compensatory effects of cougar predation on prey were addressed[a], 1970–2006.

Primary Prey Species	Study	State/Province[b]	Study Length (years)	Season of Estimation[c]	Sympatric Prey	Other Predators	Cougar Reduction/Hunting[d]	Strong Limitation (Yes/No)	Prey Relation to K	Compensation	Influences of Alternative Prey[e]
Mule deer	Bleich and Taylor (1998)	CA-NV	9	A	?[f]	Coyotes	No/No	?[g]	?[g]	Additive	?
	Lindzey et al. (1994)	UT	9	W	Elk, small mammals	Coyotes	No/No	No	Below	?	?
	Hornocker (1970)	ID	3	W	Elk, bighorn sheep, moose	Black bears, coyotes	No/Low	No	At to slightly exceeded	Partly compensatory	?
	Logan & Sweanor (2001)	NM	10	A	Bighorn sheep, javelina	Coyotes	Yes/No	No	Below, then exceeded[h]	Partly compensatory to mostly compensatory[h]	None
	Murphy (1998)	MT-WY	7	A	Elk, bighorn sheep, moose	Grizzly bears, black bears, coyotes	No/Low	No	?	Partly compensatory	?
	Shaw (1977)	AZ	4	A	Pronghorn, peccary, cattle	Coyotes	No/Yes[i]	No	?	?	Possible AC
	Ackerman (1982)	UT	2.5	A	Elk, livestock	Coyotes, bobcats	No/Yes	?[j]	Below	?	?
	Robinson et al. (2002)	BC	4	A	White-tail deer, moose, elk, caribou, bighorn sheep	Black bears, wolves, coyotes	No/Yes	Yes	Below	?	AC
	Neal et al. (1987) Neal (1990)	CA	7	A	Cattle	Black bears, coyotes, bobcats	No/No	Yes	?	?	?
	Hurley et al., unpublished	ID	9	A	Small mammals	Coyotes, bobcats	Yes/Yes	Yes	Below	Additive[k]	None
	Laundre et al. (2006, 2007)	ID-UT	15	A	Small mammals	Coyotes, bobcats	No/Yes	No	?	?	?
White-tailed deer	Kunkel et al. (1999)	MT	4	W	Elk, mule deer, moose	Wolves, bears, coyotes	No/Yes	Yes	?	Additive[l]	?
	Robinson et al. (2002)	BC	4	A	Mule deer, moose, elk, caribou, bighorn sheep	Black bears, wolves, coyotes	No/Yes	No	Below	?	None

(Continued)

Table 10.1 (Continued)

Primary Prey Species	Study	State/Province[b]	Study Length (years)	Season of Estimation[c]	Sympatric Prey	Other Predators	Cougar Reduction/Hunting[d]	Strong Limitation (Yes/No)	Prey Relation to K	Compensation	Influences of Alternative Prey[e]
Black-tailed deer	McNay and Voller (1995)	BC	9	A	?	Wolves	No/Yes	Yes	Below	Additive	?
Elk	Murphy (1998)	MT-WY	7	A	Mule deer, bighorn sheep, moose	Grizzly bears, Black bears, coyotes	No/Low	No	Fluctuated during study	Partly compensatory	?
	Hornocker (1970)	ID	3	W	Mule deer, bighorn sheep, moose	Black bears, coyotes	No/Low	No	At to exceeded	Partly compensatory	?
	Husseman (2002)	ID	3	W	Mule deer, bighorn sheep, moose	Wolves, black bears, coyotes	No/Yes	?	Below	Partly compensatory	?
	Kortello (2005)	AB	4	W	White-tail deer, mule deer, bighorn sheep	Wolves, grizzly bears, black bears	No/No	No	?	?	?
Wild horses	Turner et al. (1992)	CA-NV	5.2	S[m]	Mule deer	Coyotes, eagles	No/?	Yes	Below	?	PS
Pronghorn	Ockenfels (1994)	AZ	4	A	Mule deer	Coyotes, bobcats	?/?	Yes	?	?	?
Moose	Ross and Jalkotzy (1996)	AB	5	W	Elk, mule deer, white-tail deer, big-horn sheep	Wolves, grizzly bear, black bear	No/Yes	?	?	Mostly compensatory	?
Mountain caribou	Wittmer et al. (2005a); Wittmer et al. (2005b)	BC	20	A	Mountain goats, moose, elk, white-tail deer, mule deer	Grizzly and black bears, wolves, cougars, wolverine	No/Yes	Yes[n]	?	?	AC
Bighorn sheep	Logan & Sweanor (2001)	NM	10	A	Mule deer	None	Yes/No	Yes	Below	Partly compensatory	PS
	Ross et al. (1997)	AB	9	W	Elk, mule deer, white-tail deer, moose	Coyotes	No/Yes	No	?	Mostly compensatory	?
	McKinney et al. (2006a)	AZ	8	A	Mule deer, white-tail deer, peccary, cattle	Bobcats, coyotes, black bear	Yes/Yes	Yes	Below to at	?	PS

Study[a]	State[b]	Cougar reduction	Season[c]	Prey	Other predators				Compensatory/ additive	[e]
Kamler et al. (2002)	AZ	18.5	A	Mule deer	Coyotes, bobcats	?/Yes	Yes	?	Compensatory in SE region, additive in others	PS
Creeden and Graham (1997)	CO	0.7	S, F	Mule deer	?	?/?	Yes	Below	Additive	?
Wehausen (1996)	CA	7	W	Mule deer	?	?/?	Yes	?	?	PS
Hayes et al. (2000)	CA	5.5	A	Mule deer	?	?/?	Yes	?	?	PS
Rominger et al. (2004)[o]	NM	8	A	Elk, mule deer, cattle, pronghorn	Coyotes, bobcats, eagles	No/?	Yes	Below	?	PS
						No/?	No	At		None
Novaro and Walker (2005)	Patagonia	~20	A	European red deer, mountain vizcachas, livestock	Culpeo foxes	No/Yes	Yes	Varied across study sites	?	AC & PS
Sweitzer et al. (1997)	NV	5	W	Mule deer	None	?/?	Yes	?	Partly compensatory[p]	PS

[a]Some aspects of the following studies were inferred by authors of this chapter: Shaw (1977), McNay and Voller (1995), Sweitzer et al. (1997), Bleich and Taylor (1998), Hayes et al. (2000), and McKinney et al. (2006a).

[b]AB = Alberta, AZ = Arizona, BC = British Columbia, CA = California, CO = Colorado, ID = Idaho, MT = Montana, NM = New Mexico, NV = Nevada, UT = Utah, WY = Wyoming, Patagonia = Patagonia, South America.

[c]Season of estimation for evaluating influences on prey. A = annual, W = winter, S = summer, F = fall.

[d]Cougar reduction (Yes or No) indicates whether a study involved experimental reduction of cougar numbers. Hunting indicates whether cougars were removed through an annual state cougar hunting season (Yes or No) and, where possible, whether there was low hunting pressure.

[e]AC = apparent competition, PS = prey switching.

[f]Question marks (?) indicate a publication did not provide any information pertaining to certain aspects that are included in the summary table.

[g]Bleich and Taylor (1998) suggested that cougar predation might regulate migratory prey inhabiting arid and unpredictable environments. The study areas had experienced repeated annual droughts since 1986, but no malnutrition of deer was documented during the study. The authors acknowledged that habitat quality was not adequately known to assess the roles of nutrition and climate as factors influencing the dynamics of the deer populations.

[h]During a non-drought period (1987–88 to 1991–92), the deer population was increasing, but cougar predation seemed to be slowing the rate of increase although predation was probably partly compensatory (K. Logan, pers. comm., 2007). During the drought years (1992–93 to 1994–95), the mule deer population probably would have declined even without cougars as a primary predator, thus predation was mostly compensatory.

[i]The study was conducted between 1971 and 1975 on private ranches that were closed to cougar hunting in the fall of 1971. However, Table 1, page 24 in Shaw (1980) indicates that two to three study cougars were killed by hunters outside the study area and one cougar was illegally killed by a deer hunter on the study area.

[j]It was not clear whether cougars strongly limited the population. A population model suggested a 6 percent annual increase in the deer population, but pellet and composition counts indicated the deer population was stable in numbers over the study. Because the deer population was at a low level, the authors stated that even the limited impact of cougars may have slowed the growth of the deer population. Cougars killed 5–10 percent of the deer population annually, as estimated by the chapter authors.

[k]Cougar predation was additive during the study, but the authors felt predation was possibly compensatory over the long term.

[l]Predation was additive due to multiple predators including cougars, wolves, and bears.

[m]Study was conducted primarily in May and June, 2 weeks in September, and 1–2 weeks in January.

[n]Strong limitation was due to multiple predators, including cougars, wolves, and bears.

[o]Data reported per row represent results from two study areas, Sierra Ladron (upper row data) and Wheeler Peek (lower row data), where bighorn sheep translocation was studied.

[p]Cougar predation may have been partly compensatory when starvation of porcupines was documented in high snow years of 1988–89 and 1992–93.

responses (Figure 10.3). The type 1 response describes a relationship where the number of prey killed by the predator increases linearly as prey density increases. This relationship is unrealistic because at high prey densities, cougar handling time (search, pursue, kill, and consume) ultimately limits the predation rate and causes it to reach an asymptote, flattening at some maximum (Caughley and Sinclair 1994; Messier 1995; Mills 2007). The type 2 response reflects the more realistic situation where handling time takes up an increasing proportion of the cougar's time as prey density increases. The type 3 response also captures the upper limit to kill rate at high prey densities, but predation rate increases proportionately faster at low to intermediate prey densities (Caughley and Sinclair 1994; Mills 2007). Such a response occurs when predators develop an improved search image and handling time for prey, which might happen when a predator switches to prey item A as it becomes more numerous than prey item B (see Figure 10.5; Caughley and Sinclair 1994).

Functional responses, like numerical responses, do not necessarily occur across the full range of densities characteristic of the prey population (see Bartel et al. 2005). In general, kill rates for obligate carnivores appear fairly consistent over a wide range of prey densities (Ballard et al. 2001; Mech and Peterson 2003), but the data for cougars are too limited for such an assessment.

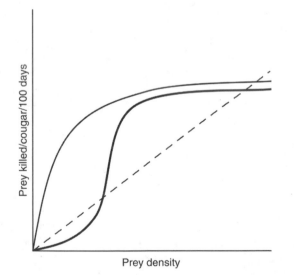

Figure 10.3 Theoretical functional responses (Holling 1959, 1965) of cougar predation rate relative to prey density. Due to constraints on handling time and because the cougar becomes satiated with feeding, predation rates should level off at some maximum. Thus, the type 1 functional response (dotted line) is probably biologically unrealistic. The type 2 functional response (thin solid line) is more realistic—the single cougar eats more as prey density increases, but the predation rate eventually reaches an upper limit (asymptote). In the type 3 functional response (thick S-shaped line), a single cougar takes few prey at low density but then predation rapidly increases before reaching an asymptote.

Total Predation Rate

Although vitally important for assessing the effects of cougars on prey species, an understanding of total predation responses of cougars at various prey densities (i.e., density-dependent response) is lacking, thus the difficulty in assessing the extent to which cougars regulate their prey. *Total predation rate, or total response,* is the product of the number of cougars present (*numerical response*) and the individual cougar kill rate (*functional response*). This results in an estimate of the number of prey killed by all cougars in an area for a given time. Total predation rates have been variously estimated by modeling cougar energetics (Ackerman et al. 1986; Laundré 2005), by calculating *cause-specific rates of mortality* among prey (Logan and Sweanor 2001; Robinson et al. 2002), or through field estimation of individual predation rates coupled with estimates of cougar population size (see Table 10.3).

Compensatory versus Additive Mortality

Even if the predation rate is high, predation that simply substitutes or compensates for other sources of mortality will not increase total prey losses (see Figure 10.4; Mills 2007). The theory of *compensatory mortality* (Errington 1946) is based on the concept that as prey population density approaches or exceeds carrying capacity (*K*), individuals increasingly compete for resources, such as forage; nutritional condition declines (McCullough 1979; Cougar Management Guidelines Working Group 2005); and birth rates fall (Ballard et al. 2001). When additional animals are produced, they tend to be removed from the population by some form of mortality, whether by starvation, hunting, disease, or predation (Sawyer and Lindzey 2002). Such mortality from any source helps lower prey population density and increases the likelihood that remaining prey survive (Cougar Management Guidelines Working Group 2005).

Conversely, when prey are well below their carrying capacity, they tend to be in good condition; individuals removed by cougar predation probably would have survived, thus predation is *additive* to other sources of mortality and increases total mortality. Cougar predation that is additive would be expected to have more of a limiting effect on prey numbers than if all of the predation mortality were compensatory.

All age classes of prey are not equally vulnerable to predation, in their compensatory or additive role, or in their importance to population growth (MacArthur 1960; Mills 2007). Predation focused on age classes with high reproductive value, such as prime age females, will especially lower growth of the prey population because offspring of these females are lost as well. Generally, cougar predation on old-age or newborn prey thus has less immediate impact on prey numbers (Murphy 1998).

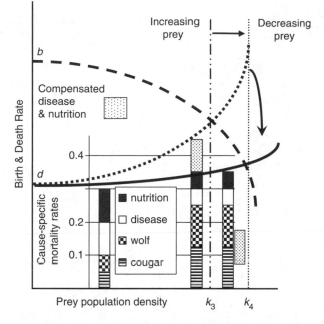

Figure 10.4 Example of how compensatory mortality changes the equilibrium point from k_3 to k_4. The figure follows from Figure 10.2, in which cougar and wolf predation were density dependent. The equilibrium k_3 occurs when mortality from all causes is additive. A new equilibrium k_4 occurs when some of the deaths due to predation were also those that would have occurred due to disease or poor nutrition. Consequently, in k_4 predation simply replaced the losses that would have occurred from disease and nutrition (dotted bar on top of middle stacked bars). In the stacked mortality bars at far right, compensated disease and nutrition (dotted bar) are now shown at the side of cougar and wolf predation to indicate that cougar and wolf kills were of diseased and poor-nutrition prey that would have died anyway. Because cougar and wolf predation was compensatory, actual total prey deaths (dark line representing d) were lower than originally assumed under additive mortality (middle-stacked mortality bars and dashed line representing d).

Apparent Competition and Prey Switching

In multiple-prey systems, changes in the density of one prey population can lead to the decline of a less abundant prey through a phenomenon known as *apparent competition* or through *prey switching* by the predator. Defining when apparent competition or prey switching occur and how long they are sustained is complicated because of time lags and because factors including prey densities, prey migration, and cougar numbers are unique to each system.

In apparent competition, cougar numbers rise (*numerical response*) because a primary prey such as white-tailed deer (*O. virginianus*) becomes very abundant, causing increased predation on and a decline in a secondary, generally less abundant prey species such as mule deer (*O. hemionus*). This increased effect of a predator on secondary prey can create the illusion of competition between prey, hence the term *apparent competition*, when in actuality predation is the main factor reducing the secondary prey (Holt 1977; Mills 2007). The proportional effect of mortality on the

secondary prey population is inversely density dependent (or *depensatory*); that is, mortality increases proportionately as numbers decline (Sinclair et al. 1998; Figure 10.5A). Apparent competition by cougars has not been well documented and clearly needs more assessment, but it has been reported (see discussion of elk and deer, mountain caribou, below).

In multiple-prey systems, apparent competition may be more common than previously recognized because it could occur any time cougar numbers track an abundant primary prey that is highly suitable to cougar predation. But cautions apply: only a few individual cougars may specialize on the secondary prey; where cougars follow migratory prey, apparent competition might not be sustained; and it is not clear when primary prey subsidize cougars or alternatively buffer the predation effects on secondary prey (see Stoddart et al. 2001).

Negative effects on alternative prey can also occur when a primary prey species *declines* to low numbers, causing cougars to increase their consumption of less abundant prey (Hamlin et al. 1984; Sinclair et al. 1998). Such *prey switching* by cougars, which increases predation on the less abundant, secondary prey (Figure 10.5B), has been reported for bighorn sheep (*Ovis canadensis*; see section on bighorn sheep below), feral horses (*Equus caballus*) in Nevada (Turner et al. 1992), and porcupine (*Erethizon dorsatum*) in northwest Nevada (Sweitzer et al. 1997). In these studies, mule deer populations declined during the same time frame that cougars switched to alternate prey. This effect should be temporary (<10 years) because cougar populations should decline in response to mule deer reductions (Logan and

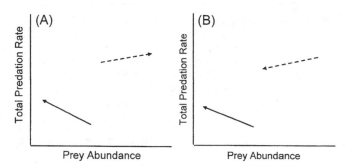

Figure 10.5 Hypothetical influence of abundant main prey (dashed line, e.g., white-tailed deer) on a less abundant secondary prey (solid line, e.g., mule deer) through apparent competition (A) or prey switching (B). In (A), what appears to be a prey decline (solid line) caused by competition between two prey is actually due to the numerical response of a shared predator to the presence of an increasingly abundant main prey. The mortality of the secondary prey increases proportionately as prey numbers decline because predator numbers are sustained, or increase, due to abundant main prey. In (B), a decline in abundant main prey results in a predator switching to a less abundant secondary prey. Due to the predator's increasing reliance on the secondary prey, mortality of the prey increases proportionately as its population declines, similarly to (A). Arrows in (A) and (B) indicate trend over time. See the text for the ecological contexts of these interactions.

Sweanor 2001; Kamler et al. 2002; Laundré et al. 2007). Seasonal migrations of primary prey outside an area of overlap with the secondary prey may define times when cougars switch to eating more secondary prey. Besides a decline in primary prey, prey switching may occur when some cougars spatially avoid areas where dominant competitors (e.g., wolves) are becoming reestablished and cause cougars to increase their overlap with vulnerable secondary prey.

Predation Rate Studies

Predation rates of individual cougars can be estimated by closely monitoring radio-collared individuals and identifying prey remains at clusters of radio location sites where cougars feed and rest, often over several days (Shaw 1977; Murphy 1998; Anderson and Lindzey 2003; Mattson et al. 2007). The time interval between kills, usually calculated as the days between when one kill is made and the day before the next kill is made, is an estimate of the cat's *individual predation rate*. Therefore, predation rate includes the days the cougar spent consuming the kill and the days the cougar hunted prior to the next kill (Figure 10.6). Sometimes a *consumption rate* that reflects the number of days an individual cougar spent consuming the kill before beginning to hunt again is calculated. Few field studies have actually measured *true daily consumption* by reweighing the prey carcass as it is consumed (but see Murphy 1998).

Field techniques, however, have limitations. Not all kills produce clusters of locations; cougars may bed elsewhere, visiting kills infrequently, or may quickly consume small prey and leave the site. Elk (*Cervus elaphus*) carcasses localize cougars for longer than do those of smaller prey. When kills are not found, their absence from the data biases kill rate estimates downward, showing fewer kills than occur; on the other hand, carcasses scavenged by cougars may be erroneously recorded as predation (Bauer et al. 2005), biasing rate estimates upward (Anderson and Lindzey 2003).

Individual rates can also be estimated from data on average daily food requirements of captive cougars, data on edible biomass of prey, and estimates of the total energy required for basal metabolism, activity, growth, and reproduction (Robinette et al. 1959; Hornocker 1970; Ackerman et al. 1986; Laundré 2005), though these require estimates and assumptions for such factors as traveling time and meat spoilage.

Individual predation rates estimated for cougars in single and multiple large prey systems in North America (Table 10.2) vary widely, ranging from as much as 74 days between elk kills (based on average daily requirements of captive cougars) to as few as 2.9 days per kill for bighorn sheep (based on field studies). Differences in methodology and sampling

Figure 10.6 Cougars may spend three to five days feeding on ungulate prey and three to five days hunting prior to killing again, resulting in a predation rate in the range of six to ten days per kill. Heavy cover and caching the prey between feedings helps slow the rate of decomposition and reduces likelihood of scavengers detecting the kill. The availability of cover to conceal prey from scavengers likely influences the kill rate, as do several other factors: temperatures that increase or retard spoilage; the size of the kill; and the age, sex, and reproductive class of the cougar. Photo © Ken Logan.

intensity, rather than biological factors, probably account for this large disparity. Among telemetry-based or snowtracking studies, individual rates of large mammal predation for all cougar social classes combined are typically in the range of 6 to 10 days per kill (nine field studies on predation on mule deer and elk, Table 10.2), supporting the common belief (e.g., Young 1946b) that cougars kill about one deer a week. Frequent monitoring of cougar activity, obtaining locations at intervals of less than four hours (Beier et al. 1995; Anderson and Lindzey 2003; Mattson et al. 2007), undoubtedly improves accuracy of individual predation rate estimates over those in studies based on a single location per day (Shaw 1977; Murphy 1998), although this bias is not apparent in two studies that located cougars infrequently (Nowak 1999; Cooley et al. 2008).

The energetics model of Ackerman and colleagues (1986) produced kill rate estimates for cougar family groups similar to field determinations, but those for solitary adults were lower compared to field studies. Recently, Laundré's (2005) recalculation of the energetics model resulted in predation rate estimates that were lower than those calculated by Ackerman and colleagues (1986) and lower than field-based

Table 10.2 Individual predation rates (see text) estimated for North American cougars.[a]

Location	Principal Prey	Season	Social Class	Predation Rate (days/mammal/cougar)	Method	Reference
Arizona	Mule deer, cattle	A	AF	10.4	Telemetry	Shaw 1977
		A	FG	5.8		
	Mule deer, elk	A	All[b,c]	7.6	GPS telemetry	Matson et al. 2007
			AM[c]	7.4		
			AF[c]	9.2		
			SM[c]	8.0		
			SF[c]	6.0		
California	Mule deer	A	AD	7.6	Telemetry	Beier et al. 1995
	Black-tailed deer, wild pig	A	FG[d]	4.8*		Hopkins 1990
	Small mammals	A	All	6.3[e]		Beier et al. 1995
Idaho	Mule deer	W	FG[f]	4.5*	Snowtracking	Horrocker 1970
		A	All	18–27	Captives	
	Elk	A	AD	10–14	Subjective	
		A	All	49–74	Captives	
Idaho-Utah	Mule deer	SM	AM	18.8	Energetics model	Laundré 2005
			AF	24.5		
			FG	9.2		
Oregon	Mule deer; elk	A	AF, FG	7.7	Telemetry	Nowak 1999
		A	AF	6.9		
		A	FG	8.0		
		SM–F	AF, FG	5.6		
		W–SP	AF, FG	9.8		
Utah	Mule deer	W	All	9.7	Snowtracking	Connolly 1949
	Mule deer	W	All	4–10	Captives	Robinette et al. 1959
	Mule deer	A	AM	8–11	Energetics model	Ackerman et al. 1986
		A	AF	14–17		
		A	FG[g]	3.3		
		A	FG[h]	4.5*	Telemetry	
	Porcupines	W	All	7.2	Snowtracking	Connolly 1949

(Continued)

Table 10.2 (Continued)

Location	Principal Prey	Season	Social Class	Predation Rate (days/large mammal/cougar)	Method	Reference
Washington	Mule, white-tailed deer	A	All	6.7	Telemetry	Cooley et al. 2008
	Mule deer	S	All	6.6		
		W	All	7.0		
	White-tailed deer	A	All	6.1		
		A	All	7.0		
Wyoming	Mule deer; elk	A	All	9.4	Telemetry	Murphy 1998
		A	AM	7.5		
		A	AF	11.1		
		A	FG	7.2		
		A	SM	11.0		
		A	SF	10.3		
	Mule deer, elk	A	All	7.0	GPS telemetry	Anderson and Lindzey 2003
		A	AM	7.8		
		A	AF	7.0		
		A	FG	5.4		
		A	SM	9.5		
		A	SF	7.3		
Alberta	Moose, various	W	AM[i]	22.7	Telemetry	Ross and Jalkotzy 1996
		W	AF[i]	38		
British Columbia	Mule deer; bighorn sheep	A	FG	2.9–5.9[j]*	Telemetry	Harrison 1990
	Mule deer; elk	W	FG	8.7–10.0	Telemetry	Spreadbury 1989

[a] Season: A = annual, F = fall, W = winter, SM = summer, SP = spring. Social Class: AD = adult (unspecified sex), AM = adult male, AF = adult female, FG = family groups, SM = subadult males, SF = subadult females, All = apparently all sex and ages of independent cougars. * = few (<3) cougars sampled.
[b] Calculated by chapter authors as the mean across average values for the social class.
[c] Rates include large mammal kills believed present, but not searched for, based on a probabilistic model.
[d] Two 5-month-old kittens.
[e] Includes opossum, cottontail rabbit, bobcat, coyote, striped skunk, and raccoon.
[f] Three 32-kg kittens.
[g] Three 13-month-old kittens.
[h] Three 6- to 8-month-old kittens.
[i] Eight females that did not kill moose were excluded; cougars of both sexes killed other large prey (excluded from calculations) during monitoring periods.
[j] Values estimated by chapter authors as total days monitored divided by the number of kills.

estimates. Differences in the models resulted from differences in techniques to determine cougar activity and associated energetic costs. Laundré (2005) monitored cougar movements every half hour, while Ackerman and colleagues (1986) inferred activity levels from variations in strength of radiotelemetry signals. Robinette and co-workers (1959) estimated individual predation rates of wild cougars at 4–10 days per mule deer kill based on the consumption rate of 2.3–5.5 kg per day they documented for wild and captive cougars. Hornocker's (1970) estimate of 18–27 days per mule deer kill for wild cougars, based on daily consumption of only 1.8–2.7 kg for captives, was probably too low because it underestimated food requirements of wild individuals.

Field studies of wild cougars support the predictions of energetics models (Ackerman et al. 1986; Laundré 2005) that predation rates differ by age, sex, and reproductive status of cougars, which in turn reflect varying body weights of individuals and nutritional requirements of cougar families. As is shown in Table 10.2, females with multiple offspring—collectively up to 145 kg of body weight—typically have the highest predation rates, making large mammal kills the least number of days apart, followed by solitary adults (females weighing 47 kg; males 72 kg) and subadults (females weighing 39 kg; males 53 kg; Murphy 1998; Anderson and Lindzey 2003). Mature males may consume large amounts of biomass, yet spend the least time on a kill (Mattson et al. 2007).

How the availability of multiple large prey and prey biomass influence predation rates is not apparent in current data. Individual rates for cougars that subsist primarily on the largest prey, such as elk, should be longer than in strongly deer-centric systems, because larger prey provide more food biomass. Predation rate comparisons among systems and across carnivore species should be standardized by incorporating prey biomass (e.g., kg killed per day, as reported by Murphy 1998).

Various factors not accounted for in models but measured in the field can influence individual predation rates. For example, where temperatures are warm (e.g., southwestern U.S. deserts), carcasses may soon putrefy and become unpalatable (Robinette et al. 1959), hastening cougar efforts to kill again. In contrast, precipitation may retard decomposition and reduce carcass depletion by microbes and arthropods, thus increasing time cougars spend at a kill (Mattson et al. 2007). Other carnivores may displace cougars from kills or feed from unattended kills (Murphy et al. 1998; Nowack 1999; Logan and Sweanor 2001; Ruth 2004a) and cause cougars to kill more frequently to compensate for food lost to scavengers (Harrison 1990). In contrast, predation rates on large prey could decline if cougars frequently scavenge or use smaller prey, as may happen during summer when

winter hibernators are active (Hornocker 1970). Murphy (1998) found that individual predation rates on ungulates declined with increasing temperatures and that small prey were more frequent in spring-fall than in winter diets of cougars, although small prey provided little food biomass. Increased costs of thermoregulation, or easy access to ungulates concentrated on winter ranges, may also have contributed to higher winter predation rates.

Nowack (1999) found the opposite trend: female cougars (both classes) had significantly higher predation rates in summer-fall than in winter-spring, possibly because newborn deer and elk (typically, young <180 days old) were available in early summer. Mattson and colleagues (2007) reported a short average interval between kills during spring drought, when prey were concentrated near water and in poor condition, and longer kill intervals during the monsoons, when prey were in improved condition and less concentrated. Cooley and co-workers (2008) found no differences in rates between winters and summers.

The total number of prey removed per unit time by the cougar population—the total predation rate—can be crudely estimated as the product of the individual rate and the number of cougars hunting (all individuals except dependent kittens). Refinements are possible if data are specific to cougar sex, age, and reproductive class (Table 10.3). The overall percentage of prey removed by cougars can then be calculated if the size of the prey population is known.

Another approach to estimating total predation rates is through monitoring radio-marked prey and documenting causes of their death (Table 10.3). Rates of mortality for each source of death (i.e., cause-specific mortality rates) are calculated for the period during which prey are monitored (e.g., monthly, annually), and they estimate the percentage of the prey killed by the predator during the time period. Thus, when the rates are expressed as a percentage, they estimate the expected total loss of the prey type to the specific mortality agent, as provided for cougars in Table 10.3.

Total predation rates have two important limitations. First, their values may fluctuate depending on mortality rates for other causes of death (Heisey and Fuller 1985) and, second, total predation is expected to vary with the relative vulnerability of prey, which may change dramatically and rapidly with environmental conditions (Mech 1977; Peterson 1977; Logan and Sweanor 2001) or with prey density. Extrapolating results—from individual studies to different locations, from single- to multiple-prey systems, or even to different periods in the same location—is risky. Although eleven of fourteen studies in Table 10.3 indicate cougar predation is ≤20 percent annually, inferences regarding the actual effects on prey populations require knowledge of births, immigration, emigration, and all other sources of mortality in the

Table 10.3 Total predation rates (percentage of prey population removed) of North American cougars estimated using individual rates (see text) and cougar population size, or cause-specific mortality rates.[a]

Location	Prey	Prey Sex/Age	Time Span/(years of study)	Total Predation Rate (%)	Method	Reference
Arizona	Mule deer	All	Annual (2)	<20; <10a	IR[b]	Shaw 1980
New Mexico	Mule deer	Adults	Annual (7)	6–28	CM	Logan and Sweanor
	Mule deer	Adult males	Annual (7)	5–47	CM	2001
	Mule deer	Adult females	Annual (7)	4–25	CM	
Colorado	Mule deer	All	Annual (1)	8–9	IR	Anderson et al. 1992
Wyoming	Mule deer	All	Annual (5)	3–5	IR	Murphy 1998
	Mule deer	Fawns	Annual (5)	3–5	IR	
	Mule deer	Adult males	Annual (5)	7–15	IR	
	Mule deer	Adult females	Annual (5)	1–3	IR	
British Columbia	Mule deer	Adults	Annual (4)	13–19	CM	Robinson et al. 2002
Idaho – Utah	Mule deer	All	Annual (12)	2b	EM	Laundré et al. 2006
				3–6b	EM	
British Columbia						
All	Black-tailed deer	Adult females	Monthly[d] (9)	0.8	CM	McNay and Voller 1995
Caycuse	Black-tailed deer	Adult females	Monthly[d] (9)	2.1	CM	
Chemainus	Black-tailed deer	Adult females	Monthly[d] (9)	2.0	CM	
Nanaimo	Black-tailed deer	Adult females	Monthly[d] (9)	0.4	CM	
Nimpkish	Black-tailed deer	Adult females	Monthly[d] (9)	0.8	CM	
California						
All	Bighorn sheep	Adults	Annual (6)	14	CM	Hayes et al. 2000
Carrizo Canyon	Bighorn sheep	Adults	Annual (6)	14	CM	
Vallecito Mts.	Bighorn sheep	Adults	Annual (6)	14	CM	
S. San Ysidro Mts.	Bighorn sheep	Adults	Annual (6)	9	CM	
N. San Ysidro Mts.	Bighorn sheep	Adults	Annual (6)	12	CM	
Coyote Canyo	Bighorn sheep	Adults	Annual (6)	26	CM	
Santa Rosa Mts.	Bighorn sheep	Adults	Annual (6)	14	CM	
British Columbia	White-tailed deer	Adults	Annual (4)	6–15	CM	Robinson et al. 2002
Arizona						
Northwest Arizona	Bighorn sheep	Adults	Annual (3)[e]	1–6	CM	Kamler et al. 2002
Southeast Arizona	Bighorn sheep	Adults	Annual (3)[e]	2–12	CM	
Southwest Arizona	Bighorn sheep	Adults	Annual (2)[f]	5–15	CM	
Central Arizona	Bighorn sheep	Adults	Annual (3)[e]	5–29	CM	
New Mexico	Bighorn sheep	Adult females	Annual (7)	0–28	CM	Logan and Sweanor
	Bighorn sheep	Adults	Annual (7)	0–25	CM	2001
New Mexico						
Sierra Ladron	Bighorn sheep	Adult females	Annual (8)	9	CM	Romminger et al. 2004
Sierra Ladron	Bighorn sheep	Adult males	Annual (8)	13	CM	
Sierra Ladron	Bighorn sheep	Adults	Annual (8)	11	CM	
Wheeler Peak	Bighorn sheep	Adults	Annual (7)	0	CM	

(Continued)

Table 10.3 (Continued)

Location	Prey	Prey Sex/Age	Time Span/(years of study)	Total Predation Rate (%)	Method	Reference
Alberta	Bighorn sheep	All	Winter (1)	13	IR	Ross et al. 1997
Wyoming	Elk	All	Annual (5)	2–3	IR	Murphy 1998
	Elk	Calves	Annual (5)	4–11	IR	
	Elk	Adult males	Annual (5)	0–1	IR	
	Elk	Adult females	Annual (5)	0–1	IR	

[a]Method: IR = predation rates of individual cougars, CM = cause-specific mortality rates of prey, EM = energetics model.
[b]Annual rates are for two study areas. Number of annual kills per cougar estimated from scats.
[c]First estimate is total predation rate before a decline in mule deer and the second estimate is after the mule deer decline.
[d]Approximate annual rate is 12 times the monthly rate.
[e]Three point estimates of mortality rate for time spans of 5–6 years.
[f]Two point estimates of mortality rate, both for 5-year time spans.

population. Total predation rates are valuable for comparing to other sources of mortality, such as human hunting or disease, but should be viewed with caution because of factors unique to each setting. For example, mortality of desert bighorn sheep due to cougars in New Mexico varied from zero in some years to a high of 25 percent for all adults in certain years (Logan and Sweanor 2001). Disease, weather, human factors, and prey switching from mule deer likely contributed to the vulnerability of the bighorn population and its subsequent decline. Logan and Sweanor's (2001) study is a good example of predation rates being documented in context with compensatory mortality, fecundity of the prey population, and environmental factors; components few other studies have adequately addressed (but see McNay and Voller 1995). We discuss cougar predation by prey categories in the following sections, and invoke, where possible, specifics such as spatial and temporal changes and climatic effects particular to each study.

Influences on Native Prey

At present, few studies on cougar-prey relationships have lasted long enough (>10 years) to include the full range of population fluctuations of cougars and their prey (but see Laundré et al. 2006, 2007). Thus studies have not fully documented the extent to which cougar predation limits or regulates prey populations and the compensatory or additive nature of cougar predation (Logan and Sweanor 2000). Clearly, the effects of cougar predation may be different in various systems and may depend on conditions present during the study period. Here, our review is focused on both cougar-centric and prey-centric studies that addressed the

effect of predation as a primary study objective (see also Table 10.1).

Elk and Deer

Across North America, elk and deer constitute major prey of cougars and are also economically important to humans. Understanding the influences of predation on these prey species is highly important to managers and the public, especially hunters.

A number of studies finding cougar predation as an important mortality factor of prey did not address the limiting, regulatory, or compensatory nature of cougar predation (see Tables 10.2, 10.3; Ballard et al. 2001), and few have addressed the prey population's status relative to carrying capacity (Ballard et al. 2001). In their review of thirty studies on cougar, coyote (*Canis latrans*), and wolf (*C. lupus*) predation on mule, white-tailed, and black-tailed deer (*O. h. columbianus*), Ballard and colleagues (2001) concluded that deer populations well below carrying capacity were more likely limited by predation and that in such instances, deer survival increased significantly when predator numbers were reduced. In contrast, deer populations at or near carrying capacity were food limited and did not respond to predator removal experiments (Ballard et al. 2001). Our review of studies (Table 10.1) generally supports findings by Ballard and co-workers (2001).

While cougars kill mainly elk and deer where these prey are the most abundant, cougar predation has not been implicated as a strong limiting factor of elk, and approximately half of the studies indicated that cougar predation did not strongly limit deer populations. Stable numbers of resident cougars did not prevent elk and mule deer from increasing in Idaho (Hornocker 1970) and Utah (Lindzey et al. 1994). Predation on radio-collared adult elk (Evans et al. 2006) and calves (Smith and Anderson 1996; Singer et al. 1997; Barber-Meyer 2006; Smith et al. 2006) was minimal across the Greater Yellowstone Ecosystem. In northern Yellowstone, an increasing resident cougar population did not strongly limit the numbers or growth rates of elk and mule deer during a seven-year study (Murphy 1998). These prey commonly used open grassland and agricultural fields that lacked sufficient cover for cougars. Although cougar predation was a major source of mule deer mortality in two Arizona study areas (<10 percent and <20 percent annual removal), cougar predation alone did not prevent mule deer populations from increasing (Shaw 1977, 1980). Recently, Laundré and co-workers (2006, 2007) suggested that not only do cougars have little effect on deer populations, even at relatively high cougar densities, but cougar numbers were influenced by cycles in deer abundance that were driven by winter snowfall in northern systems. A caveat is that their estimates of individual cougar predation rates as calculated in Laundré (2005) and resulting estimates of total predation impacts on deer were low compared to field studies.

Some studies indicate strong limiting effects on white-tailed deer, black-tailed deer, and mule deer. In British Columbia, cougar and wolf predation were found to reduce numbers of resident black-tailed deer in isolated stands of old growth forests during late March, a time migratory deer departed winter ranges (McNay and Voller 1995). Cougar predation was not sufficient to strongly limit the migratory segment of the deer population. Because all study areas were extensively logged, the authors hypothesized that forest harvest and resulting road building had isolated winter habitats, possibly intensifying predation on the resident deer population. Habitat fragmentation was suggested as the ultimate factor initiating declines in the deer. In a California study, cougar predation was the largest single cause of mortality of mule deer fawns (Neal 1990). Although Neal (1990) suggested that removal of cougars would reverse a long-term downward trend in the deer population, survival of adult deer and possible effects of climate and range conditions were not studied, nor was compensatory predation addressed. Also documented, but not evaluated, was predation by coyotes, bobcats (*Lynx rufus*), and bears.

Strong limiting effects of cougar predation may be related to increases in primary prey or seasonal migration of primary prey that trigger apparent competition with secondary prey. In large portions of northeast Washington, timber harvest, severe winters and drought, and competition with livestock are factors suggested to have influenced declines in mule deer. A combination of these factors and agricultural practices may have increased numbers of white-tailed deer and their expansion into traditional mule deer ranges (Robinson et al. 2002, Cooley et al. 2008). Currently, white-tailed deer outnumber mule deer by estimates of 2:1 to 4:1 (Robinson et al. 2002; Cruickshank 2004). Recent research by Cooley and colleagues (2008) indicates that cougars primarily preyed on white-tailed deer and followed them to higher elevations during summer. Thus, spatial overlap between cougars and mule deer increased; cougars then selected for mule deer on the summer range. In south-central British Columbia, cougars responded positively to the numbers of their primary prey (white-tailed deer), and predation on mule deer (secondary prey) was considered to be *depensatory*, or *inversely density dependent*, causing a decline in the mule deer population concurrent with an increase in white-tailed deer (Robinson et al. 2002). However, this interpretation is suspect because of a single data point that disproportionately influenced the mule deer trend line, which leaves the interpretation of inverse density dependence uncertain (see Figure 7 in Robinson et al. 2002). The case for increased cougar predation rate with increased density of white-tailed deer (positive density dependence) seemed reasonable.

In an evaluation of deer survivorship and cause-specific mortality across five winter ranges, Bleich and Taylor (1998) suggested that cougar predation might regulate mule deer herds in the stressful environments of the western Great Basin, California, and Nevada. Yet they did not evaluate whether cougar predation was density dependent or density independent, and thus the extent to which it might regulate mule deer. At the time of the study, the Great Basin had experienced repeated droughts, which may have reduced ecological carrying capacity. A harsh winter during 1992–93 killed many deer. The researchers suggested cougar predation was primarily additive to other sources of mortality because cougar predation was the leading cause of deer mortality, and evidence that some deer died of malnutrition was found in only two of five study populations. The authors acknowledged that fawn survival and the relative roles of nutrition, predation and climatic factors were not fully evaluated.

Determining whether cougar predation strongly limits or regulates prey populations is difficult without experimental manipulation of predator numbers (Ballard et al. 2001) through translocations or controlled hunting, although this was accomplished in two studies on mule deer (Logan and Sweanor 2001; Hurley et al., Idaho Department of Fish and Game, unpublished data). In an effort to assess predation effects of cougars, a portion of a cougar population was translocated from a treatment section of a New Mexico study area in conjunction with monitoring mule deer survival and population trends on both the treatment and reference (nonremoval) sites. The desert mule deer population increased despite increasing cougar density, at least until the onset of a drought (Logan and Sweanor 2001). During the drought, fawn production declined and predation rates on deer aged one year old or older increased. The deer population declined; thus predation was inversely density dependent. Although cougar predation contributed to the deer population decline, it was considered to be mostly compensatory because the drought reduced ecological carrying capacity (Logan and Sweanor 2001).

In Idaho (1997–2003), where mule deer populations were below carrying capacity and cougar predation was mostly additive, removal of cougars through hunting increased deer survival and fawn production and had a weakly positive effect on the rate of population increase (Hurley et al., unpublished data). However, the biologists suggested that in a longer time frame, cougar predation might be partly compensatory. Removal of cougars potentially decreased the fitness of the deer population due to an increased survival of older animals, ultimately lowering recruitment and potentially increasing vulnerability of adults to weather events and other predators (Hurley et al., unpublished data). Although cougar and coyote predation were significant limiting factors, weather conditions that influenced habitat, not predation, were thought to regulate mule deer populations in eight experimental areas.

Other large carnivores and humans may complicate cougar-prey relationships through their effects on cougar predation rates and on cougar and prey population size. Human hunting and cougar and wolf predation were additive sources of mortality and appeared to limit white-tailed deer in the Flathead River Drainage, Montana and British Columbia (Kunkel et al. 1999). On a ~340 km² winter range in Banff National Park, cougar predation had minor demographic impact on elk, but wolf predation in winter appeared to perpetuate a decline in the elk population (Kortello 2005) after park managers translocated 0.4 to 26 percent of the elk during three years of the study. Within two years, and coincident with a period of increasing wolf and decreasing elk numbers, cougar diets shifted from predominantly elk to alternative prey consisting primarily of bighorn sheep and mule deer.

In summary, the effect of cougars on deer and elk populations is influenced by prey numbers relative to carrying capacity, weather, compensation with other sources of mortality, and prey birth rates. Similar to findings by Ballard and colleagues (2001), in studies where cougars had weak limiting influences on elk and deer, prey were at or near carrying capacity and predation was partly compensatory to other sources of mortality (see Table 10.1; Hornocker 1970; Murphy 1998; Logan and Sweanor 2001). Alternatively,

where cougar predation did strongly limit deer, populations were below carrying capacity and predation appeared additive to other sources of mortality (McNay and Voller 1995; Hurley et al., unpublished data). In general, studies have been hampered by a lack of information about these factors and by a lack of controlled experimentation.

Moose and Wild Horses

The large size of adult moose (*Alces alces*) and adult wild horses, the most reproductively valuable component of prey populations, impedes predation by cougars. Thus cougars are expected to have weaker limiting effects on these species than on smaller prey. Although a number of studies (Ross and Jalkotzy 1996; Murphy 1998; Kunkel et al. 1999; C. R. Anderson and Lindzey 2003) have documented moose in the diet of cougars, only one study has assessed the effect of cougar predation on a moose population. In southwestern Alberta, cougars avoided predation on adult moose and selected calves over adults and yearlings (Ross and Jalkotzy 1996). Male cougars preferred moose calves to other species of prey, including elk, mule deer, bighorn sheep, and white-tailed deer. An estimated 16–30 percent of the early winter calf crop was killed by cougars, but cougar predation was largely compensatory to other moose mortality. Femur marrow fat of calf and yearling moose averaged 19.9 percent, suggesting malnutrition, and more than one-third of calves and yearlings taken had marrow fat levels of less than 10 percent, a level indicating that death was likely, even without predation (Franzmann and Arneson 1976).

In central California and western Nevada, all wild horses killed by cougars were less than six months old, and almost half of the annual foal losses, most due to cougar predation, occurred during the peak foaling period of late May through June (Turner et al. 1992). Cougars switched to mule deer as their primary prey once foals attained a larger size between April and October. The horse population was apparently not limited by forage availability. In this case, cougar predation on foals was sufficient to limit strongly the growth of the wild horse population and had beneficial stabilizing effects on wild horse numbers (Turner et al. 1992).

Mountain Caribou

Predation, including that by cougars, was linked to the decline of mountain caribou (*Rangifer tarandus caribou*) in the United States and Canada (Schaefer et al. 1999; Wittmer et al. 2005a, b). Predation on caribou varied annually; was influenced by elevation shifts and population increases of alternative prey (moose, white-tailed deer, elk), leading to apparent competition; and had the greatest influence on small populations of caribou in more southerly areas (Wittmer et al. 2005a, b). The loss of mature forests, and concomitant changes to young-aged forest stands, could compromise a common predator avoidance strategy of caribou: that is, separating themselves from alternate prey and their predators. In British Columbia, predation by cougars, wolves, and bears was the primary (≥50 percent) cause of mortality in eleven of thirteen subpopulations of woodland caribou and precipitated a caribou decline. In this study, cougar and other predator numbers were subsidized by the presence of alternate ungulate species (deer, elk, and moose), which caused small caribou populations to suffer proportionately greater predation mortality. Hence, predation effects were inversely density dependent. This was particularly apparent in southern areas, where caribou populations were smaller and had low rates of increase. Although the proximate cause of population decline was predation on adult caribou, the loss of mature forests was suggested to increase vulnerability of caribou. The study did not evaluate adult caribou condition, leaving the possibility that predation was compensatory.

Katnik (2002) also linked increased cougar predation on mountain caribou to a seasonal shift in elevation by their primary prey, white-tailed deer, resulting in greater spatial overlap between cougars and caribou in the Selkirk Mountains, Washington, Idaho, and British Columbia. However, not all cougars overlapped with and had the opportunity to prey on caribou. Therefore, Katnik (2002) proposed that any removal of cougars (translocation or killing) should be targeted at specific individuals.

Bighorn Sheep

Cougar predation on bighorn sheep is of considerable interest because bighorns often occur in small groups, have a naturally fragmented distribution (Bleich et al. 1990), and have declined in several areas of the western United States (Hayes et al. 2000; Krausman and Shackleton 2000; Kamler et al. 2002). The potential for population-level effects of cougar predation is greatest in small, isolated sheep populations (Ross et al. 1997; Hayes et al. 2000; Logan and Sweanor 2001). Cougar predation on bighorn sheep may be sporadic, varies annually, and is often a result of specialized predation by a few individual cougars (Ross et al. 1997; Logan and Sweanor 2001).

In many areas, predation on bighorn sheep occurred during winter (Wehausen 1996; Ross et al. 1997; Hayes et al. 2000; Schaefer et al. 2000). Hayes and colleagues (2000) documented cougar predation in all months except June, but 62 percent of cougar kills occurred between December and March. In southwestern Alberta, cougar predation varied greatly from year to year, causing 0–57 percent of

the winter bighorn mortality (Ross et al. 1997). However, this study found no long-term relationship between cougar predation and bighorn numbers.

Prior to the 1990s, few studies in western North America reported negative impacts of cougar predation on bighorn sheep populations (Munoz 1982, cited in Kamler et al. 2002). For example, cougar predation had an insignificant effect on a population of approximately 125 bighorn sheep in the central Idaho wilderness, where elk and mule deer were the primary prey (Hornocker 1970). However, since the 1990s, cougar predation has increasingly been documented as a substantial mortality factor in bighorn sheep populations in western North America (Kamler et al. 2002), particularly the southwestern United States. Cougar predation on bighorn sheep is apparently greatest where bighorn and mule deer are sympatric, and increased cougar predation corresponded with declines in mule deer populations during the 1990s (Hayes et al. 2000; Schaefer et al. 2000; Ballard et al. 2001). In the Mojave Desert of California, a small population of bighorn (about a dozen) sympatric with mule deer experienced low annual survival (1989–92), mainly due to cougar predation (Wehausen 1996). However, during 1993–96 there was no cougar predation, and the bighorn population increased 15 percent annually. Cougar predation also resulted in low survival and prevented recovery of an endangered California bighorn population (Hayes et al. 2000). In contrast, no evidence of predation by cougars on desert bighorn was found in the Sonoran Desert, possibly because of the low density of mule deer and cougar (Andrew et al. 1997, cited in Schaefer et al. 2000).

Although some efforts to reestablish or augment bighorn sheep populations through translocation have been successful, others have failed or been hindered due to cougar predation. In addition to cougar predation, the presence of alternative prey (apparent competition), declines in alternative prey (prey switching), and other factors potentially influence the vulnerability of sheep to predation and hence the success of translocation efforts. Group sizes of released bighorn, habitat quality and quantity, amount of escape terrain at release locations, and the abundance of mule deer and their proximity to well-established cougar populations influenced cougar predation on desert bighorns in Arizona (McKinney et al. 2006b). Cougars were responsible for declines in other translocated bighorn populations (Kamler et al. 2002), where cougars switched to bighorn as alternate prey four years after mule deer numbers declined. In New Mexico, cougar predation was assumed to limit growth of one population but did not limit the growth of a second population of translocated bighorn (Rominger et al. 2004). At the study site where cougar predation limited translocated bighorn, the researchers hypothesized that the presence of cattle near bighorn habitat subsidized cougar diets and enabled cougars to remain at higher densities than when natural prey densities were low (Rominger et al. 2004). However, the researchers did not provide data on densities of cattle, subsequent response of cougar numbers to cattle densities, or whether cougar predation was compensatory to other sources of mortality.

Experimental removal of cougars, in conjunction with monitoring of bighorn sheep and cougar populations, indicated that cougar predation was a limiting factor during a fourteen-year study in Arizona (McKinney et al. 2006a) and a ten-year study in New Mexico (Logan and Sweanor 2001). As in other areas, an increase in predation on bighorn corresponded with a drought-influenced decline in mule deer. In the Arizona study, reduction of cougar numbers through hunting reduced predation by cougars on bighorn. However, nutritional status (influenced by winter rainfall) and cougar predation were both limiting factors affecting growth of the sheep population during a drought (McKinney et al. 2006a). In the New Mexico study, cougar predation and other factors limited a small population of bighorn sheep (Logan and Sweanor 2001). A combination of historic human-induced isolation of bighorn sheep, disease, low reproductive rate, drought, and cougar predation ultimately played a role in bighorn extinction.

Cougar predation may compensate for other sources of mortality in bighorn sheep, or it may be additive. Cougars selected for lambs, the nonreproductive portion of the sheep population, and one-third or more of bighorn sheep had behavioral or anatomical disabilities prior to being killed by cougars in Alberta (Ross et al. 1997). Approximately 30 percent of cougar predation on desert bighorn was considered compensatory in New Mexico because many bighorn killed were more than twelve years old, emaciated, or had severe scabies infestation (Logan and Sweanor 2001). But in a Colorado study, increased cougar predation on adult and subadult bighorn was additive to other sources of mortality and, in combination with low lamb recruitment, was suspected to be strongly limiting (Creeden and Graham 1997). In Arizona, increased cougar predation may have been compensatory to other sources of mortality in the southeastern region, yet additive in the central and southwestern regions (Kamler et al. 2002).

Specialized predation by individual cougars, possibly a learned behavior, may impact bighorn populations independently of cougar density. Not all cougars kill bighorn sheep repeatedly, if at all, and where predation occurs, it typically varies from year to year (Hornocker 1970; Ross et al. 1997; Logan and Sweanor 2001). Although individual cougars may have substantial impacts on small, isolated bighorn populations, targeted removal of individuals can reduce predation on bighorn while minimizing impacts to cougar populations (Ernest et al. 2002).

Ross et al. (1997) recommended that managers should expect highly variable cougar predation rates on bighorn populations of fewer than 200 individuals. Sawyer and Lindzey (2002) suggested that the potential for cougar predation to have population-level effects appeared greatest in bighorn populations of fewer than one hundred inhabiting desert environments. In general, predation is more likely to be a limiting factor of bighorn sheep populations inhabiting ranges without adequate escape terrain (Hass 1989, cited in Sawyer and Lindzey 2002) and is less important in habitats where they can escape into cliffs and other rugged terrain. Conversely, Laundré and colleagues (2002) argued that what has traditionally been considered escape terrain is also ideal hunting habitat for cougars. They suggest that evaluation of escape terrain relative to other habitats needs further investigation. Bighorn populations within or adjacent to areas with declining mule deer populations may also be limited by cougar predation until cougar numbers decrease in response to declining mule deer numbers (Kamler et al. 2002).

Pronghorn

Cougar predation is typically not important on pronghorn (*Antilocapra americana*) because cougars are unlikely to stalk and catch adult pronghorn consistently in open grassland or sage steppe (Byers 1997). However, where pronghorn use more rugged terrain, their vulnerability increases (Ockenfels 1994; Logan and Sweanor 2001, C. Anderson, pers. comm., 2004). In Yellowstone National Park, ten pronghorn were killed by four female cougars in steep and rocky terrain with sagebrush and tree cover (Ruth and Buotte 2007). One female killed seven of the pronghorn, four within fourteen days and the remaining three over two years (Ruth and Buotte 2007), suggesting specialized killing behavior by an individual similar to that seen with bighorns. In Arizona, Ockenfels (1994) suggested that cougar predation on reproducing female pronghorn stabilized or decreased the pronghorn population in rugged, brushy terrain. Urbanization and fragmentation by an interstate highway may have introduced disturbance-related effects that caused changes in habitat use by pronghorns that were supportive of cougar predation (H. Shaw, pers. comm., 2007).

Guanacos

Historically, guanacos (*Lama guanicoe*) were the main native prey of pumas in Patagonia, but increased exotic game ranching, overgrazing by livestock, hunting, and habitat degradation have drastically altered relationships between native prey and carnivore populations (Novaro and Walker 2005). Exotic wild prey (such as European red deer, *Cervus elaphus*, and European hares, *Lepus europaeus*) and domestic livestock compete with native guanacos for resources and sustain high numbers of pumas and culpeo foxes (*Pseudalopex culpaeus*). Both carnivores were found to limit guanacos strongly at low densities and to prevent recovery of guanaco populations in parts of Patagonia. In addition, pumas may switch and accelerate their predation on guanacos if more common exotic prey decline. Puma food habits examined through scat analysis at different sites and at a range of guanaco densities indicated that puma predation was density dependent and suggested a type 3 functional response (see Figure 10.2). Accelerated predation rate and, thus, the potential to prevent guanaco population growth, occurred only at low guanaco density (known as the Allee effect; see Allee 1931). The potential to prevent guanaco population growth occurred below a threshold density of about 8 guanacos/km² (Novaro and Walker 2005, 278–79). With guanacos occurring at an average density lower than 1/km² (Novaro et al. 2000) in many areas of their range, and pumas widespread and increasing, Novaro and Walker (2005) predicted that puma predation could increasingly prevent guanaco recovery, particularly as domestic sheep are removed from many areas (see Chapter 7 for fuller discussion).

Small Mammals

Because small mammals (<15 kg) constitute an important component of annual cougar diets in southern latitudes and summer diets in northern latitudes (see Chapter 9), cougar predation may influence population dynamics of small prey, which may in turn influence larger prey. In southeastern Idaho during the winters of 1997 to 2002, mule deer mortality due to coyote predation was highest in February and March, but declined significantly in April and May after small mammals emerged and coyotes switched to this prey (Hurley et al., Idaho Department of Fish and Game, unpublished data). In the Great Basin Desert, increased cougar predation was linked to the decline of a porcupine population from over eighty to less than five individuals in three years (Sweitzer et al. 1997). Grazing by domestic and feral livestock in the 1800s altered plant communities and eventually led to an irruption of mule deer and an accompanying increase in cougars (Berger and Wehausen 1991). Drought and a decline in mule deer density during 1987–90 were believed to have caused cougars to switch to preying on porcupine (Sweitzer et al. 1997). More extensive research would provide insights into cougar predation influences on both small and large prey.

Patterns in Cougar Predation

Our review of studies illustrates the complex and dynamic nature of cougar-prey relationships. Prey and cougar population density as well as duration and location of study (e.g., conducted entirely on a winter range) may all affect conclusions (Table 10.1). While this reinforces caution in comparing studies, strong limitation appears to occur when prey are at low density, in fragmented groups, or in harsh and unpredictable environments, particularly desert environments with vegetation sensitive to drought. Persistent drought reduces habitat carrying capacity on deer ranges, causes poor body condition and low neonatal survival (Ballard et al. 2001), and exposes offspring and adult deer to increased mortality from predation (Logan and Sweanor 2001). Moreover, strong limitation of low-density secondary prey may occur when seasonally abundant primary prey subsidizes cougar predation or when primary prey declines and cougars switch and kill more of the less abundant, secondary prey. However, the conditions in which the primary prey subsidizes, rather than buffers, the effects on a secondary prey need further evaluation, including the time and spatial scale over which a subsidy is likely to occur.

Alternatively, predation by cougars may combine with environmental factors and other predators to limit prey that were initially at high densities (Messier 1994; Ballard et al. 2001). Predation may not contribute to prey population declines unless other factors initially tip the scale—severe and potentially prolonged weather events, human overharvest, or human alteration of habitat (Ballard et al. 2001; Logan and Sweanor 2001; Laundré et al. 2006)—after which cougar predation may exacerbate the decline in prey. In some of these cases, however, cougar predation is more likely compensatory with other mortality factors.

Trends in cougar density and seasonal distribution can mislead assessment of the influence of predation if data are not coupled with knowledge of individual predation behavior. Regardless of cougar population density or trend, only a few individuals may overlap spatially with prey species of concern (mountain caribou or bighorn sheep) or may selectively kill certain prey (bighorn sheep, pronghorn), yet these individuals could have a significant influence on prey numbers. Thus, knowledge of the spatial relationship and predatory behavior of individual cougars can be an important consideration when developing publicly acceptable management goals that may include cougar removals through translocation or killing (Katnik 2002).

Habitat use patterns and migratory behavior of prey can also affect the nature and impact of cougar predation. Cougars may have less effect on prey that use open habitats than on prey that frequently use habitats providing stalking cover for cougars (Ockenfels 1994; Byers 1997; Murphy 1998).

Concentration of prey on winter ranges may make them vulnerable to cougars. Similarly, migratory prey that move to habitats seasonally unavailable to cougars are potentially impacted less than nonmigratory prey that live in cougar habitat yearlong (McNay and Voller 1995; Murphy 1998). As has been found through simulation modeling, this pattern is similar to that for ungulates in East Africa, for which migratory patterns decreased the influence of predation on prey (Fryxell et al. 1988).

Human development can change the impact of cougar predation. In the San Bernardino Mountains of California, urbanization reduced high-quality winter habitat for mountain sheep and mule deer (Nicholson 1995, cited in Schaefer et al. 2000), resulted in increased seasonal densities and spatial overlap between sheep and deer, and was thought to increase risk of predation for both species. Other human-induced habitat changes contribute to predator-prey disequilibriums (Berger and Wehausen 1991; Novaro and Walker 2005). In the Great Basin of the western United States, mule deer rapidly increased in response to habitat changes caused by livestock grazing, which apparently resulted in increased cougar densities and a significant reduction of bighorn sheep by cougars (Berger and Wehausen 1991). However, as earlier noted, collective studies to date indicate that a range of different factors could explain bighorn declines in specific areas, including specialized predation by individual cougars, effects of drought, disease, and prey switching due to mule deer declines (see Sawyer and Lindzey 2002). Whichever factors are operating, the human-induced changes described may result in a more predictable and higher density prey base, which contributes to high risk of cougar predation for bighorn (Schaefer et al. 2000).

The body size of ungulates may be important in determining the impact of predation: small ungulates may experience strong predation effects related to their size, whereas large ungulates are less vulnerable to solitary predators such as cougars. For example, Murphy (1998) suggested that cougars may have only weak limiting effects on ungulate prey when landscapes support adult prey that, on average, are too large to be killed by cougars. In the Serengeti of Tanzania, large ungulates had few natural predators above a threshold of about 150 kg and were primarily food limited (Sinclair et al. 2003).

When predation by cougars has significant impacts or threatens the persistence of native prey, particularly those in small, naturally fragmented groups (e.g., bighorn sheep, pronghorn), then removal of individual cougars may be necessary to maintain these populations. Using a simulation model, Ernest and colleagues (2002) found that both population-wide reduction of cougars and removal of one to two individuals per year reduced extinction risks for bighorn sheep in California. Targeting individuals responsible

for predation achieved effects similar to population-wide removals but averted unnecessary removal of cougars that did not specialize on bighorn. Similarly, prior to removing cougars, careful consideration should be given to habitat (Hayes et al. 2000; Bowyer et al. 2005) and human factors that influence prey vulnerability and predation responses. For example, bighorn sheep in North America have lost more than half their range due to human disturbance and habitat loss. In view of declines in mule deer populations, cougar predation may be greater today than historically (Kamler et al. 2002). Cougars removed an estimated 3 to 5 percent of mule deer annually compared to 6 to 12 percent of deer removed annually by human hunting on the northern range of Yellowstone (Murphy 1998). Thus, the influence of human hunting in addition to other factors should be considered when evaluating cougar impacts on prey and various management approaches. Because removal of cougars could be a prolonged and costly effort, and potentially unacceptable to the public, long-term costs and benefits should be compared with other alternatives before embarking on a removal program. Ballard and co-workers (2001) provide helpful criteria to assist in evaluating when predation is an important mortality factor and whether cougar population control or other management approaches should be considered.

Indirect Effects of Predation

In addition to direct effects on prey, cougar predation may also have indirect effects. These include influences on behavior (e.g., vigilance) and foraging patterns of prey, sexual segregation of prey in response to predation risk, and effects that cascade down from upper to lower trophic levels. For example, bighorn sheep vigilance to predation risk, which may incur foraging costs, decreased as group size increased above five individuals (Mooring et al. 2004). In response to predation risk, prey species may alter their use of habitats, trading a reduction in forage quality or quantity for increased security, and such changes in habitat use may be seasonally dependent (Hernández and Laundré 2005; Mao et al. 2005). Recent research indicated that elk were likely to occupy foraging sites in open grasslands when wolves were absent but quickly moved into coniferous forest when wolves were present (Creel et al. 2005; Hernández and Laundré 2005). Hornocker (1970) first noted that elk and deer responded to cougar predation risk by moving to different areas within a spatially limited winter range. Altendorf and co-workers (2001) demonstrated that mule deer tended to avoid edge habitat where predation rates by cougars were higher (Laundré and Hernández 2003). Predation by cougars may have caused a small population (≤60) of bighorn sheep to abandon use of low elevation winter range throughout the southern and central Sierra Nevada (Wehausen 1996).

Antipredator behavior of prey may be specific to certain habitats that are riskier than others. Thus prey response to predation risk may not result in changes in prey distribution or behavior at a landscape level, as recently found for elk during winter in Yellowstone (Kaufman et al. 2007). Access to critical winter forage coupled with the ability of elk to avoid wolf predation by moving 1–2 km to safe habitats most likely explained why elk have not made broad-scale changes in winter habitat to avoid encounters with wolves (Fortin et al. 2005; Mao et al. 2005; Kaufman et al. 2007). Similarly, in other areas prey did not always respond strongly to predation risk from cougars (Logan and Sweanor 2001; Pierce et al. 2004).

Response to predation risk is not always equal across prey gender. In Montana, individual bull elk faced wolf predation risk 6.3 times higher than that of individual cow elk, yet bulls were less responsive in selecting woodland habitat in the presence of wolves (Creel et al. 2005). In this system and in Yellowstone National Park, male elk may be nutritionally constrained and less able to pay foraging costs associated with antipredator behavior (Laundré et al. 2001; Creel et al. 2005). How prey genders respond to cougar predation likely varies by prey species, habitat types, and climatic regime, but few studies addressing this aspect of predation are available. Pierce and colleagues (2004) detected no differences in habitat selection between male and female mule deer during winter; both sexes selected habitats that had a relatively high proportion of bitterbrush. Adult bighorn rams (class three-quarters to full curl) occurred in smaller groups, were much more vigilant than other sex and age classes of sheep, and were at the greatest risk of cougar predation (Mooring et al. 2004). Due to their smaller body size and the need to minimize risk to their offspring, female bighorns and their young used habitats with fewer predators and had greater opportunities to evade predation than did mature males (Bleich et al. 1997).

In combination with numerical effects of predation on prey, risk-driven shifts in habitat selection may also alter population and community dynamics (Hornocker 1970; Estes 1996; Creel et al. 2005) and, through trophic cascades, influence biodiversity (Ray et al. 2005). Predation by cougars (Ripple and Beschta 2006) and other carnivores may increase biodiversity by preventing primary prey from becoming overly abundant, or by reducing foraging pressure on certain habitats (top-down influences), thus promoting coexistence among prey competitors (Estes 1996) and having positive influences on other species in the community (Figure 10.7). Recent studies indicate that ecosystems can be altered by ungulates in the absence of top carnivores

(Berger et al. 2001; Ripple and Beschta 2006) and following reintroduction of those carnivores (Ripple et al. 2001; Fortin et al. 2005). Ultimately, the existence and strength of behaviorally influenced trophic cascades will also be influenced by how prey perceive and manage the trade-off between food and safety at local scales, and if resultant changes translate to broader scales (Kauffman et al. 2007). Finally, a yet unstudied hypothesis is that a reduction of cougars could lead to an increase in midsized carnivores such as coyotes, a process referred to as *mesopredator release*, and that could in turn lead to increased predation on small prey (see hypothetical example in Figure 10.7).

Cougar Predation on Domestic Prey

Loss of domestic animals to cougars was historically a primary motivation for eradication efforts and remains contentious. Incidents of cougar predation on domestic prey (termed *depredation*) have increased over the last thirty years in the western United States. Cougars have killed cattle, domestic sheep, goats, pigs, immature horses, llamas, alpacas, emus, chickens, ducks, geese, dogs, and cats (Cougar Management Guidelines Working Group 2005), creating conflict on both public and private lands (Wilson et al. 2006). Concern exists not only about loss of livestock but also about the impact livestock grazing has on carnivores and native prey. In many areas of Latin America (see Chapter 6; Hoogesteijn 2000) destruction of habitat and human persecution of pumas and jaguars as a result of documented and perceived depredation negatively affects these carnivores. Livestock grazing directly affects cougar numbers, because depredating individuals may be killed, and stock also may compete with native ungulates, thereby influencing predation patterns on native prey. Livestock grazing has been found to lower the density of grasses, result in competition with native prey for forage, reduce the cover used by prey to hide from predators (e.g., pronghorn fawns, Lee et al. 1998), and increase transmission of livestock-borne diseases to native prey (Fleischner 1994; Lee et al. 1998).

Factors that potentially influence depredation by cougars include high vulnerability of domestic animals, degree and seasonality of their spatial overlap with cougars, husbandry practices, abundance and distribution of native prey, changes in land use, and divisions of space and resources among sympatric carnivores. Relative importance of these factors has not been thoroughly investigated (but see Wilson et al. 2006).

In preying on domestic animals, cougars may kill more prey than they can eat. Cougars have killed five to ten sheep at one site, with generally only one or two eaten (Shaw 1983).

In one incident in Colorado, 350–400 sheep were killed or injured over a two-week period (Anderson and Tully 1989). Similar surplus killing has been reported in other carnivores (Kruuk 1972; Mech and Boitani 2003, 144–45, 306–7; Mills 2007). Because domestic livestock lack ancestral antipredator behaviors, they are easily killed, especially when in fenced areas; for cougars, killing such prey carries little risk of injury and there is no adaptive reason to stop killing (Linnell et al. 1999; Mills 2007). Many small carnivores catch more prey than they can eat at one time and cache them for later use. Thus surplus killing in large carnivores has been proposed as an extension of "multiple killing" behavior exhibited by small carnivores (Linnell et al. 1999; Mills 2007). Surplus killing of native prey by cougars seems rare, except for infrequent reports of predation on mother-offspring or sibling pairs of prey (see Shaw 1977; Murphy 1998; Ruth and

Buotte 2007). The keen antipredatory adaptations of most native prey make kills of native prey more difficult to repeat before prey escape the area (Mills 2007).

Because young carnivores may lack refined hunting skills and feed on easily killed prey, they and very old adult carnivores are thought to prey on livestock to a greater degree than do resident adult individuals (see review by Linnell et al. 1999). Information is limited regarding ages or gender of cougars responsible for livestock depredation. Male cougars were more frequently involved in depredation on livestock than were female cougars in Arizona (Shaw 1977; Cunningham et al. 1995), California (Torres et al. 1996), Nevada (Suminski 1982), and Montana (Aune 1991), and this gender bias holds for other carnivores as well (Linnell et al. 1999). Six of nine cougars involved in Montana depredations during 1989–90 were two years old or younger,

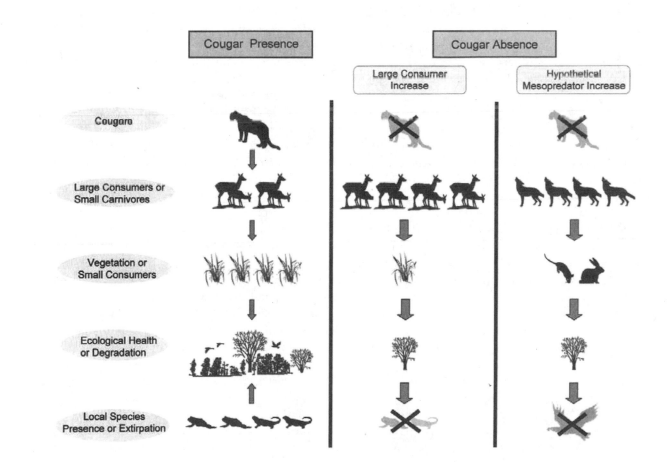

Figure 10.7 Examples of cougar effects on ecological processes. In North Creek, Zion National Park, Ripple and Beschta (2006) documented long-term cougar presence, lower numbers of large consumers (deer), high numbers of cottonwood trees, and a high diversity of amphibians and lizards compared to a second site where cougars were rare. Rarity of cougars in the park's Zion Canyon was linked to higher abundance of large consumers (deer), reduced abundance of cottonwood trees, increased stream bank erosion, and decreased riparian biodiversity, including fewer amphibians, lizards, and butterflies (Ripple and Beschta 2006). A yet unstudied hypothesis is that a reduction in or an absence of cougars can also lead to an increase in the number of mesopredators (e.g., coyotes), which could result in fewer rodents and rabbits and cause reduction or local loss of birds of prey. Down arrows indicate trophic influences with thin arrows indicating more equable influences and thick arrows indicating greater influences. The up arrow indicates species diversity is maintained within an ecological community.

and a majority of depredations coincided with the period (April through October) when many twelve- to twenty-four-month-old cougars become independent from their mothers and disperse from their natal area (Aune 1991). However, older cougars were also occasionally involved in depredations on livestock (Aune 1991). The mean age of cougars killed in depredation cases in southeastern Arizona was fifty four months old (Cunningham et al. 1995) In California, 33 percent of depredations were made by subadult cougars of two years old or less (Torres et al. 1996).

Based on the spatial clustering of sheep depredations, Anderson and colleagues (1992) suggested that a few individual cougars may have killed sheep repeatedly. If individual cougars are responsible for killing livestock, then nonselective programs of cougar elimination may not help diminish depredation problems as efficiently as selective removal (Hoogesteijn 2000). Cattle ranches that removed cougars over twenty-five to forty years still experienced loss of calves to cougars, and Shaw (1981) hypothesized that moderate removal of cougars may actually increase the presence of young cougars. In addition, recent research found that widespread and long-term government-subsidized predator control efforts failed to prevent a decline in the sheep industry; thus, such predator removal programs may have substantial effects on local populations of cougars and other carnivores with little positive impact on the economic viability of this industry (Berger 2006).

Cougar predation on domestic prey varies annually and both across and within states or provinces (Torres et al. 1996). Depredation incidents are typically most common where livestock or pets range within or adjacent to cougar habitat (Cougar Management Guidelines Working Group 2005; Torres et al. 1996). In North America, livestock grazing occurs on about 91 percent of all federal lands in eleven contiguous western states (Armour et al. 1991). In Montana, cougar encounters with livestock occurred during May–October when domestic animals were moved onto summer ranges that overlapped cougar habitat, often on federal lands (Aune 1991). From 1970 to 1990, 68 percent of reported cougar encounters were depredations on livestock east of the Continental Divide, where human densities were low and livestock were widely distributed; west of the Continental Divide 73 percent of reported cougar encounters involved people rather than livestock (Aune 1991). Between 1977 and 1981, an average annual loss of 375 sheep (62 percent lambs) to cougar predation was reported in Nevada; almost all sheep in the state were grazed on summer range in the high mountains (Suminski 1982). The southwestern United States, particularly Arizona, have historically reported a greater amount of depredation than northern states (Shaw 1981). But in these more southerly areas, the season of greatest depredation was not as easily defined because the extent of seasonal overlap between livestock and cougars varied across the region.

Cougar predation on cattle has been linked to the relative abundance of native prey and to cattle management practices that situate calves in cougar habitat. Cougars mainly kill calves and yearling cattle, and predation appears greatest where calves are born in cougar habitat (Shaw 1977; Cunningham et al. 1995) and where deer densities are low (Shaw 1977). In central Arizona, cattle comprised 37 percent of kills (Shaw 1977), but 93 percent of depredated cattle were less than one year old (Shaw 1983). Kills of mature animals weighing more than 300 pounds were infrequent (Shaw 1977, 1983). Cunningham and colleagues (1995) speculated that domestic calves were selected over deer in southeastern Arizona because calves were more vulnerable to predation than were deer. In contrast to high relative availability of and predation on domestic calves on the Spider–Cross U study site (calf:deer ratio of 43:57), a second Arizona study site on the North Kaibab had a low relative availability of calves (calf:deer ratio of 8:92), and few cattle were documented in the diet of cougars (Shaw 1981). Calving away from cougar habitat and allowing calves to gain weight reduced livestock losses for ranchers who were able to implement this management strategy (Ackerman 1982; Shaw et al. 1988). Thus the nature of a livestock operation is relevant to the extent of depredation: losses can be reduced if yearling and adult cattle only are allowed to roam in cougar habitat, and if the numbers of cattle are maintained at a low ratio relative to native prey.

Depredation in the llanos of Venezuela mirrors these findings. A ranch and wildlife preserve of 800 km² experienced livestock losses of 13 percent over ten years attributed to pumas and jaguars (Polisar et al. 2003). Calves between one and thirty days old were the primary domestic prey killed, and losses were greatest during peak calving months (August–October). Cattle maternity pastures were located on high and dry sites during the wet season, and pumas tended to use the adjacent forests more frequently than jaguars and also killed more livestock than jaguars (Polisar et al. 2003). The frequency of cattle depredation was inversely related to the availability and vulnerability of natural prey (Polisar 2000, cited in Hoogesteijn 2000; but note that livestock losses in Latin America vary sharply across regions and types of stock, as described in Chapters 6 and 7).

Little studied is the degree to which livestock availability and density may facilitate a numerical response in cougars, and thus increase cougar predation on native prey (apparent competition or prey switching). Shaw (1981) and Rominger and colleagues (2004) proposed that availability of young calves during livestock calving operations may support cougars through periods of low deer availability, which in turn may result in a greater impact on deer and bighorn sheep at other times of the year. As already noted, in parts of Patagonia cougar predation is preventing the growth of guanaco populations where abundant exotic prey or livestock support high cougar densities (Novaro and Walker 2005). Because cattle constitute an important part of cougar diets in many areas of Latin America, this area of research may provide new insights into whether domestic prey influence cougar predation impacts on native prey (see Chapters 6 and 7).

Large-scale alterations in land use and habitat can influence cougar depredation on livestock. Deforestation can diminish some populations of wild prey, thus increasing depredation of cattle in the absence of natural prey. Deforestation may also predispose livestock to predation because cougars move to more marginal habitats where they potentially have greater access to livestock (Hoogesteijn 2000). Similarly, exurban development may attract deer and other native prey to watered lawns, which exacerbates overlap between cougars and domestics, including pets. The Cougar Management Guidelines Working Group (2005), Hoogesteijn (2000), and Polisar and co-workers (2003) provide valuable suggestions for livestock husbandry practices and protection of native prey populations to reduce cougar depredation problems.

Conclusion

Cougar-prey relationships are complex and variable. Although the theoretical basis for understanding the numerical effects of predators is now reasonably well established, further study of cougars in a broad range of prey systems is still badly needed. Data for many carnivore species—including coyotes, wolves, African lions, and cheetah—indicate that carnivore density is positively correlated with prey density (see review in Fuller and Sievert 2001), although social forces may act to regulate carnivore numbers at high prey densities (Hornocker 1970; Messier 1994, 1995; Lindzey et al. 1994). For cougars, we have not yet achieved a sufficient understanding of these relationships (see Logan and Sweanor 2001; Pierce et al. 2000b), chiefly because we often lack the necessary field data. Nonetheless, knowledge of cougar population responses to varying prey densities in multiple and single-prey systems as well as the cat's influence on top-down processes and biodiversity are critical for management and conservation of cougar-prey systems.

Cougar predation has commonly been viewed as a factor associated with declines in mule deer across the western United States. But differentiating the effects of multiple-prey versus single-prey systems has received little attention.

Recent studies have identified apparent competition and prey switching in some predator-prey systems, yet we have little understanding of the time frames over which these processes are sustained or clear support for their subsidizing effects on cougar numbers. Equally important is understanding how sources of mortality among prey change for populations above, at, and well below ecological carrying capacity (Bowyer et al. 2005). As in Novaro and Walker's (2005) analysis of guanacos, we may be able to identify prey density thresholds that suggest when cougar predation influences on prey require adaptive management. As noted, if cougar populations naturally decline within four to five years after prey populations decline (Logan and Sweanor 2001; Laundré et al. 2007), measures to control cougars may be unnecessary. A broader view that incorporates interactions among prey species might lead to management of alternative prey species or manipulating habitat rather than simply reducing numbers of cougars.

Hornocker (1970) was the first to note that cougar predation may act on lower trophic levels. That is, by redistributing ungulates on winter ranges, cougar predation may have beneficial ecological effects on vegetation and other animal species. Although information regarding top-down/bottom-up regulation and trophic cascades within ecosystems has increased (see Ray et al. 2005), few studies have focused on the role cougars play in structuring communities and influencing biodiversity (but see Novaro and Walker 2005 and Ripple and Beschta 2006). Rather than continuing to focus heavily on prey numbers, we should be striving to understand predation effects on broader ecosystem processes and for insight into where cougars play ecologically significant roles and at what densities they affect trophic interactions (see Ray et al. 2005, 425).

For example, the extent to which carnivores select diseased prey, or those with heavy parasite loads, is of strong interest to managers. Helminthes (blood parasites) were found to increase vulnerability of snowshoe hares and red grouse to predation (Hudson et al. 1992; Murray et al. 1997). Schaller (1972) suggested that predators preferentially selected diseased prey. Do cougars similarly eliminate the most infectious individuals from the prey population, thereby reducing the spread of parasites or disease? If so, population control of cougars and other predators that key on vulnerability in prey may have unintended and harmful consequences (Packer et al. 2003). Another example of improving our understanding of energy flow and trophic interactions would be gaining a clearer grasp of the extent to which cougars provide carrion for avian and mammalian scavengers and how carrion influences detritivores and contributes to soil health (DeVault et al. 2003; Wilmers et al. 2003; Jedrezejewska and Jedrezejewska 2005).

More broadly, the flexibility in cougar and prey behavior and their responses to ecological changes are key components of their coevolution and their role in wildland communities. Anthropogenic influences such as logging, housing development, and livestock grazing can influence prey numbers, behavior, and distribution and thus need consideration. Few predator-prey interactions now occur in the absence of such anthropogenic influences. Carnivores live in increasingly human-dominated landscapes that support altered predation processes on regional and landscape scales. Therefore, the greater the number and type of wildland ecosystems that can be conserved, the greater the potential to conserve natural processes, including the range of interactions between cougars and prey (Mills 2005), which in turn promote genetic health and diversity and ultimately help reduce ecological degradation

Chapter 11 Competition with Other Carnivores for Prey

Toni K. Ruth and Kerry Murphy

IMAGINE STUMBLING ACROSS a deer carcass while hiking in the fresh spring air of Montana. The carcass is cached next to a tree and much of it is covered with grass and pine needles. It looks to be a cougar kill, but you see no cougar and any that had been present probably left the area when you approached. The detective in you is drawn to investigate the scene: to find the pieces to the puzzle and the story behind the death. Suddenly, the snap of a stick sets off warning bells in your brain and reminds you to be cautious: to stop, listen, look over your shoulder, and make noise. Why? Bears. Similarly, cougars also must be ever vigilant for other carnivores that could steal their kill or harm them. In this chapter, we synthesize current knowledge of interactions between cougars and other carnivores and how these interactions may affect predation and prey.

Cougars (*Puma concolor*) are efficient hunters of a wide range of vertebrate prey animals and they inevitably compete to some extent with other predators for prey. Cougars also provide prey in the form of carrion to these other predator species. Fellow members of the predator guild potentially relying on similar prey resources are wolves (*Canis lupus*), bears (*Ursus arctos* and *Ursus americanus*), coyotes (*Canis latrans*), foxes (*Vulpes vulpes, Urocyon cinereoargenteus, Cerdocyon thous*), bobcats (*Lynx rufus*), lynx (*Lynx canadensis*), jaguars (*Panthera onca*), and wolverines (*Gulo gulo*). A guild is a group of species exploiting the same class of environmental resources in a similar way (Root 1967). In doing so, members of carnivore guilds interact directly and indirectly and may suppress populations of other guild members and alter behavior and hunting patterns of animals at the same trophic level through interspecific competition. There is growing recognition that cougars and other carnivores are often engaged in a mixture of competition and predation known as *intraguild predation* (Polis and Holt 1992; Holt and Polis 1997; Creel et al. 2001). More specifically, intraguild predation is the relationship between two or more species where they compete for the same prey, but at least one of the species also preys on the other (Polis et al. 1989). Such interactions may affect predator-prey relationships in complex and dynamic ways (Polis and Holt 1992; Linnell and Strand 2000).

In general, predator diversity and intraguild predation have effects that can indirectly propagate through the food web (termed *trophic cascades*; Beckerman et al. 1997; Pace et al. 1999) and can ultimately affect ecosystem function and species diversity (see Terborgh 1988, 403; Finke and Denno 2005). For example, Finke and Denno (2005) found that increasing predator species richness influenced trophic cascades in a salt marsh food web, but the magnitude and direction of the effects depended on the composition of predators present. Specifically, an increase in numbers of three hunting spiders (*Pardosa littoralis, Clubiona saltitans*, and *Marpissa pikei*) promoted aggressive interactions between these intraguild predators and resulted in a higher density of herbivorous planthoppers (*Prokelisia* spp.) and lower plant productivity. In southern latitudes, pumas and jaguars may control the numbers of peccary (*Tayassu* spp.), paca (*Agouti paca*), and agouti (*Dasyprocta punctata*), which feed on fallen seed from large-seeded trees, and, in turn, may regulate the balance between large- and small-seeded tree species in tropical forest regeneration (Terborgh 1988, 403).

Biologists have traditionally framed cougar population and prey relationships in a single-species context (Linnell and Strand 2000), as is also the case for other large carnivores. However, recent advances in technology, such as use of global positioning system (GPS) transmitters and remote cameras, and collaborations between carnivore researchers (e.g., Ruth et al. 2004) have allowed a shift in emphasis to understanding interactions between carnivores and their relevance to community structure and ecosystem processes (Estes 1996; Linnell and Strand 2000). Because research on multispecies interactions across trophic levels is unwieldy for detailed investigation (Polis 1994), examination of interactions within a guild has been a more common approach (Singer and Norland 1994; Karanth and Sunquist 1995; Creel and Creel 1996; Creel et al. 2001; Ruth 2004b; Ruth et al. 2004; Kortello 2005). Few studies in North, Central, or South America have documented population-level interactions between cougars and other carnivores due to their secretive nature, the large scale at which large predator assemblages operate, and because few places have retained their entire complement of native carnivores. In addition, large carnivores and their prey are often influenced by legal and illegal human hunting, which may cause interactions and predation to deviate from natural patterns.

Competitive Interactions among Carnivores

Two main types of competitive interactions are commonly identified by ecologists. The first are direct, aggressive interactions, known as *interference competition*, leading to the immediate exclusion of individuals from a resource (e.g., food stealing, also termed *kleptoparasitism*). In contrast, *exploitation competition* between species is based on differential efficiency in using shared resources without direct interactions. One species negatively affects the other species' access to and use of resources (Caughley and Sinclair 1994; Mills 2007), sometimes by depletion of the resource used in common. Alternatively, species may actively avoid exploitation or interference competition through shifts in activity, space or habitat use.

Interference and exploitation competition can lead to intraguild predation, the strength of which is influenced by similarity of niche use, abundance of main prey species (i.e., nonguild prey), and the ratio of guild to nonguild prey (Figure 11.1; Polis et al. 1989). In carnivorous mammals, direct interactions are often associated with scavenging or kleptoparasitism at kill sites and frequently (but not always) lead to one carnivore killing another. Species that compete for prey indirectly may have strong competitive influences, yet may avoid direct interactions by partitioning space or being active at different times of the day. Thus, intraguild predation is weak or potentially absent in such situations. Strong intraguild predation can eventually weaken exploitation competition between species (Polis et al. 1989) because predator densities are reduced and the ratio of guild to nonguild prey decreases. Coexistence is also possible if the species more vulnerable to predation from a competitor is superior at exploiting resources (Polis et al. 1989). Intraguild predation is usually asymmetric; that is, interactions do not have equal impact on each competitor, and the dominant competitor may kill the subordinate competitor or exclude

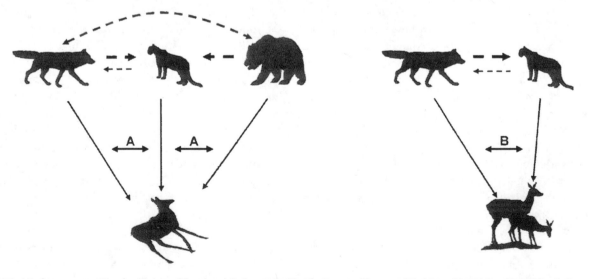

Figure 11.1 Interference competition for dead prey (A) and exploitation competition for live prey (B) can result in intraguild predation (dashed lines) where subordinate cougars are killed by dominant competitors (thick dashed lines). Cougars may occasionally kill a typically dominant competitor, such as a solitary wolf (thin dashed line), but more likely cougars are killed by wolves and bears. Solid arrows indicate that one entity (point of arrow) is eaten by another entity (base of arrow). Intraguild predation may increase as the abundance of nonguild prey (elk, deer) decreases; as the ratio of guild to nonguild prey increases; or with similarity of resource and niche overlap between cougars and competitors (see Polis et al. 1989).

it from certain resources. A host of factors influence the outcome and magnitude of dominance hierarchies between cougars and other carnivores: environmental changes, body size, prey availability and density (and hence dietary overlap), sociality or grouping behavior, habitat and temporal overlap, carnivore population densities, and predatory behavior (Palomares and Caro 1999; Creel et al. 2001; Donadio and Buskirk 2006).

Interference Competition

Because cougars are solitary and specialized at killing disproportionately large prey (Hornocker 1970; Kleiman and Eisenberg 1973), they may require extended periods, typically three to six days, to consume a kill (Shaw 1977; Murphy 1998; Ruth and Buotte 2007). During this time cougars attempt to conceal their prey from scavengers. This pattern differs from communal feeding in social carnivores such as wolves, which generally feed for up to one day per kill (Peterson and Ciucci 2003; Smith and Bangs 2009). Murphy 1998 estimated that up to twenty-one independent adult and subadult cougars annually provided about eight hundred dead elk (*Cervus elaphus*) and mule deer (*Odocoileus hemionus*) of various age classes on the northern range of Yellowstone National Park. Thus, cougar kill sites may function as potential epicenters of carnivore interaction to a greater degree than do kills of smaller felids (i.e., bobcats and ocelots, *Leopardus pardalis*), which eat smaller prey, or social canids (i.e., wolves and coyotes), which tend to consume prey more quickly. Nonetheless, cougars evolved in environments with carnivorous competitors and have adapted a variety of strategies to minimize detection of their kills and to optimize escape when direct encounters occur.

One behavioral strategy cougars employ is to make or place their kills in areas that minimize detection by vocal avian scavengers, such as ravens and magpies, the activity of which alerts mammalian scavengers like coyotes, wolves, and bears and assists these species in detecting carcasses (Stahler 2000). Cougars typically kill their prey in areas with ample ground or tree cover, or drag killed prey to such places. They also cover or "cache" their kills with debris of conifer duff, grass, and sheared hair from prey, thus minimizing detection of their kills by scavengers (Figures 11.2, 11.3; Connolly 1949; Beier et al. 1995; Bank and Franklin 1998). If a direct encounter occurs, cougars can escape by running into rock outcrops or patches of downed trees or by quickly climbing trees. They also use or remain near habitats with vertical and horizontal cover that provides such escape options (Beier et al. 1995).

Although these adaptations may be effective, other carnivores do detect and encounter cougars at cougar-killed prey.

Figure 11.2, 3 Small ungulate prey, such as deer (top), are easily cached in heavy cover and are more quickly consumed by cougars, thus reducing detection by avian and mammalian scavengers. Because they are more difficult for cougars to kill, large ungulate prey like bull elk (bottom) may be killed in open habitats more commonly than smaller prey are. In such instances, cougars cannot move or conceal the carcass as easily, and larger prey are more readily detected by scavengers and competitors, such as wolves and bears.
Top photo © Ken Logan; bottom photo by Terry Hofstra, National Park Service.

Wolves visited or scavenged 22.5 percent and 33 percent of cougar kills and directly usurped between 6.3 percent and 14 percent of cougar kills in Yellowstone and Banff national parks, respectively (Kortello 2005; Ruth and Buotte 2007). Similarly, wolves visited or scavenged 18 percent of cougar kills in the central Idaho wilderness (Akenson et al. 2005) and 20 percent of cougar kills near Glacier National Park (Kunkel et al. 1999). Both black bears (40–200 kg, Wilson and Ruff 1999) and grizzly bears (98–270 kg, Schwartz et al. 2003) outweigh cougars and may rely heavily on carcasses prior to appearance of vegetative forage during spring and during fall prior to winter denning (Green et al. 1997). Detection of carcasses by bears is probably enhanced by their refined olfactory senses, and bears have arrived at cougar kills within four to six hours after the kill was made (Ruth and Buotte 2007). In Glacier and Yellowstone national parks, both species of bears visited cougar kills (13 percent in Glacier, 33 percent in Yellowstone) and often displaced the cougars (four of eight carcasses in Glacier, seven of nineteen in Yellowstone; Murphy et al. 1998). Cougars occasionally scavenge kills of other cougars or animals that have died of natural or other causes (e.g., road-killed mammals; Logan and Sweanor 2001; Bauer et al. 2005; Mattson et al. 2007), but scavenging the kills of dominant competitors may carry greater risk of injury or death and has been less frequently reported (but see Akenson et al. 2005, 182; Kortello 2005, 32, Table 1).

Cougars can incur injuries or be killed when defending their kills, and the outcome of encounters may depend not only on body size but also on group sizes of competitors. When opponents in interspecific interactions are similar in body size, fighting may be avoided because attack carries a high risk of injury or death (Donadio and Buskirk 2006). Operating as a group or pack has advantages in the outcome of interactions, and carnivores that group can kill species that weigh up to twelve times their own body mass (Palomares and Caro 1999). Although cougars (35–85 kg) and wolves (32–64 kg) are similar in size, the social nature of wolves enhances their competitive ability, and in direct interactions with cougars they tend to be dominant. Wolves have directly killed kittens, subadults, and adult cougars in Glacier (White and Boyd 1989; Boyd and Neal 1992; Ruth 2004b) and Yellowstone national parks (Ruth 2004a; Ruth and Buotte 2007). Reports of cougars killing wolves are less common. In Alberta, an adult female wolf was killed by a cougar in March 1985 (Schmidt and Gunson 1985), and near Yellowstone National Park cougars killed solitary subadult wolves in two separate instances (Yellowstone Science 2004). In direct encounters that resulted in death, biomass of wolf packs outweighed cougars by an average of almost 13:1, whereas the biomass ratio was 1:1 when a cougar killed a wolf (Ruth and Smith, unpublished data). Pumas are occasionally killed by jaguars, their main competitor in more southerly latitudes (Quigley 1987; Rodrigo Nuñez, pers. comm., 2007).

Felids and canids of smaller body mass are often subordinate to larger carnivores. In their interactions with cougars, smaller carnivores trade the risk of being killed against the energetic benefits of scavenging. There are numerous reports of cougars killing bobcats, coyotes, and foxes near cache sites of cougar-killed prey (bobcats—Koehler and

Hornocker 1991, Beier and Barrett 1993; coyotes—Boyd and O'Gara 1985, Koehler and Hornocker 1991, Murphy et al. 1998, Arjo and Pletscher 1999, Ruth and Buotte 2007; foxes—Logan and Sweanor 2001, Ruth and Buotte 2007). This often occurred when subordinate carnivores scavenged at cougar kills (Figure 11.4); Koehler and Hornocker (1991) found that 40 percent of thirty-three cougar kills were visited by coyotes. The degree to which wolverines scavenge from and interact with cougars at cougar kills is not known, but it may be rare due to use of habitats at different elevations, particularly in winter (B. Inman, pers. comm., 2007); cougars have nevertheless occasionally killed wolverines (Hornocker and Hash 1981; Krebs et al. 2004). Smaller felids and canids can kill cougar kittens of similar or smaller body mass. In New Mexico, a six-month-old cougar cub was killed by coyotes (Logan and Sweanor 2001).

Cougars vary in their consumption of other carnivores (Koehler and Hornocker 1991; Beier and Barrett 1993), and consumption can be influenced by prey density and the availability of alternative prey (Palomares and Caro 1999). In some areas, small carnivores are part of the cougar diet and are more consistently consumed. Puma and jaguar predation was found to account for up to half of all mortality to adult white-nosed coatis (*Nasua narica*) in Mexico (Aranda and Sánchez-Cordero 1996; Hass and Valenzuela 2002). In the Swan Valley of Montana, cougars accounted for fourteen out of fifteen lynxes killed by predators from 1998 to 2006 (J. Squires, pers. comm., 2007).

When their kills are frequently detected, cougars are likely to suffer adverse energetic costs when displaced from kills

Figure 11.4 Numerous scavengers may detect and feed at cougar kills, including coyotes and red foxes. These smaller and typically subordinate carnivores may be killed by cougars defending their prey. Photo by Kerry Murphy.

by a dominant competitor (Murphy et al. 1998). Estimates of cougar food loss to other carnivores are extremely rare; however, when cougars are displaced from their kills, they may compensate for losses by killing additional prey, thus increasing their rate of predation to offset energetic costs of displacement. Although scavenging by bears in Yellowstone National Park was sporadic and did not increase cougar predation rate overall, the ungulate biomass available was 71 to 113 percent of the daily caloric requirements of bears, while the losses amounted to 17 to 26 percent of cougar daily energy requirements (Murphy 1998). In British Columbia, Harrison (1990) reported that coyotes displaced cougars from their kills less frequently where coyote numbers were controlled by humans. Where coyotes were not reduced by humans, a maternal female cougar killed more frequently than did a maternal female living where coyotes were controlled (Harrison 1990). Frequency of displacement for individual cougars may vary substantially within a system due to differences in spatial overlap with other carnivores (Ruth 2004b; Kortello 2005).

Little is known about whether frequent loss of kills to dominant competitors has adverse effects on cougar reproduction and survival at a population level. However, in the North Fork of the Flathead River in Montana, Ruth (2004b) hypothesized that interactions with and loss of food to bears during spring may have accentuated conditional stress for cougars that lost kills during winter to wolves and to bears active outside dens (Ruth and Gniadek 1996). In addition to three cougars killed by wolves and one by a bear, the deaths of five adult cougars (four to nine years old), representing 23 percent of cougar mortalities between 1993 and 1996, were due to starvation. Based on body condition indices, cougars dying from causes other than starvation were also in poor condition (Ruth 2004b). In Banff National Park, Kortello (2005) documented low annual survival rates for cougars (0.51; 95 percent C.I. = 0.30, 0.88); found evidence of both interference and exploitation interactions between cougars and wolves; and suggested that food limitation due to an elk decline (deriving from both human relocation of elk from the study area and wolf predation) contributed to cougar mortalities during the winters of 1999 to 2003.

Exploitation Competition

Exploitation competition typically occurs when carnivores share the same food resources. In many reported cases where intraguild predation occurs, substantial dietary overlap exists between the dominant and subordinate predators (Polis et al. 1989). Therefore, in systems where cougars interact with other carnivores at kill sites, they are also likely engaged in exploitation competition for prey, space, and other resources. For example, cougars and wolves are

similar in size, are mobile within large home ranges (72–826 km^2 for cougars; 69–2,600 km^2 for wolves; Logan and Sweanor 2000; Mech and Boitani 2003), and both rely primarily on ungulate prey larger than 25 kg. Where overlap was high in winter cougar and wolf diets (elk, mule deer, and white-tailed deer, *Odocoileus virginianus* prey), exploitation competition and intraguild predation between cougars and wolves were also documented (Table 11.1; Ruth 2004b; Kortello 2005). In Yellowstone National Park, elk constituted 80 percent of annual cougar diets and 90 percent of wolf winter diets (Smith 2005; Ruth and Buotte 2007). Cougars and wolves rely on large ungulate prey throughout the year, although access to and use of smaller prey probably relaxes competition for larger prey to some degree. Determining the degree of dietary overlap between these two carnivores, particularly during summer, is logistically difficult for biologists (D. W. Smith, pers. comm., 2005).

Various factors influence dietary overlap and the potential for exploitation competition between species; and the magnitude of interactions varies temporally and geographically. Dietary overlap is lower between cougars and smaller carnivores (coyotes and bobcats, Table 11.1) as diets of the latter generally include more small prey of less than 10 kg (Hornocker 1970; Koehler and Hornocker 1991). Thus, exploitation competition between these smaller species and cougars may be less than between cougars and other large carnivores, but this, too, likely varies by season and latitude (see later discussion). Changes in prey availability also influence the degree of dietary overlap and potential for exploitation competition between species. In Big Bend National Park, dietary overlap between cougars, coyotes, and bobcats was lower during 1980–81 than 1972–74 (Table 11.1) due to a decline in desert mule deer populations during the later years, which altered diets of all three carnivores (Leopold and Krausman 1986). The long growing season and availability of vegetative forage in South Florida may explain the infrequent use of large vertebrate prey by black bears and hence low dietary overlap (Table 11.1) between bears and panthers in Florida (Maehr 1997b).

In certain systems, predation on elk calves by bears, wolves, and coyotes could reduce the availability of this important prey to cougars. Near Jackson Hole, Wyoming, 68 percent of newborn elk mortality resulted from predation by black bears and coyotes (Smith and Anderson 1996). In north-central Idaho black bears caused 94 percent of the total mortality (Schlegel 1976). Prior to wolf reintroduction in Yellowstone National Park, bears accounted for 23 percent and coyotes for 17 percent of confirmed elk calf deaths (Singer et al. 1997). After wolf reintroduction, bears accounted for 58–60 percent, wolves 14–17 percent, and coyotes 10–11 percent of newborn elk (Barber-Meyer 2006). Cougars killed newborn elk infrequently, 0–3 percent, until

Table 11.1 Dietary overlap (by percentage occurrence of prey items in diet) between cougars and sympatric carnivore species.

Carnivore Pair	Primary Prey Species	Dietary Overlap[a]		Sample Type (Size)[b]	Season[c]	Study
		By Species	By Age & Species			
Cougar-Wolf	White-tailed deer, elk	RI = 0.91 HI = 0.07	RI = 0.60 HI = 0.06	Kills (302)	W	North Fork of the Flathead River, Montana (Kunkel et al. 1999, Ruth 2004a)
	Elk, mule deer	RI = 0.96 HI = 0.99	RI = 0.83 HI = 0.96	Kills (218)	W	Panther Creek, Idaho (Husseman et al. 2003)
	Elk, mule deer	RI = 0.44–0.90	n/a	Kills (255)	W	Bow Valley, Alberta (Kortello 2005)
Cougar-Coyote	Elk, mule deer	RI = 0.49 HI = 0.74	n/a	Kills (57)	W	Big Creek, Idaho (Koehler and Hornocker 1991)
	Deer, collared peccary, rabbits	RI = 0.36 & 0.26[d] HI = 0.59 & 0.47	n/a	Scats (406 and 682)	A	Big Bend National Park, Texas (Leopold and Krausman 1986)
Cougar-Bobcat	Elk, mule deer	RI = 0.45 HI = 0.65	n/a	Kills (40)	W	Big Creek, Idaho (Koehler and Hornocker 1991)
	Deer, collared peccary, rabbits	RI = 0.39 & 0.36[d] HI = 0.67 & 0.52	n/a	Scats (289 and 488)	A	Big Bend National Park, Texas (Leopold and Krausman 1986)
	Feral hog, white-tailed deer	RI = 0.12 PI = 0.13	n/a	Scats (n/a)[e]	A	South Florida (Maehr 1997)
Cougar-Black Bear	Feral hog, white-tailed deer	RI = 0.05 PI = 0.015	n/a	Scats (n/a)[e]	A	South Florida (Maehr 1997)
Cougar-Jaguar	Livestock, white-lipped peccary, capybara, deer	RI = 0.50–0.55 PI = 0.70	n/a	Scats (84) Kills (80)	A	Hato Piñero, Venezuela (Maxit 2001; Scognamillo et al. 2003; Polisar et al. 2003)
	White-tailed deer, coati, collared peccary, armadillo	RI = 0.68 PI = 0.84	n/a	Scats (115)	A[f]	Chamela–Cuixmala Biosphere Reserve, Jalisco, Mexico (Nuñez et al. 2000)
	Collared & white-lipped peccary, white-tailed and brocket deer, coati, agouti	RI = 0.34 & 0.36[g] PI = 0.30 & 0.33	n/a	Scats (221)	A[h]	Maya Biosphere Reserve, Guatemala (Novack et al. 2005)
	Brocket deer, rabbits, armadillos, collared- white-lipped peccaries	RI = 0.65	n/a	Scats (201)	A	Chaco, Parguay (Taber et al. 1997)
	Collared peccary, deer, coati, armadillo	RI = 0.45 HI = 0.65	n/a	Scats (52)	A	Calakmul Biosphere Reserve, Campeche, Mexico (Aranda and Sánchez-Cordero 1996)
Cougar-Ocelot	Agouti, paca, red brocket deer, collared peccary, sloth	RI = 0.63 PI = 0.72	n/a	Scats (278)	A	Barro Colorado Island, Panama (Moreno et al. 2006)

[a]RI = Renkonen's Index (Renkonen 1938, cited in Krebs 1989), HI = Horn's Index (Horn 1966), and PI = Pianka's Index (Pianka 1973) of dietary overlap range from 0 (no overlap) to 1.0 (complete overlap). RI is a simple measure of percentage similarity and is not sensitive to how resource categories are divided, while HI is more sensitive to bias associated with how resource categories are divided and is less sensitive to differences in sample sizes (Krebs 1989). PI is another commonly used index of dietary or niche overlap (Krebs 1989). RI and, where possible, HI were calculated by chapter authors for the following studies: Aranda and Sánchez-Cordero (1996), Husseman et al. (2003), Koehler and Hornocker (1991), Leopold and Krausman (1986), Moreno et al. (2006), Novak et al. (2005), Nuñez et al. (2000), Scognamillo et al. (2003).
[b]Dietary overlap was based on sampling kills or scats.
[c]W = winter (generally November through April); A = annual.
[d]Diets of cougars, bobcats, and coyotes were determined during two time periods, 1972–74 and 1980–81, and dietary overlaps and scat sample sizes are presented in chronological order to reflect these two time periods. The authors averaged relative frequency across the seasons presented in Leopold and Krausman (1986, 293, table 3) in order to calculate annual dietary overlap.
[e]Percent frequency of food items in diets was summarized from several studies (see Maehr 1997, 28).
[f]Scats were mainly (85 percent) obtained during the dry season of November through April (Nuñez et al. 2000).
[g]RI and PI were calculated for an area with subsistence hunting (PI = 0.33) and without subsistence hunting (PI = 0.30).
[h]Samples were mainly (95 percent) obtained during December to July (Novack et al. 2005).

calves became more mobile (0 percent, Smith and Anderson 1996; 1–2 percent, Singer et al. 1997; 2–3 percent, Barber-Meyer 2006). Barber-Meyer (2006, 70) documented that mean calf age at death was 10 days for bear kills, 29 days for coyote kills, 35 days for wolf kills, and 107 days for cougar kills.

Using dietary overlap to measure exploitation competition without benefit of other data on resource partitioning is problematic. For example, although high observed overlap may imply competition for resources, it may also indicate the absence of competition, if we assume that persistent competition should already have reduced resource overlap (Putman 1996). Therefore, the magnitude of competition, and whether it results in one species excluding another or in coexistence, depends not only on overlap in forage consumed by the two species but also on the degree of habitat and spatial overlap, degree of interference competition, and whether the availability of food, habitat, or other resources is limited (Wiens 1989; de Boer and Prins 1990). Additionally, while dietary overlap in use of prey species may exist, differential hunting strategies and hunting success rates between carnivores suggest that prey selection at a finer scale should differ by age (and thus size), sex, and condition between stalking predators (cougars) and coursing predators (wolves, coyotes). This could minimize the degree of dietary overlap and could influence combined effects on prey. Slightly lower indices of overlap emerge if the proportion of prey age classes in cougar and wolf diets are included than in comparisons of prey without age specificity (Table 11.1).

Both cougars and wolves selected disadvantaged prey in Idaho, but wolves exhibited a greater overall tendency to select juveniles and malnourished deer and elk than did cougars (Husseman et al. 2003). In other systems, researchers found little support for differential prey selection due to differences in hunting behavior between cougars and wolves (Kunkel et al. 1999) and between cougars and coyotes (Pierce et al. 2000). Kunkel and colleagues (1999) found that both cougars and wolves preferred the most abundant prey, white-tailed deer, over less abundant elk and moose (*Alces alces*). Yet, there was no difference in the age or sex ratio of deer killed by cougars and wolves. In the Sierra Nevada, California, cougars selected young mule deer as prey, but among adult mule deer, cougars and coyotes both selected older age classes, and predation was not associated with especially poor body condition (Pierce et al. 2000). In instances where prey selection differed little between sympatric canids and felids, predatory behavior by canids was thought to be influenced by dense vegetation that permitted them to approach and kill prey in a felid-like manner (Murray et al. 1995; Kunkel et al. 1999).

When sympatric with jaguars, puma showed a preference for smaller to medium-sized prey of up to 15 kg, while jag-

uars showed a preference for prey larger than 15 kg (Taber et al. 1997; Polisar 2000; Maxit 2001; Polisar et al. 2003; Scognamillo et al. 2003). Dietary overlap ranged from 45 to 70 percent (Table 11.1). Both pumas and jaguars selected for collared peccary (*Pecari tajacu*) in Venezuela, but pumas primarily killed juvenile peccaries (Maxit 2001; Scognamillo et al. 2003), and puma dietary breadth exceeded that of jaguar (Maxit 2001). In Mexico, white-tailed deer made up 54 percent and 66 percent of jaguar and puma diets, respectively, and although there was high overlap (Table 11.1), pumas had a broader food niche than jaguars (Nuñez et al. 2000). In these studies, differences in prey use and selection were explained by body size differences between pumas (26–51 kg) and jaguars (52–87 kg) and jaguars' greater use of habitats near water and deeper in the forest (>500 m from the forest edge; Scognamillo et al. 2003). Because pumas take a greater range of prey sizes, they may be superior exploitative competitors to jaguars and have an advantage when faced with human-induced prey and habitat changes (Nuñez et al. 2000; Maxit 2001).

Results from these studies suggest some general patterns related to size, abundance, and diversity of prey, and to habitat structure. First, there may be little to no differential prey selection between cougars and wolves or cougars and coyotes where smaller ungulates are the primary prey (i.e., deer are most abundant) or where landscape and vegetation features reduce opportunities for coursing by the canids. Conversely, greater differences in prey selection might be expected when larger prey (elk) are more abundant or where the prey's frequent use of open habitats enhances the coursing predator's ability to select injured or nutritionally stressed prey. Finally, dietary overlap may increase when primary prey is highly abundant (Colwell and Futumaya 1971), may decrease due to prey switching when abundance of primary prey decreases (see Kortello 2005), or may decrease due to diversity of available prey promoting greater dietary breadth for one predator or both (Nuñez et al. 2000; Maxit 2001). Research in various environments, including more southerly latitudes, could enhance understanding of these possible patterns.

Being dominant to some species of carnivores but subordinate to others, cougars can have influences both upward and downward in the carnivore guild and across trophic levels, the extent of influence varying with carnivore and prey diversity and environmental conditions. For example, while loss of prey to other carnivores can negatively affect cougars, access to cougar-killed prey may directly influence the behavior and survival of other carnivores. Meat from ungulates becomes more important to grizzly bears during years with poor whitebark pine (*Pinus albicaulis*) cone crops in the Greater Yellowstone Ecosystem (Felicetti et al. 2003). In addition, carcasses available to bears from cougar and other carnivore kills probably increase in importance during these periods, as well as during spring seasons in years when winter-killed ungulate carcasses are less available (Green et al. 1997).

Similar to the positive effects of wolf presence on red fox (*Vulpes vulpes*) populations (Carbyn et al. 1993), competition from cougars may also increase population size and species richness among smaller predators (Terborgh et al. 1999; Cougar Management Guidelines Working Group 2005). Coyotes and cougars compete directly for mule deer in the eastern Sierra Nevada of California. As cougar populations decreased, predation on mule deer by coyotes increased (Pierce et al. 1999, cited in Cougar Management Guidelines Working Group 2005), suggesting that cougars potentially limit coyote numbers. Since coyotes probably compete more directly with bobcats and foxes (Ralls and White 1995; Crabtree and Sheldon 1999), the presence of cougars and their influence on coyotes may benefit these smaller carnivore species (Cougar Management Guidelines Working Group 2005).

Some Factors Influencing Interactions

In northern latitudes, seasonal changes in climate and prey distribution greatly influence the potential for interference and exploitation competition. Both cougars and wolves rely on migratory elk and deer that become concentrated on low-elevation winter ranges due to accumulating snow (Murphy 1998; Kunkel et al. 1999, Ruth 2004b; Ruth and Buotte 2007). In the North Fork of the Flathead River in Montana and British Columbia, cougars and wolves had greater spatial overlap and winter dietary overlap (40–55 percent spatial overlap in home ranges, 13–20 percent in core areas; Ruth 2004b; Kortello 2005). During winter in Idaho and Montana, there was little to no difference in the spatial distribution of wolf kills versus cougar kills (Kunkel et al. 1999; Husseman et al. 2003). Wolves are highly mobile, and Ruth (2004b) hypothesized that because of increased overlap during winter, traveling and hunting wolves are more likely to detect and encounter cougar-killed prey than during summer. Overall, more carnivore interactions have been documented during the cold season (Koehler and Hornocker 1991; Palomares and Caro 1999), and all reported cases of intraguild predation between cougars and wolves were during winter (November 1–April 15; White and Boyd 1989; Boyd and Neale 1992; Ruth 2004b; Kortello 2005; Ruth and Buotte 2007). Overlap in food and habitat use among cougars, bobcats, and coyotes was significant during winter, and cougars killed five bobcats and two coyotes near kills (Koehler and Hornocker 1991).

Interactions with bears are the exception to this pattern due to the bears' winter denning behavior and seasonal foraging, which limit interactions at kill sites to spring, summer, and fall.

During summer, changes in the availability of prey types, distribution of prey on the landscape, vegetation, and carnivore reproductive behaviors affect carnivore distribution and prey selection. Cougars often follow prey species to higher elevations (Kunkel et al. 1999, Pierce et al. 1999; Ruth 2004b), whereas wolves tend to restrict their movements to denning and rendezvous areas in valley bottoms when provisioning young offspring (Arjo 1998). Ruth (2004b) documented that overlap in cougar and wolf home ranges relaxed during summer months (30 percent in home ranges, 7 percent in core areas). Compared to winter, there was less overlap in use of topographical and habitat features by cougars, bobcats, and coyotes during summer, and cougars used higher elevations more than either bobcats or coyotes (Koehler and Hornocker 1991). This dichotomy in behavior and the greater availability of smaller and alternate types of prey may relax competition and enhance coexistence between cougars and certain species of carnivores over the long term. For other species of carnivores, spatial overlap with cougars may increase during summer months. In the Swan Valley of Montana, lynx became more vulnerable to predation during the snow-free months when cougars and other predators were present in high-elevation lynx habitat (J. Squires, pers. comm., 2007).

In South America, changes in prey movement patterns and habitat availability between the dry and wet seasons may also affect the degree of interference and exploitation competition. During the dry season, pumas were more active than jaguars during both night and daytime; but in the rainy season, jaguars were more active than pumas by day (Scognamillo et al. 2003). Pumas more frequently used areas bordered by dry forest (Maxit 2001), and jaguars used more mesic habitat (Scognamillo et al. 2003). Novack and colleagues (2005) noted that during the dry season, armadillos (*Dasypus novemcinctus*) and coatis—the primary prey species of jaguar in Maya Biosphere Reserve—foraged in areas of greater soil moisture, which were more mesic sites preferred by jaguars, a pattern that probably reduced dietary overlap between jaguar and puma. At two study sites in Panama, the diet of ocelots and pumas was most similar in the wet season when both carnivores frequently ate agoutis (Moreno et al. 2006).

Habitat structure and complexity may influence detection of kills and active avoidance of one species by another. Visibility can exacerbate competition between carnivores, and open habitats appear to intensify competitive interactions

(Creel et al. 2001). Dense shrub and conifer vegetation along the North Fork of the Flathead River Valley in Montana and British Columbia could reduce detection of cougar-killed prey by wolves as compared to more open habitats (e.g., grasslands) of Yellowstone National Park.

Prey density probably also influences competition between carnivores. For example, abundant prey characteristic of much of the Greater Yellowstone Ecosystem could reduce competition for prey and kleptoparasitism among its multiple large carnivores. Competition theory predicts that increasing prey density will weaken competition due to increased food availability, but it assumes that live prey are the resource for which carnivores compete (Creel et al. 2001). Creel and colleagues (2001) suggested, in contrast, that if competition is for carcasses and not live prey, then increasing prey density may support a higher density of competitors, and the risk of subordinate species experiencing food loss and predation may increase. Collaborative research in the Greater Yellowstone Ecosystem is currently assessing interspecific competition among cougars, wolves, and bears at ungulate carcasses and how landscape and carnivore diversity, density, and behavior interact to drive that competition (H. Quigley, pers. comm., 2006).

Influences on Prey

In North America, studies of predation in multicarnivore systems have suggested that predation by multiple carnivores may have strong limiting effects on prey (Bergerud 1992; Kunkel 1997). Predation by cougars, wolves, and bears in the North Fork of the Flathead River was thought to be largely additive (Kunkel 1997). Mech and Peterson (2003, 157) felt that the combined effects of wolf and bear predation could be sufficient to reduce primary prey populations to levels below that which could be supported by their forage base. Based on these studies, we might assume that the combined influence of cougar, wolf, and other carnivore predation on prey is likely to have a more suppressing effect on the numbers of prey compared to levels in environments with cougars alone. Yet, in multipredator systems, the interplay of competitive influences can also affect carnivore density and distribution and may result in more complex and variable predatory influences on prey species (Polis and Holt 1992; Palomares and Caro 1999; Gunzburger and Travis 2005). Intraguild predation can produce a combined predator effect that is less than additive since predators reduce their own total density by treating one another as alternative prey (Holt and Polis 1997); therefore, intraguild predation can operate to stabilize the predator trophic level and effects of predation on prey, and it may dampen trophic

cascades (Finke and Denno 2005). In southern Spain, for example, small predators avoid areas occupied by Iberian lynx, resulting in an increase in rabbit density in areas of lynx activity (Palomares et al. 1995). Competition between cougars and coyotes for similar prey (Pierce et al. 1999) has unknown yet possibly stabilizing effects on predation. On the northern range of Yellowstone National Park, increased intraspecific strife in the wolf population appeared related to high total predator densities of 121–129/1,000 km^2 for cougars, wolves, and grizzly bears between 2000 and 2003 (calculated from Ruth et al. 2004; Ruth and Buotte 2007; Smith and Bangs 2009). The situation may also explain declines in cougar (since 2003; Ruth and Buotte 2007) and wolf densities; yet, as compared to other northern systems, elk numbers have remained relatively high at 7,000–9,000 elk/1000 km^2 (Smith and Bangs 2009).

Carnivores can shift from one prey species to another because the densities and sex and age composition of different prey populations may change from year to year in uncorrelated patterns (Soulé et al. 2003). During declines in primary prey abundance, the degree of dietary overlap and interspecific competition between cougars and wolves may be highly dependent on the availability of alternative prey (Ruth 2004b; Kortello 2005). Due to their ability to stalk close to prey in rugged terrain, cougars are better adapted than wolves for predation on bighorn sheep (*Ovis canadensis*), and in some systems cougars are capable predators on pronghorn (*Antilocapra americana*) as well, particularly during the fawning season (Ockenfels 1994; Ruth and Buotte 2007). Declines in elk numbers in Banff National Park and exploitation competition for food between cougars and wolves appeared to elicit a dietary shift in cougars from elk to mule deer and bighorn sheep (Kortello 2005). Diet overlap in prey species between cougars and wolves was 90.9 percent in 1999–2000, dropped to 72.6 percent in 2000–2001, fell to 43.6 percent in 2001–2002 (when elk numbers were at their lowest), and increased again to 78.9 percent in 2002–2003. While wolves also demonstrated a dietary shift from elk to alternate prey, wolf dietary response lagged behind that of cougars by one year (Kortello 2005). Dietary switching may have consequences for alternative prey populations, such as bighorn sheep or pronghorn; these are species that tend to occur at lower densities and in smaller groups and for which cougar predation has been shown to have population-level impacts (Wehausen 1996; Ross et al. 1997; Hayes et al. 2000; Logan and Sweanor 2001).

If an antipredator behavior that is effective against one species dramatically increases predation from another species, the effects of combined predators can be greater than expected (Soluk 1993). Wolf predation may facilitate cougar access to and predation on some prey species in certain areas (see Atwood et al. 2007). In the North Fork of the Flathead River in Montana, white-tailed deer selected denser canopy and hiding cover in the presence of wolves, which may have placed them at greater risk for predation by cougars (Kunkel 1997). In the Greater Yellowstone Ecosystem, elk avoided areas of high wolf density in summer and selected higher elevations, less open habitat, steeper slopes, and more burned forest after wolf recovery compared to before wolf recovery (Mao 2005). On a finer scale, elk moved into the protective cover of wooded areas when wolves were present (Creel et al. 2005). Selection of more structurally complex habitats by elk might facilitate predation by cougars (Atwood et al. 2007) as cougars generally use steeper slopes (Ruth 2004b), make kills on steeper slopes (Husseman et al. 2003), and hunt in less open habitats than wolves. In contrast to summer, Mao (2005) found that elk were unable to separate themselves spatially from wolves in winter and selected more open habitats after wolf recovery. Thus, an alternative scenario may exist during winter: wolf presence may impede cougar predation on elk at a time when the two carnivores and elk are overlapped on winter range. Considerable variation should exist among ungulate prey in the way they respond to predation risk (Creel et al. 2005), especially considering the diversity of carnivore hunting strategies and habitat features in ecosystems inhabited by cougars.

Conclusion

The examples given emphasize that management and conservation of prey require knowledge not only of their predators but also of their predators' predators (Linnell and Strand 2000, 172). Competition among carnivores is an expanding area in cougar research, raising new questions about how different predator combinations affect both the carnivores and the prey populations in particular settings. The cougar's broad geographic distribution and adaptability lend considerable variability to its interactions with prey and other carnivores. Simple conclusions are elusive. Some cougars must compete with other large carnivores, while others coexist only with smaller predators; some live in equable climates with a high diversity of prey, whereas others must deal with extreme winters and lower prey diversity. Today cougars and other carnivores are often confined to protected areas to a greater extent than other animals (Creel et al. 2001), and intraguild interactions may play out differently in such areas than in places where carnivore and prey densities are heavily affected by human hunting and development. In planning for conservation of cougars and their prey, assessment of how cougar predation is influenced by competition and intraguild predation can provide a better understanding of the role of cougars in structuring communities and ecosystem processes.

Part IV

Conservation and Coexisting
with People

Death of a Towncat

Harley Shaw

SEVENTEEN YEARS after I left the Arizona Game and Fish Department, I became involved in a cougar incident in my neighborhood. As a game biologist, I'd occasionally responded to urban cougar incidents. In Yarnell, for example, a woman reported a cougar eating from a dish on her porch. I was skeptical, but found tracks of a female cougar with kittens in the lady's yard and all over town. I recommended that townsfolk give the cat time to go away. Apparently, it did.

I also investigated reports in Chino Valley, where I lived. These reports had cougars sunning on people's roofs and denning in their yards. None held up to scrutiny, stemming from untraceable rumors. I learned to maintain healthy skepticism but not dismiss reports out of hand. Investigation on the ground was essential, and the best investigations could never completely disprove a sighting. I learned to listen respectfully, even to incredible stories.

So, when I heard that Mrs. Huber, six houses down my street after I had moved to New Mexico, had a cougar in her yard, I assumed my old professional skepticism. Something had killed a feral cat and a raccoon near a feeding station she maintained. She said they had been partially buried, which certainly suggested cougar or bobcat. But she cleaned up the remains before anyone saw them.

At four the next morning, a vehicle stopped in front of our house. The driver was our local forest ranger.

"Mrs. Huber has a lion in her back yard!"

At her home, Mrs. Huber, her son, and the ranger's son were gathered at a window watching a cougar munch on a domestic cat. Mrs. Huber was anxious but admirably reasonable. Her main concern involved her son Chip, a young man with Down's Syndrome. He sometimes wandered outside at night. He seemed unlikely cougar prey but, with an animal already behaving strangely, who could say. So far, the cougar hadn't threatened anyone, and the cats it had eaten were feral. Personally, I thought it was performing a service by reducing these half-wild bird eaters.

The cougar finished its snack and departed. I advised Mrs. Huber to contact the New Mexico Game Department. A New Mexico Game and Fish officer arrived at midday. He recommended a wait-and-see approach, hoping that the cat would not reappear, but the next morning Mrs. Huber reported that the cougar had again passed through her yard.

John Laundré and family arrived at our house that day for a visit. John has studied cougars for twenty years in Idaho and Mexico. He and I wandered the dirt streets of Hillsboro and outlying washes, hoping to detect a pattern in the cougar's movements. We found no sign! For me, this was humbling—two people with fifty years of cougar experience between them unable to confirm the presence of an animal that I had actually seen. If I had been the investigating officer, without the testimonies of a cougar biologist and a forest ranger, I would have discounted Mrs. Huber's sighting.

I heard no new reports for a day or two. I walked Percha Creek through town, hoping to find a track in the sand. No luck. I learned that Game and Fish had brought in two federal Wildlife Services agents. They, too, found no sign. I began to hope the animal had moved on.

But four nights later, town dogs began to raise a ruckus. I was sleeping on our screened porch and could hear the disturbance advance along a road that runs behind our house. I heard a domestic cat yowl nearby, followed throughout the night by a cougar's raspy *ouch* vocalization. The next morning, I found no tracks. Nonetheless, I contacted Game and Fish and suggested that the cougar was becoming habituated to humans and should be removed.

Over the next few days, two townspeople reported seeing a cougar near the creek. I still found no tracks, in spite of soft sand and excellent tracking conditions. A night later, the town dogs were again upset and, come daylight, I quickly found cougar tracks along streets near Mrs. Huber's house. I emailed Game and Fish, again suggesting that the cat might be dangerous, especially to children.

Ron Thompson, another Arizona Game and Fish retiree with cougar experience, stopped by the next day, and I discussed the cougar's behavior with him. Ron was assisting with an urban cougar study near Tucson, and I asked him to compare behavior of the Hillsboro cat with their nine radio-collared cougars. None of their cats was spending similar time near residences.

For the next three nights, dog noise suggested that the cat was still around. A supervisor from the Las Cruces office of the New Mexico Game and Fish Department called to discuss the options for taking the cat. By this time, the agency agreed that the animal should be considered a risk.

Twenty-five days after the initial incident, a local EMT called me from Mrs. Huber's—this time during daylight. The cat was back. Mrs. Huber was in hospital for surgery (unrelated to the cougar), and the EMT was staying with her son. When I arrived, the cougar was pacing the yard, and I could see that it was thin and weak. Either it was diseased, or its fare of feral cats and raccoons was proving inadequate. It trotted feebly down the street toward Percha Creek. I called Game and Fish, and they dispatched an officer and alerted Wildlife Services.

The Wildlife Services houndsman arrived about noon and released his dogs where we had last seen the cougar. They jumped it from the shade of a creekside cottonwood, the chase was short, and the cougar climbed a Chinese elm. The houndsman dispatched it with three rapid shots from his rifle—a procedure necessary to assure the animal was dead before it hit the ground. It was a badly emaciated male, perhaps two-and-a-half years old.

Some townspeople questioned the need for the cat to die. It had not directly threatened anyone, and its diet had been feral cats and raccoons. Under the circumstances, I believe it had to be killed. Moving cougars does not work. I moved two cougars in northern Arizona; both turned up a long distance from where we placed them, and both ended up in trouble—killing cattle or sheep. Of fourteen cougars Logan and Sweanor moved from southern to northern New Mexico, two returned home. Several of the others died soon after relocation—killed by resident cougars in the new area, hit by cars, or shot.

Cougars regularly wander the landscape near Hillsboro, but they do not habitually occupy the town or attack small, semidomestic prey. A cougar showing so little wariness of humans represents a risk. Children running about at play or joggers in cougar habitat can trigger an attack—rather a cat-and-mouse phenomenon. While human attacks are rare, they attract disproportionate media attention compared with auto accidents and other ways humans die. Should an attack occur, an agency that fails to respond to a habituated cat can become the target of litigation and is vulnerable to opportunistic politicians. Removing a misbehaving cougar may preclude more serious problems for the species as a whole. And in a state like New Mexico, where perhaps 300 cougars are killed by sport hunters or ranchers every year, one more dead cougar has little impact on the population.

Chapter 12 A Focal Species for Conservation Planning

Paul Beier

CONSERVATION PLANNING HAS been around since early humans protected the first sacred grove or royal game reserve for future generations. Although conservation planning focuses primarily on identifying land for conservation, the discipline also includes recommendations for managing land and populations outside reserves. Saving large charismatic mammals was probably one of the earliest priorities in conservation planning, although in much of the world, large carnivores like cougars (*Puma concolor*) were not considered charismatic until the mid-twentieth century. With Shelford (1926), planning goals started broadening to include conservation of representative vegetation communities, endemic and endangered species, large wildlands with ecosystem functions intact, areas with high evolutionary potential, and wildland networks robust enough to allow animal movement in response to climate change. Since about 1980, conservation planning has increasingly emphasized efficiency—that is, how to achieve these ambitious goals with limited financial resources, limited political will, and incomplete knowledge of how biodiversity is distributed.

These changes in conservation planning are best thought of as evolution rather than revolution. The old emphasis on large charismatic mammals still has a role to play. In particular, the need for efficiency bestows priority on species that can attract dollars and political will, and on well-understood species needing such extensive habitat that conserving them provides benefits for a spectrum of other species, thus conserving biodiversity. The cougar fits the bill.

Cougars are important in several contexts in conservation planning: conservation of isolated cougar populations at risk of extinction; as a species to be recovered or reintroduced into areas where it has been extirpated; and as a focal species for designing a regional conservation network. These three contexts are the central themes of this chapter, and they are not mutually exclusive—the Florida panther (*Puma concolor coryi*), for instance, is prominent in all three.

Understanding habitat use by cougars is crucial to all three of these planning contexts with cougars as a focal species. Therefore, after outlining reasons to use cougars in conservation planning, I summarize what is known about habitat use at spatial scales from the geographic range of the species through fine-scale movements. The last three sections deal with the main contexts in which cougars are currently used in conservation planning and include an important caution about using cougars (or any focal species) as an "umbrella" species for designing a wildland network.

Cougars are also a focus of efforts to restrict hunting, which might be considered a fourth way in which the species enters conservation planning. Human persecution was a major factor in historic declines of cougars and remains a limiting factor in some parts of Latin America (see Chapters 6, 7). This is less a conservation planning issue than a matter of how people value carnivores and hunting.

Why Cougars Are Useful for Conservation Planning

Cougars are useful in conservation planning for three reasons. First, if the planning goals include conservation of communities, ecological functions, and evolutionary relationships, a conserved area lacking cougars will fail because it will lack an important regulating force. Second, cougars demand such large, well-connected natural areas that an area conserving cougars will meet the space requirements for most other species, even if it does not meet their habitat requirements. Third, cougars can garner public support for broader conservation efforts.

Ecological Role as Top-Down Regulator

By preying on ungulates, cougars are important top-down ecosystem regulators. The extirpation of cougars from the eastern and midwestern United States almost certainly played a significant role in today's large populations of white-tailed deer and the resulting deer damage to natural vegetation in that region. Unfortunately, these major ecosystem changes provide anecdotal support for top-down regulation by cougars but fall short of scientific proof, because cougar extirpation, the near-extirpation of white-tailed deer (*Odoco lieus virginianus*), and the subsequent recovery of deer and vegetation took place over a century that included other changes in wildlife and land use. Furthermore, there was no "control area," or comparable landscape where cougars remained.

Two recent studies provide strong evidence for cougar-driven ecosystem regulation and increase the plausibility of cougars being important top-down regulators throughout their range. These two studies quantified changes in trophic levels during decline or extirpation of cougars and compared these results to simultaneous conditions in areas that retained cougars. Ripple and Beschta (2006) report that increased human visitors in Zion Canyon, Utah, apparently reduced cougar densities relative to other parts of Zion National Park. This led to higher mule deer densities and higher browsing intensities, which in turn reduced recruitment of riparian cottonwood trees (*Populus fremontii*), increased bank erosion, and diminished both terrestrial and aquatic species abundance (see Figure 10.8). Terborgh and colleagues (2001, 2006) observed changes in herbivores, small predators, and vegetation when a hydroelectric project in Venezuela flooded a large valley in 1986, creating fourteen islands that lacked three top predators (harpy eagle, *Harpia harpyja;* jaguar, *Panthera onca;* and cougar). Compared to the mainland areas where these top carnivores persisted, the islands had tenfold to hundredfold increases in herbivore populations, leading to loss of saplings, followed by an understory lacking foliage or leaf litter, and ultimately leading to dominance by herbivore-resistant lianas that smother and kill the tree canopy.

Cougars probably have additional ecosystem effects by killing smaller carnivores. They regularly kill coyotes (*Canis lantrans*), bobcats (*Lynx rufus*), skunks (family *mephitidae*), and raccoons (*Procyon lotor*), (Boyd and O'Gara 1985; Koehler and Hornocker 1991; Beier and Barrett 1993;

Logan and Sweanor 2001; see also Chapter 9, and Tables 2.1, 9.1–9.5, 11.1, this volume). Such kills sometimes occur in settings that suggest the small carnivores were scavenging cougar-killed ungulate carcasses and others occur away from ungulate carcasses. Cougars consume the smaller carnivore in some but not all cases. Thus, small carnivores experience both predation and interference competition from cougars, and these interactions probably affect niche relationships, the structure of the community of carnivores, and habitat use by subordinate carnivores, with cascading ecosystem effects (see Chapter 11). Although no research has addressed this issue, ecological theory and evidence that cougars kill other carnivores justify the assumption of such relationships until proven otherwise.

Advocates for cougars readily accept the idea that cougars should be conserved because they are important top-down regulators. But when confronted with a proposal to reduce cougar numbers temporarily to benefit a struggling bighorn or pronghorn population, some of these same advocates deny any significant top-down role for cougars. Although cougars prey disproportionately on young, old, or weakened ungulates with low reproductive value (Hornocker 1970; Kunkel et al. 1999; Logan and Sweanor 2001; Husseman et al. 2003), cougars kill more than the "doomed surplus," both in widespread herbivore populations (Terborgh et al. 2001, 2006; Ripple and Beschta 2006) and in populations of secondary prey, such as bighorn sheep or pronghorn. Specifically, cougar predation can be a proximate factor endangering or eliminating populations of bighorn sheep and may preclude successful bighorn restoration (Wehausen 1996; Rominger et al. 2004; McKinney et al. 2006). In another example, following a decline in the mule deer population in the Great Basin of Nevada, cougar predation nearly extirpated a local population of porcupines (*Erethizon dorsatum*)(Sweitzer et al. 1997). In many such cases, the impact of predation is exacerbated by disease (e.g., scabies and pneumonia in bighorn; Logan and Sweanor 2001), woody plant invasion due to overgrazing or fire suppression (Sweitzer et al. 1997; Rominger et al. 2004), or artificially high cougar populations "subsidized" by year-round livestock operations (Rominger et al. 2004). Regardless of those interacting factors, short-term cougar reductions may be an appropriate part of a long-term program to address all the factors endangering a particular herbivore population.

Focal Species

A *focal species* is one that demands so much of an ecosystem resource that a conservation plan providing enough of that resource for the focal species will also provide enough for other species (Lambeck 1997). Examples of ecosystem resources are land area, abundance of large snags, or absence of artificial night lighting. Large carnivores such as grizzly bears (*Ursus arctos*), wolves (*Canis lupus*), and cougars make ideal focal species for the adequacy of habitat area because they exist at low density and require large areas (Beier 1993; Servheen et al. 2001). As habitat areas are fragmented and lost due to increasing human use of the landscape, these are the first species to suffer. Of these, only the cougar plays a significant ecological role in much of North America. Thus, in many locales, the cougar is the best focal species for ensuring that a conservation plan will conserve enough habitat area.

In addition to being highly sensitive to an area, cougars are sensitive to human activities that can degrade the utility of wildland blocks, buffer zones, and corridors. For example, although cougars use culverts and other road-crossing structures, Beier (1995) reported that in five out of seven overnight monitoring sessions, a cougar walking in a drainage bottom climbed out of the canyon to cross a two-lane road at grade and then reentered the drainage. In the other two sessions, the animal used box culverts 1.8 m in height to pass under two-lane roads. Thus, cougars can be a useful focal species to ensure that a conservation plan includes roadside fencing that funnels animals to crossing structures (Figures 12.1, 12.2). Because dispersing cougars are sensitive to artificial night lighting (Beier 1995), the cougar is likewise a useful focal species for minimizing artificial night lighting in wildlife corridors. Cougars may also be a useful focal species for minimizing the secondary and tertiary effects of anticoagulant poisons. Sauvajot and colleagues (2006) reported that two mountain lions in southern California died from anticoagulant poisoning apparently acquired by preying on coyotes that had eaten poisoned rodents.

Figure 12.2 A Florida panther enters a culvert. Panthers prefer crossing structures with natural substrate but will use hard-bottom culverts. Various kinds of crossing structures are needed to promote movement and gene flow for all species in a landscape. Photo courtesy of Florida Fish and Wildlife Conservation Commission.

Cougars are obviously an appropriate focal species for efforts to manage an endangered population, reintroduce cougars, or facilitate recolonization. There are advantages to using cougars as a focal species for designing wildland networks and linkages. However, as discussed later, it is not appropriate to use cougars as the sole focal species in this context.

Flagship Species

Most Americans have a positive, protective attitude toward cougars, wolves, and grizzly bears and have fewer negative associations with cougars than with the other two large carnivores (Kellert 1996). One measure of the cougar's appeal is sale of special vehicle license plates by the state of Florida. Since 1990, the $25 fee for panther plates has raised about $3 million per year, of which 85 percent is to be spent on panther research and management (but not habitat acquisition). In 1990, California voters approved a ballot measure that prohibited sport hunting of mountain lions and required public agencies to spend $30 million per year principally for the acquisition of habitat for cougars and deer, and rare, endangered, and threatened species (see Appendix 5).

Positive feeling toward cougars is also evident in public and media reactions to cougar attacks on humans. Since my review paper on this topic (Beier 1991), I have often been contacted by the media after an attack has occurred, and I follow the media coverage of these attacks. Although some news reports, especially on television, are luridly sensational, most newspaper reporters ask me for (and print) statistics on the rarity of attacks and advice on coexisting with cougars. Follow-up editorials invariably emphasize

Figure 12.1 Interstate 75 in Florida is fenced for the entire right-of-way between wildlife undercrossings. These fences funnel animals toward the undercrossings and are critical to achieving the 95 percent reduction in numbers of large mammals killed by collisions in Alligator Alley. Electronic sensors in the fence alert authorities to breaks in the fence so that repairs can be made promptly. Photo courtesy of Florida Fish and Wildlife Conservation Commission.

both the benefits and risks of cougars. Most strikingly, when a cougar is killed after an attack on a human, most letters to the editor criticize the management agency for removing the cougar. Letters expressing sympathy for the human victim or calling for control of cougars are rare. Although letters to the editor are not statistical polls, these letters and media coverage of cougars at their worst suggest that the cougar is an effective flagship to garner public support for conservation. As described later, the flagship value of cougars may vary with geographic region (strongest in areas where cougars occur) and the conservation context (weaker for reintroduction efforts).

Cougar Habitat Use Relevant for Conservation Planning

Cougars are cosmopolitan, occurring across many countries and from sea level to 4,000 m elevation in vegetation types from deserts to tropical and temperate rain forests. The three essential habitat requirements are freedom from excessive human interference, adequate prey including large ungulates, and ambush or stalking cover (Seidensticker et al. 1973; Currier 1983; Lindzey 1987; Koehler and Hornocker 1991). Cougars have relatively broad tolerance with respect to each factor. For instance, they survive in areas where they are subject to moderate levels of predator control, hunting, and road mortality, and even in areas with high mortality if they are connected to source populations (Stoner et al. 2006; see source-sink discussion in Chapters 4, 5). In any given region, cougars consume ten or more species of prey, but they are resident only in areas with sufficient density of large ungulate prey (Hornocker 1970; Seidensticker et al. 1973; Hemker et al. 1984; Pierce et al. 2000b; Logan and Sweanor 2001). In the San Andres Mountains of New Mexico, no cougar home range extended more than 2 km from mountainous terrain because the desert basins had few native ungulates (Sweanor et al. 2000). Because cougars cannot run down prey, they hunt by stalking or ambushing, and stalking cover can be provided by trees, shrubs, or broken terrain (Young 1946b; Hornocker 1970; Seidensticker et al. 1973; Koehler and Hornocker 1991).

Habitat selection is a hierarchic process (Johnson 1980) that includes establishing a geographic range within continental areas, establishing home ranges within a larger landscape, choosing point locations within home range, and making decisions about large dispersal movements and fine-scale movements during daily activities. Despite their broad habitat tolerances, cougars prefer some and avoid other habitat features at each scale of selection. These preferences are relevant to conservation planning and are detailed in the segments that follow.

Geographic Range

Although there are no reports of reproducing cougar populations in desert flats, agricultural landscapes, urban areas, or large grasslands, cougars do use almost every vegetation type. They are known to use coniferous and deciduous forests, woodlands, swamps, savannahs, chaparral, gallery riparian forests, desert canyons and mountains, and semi-arid shrublands (Hansen 1992). Absence of cougar populations is probably due to human disturbance, historic persecution (eastern United States), or low prey availability rather than to the direct influence of vegetation type.

Home Ranges

Within a regional landscape, cougar home ranges tend to occur in areas with more forest or woodland cover, less human disturbance, lower density of paved roads, and little grassland cover (Van Dyke et al. 1986; Laing 1988; Belden and Hagedorn 1993a; Dickson and Beier 2002; Kautz et al. 2006; Arundel et al. 2006). In Florida, panther home ranges tended to encompass relatively small (<50 ha) forest patches surrounded by natural vegetation (Kautz et al. 2006). Cougar home ranges in Arizona avoided forest areas logged within the previous six years (Van Dyke et al. 1986).

In Arizona, cougar home range edges tended to coincide with paved roads such that individuals rarely crossed paved roads (Arundel et al. 2006). In southern California habitat of oak savannah, woodland, and shrubland, no cougar home range overlapped an area with a paved road density of 1.0 to 1.5 km/km^2 (Dickson and Beier 2002). The roads had been created to access lots for rural residences, but few lots had homes at the time of the study, suggesting that this density of lightly used roads was incompatible with cougar occupancy.

Locations within the Home Range

Most habitat-use studies describe cougar locations during daylight hours, when cougars are least active (Beier et al. 1995). Cub-rearing sites tended to occur in nearly impenetrable vegetation away from cougar travel routes (Young 1946b; Maehr et al. 1989; Beier et al. 1995). Daybed locations tended to be in relatively dense natural vegetation and almost never in urban or human-disturbed areas (Williams et al. 1995; Maehr and Cox 1995; Dickson and Beier 2002; Kautz et al. 2006; Cox et al. 2006; Arundel et al. 2006). In arid southern California, daybeds tended to occur in riparian areas (Dickson and Beier 2002). Cougar locations in dense vegetation were often closer than expected to more open vegetation types, such as grasslands and swamps

(Laing 1988; Williams et al. 1995; Kautz et al. 2006; Cox et al. 2006), suggesting that open habitats are important for prey, and woodland-grassland edges are important stalking areas. This interpretation is supported by the finding that during winter, cougars killed mule deer (*Odocoileus hemionus*) within 20 m of forest-grassland edges about 2.5 times as often as expected based on availability of edges (Laundré and Hernández 2003). Cougars selected areas with steep slopes in Wyoming (Logan and Irwin 1985), Utah (Laing 1988), and Arizona (Arundel et al. 2006) but not in southern California (Dickson and Beier 2002); these differences are probably due mostly to differences in scale and methods. They avoided recently logged areas within their home range in Arizona (Van Dyke et al. 1986) and California (Meinke et al. 2004). In Arizona, cougars selected sites relatively close to water, and males strongly selected lands where hunting was not allowed, especially during fall and winter (Arundel et al. 2006).

Although southern California cougars tended to locate home ranges in areas with few roads, they did not avoid areas near the paved roads in their home range (Dickson and Beier 2002). This doubtless contributes to the high incidence of cougar mortality on roads in southern California. In contrast, cougars avoided areas within about 2 km of highways in Arizona (Arundel et al. 2006). The Arizona study area had significantly fewer miles of paved roads than the southern California study area. Thus, this difference may reflect habituation to roads in areas with greater road density, or the fact that it is easier to avoid paved roads when there are not as many of them.

Dispersal Movements

Dispersal plays a crucial role in cougar population dynamics because recruitment into a local population occurs mainly by immigration of juveniles from adjacent populations, while many of the population's own offspring emigrate to other areas (Beier 1995; Sweanor et al. 2000). Regardless of population density, almost all male offspring and 20–50 percent of female offspring disperse (Sweanor et al. 2000). Juvenile dispersal distances averaged 32 km (range 9–140 km) for females, and 85–110 km (range 23–274 km) for males (Anderson et al. 1992; Sweanor et al. 2000; Maehr et al. 2002; but see two extreme records in recolonization discussion, this chapter). Dispersing cougars in New Mexico crossed up to 100 km of unsuitable habitat; such movements were generally direct and quick (Sweanor et al. 2000). Dispersers crossed rivers and freeways more readily than did adults (Beier 1995; Sweanor et al. 2000; Maehr et al. 2002) but experienced high mortality on roads (Beier 1995).

Urban areas are impenetrable to movement of both adult and dispersing cougars (Beier 1995). Large expanses

of agriculture, such as in California's central valley, and major freeways in California, Arizona, and New Mexico coincide with significant gradients in gene flow (Ernest et al. 2003; McRae et al. 2005). Ameliorating barriers (highways, canals, urbanization) and minimizing human-caused mortality (depredation, shooting, road kill) are the key issues in promoting successful dispersal and are thus key to conservation planning (Beier 1993, 1995, 1996; see Chapters 5 and 8 for further discussion of dispersal).

Fine-Scale Movement

Cougar tracks in snow suggested that movement routes in Utah were associated with woody cover (Laing 1988). In California, radio-tagged cougars monitored overnight at fifteen-minute intervals showed weak preferences among vegetation types, although rank-order was similar to that observed during daylight hours (Dickson et al. 2005). Travel routes of California cougars tended to follow canyon bottoms and gentle slopes, a finding that was not the result of correlation of these topographic positions with preferred vegetation types (Dickson and Beier 2007). Cougars showed a strong aversion to crossing freeways in California (Beier 1995; Dickson et al. 2005) and Arizona (Arundel et al. 2006). In the Arizona study, they also crossed a low-traffic paved road through a large wildland much less often than expected based on proximity to the road. Sweanor and colleagues (2000) documented seven successful at-grade crossings of a four-lane highway in New Mexico; after the road was widened to six lanes, they documented two attempted crossings, both of which resulted in fatal collisions with vehicles. Highway underpasses and overpasses have successfully facilitated cougar movement across freeways in Canada (Clevenger et al. 2001; Clevenger and Waltho 2005), California (Beier 1995), and Florida (Figure 12.3; Foster and Humphrey 1995). One young male cougar in California successfully used a culvert and an underpass to cross a freeway at least twenty-two times in nineteen months (Beier 1996).

Conservation of Isolated Cougar Populations

A population is a group of interbreeding individuals of one species living in a given area. One population can be linked to others by dispersal. In North America, dispersal among populations occurs frequently enough that most cougar populations form a single metapopulation (see Chapters 3, 5, 8; Culver et al. 2000a). However, three cougar populations have been isolated long enough to diverge genetically from other North American populations (Culver et al. 2000a): the Florida panther; Vancouver Island cougars,

Figure 12.3 Large crossing structures like this one in Florida should be placed at intervals of about 1 km. Placement of crossing structures must be integrated with land use planning to ensure that human land uses are compatible with landscape-scale wildlife movement. Photo by Mark Lotz.

(*P. c. vancouverensis*); and cougars on the Olympic Peninsula of Washington (*P. c. olympus*). Two other populations—in the Black Hills of South Dakota and the Badlands of North Dakota—are relatively isolated from the nearest larger cougar populations. Different conservation interventions may be appropriate for each isolated population. The overarching conservation lesson is that it is better to avoid isolation than to try to manage isolation after it has occurred.

Florida

The Florida panther population in south Florida is separated by about 1,000 land miles from the nearest confirmed cougar population in Texas. It has been recognized as a distinct subspecies, which once extended as far west as western Arkansas and Louisiana. By the mid-twentieth century, some observers feared the population was extinct, until a small remnant population of fewer than thirty-five animals was confirmed in south Florida in 1973 (Nowak, cited in Pritchart 1976). The population probably remained below fifty animals from 1980 through 1995 (Pimm et al. 2006). One of the first mammals in North America to be listed as endangered under the U.S. Endangered Species Act (ESA), the panther has been the subject of massive research and recovery efforts, including habitat acquisition, genetic augmentation, and research to support establishment of new panther populations.

Although persecution caused the near-extinction of the Florida panther, by the late twentieth century the population

was at risk due to three habitat-related problems: limited habitat, continuing loss and fragmentation of habitat due to urbanization and conversion to agriculture, and urban and geographic barriers to northward range expansion. It was also at risk due to a problem common in captive and domestic animals but rarely observed in wildlife populations—the deleterious effects of low genetic variation. Loss of genetic variation is an inevitable consequence of a small population persisting for a long time, which is known as a genetic bottleneck; for the panther, this was probably the case for several decades. High prevalence of cowlicks and kinked tails was the first evidence that low genetic variation was affecting the panther phenotype, and the trivial nature of these traits made for an interesting story but did not alarm most observers in the 1980s.

But research soon brought much more attention to panther genetics. Compared to other North American cougar populations (including on Vancouver Island and the Olympic Peninsula), Florida panthers by the early 1990s had the lowest value for every measure of genetic variation, typically one-half to one-eighth that of other populations (Roelke et al. 1993; Culver et al. 2000a; see Chapter 3 for further discussion of the Florida panther and genetics). The strong preponderance of evidence suggests that Florida panthers lost most of their genetic variation after about 1890–1920 (Beier et al. 2003). Simultaneously, prevalence of cowlicks, probably genetically controlled, increased from 33 percent—two of six animals—during 1896–98 to over 94 percent in modern specimens (Wilkins et al. 1997). Similarly, cryptorchidism—failure of one or both testes to descend into the scrotum, a trait almost certainly under genetic control—increased fourfold during 1970–92 (Roelke et al. 1993). By 1993, motile sperm per ejaculate in the Florida panther was 18–38 times lower than in other cougar populations (Roelke et al. 1993a), and cryptorchidism affected 49–56 percent of male panthers (Roelke et al. 1993; Mansfield and Land 2002) compared to 3.9 percent in other puma populations (Barone et al. 1994). Some males were sterile (Roelke et al. 1993). Heart murmurs consistent with atrial septal defect occurred in 80 percent of Florida panthers compared to 4 percent of other cougars, and atrial septal defects caused the deaths of at least two panthers (Roelke et al. 1993).

These alarming results caused an emergency amendment to the Florida panther recovery plan to introduce new genetic material to the population. To emulate natural gene flow between adjacent populations, managers released eight female cougars from Texas into apparently vacant panther habitat in 1995 as part of the Florida Panther Genetic Restoration Program. The Texas population historically interbred with *P. c. coryi* and both populations probably belong to the same subspecies (see Chapter 3). Within five years, the target of 20 percent introgression of Texas genes into the Florida population had been met (Land and Lacy 2000).

The results were dramatic. None of the progeny resulting from genetic restoration efforts has been cryptorchid (Mansfield and Land 2002). Rates of heart murmurs and opportunistic infections have plummeted in hybrid individuals with a mixture of Florida and Texas genes (Beier et al. 2003). The hybrid animals have a kitten survival rate of 72 percent compared to 52 percent for pure Florida panther kittens observed during the same time period (Shindle et al. 2001), which translates into a threefold greater probability of surviving to reproductive age (Pimm et al. 2006). Panthers now breed in swamp-dominated areas that pure Florida panthers had not permanently occupied, and the population probably exceeds one hundred animals (Figure 12.4). Introgression seems to have reversed the deleterious effects of genetic erosion. The 20 percent level of introgression carries little or no risk of losing genes adapted to the local environment (Hedrick 1995).

Genetic restoration was a necessary step for panther recovery. But it does not solve the fundamental habitat-related problems. Securing panther habitat in south Florida will require conservation of over 300,000 ha of land that is now in private ownership (Maehr 1990; Logan et al. 1993; Main et al. 1999). Some land has been acquired specifically for panthers, including the 26,400-acre Florida Panther National Wildlife Refuge, established in 1989. Florida voters have passed over $6 billion in bond measures to acquire habitat for panthers, bears, and other wildlife, but owners of some key panther habitat are not willing to sell, and the panther's share of bond money will not buy all the land needed. Thus, innovative collaborative agreements with private landowners will be needed to conserve panther habitat (Maehr 1990; Main et al. 1999).

Zoning by county and municipal governments, and consultations with the U.S. Fish and Wildlife Service (USFWS) required by the ESA, can minimize and mitigate adverse impacts of development projects on panther habitat. In its calculation of potential adverse impacts and appropriate mitigation, the USFWS relied for years on a Panther Habitat Evaluation Model (Maehr and Deason 2002) that was found to be scientifically unreliable (Beier et al. 2006) and was formally abandoned by the USFWS in 2005. Decisions regarding the impact of land use changes on panthers, and the appropriate ways to minimize and mitigate them, remain contentious.

Formal designation of critical habitat for the Florida panther is required by the ESA and could give the USFWS greater clout in consultations, but it could also make landowners less likely to cooperate voluntarily with conservation agencies. For about fifteen years, the USFWS has declined to designate critical habitat for any species except under court order, claiming that procedural difficulties, cost, and political ill will outweigh the benefits of designation. Removing the panther from the list of endangered species requires three viable populations within historic range (USFWS 1995, 1999). However, urban and agricultural barriers in central Florida probably preclude natural colonization of suitable habitat outside south Florida (Maehr et al. 2002); therefore, reintroductions will be required to establish these new populations (see later discussion).

Vancouver Island

Vancouver Island is the largest island on the West Coast of North America, at over 32,000 km². It is separated from mainland British Columbia by the saltwater Johnstone Straight, which is at least 2 km wide for most of its 200 km length. The saltwater barrier has affected the mammal community, which consists of only eighteen of the forty-six mammals of mainland British Columbia and includes endemic races of cougar, mink (*Neovision vison*), and wolf and an endemic species of marmot (Family *Marmota*).

A conservative estimate of one adult per 100 km² suggests that there are at least 300 adult cougars on Vancouver Island. The island cougar population is one of only three cougar populations (along with the Olympic Peninsula and Florida) with no variation in a large proportion of its genes. There are no reports of physiological or demographic problems related to low genetic diversity of

Figure 12.4 Since 1994, when Texas cougars were introduced to increase genetic diversity, the Florida panther population has expanded its range and numbers, and the animals now breed in swamp-dominated areas. Some genetically based medical conditions, such as heart murmurs and male reproductive abnormalities, have greatly decreased. Photo by Mark Lotz.

this population. The population is hunted and is not considered at risk by management agencies. Wilson and colleagues (2004) attributed smaller litter sizes to lower food availability. Because this population is isolated by a natural saltwater barrier and the population is relatively large and stable, intervention to counteract isolation is not needed. The Vancouver Island cougar population may provide a useful benchmark for understanding genetic and demographic changes in smaller, more recently isolated cougar populations.

Olympic Mountains

The Olympics Mountains at the north end of Washington's Olympic Peninsula have at least 6,200 km^2 of remote, rugged habitat suitable for large forest carnivores (Singleton et al. 2001), and probably a larger area suitable for cougars. Although surrounded by water on three sides, the peninsula is wide (45 km) and, prior to settlement by European-Americans, forest vegetation was well distributed in the lowlands of the south peninsula and southwestern Washington. Thus it was probably part of a large cougar population extending through the Cascade Mountains and beyond. Since European-American settlement, much of the lower peninsula has been converted to farms and urban uses; several highways, most notably Interstate 5, cross through or near the base of the peninsula. The Olympic range is about 115 km from the nearest large block of prime habitat in the South Cascades. Singleton and colleagues (2001, 593) speculated that the current "landscape is an effective barrier for lynx (*Lynx canadensis*), wolverine (*Gulo gulo*), wolves, and grizzly bears." The same landscape patterns may also limit but probably do not preclude cougar movement. Elk and deer are common through most of this 115-km-long area, which also generates occasional cougar complaints (G. Koehler, Washington Department of Fish and Game, pers. comm., 2007). Although no population estimates exist, a density of one adult per 100 km^2 yields a crude estimate of at least sixty-five breeding animals in prime habitat of the Olympic Mountains.

This cougar population has a large proportion of genes lacking variation and a unique and fixed mutation in mitochondrial DNA (Culver et al. 2000a), suggesting a genetic bottleneck at least as severe as that experienced by the Vancouver Island population, which has probably been isolated since sea level rose to its current level about 10,000 years ago. The severity of a genetic bottleneck depends on both its narrowness (population size over the years) and its length (how many years the population was at a low ebb). It is surprising that a population of at least sixty-five breeding adults would exhibit such severe genetic erosion in a short time. Possible explanations include (a) human-caused iso-lation began with agricultural conversion a century ago or earlier—well before suburbs and highways became factors; (b) the population was persecuted to extremely low numbers and recovered from a few founders; (c) peninsularity affected population genetics long before European-American settlement; and (d) the genetic samples do not adequately represent existing variation. A combination of these factors may apply.

The Olympic population is probably large enough to maintain demographic stability on its own (Beier 1993), but if a disaster occurs, natural recolonization may be difficult. Furthermore, if there is no gene flow from other cougar populations, genetic erosion will continue, eventually impacting population fitness. Planning and implementation of corridors in this urbanizing area are sorely needed. However, there are apparently no management efforts to conserve and enhance wildland connectivity between the Olympic and Cascade mountain ranges. Before cougars are translocated into the area to enhance genetic diversity, a genetic analysis based on a large sample of cougars from the Olympic Peninsula and nearby areas should be conducted.

Black Hills

The Black Hills of South Dakota and Wyoming (8,400 km^2) are separated by over 150 km of grassland plains from the nearest cougar populations in the Laramie Mountains 160 km to the southwest and the Bighorn Mountains 200 km to the west (Fecske 2003). Although early reports indicated that cougars were widespread throughout South Dakota, they were apparently extirpated by 1906 (Fecske 2003). There were no records of cougars in South Dakota from 1906 to 1930, until one male and one female were shot in the western Black Hills. Anderson and colleagues (2004) surmised that a remnant population may have persisted during 1906–30. Genetic patterns today suggest gene flow from both the Laramie and Bighorn Mountains (Anderson et al. 2004). If extirpation did occur, colonists apparently came from both the nearby populations.

Cougar numbers and habitat area in the Black Hills are similar to those in the Olympic Mountains. Fecske (2003) estimated the population at 58–78 adult cougars (>2 years old) and concluded this was large enough to ensure that the population would persist even without immigration. However, immigration may have rescued the Black Hills population once before and would be important for recolonizing the Black Hills if another catastrophic decline were to occur. Immigration is also important to conserve genetic diversity. Northeastern Wyoming is still predominantly rural, and natural habitat differences (flat grasslands) are the main impediment to immigration. Nonetheless, landowner tolerance, highways, and land use changes could affect ability of cougars and other mountain species to immigrate to the Black Hills.

It is unlikely that 160 km grassland corridors to the Laramie and Bighorn mountains can conserve genetic diversity or allow recolonization of the Black Hills. Thus, in this landscape, maintaining permeability across northeastern Wyoming is a better conservation strategy than planning for corridors, which are more appropriate in smaller landscapes at imminent risk of being fragmented by human activities. Landscape-scale permeability can be maintained by ensuring that all highways have at least one wildlife crossing structure per mile, retaining pastoral land use, minimizing ranchette development, and educating people to tolerate predators.

Other Isolated Populations

Development of transportation networks, increased border security, and urbanization threaten to isolate other cougar populations. The case of the Florida panther illustrates the seriousness of problems arising from isolation and should motivate management agencies and conservationists to prevent isolation. The final section of this chapter discusses such preventive strategies for small populations at risk of isolation, like those in the Santa Monica Mountains of southern California. As the panther example illustrates, trying to manage isolation after it occurs is expensive and politically contentious, and it may fail.

Cougars as the Focus of Reintroduction or Recolonization Efforts

Recolonization is the natural process of animals establishing breeding populations in areas where they have become locally extinct. Reintroduction is the deliberate release (by humans) of animals to establish a population in the animal's former range. Because cougars have been extirpated from more than half of their North American range (see Range Map), many areas are available for recolonization and reintroduction. Although agricultural conversion, urbanization, and high road density have made most of the eastern and midwestern United States unsuitable for self-sustaining cougar populations, there remain several large areas with sufficient ungulate prey and relatively low human density. Recolonization is most likely to occur first in one or more of three large areas: the forests of northern Minnesota, Wisconsin, and Michigan; the Ozarks of Arkansas and Missouri; and the pine and bottomland hardwoods of Louisiana, Mississippi, Alabama, south Georgia, and north Florida. Farther east, the best block of potential habitat is the forest extending from upstate New York through Maine and into eastern Canada.

Recolonization of the Midwestern and Eastern United States

Cougars are already on the move. Since about 1990, cougars have been killed by vehicles or hunters, or captured on film in areas where the species had not been reported for fifty years. These confirmed occurrences extend as far east as southeastern Minnesota, eastern Iowa and Missouri, central Arkansas, and east Texas (http://easterncougarnet. org). A male tagged in the Black Hills of South Dakota dispersed a straight-line distance of 1,067 km to central Oklahoma (Thompson and Jenks 2005). Another radio-tagged young male moved over 800 km from the Black Hills through North Dakota and into northwest Minnesota (Fecske 2006).

Although some photographed cougars could have been captives, the animals killed by cars or hunters were apparently naturally occurring animals. Most of these deaths were near major river corridors, and most were males. Given the stronger dispersal tendencies of males, many of these were probably dispersing young males from known populations in the west using rivers as travel routes. But some females are also showing up far east of known breeding areas, including one female shot in southeastern Missouri in December 1994 and another shot 160 km east of Dallas, Texas, in January 1991. Unless otherwise cited, the details on midwestern animals are from the Cougar Network Web site (see Cougar Network 2009). For each purported occurrence, the organization archives and posts supporting information, such as photographs, news articles, names of agency personnel, dates of communications, and any evidence suggesting whether the cougar might be a captive or released animal. A few samples of cougar fur and scat have been confirmed in New England, New Brunswick, and Quebec. Because these are so far from any known breeding population, and because no carcass of a noncaptive animal has been observed in this area for a century, I suspect all these samples were from captive animals.

The eastward movement of cougars has resulted in at least one recolonization event. After no confirmed cougar reports in North Dakota from 1902 to 1958, cougars had recolonized the Little Missouri River badlands by the early 2000s, and North Dakota held experimental cougar hunts in 2005 and 2006. One cougar kitten was killed by a North Dakota hunter in 2006 (*Bismarck Tribune*, November 9, 2006). This population was most likely founded either by dispersers from the Black Hills, over 200 km south, or dispersers from Montana, about 300 km west. If the Black Hills population also is the result of recolonization (rather than a remnant population), then within a few decades cougars have moved in steppingstone fashion to establish populations across gaps of at least 160 and 200 km. At least two males have explored about 1,000 km of apparently unoccupied habitat.

The next steppingstones that could support cougar populations are probably the north woods of Minnesota (about 1,000 km from the North Dakota badlands) and the Ozarks of Arkansas and southern Missouri (about 1,500 km from western Colorado or Texas). Establishing a population in one of these areas would require at least one male and one female cougar to disperse this distance at about the same time, find each other, and breed. Within a few years, additional dispersal into the nascent population would be needed to avoid inbreeding. I suspect this chain of events will happen soon if it has not happened already.

The Cougar Network was founded in 2002 (as Eastern Cougar Network) to document this eastward movement, anticipate where it may occur, and use science to facilitate "the continued recovery and long-term conservation" of "North America's apex predator" (http://easterncougarnet. org). Besides documenting cougar occurrences east of known range, the Cougar Network is mapping potential habitat in the Midwest and sharing information on cougars and other wild cats through the online publication *Wild Cat News*. Emphasis is on objective reporting and credibility (e.g., the account of North Dakota's experimental hunting seasons was written by the state's furbearer biologist), allowing the network to become a trusted source where agency biologists are comfortable sharing information.

Fourteen states appear to be regularly receiving cougars dispersing from known breeding populations, or can soon expect to receive dispersers. Policy in each of these states will have a strong impact on the ability of cougars to recolonize that state or states farther east. Of these states, two (Arkansas, Louisiana) are in the historic range of the Florida panther, and two (Illinois, Michigan) are in the historic range of the eastern North American cougar (*P. c. couguar*). Both *coryi* and *couguar* are listed under the ESA. Recolonized populations, however, would be from non-endangered cougar lineages and would not be protected under the ESA (letter from USFWS Director J. Clark to Eastern Cougar Foundation, June 21, 2000). In January 2007, the USFWS initiated a review of the status of *Puma concolor couguar* (Federal Register 72 [18]:4018–19). Based on evidence that all U.S. races are a single subspecies (see Chapter 3; Culver et al. 2000a), this review may result in removal of *couguar* from the list of endangered species.

Each of the other ten states (Iowa, Kansas, Minnesota, Missouri, Nebraska, North Dakota, Oklahoma, South Dakota, Texas, and Wisconsin) lies entirely or mostly within the ranges of cougar subspecies that are not listed under the ESA. Some of these states (Texas, North Dakota, and South Dakota, at a minimum) harbor cougar populations but also include large

areas that dispersers would have to cross during eastward expansion.

Of these fourteen "frontline" states, cougars are protected in Louisiana, Michigan, and Wisconsin. They are designated as a game species with current harvest quota of zero animals in Nebraska and Oklahoma and a game species with regulated seasons in North and South Dakota. In Arkansas, Kansas, Missouri, and Minnesota cougars are categorized as nongame wildlife, meaning animals can be killed only to protect property or under special permit. There are no restrictions on killing cougars in Texas, Iowa, or Illinois. A change to game, nongame, or protected status in these states would facilitate eastward movement because dispersing animals could no longer be legally killed without a permit. States may prefer game status, perhaps starting with a harvest quota of zero animals, because this offers more management flexibility than stricter prohibitions on killing cougars. Eastward dispersal to or through these fourteen states would also be facilitated by management plans to respond to presence of dispersing cougars or new populations. For instance, Nebraska's 2004 Mountain Lion Response Plan established procedures for responding to cougar occurrences and created a state Mountain Lion Work Group. Although Nebraska's Work Group includes only state employees, such work groups could constructively engage hunters, ranchers, farmers, and wildlife advocates in dialogue about managing cougar dispersal or colonization of the state.

Following the lead of the Cougar Network, advocacy groups should focus on broad goals such as creating a viable population in one specific area or several areas, with suitable habitat for dispersal between population centers. Advocates should be flexible and work collaboratively on the specific details, such as conservation arrangements with private landowners and rules for hunting or depredation. Scientific polls on how people in these frontline states feel about the return of cougars are apparently lacking. In both rural and urban areas, these feelings surely include excitement that a top predator is returning home and relief at a second chance to consider the fate of a species that humans had extirpated from the state. These feelings also include fear about risks to humans, pets, and livestock. Perhaps the most important step to facilitate expansion is to educate the public about cougars. Surveys to learn what people know and feel about cougars can help guide the content and identify target audiences for an educational campaign. Appropriate education approaches include Web sites, news stories, letters to the editor, and brochures for areas where cougars are expected to occur. Brochures will be most helpful if they are factual and comprehensive (addressing both benefits and risks) and if the official sponsors include the state wildlife agency.

To emphasize the potential for an unfortunate public response, the next section describes public relations problems related to an experimental panther reintroduction. Natural recolonization by a nonendangered species will probably not be as controversial as deliberate reintroduction to facilitate release of an endangered species. Nonetheless, whether recolonizing cougars are welcomed or cursed will depend more on public education than on scientific research.

Reintroduction

As of 2008, reintroduction of cougars is being considered only for the Florida panther. Under the ESA, getting the panther to the point where it becomes a candidate for delisting requires three populations within historic panther range, which is two more populations than currently exist. These new populations will not occur via natural recolonization (Maehr et al. 2002). Reintroduction has therefore long been part of the recovery plan.

Experimental releases of sterilized cougars in north Florida during 1988–89 and 1993 provided useful information on home range establishment, (see plate 18) response to deer hunting seasons, influence of roads, size of release area, and use of wild versus captive stock (Belden and Hagedorn 1993a; Belden and McCown 1996). Jordan (1994) identified fourteen potential reintroduction sites. In a statistical approach using remotely sensed data, Thatcher and colleagues (2006) identified nine areas that might be of sufficient quality and size to support a panther population—a subset of the fourteen identified by Jordan—and recommended field work and sociological surveys to prioritize further among the nine.

Belden and McCown (1996) documented that the biggest hurdles to reintroduction are not biological but political. Opinion surveys showed that only 7 percent of Florida residents opposed panther reintroduction, and over 70 percent (both statewide and in the release area) supported reintroduction in or adjacent to their county. Despite these favorable numbers, the Florida wildlife agency wisely anticipated the need to educate local people before the 1993 experimental release. They developed a list of local leaders and met with them early in the process, before news of the planned release could circulate by word of mouth. They distributed a brochure targeted at concerns of landowners and hunters, and had procedures in place to reimburse for livestock killed by cougars. They invited fifty-eight hunt clubs to a special meeting before the release. However, when three calves were killed by released cougars about a year after release, a "not in my backyard" group emerged. The opposition was politically astute, using the media and well-connected allies to create a local atmosphere hostile to the experiment. Public expressions of concern about human safety, safety of pets and livestock, landowner rights, and impact on deer populations increased markedly. Released cougars were then blamed for all depredation incidents, even when evidence clearly implicated dogs. Some experimental cougars were illegally shot with impunity. Despite well-planned educational efforts that started early, the political fallout has deflated the state's enthusiasm for reintroduction for over a decade. Clearly, education effort for any future reintroduction should start even earlier, be even more aggressive about involving potential opponents, and quickly respond to changing developments.

By 2010, the USFWS will initiate a status review of *P. c. coryi* that may well result in removing *coryi* as a listed subspecies. If this were to occur, the existing Florida panther population in south Florida would probably be listed as a distinct population segment with full ESA protection. However, the range of a distinct population segment is defined by current conditions, and therefore "historic range" would have no meaning. Thus, the recovery plan's goal of establishing two new populations "within historic range" would become meaningless. Federal policy on distinct population segments (USFWS-NOAA 1996) does not mention the possibility of creating new distinct population segments by reintroduction. It seems doubtful that an amended recovery plan would include reintroduction plans, especially because natural recolonization from the west seems plausible.

Cougars as a Focal Species in Reserve Design

A reserve design is often identified with one particular product—a map depicting a network of cores (wildland blocks) connected by linkages and buffered from areas of intense human use. But, in fact, reserve design is not a product but a *process* that unfolds over many years, during which "the map" continually evolves. The process starts by engaging multiple stakeholders, including land management agencies, wildlife agencies, conservation NGOs, transportation agencies, and local land use planners. It continues by identifying conservation targets (such as connected populations of focal species), developing strategies for meeting those targets, and implementing those strategies. During implementation, stakeholders adjust the targets and strategies in responses to changes in the landscape, scientific understanding, and financial and political resources. Examples include the Yellowstone-to-Yukon Conservation Initiative (2006), the Florida Ecological Network (Hoctor et al. 2000), and the Wildlands Network in southern California (Chapter 15; South Coast Wildlands 2006). In most linkage design efforts, focal species are used to help define conservation targets, create maps, develop strategies, and evaluate proposed adjustments to targets, maps, and strategies.

As mentioned earlier, cougars are an appropriate focal species for the size of core areas, for road-crossing structures and associated highway fencing in linkage areas, and for restrictions on artificial night lighting and rodenticides in cores, corridors, and buffer areas. The fact that cougars have been documented to use narrow habitat corridors and peninsulas of habitat up to 7 km long in urban areas (Beier 1995) further justifies their utility as a focal species for corridors.

Cores

Cores should include at least 1,000–2,200 km² of suitable cougar habitat (depending on habitat quality) to ensure the persistence of a cougar population for several decades (Beier 1996). Much larger areas are needed to ensure longer-term demographic stability and to prevent loss of genetic diversity. Smaller patches will not support cougars without linkages to larger source populations. Belden and Hagedorn (1993a) recommended that a reintroduction site be at least 2,590 km², with a deer density >1 deer/36 ha, and minimal human population, to support 50–60 panthers (2–3/100 km²).

Ungulate prey are an essential component of suitable core habitat. Where highways cross a core, large crossing structures such as bridges, viaducts, or overpasses should be located no less than 1.5 km (0.94 miles) apart (Clevenger and Wierzchowski 2006). Density of small paved roads should be less than 1 km/km². Some, but not necessarily all, core areas in a network should be managed as strategically located refugia, where mortality of cougars would be limited to natural causes (Logan and Sweanor 2001). Each refugium should contain at least 3,000 km² of high-quality cougar habitat so that it can serve as a source population in the network and provide dispersers to restock adjacent populations depleted by hunting or other causes. (see discussion of source populations, Chapter 5)

Corridors

Habitat corridors for cougars should have nearly continuous woody cover, underpasses integrated with 3-m roadside fencing at crossings of high-speed roads, little or no artificial night lighting, and less than one human dwelling per 16 ha (Beier 1995; Foster and Humphrey 1995). Beier (1995) suggested that gaps in woody cover should not exceed 400 m. The approach to a road-crossing structure should give the animal a good view of the landscape on the other side of the road. Dirt roads or trails probably do not impede cougar movement and could help guide cougars through a corridor (Beier 1995; Dickson et al. 2005). In xeric landscapes, riparian corridors and gentle terrain are favored movement routes (Dickson et al. 2005; Dickson and Beier 2007). Unrestrained pets should be prohibited from corridors to

prevent depredation and subsequent demands to remove the offending cougar. Minimum width should be at least 400 m for corridors up to 7 km long (Beier 1995). Small patches of suitable habitat along a linkage can provide enough space for a breeding female, transient subadult, or disperser, and thus can enhance connectivity between larger patches, even if the small patch is not continuously occupied (see Figure 12.3; Beier 1996). The utility of small patches of stopover habitat is also suggested by panther movement in south Florida (Maehr et al. 2002; Kautz et al. 2006).

Flagship, Umbrella, and Focal Species for Reserve Design

Clearly, cougars are a useful focal species for designing core areas and corridors. In my experience advocating for regional conservation plans based on multiple focal species in the western United States (Beier et al. 2006), the cougar makes a great flagship, too. After presentations emphasizing how each of several focal species will benefit from a linkage design, I have occasionally been surprised by someone commenting that cougar conservation is the only reason—and a sufficient reason on its own—to conserve the linkage. Some people also ask about the risk that a corridor may bring cougars close to humans, but usually the concern is that such proximity will be fatal to cougars, and concerns about human safety never rise to the level of an objection. Cougars are an excellent flagship species for getting people excited about conserving a core or corridor.

But the cougar should never be the *sole* focal species for a reserve design or a corridor design; in other words, it is not a good umbrella species. Although cougars may need more habitat than other species, cores based on cougars do not cover endemic species or communities well (Thorne et al. 2006). The cougar may be the first species to feel the impact of the loss of connectivity, but many other species need linkages to maintain genetic diversity and metapopulation stability. Furthermore, as habitat generalists, cougars can move through marginal and degraded habitats, and a corridor designed for them does not serve most habitat specialists with limited mobility (Beier et al. 2009). Indeed, successful implementation of a single-species "cougar corridor" could have a negative umbrella effect if land use planners and conservation investors become less receptive to subsequent proposals to provide corridors for less charismatic species.

The cougar best serves biodiversity if it is one of many focal species used to design a network. For instance, before developing linkage designs for fifteen key wildlife corridors in southern California, Beier and colleagues (2006) invited agency and academic biologists familiar with each corridor planning area to identify species that would be a useful umbrella for other species sharing particular traits. Examples would be species that required intercore dispersal for metapopulation persistence, had short or habitat-restricted dispersal movements, represented an important ecological process (e.g., predation, pollination, fire regime), needed connectivity to avoid genetic divergence of a now-continuous population, might change from being ecologically important to ecologically trivial if connectivity were lost, or were reluctant to traverse barriers (e.g., culverts under roads). Each of the resulting linkage designs had 10–20 focal species, often including reptiles, fish, amphibians, plants, and invertebrates. By serving as an excellent flagship for this suite of umbrella species and the most appropriate focal species for specifying some attributes of corridors and cores, cougars will play their best role in reserve design.

My recommendations about quality and size of cores and corridors for cougars should be used to develop a formal habitat model that drives reserve design, not as rules of thumb that planners can use to delineate cores and corridors in an informal way. Less formal mapping is subjective and may not identify the best cores or corridors, especially when the needs of many species must be considered. At worst, such seat-of-the-pants maps can be abused to rationalize a bad plan and avoid hard decisions to conserve the best habitats. Instead, I advocate a rigorous and transparent approach (Beier et al. 2008) that enables all stakeholders to evaluate how well a proposed corridor or reserve design serves each focal species.

Conclusion

Conserving the cougar is a goal within our grasp. Fortunately, cougars are widespread and resilient, and in many areas conservation requires only tolerant human attitudes and a modicum of human restraint in transforming landscapes. The Florida panther has dodged genetic implosion, and may be one successful reintroduction away from recovery. Driven by their impressive dispersal abilities, cougars are capable of restoring themselves to midwestern and eastern North America with only a little human assistance. Cougars also can inform conservation strategies and motivate conservation efforts that benefit many other species. They can help planners estimate the minimum size of habitat cores and set limits on human activities that can degrade wildland blocks, buffer zones, and corridors. Finally, cougars can garner public support for efforts to create wildland networks for all biodiversity.

This chapter began with some reasons to conserve cougars: they are important top-down regulators of ecosystems; they play a significant role in structuring herbivore communities and controlling patterns of herbivory; and they are part of the evolutionary environment throughout the Western Hemisphere. Perhaps more important than any of these reasons, however, is the cultural reason: human life is enriched wherever big predators are part of the landscape. One cultural benefit of living with large predators is the reminder that humans are part of nature—and not always the dominant part. My home region of Arizona once had three such big predators, and we may be irreversibly impoverished by the loss of one of them, the grizzly bear. Our lives have recently been enriched by the return of another, the Mexican wolf. But through all centuries since the last glaciers receded, our third big predator, the cougar, has always been here to remind us that we are plain citizens of the biotic community. I trust it will always be so.

Chapter 13 Cougar-Human Interactions

Linda L. Sweanor and Kenneth A. Logan

O N JANUARY 8, 2004, a two-year-old, 110-pound male cougar (*Puma concolor*) killed a man cycling alone on a mountain biking trail in Orange County, California. Within hours of killing the man, the cougar attacked and seriously wounded a woman cyclist as she rode with a companion down the same trail. The man's death is the more recent of two that were attributed to cougars in the present decade in the United States and Canada. From 1890 through 1990, there were 63 verified cougar attacks on humans in these two countries, resulting in ten human deaths (Beier 1991; Fitzhugh et al. 2003). Another 54 attacks resulting in nine human deaths occurred in North America between 1991 and 2005 (Fitzhugh et al. 2003; Lewis 2006)—almost as many in 16 years as in the previous 100 years. Cougar attacks are extremely rare (nineteen human deaths in 116 years), and an analysis by Mattson (2007a) suggests a recent, slight decline in the annual number of cougar attacks on humans, down from about 3.7 per year in the 1990s to about 3.5 attacks per year during the present decade (2000–2006; Figure 13.1). Nevertheless, the overall increasing trend in cougar attacks has become a growing concern for wildlife agencies responsible for cougar management (Cougar Management Guidelines Working Group 2005). Recent well-publicized accounts of attacks have also intensified public apprehension about the dangers cougars might pose to human safety (Deurbrouck and Miller 2001; Etling 2001; Baron 2004).

No matter how rare incidents are, no one wants to be attacked by a cougar. Attacks also often provoke demands for retribution and sometimes reductions of local cougar populations. At the extreme, allowing any cougars at all in areas frequented by people is brought into question. However, wildlife and public land managers have the dual responsibility to provide for public safety while also maintaining viable wildlife populations, including cougars. Additionally, public attitudes toward cougars are not all negative; many value the cougar for its aesthetics, recreational value, and ecological role (Chapters 14, 15; Cougar Management Guidelines Working Group 2005). The ability to minimize the risk of negative cougar-human interactions can allay human fears and help ensure the long-term viability of cougar populations. Consequently, people need a better understanding of the factors that lead to cougar attacks and of ways for people to reduce risk. In this chapter, we first summarize information from recent studies that have examined cougar behavior around people. We then apply some of this information and observations from records on cougar-human interactions (particularly attacks) to an assessment of human risk of attack. We discuss behaviors and actions by individuals and responsible management agencies that may reduce human risk. Finally, we discuss avenues for future research that would help us better understand cougar behavior and further reduce the chance of a negative cougar-human interaction.

Research on Cougar Behavior around People

Recent works have documented, determined the validity of, and analyzed cougar attacks on humans (Beier 1991; Etling 2001; Fitzhugh et al. 2003; Mattson 2007a). Other studies have provided some information on cougar activity

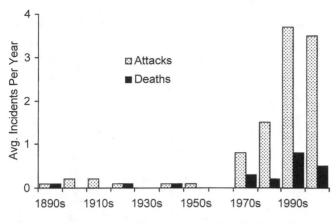

Figure 13.1 Annual number of cougar attacks and resulting human deaths, averaged per decade, for the United States and Canada (from Mattson 2007a).

patterns (Ackerman 1982; Van Dyke et al. 1986; Sweanor 1990; Beier et al. 1995; Ellingson 2003). Limited scientific information also suggests cougars may modify their behavior when exposed to human disturbance, primarily by using, or establishing home ranges in, areas subject to less human disturbance (Van Dyke et al. 1986; Janis and Clark 2002). But information on temporal avoidance of human use areas, or alteration of behavior to avoid direct contact with humans, has been restricted to a handful of individual cougars (see Van Dyke et al. 1986; Ruth 1991; Ellingson 2003). Only recently has an attempt been made to employ cougar movements and habitat-use patterns to determine the probability of cougar activity and human exposure to cougars (Arundel 2007).

These data gaps hamper our ability to understand cougar behavior around people and minimize human exposure to risk. Hence, we discuss findings from three recent studies that provide new information on three different scales. The first quantifies cougar behavior during episodes when people encounter cougars but the cougars do not attack (Sweanor et al. 2005). Prior analyses of rare cougar attacks provided important insights, but recording nonattack encounters as thoroughly as we document attacks may allow us to create more hypotheses on cougar and human behaviors and interactions that lead to attacks (Fitzhugh et al. 2003). The second study examines temporal relationships between cougars and people in shared environments (Sweanor et al. 2007). More detailed information on temporal relationships between cougars and people in shared environments can increase our ability to assess and reduce the potential for negative cougar-human encounters. The third study discusses the development of explanatory models using cougar behavioral, movement, and habitat selection data to obtain landscape-level predictions of cougar activity (Arundel et al. 2007). In concert with cougar attack data, these studies provide information that increases our ability to assess and

reduce the potential for negative cougar-human encounters, identify high risk areas, and provide insights into the impacts of human activities on cougar populations.

Cougar Responses to Human Approach

From 1985 to 1995, researchers approached and visually observed cougars in a remote New Mexico study area to gather data on cougar population biology and predation and to address a series of scientific hypotheses (Logan and Sweanor 2001). We later analyzed detailed field notes recorded on the range of responses that radio-collared cougars exhibited to those approaches (Sweanor et al. 2005). In contrast to reports in the literature, where the cougar probably initiated most of the encounters that led to attacks (e.g., Beier 1991; Etling 2001; Fitzhugh et al. 2003), our research team instigated the cougar-human encounters. Because our 2,059 km² study area was almost completely within White Sands Missile Range, human activity was minimal and year-round human density was low (about 0.65 persons per 100 km²). Cougars living there had few opportunities for human encounters, except in the context of cougar research. We hoped our observations would provide further insights into cougar behavior and increase our understanding of the variation in behaviors expected from wild cougars.

Our research team captured 126 cougars from one to six times each and fitted them with radio collars. Of the 126 animals, we approached and obtained visual observations on 67 independent cougars from one to twelve times each (mean = 4.7 times) and of eight weaned kittens once or twice each. We typically approached radio collared cougars using routes that provided the best concealment and kept us above the cougar. We approached family groups (i.e., females with nursing or weaned kittens), breeding pairs, solitary independent males and females, and weaned kittens in the absence of their mothers; however, our approaches more frequently involved female cougars and their young and thus were not a random or systematic sample of the cougar population. For each observation, we recorded the cougar's status, researcher approach distance, the number of people involved, and the cougar's behavior.

Based on our observations of cougar behaviors, we developed four response categories.

1. Cougar detected the researchers, as expressed by apparent visual contact, and moved away.
2. Cougar stayed at first observed location, visually detected the researchers, and watched.
3. Cougar appeared to ignore or did not appear to detect the researchers.
4. Cougar displayed a threat response toward the researchers.

Threat responses included vocalizations such as hissing, spitting, growling (with ears forward or down and back), deliberate approaches or charges toward the observers, or a confrontational stance. A confrontational stance consisted of the cougar quickly moving a short distance out from cover and in the direction of the observer, where it would stand stiff-legged in clear view of the observer.

No cougar exhibited a threat response during 84 observer approaches to distances of over 50 m. Most cougars (60 percent) simply left the area. Another 18 percent stayed without giving a threat, and 23 percent either ignored or failed to notice the observer(s). During 172 close approaches (2–50 m), most cougars (66 percent) left the area; another 24 percent remained in place without giving a threat response (Figures 13.2, 13.3). However, 16 close approaches resulted in threat responses (9 percent; Table 13.1). Threats occurred at a mean approach distance of 17 m (range = 2–50 m), whereas cougars that left or stayed without exhibiting threats were approached to a mean distance of 22 m (range = 3–50 m).

Mothers with cubs exhibited the most threats (14 threats during 79 close approaches); consequently, they were probably the most dangerous class of cougar to approach. A similar percentage of mothers (~17 percent) exhibited threats whether they had nursing or weaned cubs, and at an average approach distance of 16 m (range = 3–35 m). The remainder either left (58 percent) or stayed (24 percent) without giving a threat. In all 14 instances when the cougar

Figure 13.2 In New Mexico, a cougar's most common reaction to researcher approach (66 percent of 172 approaches to within 50 m) was to leave the area without giving a threat response. Photo © Ken Logan.

mother gave a threat response, she vocalized. Vocalization was the full extent of the threat for nine mothers; another two mothers approached, two charged, and one gave a confrontational stance before fleeing. We interpreted the mothers' threats to be in defense of their cubs. Females with cubs also were more likely to stay in place without threat

Figure 13.3 During 24 percent of 172 occasions when New Mexico researchers approached to within 50 m, the cougar remained in place without giving a threat response. Photo © Ken Logan.

Table 13.1 Cougar responses to human observers that approached to 2–50 m, San Andres Mountains, New Mexico, 1985–1995.

Puma Class[a]	Number of Approaches	Median (range) Approach Distance (m)	Median (range) Number of Observers	Cougar Response Left	Stayed[b]	Threat
Family	79	20 (3–50)	2 (1–5)	46	19	14
Solitary	82	20 (2–50)	1 (1–8)	62	19	1
Breeding pair	3	50 (10–50)	1 (1–2)	0	2	1
Cubs	8	20 (6–50)	2 (1–3)	6	2	0
Total	172	20 (2–50)	2 (1–8)	114	42	16

[a]Family includes female with nursing or weaned cubs; solitary includes adults and subadults; cubs includes weaned cubs 11–52 weeks old and in the absence of their mother.
[b]Includes five observations when cougar either ignored or did not detect human observer(s).

(42 percent stayed) than were solitary cougars and cubs in the absence of their mother (24 percent and 25 percent stayed, respectively).

Only two other close approaches resulted in threats, including one of 82 approaches of solitary cougars and one of three approaches of breeding pairs. A solitary adult male vocalized and then fled after he was approached within 2 m by a single observer. A breeding male charged out from behind a rock outcrop, made a confrontational stance and subsequent eye contact with a single observer, then turned and fled after the observer took one step upslope, diagonally toward the male cougar so that he could easily be seen. No dependent, weaned cubs gave threat responses during eight times they were approached while their mothers were away.

Researchers found variation in threat response between and within individual cougars. Only 14 (2 males, 12 females) of 67 independent cougars exhibited threats when approached (Figure 13.4). Each cougar gave its first threat after being visually observed without threat on up to six other occasions (mean = 1.8). Two females gave threat responses during two different approaches. However, 9 of 14 cougars that threatened observers were approached on subsequent occasions when they did not give a threat. These same 9 cougars chose to leave, without giving threats, during 25–83 percent of approaches (mean = 59 percent), and stay without threat during 0–50 percent of approaches (mean = 16 percent). Consequently, researchers could not predict whether a cougar that behaved one way during an approach would behave in a similar way during a subsequent approach. Regardless, most cougars showed great restraint, considering that human observers (often singly) initiated the encounters. The majority of cougars left the area when they were closely approached. Many other cougars remained in place without exhibiting threat behavior. Sometimes cougars appeared to relax after tense moments associated with initial human-to-cougar eye contact. Others

tended to ignore the observers, or appeared to react as though observers could not see them.

Clearly, the majority of study animals avoided direct human encounters. Because researchers instigated the encounters, cougars they approached were probably placed in a more subordinate position, and the threats they gave were likely more defensive than aggressive in nature. The observers' dominant behavior as well as careful approach tactics and angle (typically from above) also may have reduced the chance of a threat. Although the threshold distance for a threat response was 50 m, approaching to within that distance did not guarantee that a threat would occur. In fact, during 85 percent of approaches to with 10 m, there was no resulting threat.

Cougar Behavior toward Human Activity Areas

Cuyamaca Rancho State Park (CRSP) in southern California became a focus of research on cougar-human interactions in 2000 (Figure 13.5; Sweanor et al. 2004, 2007). Interest in such research was partly motivated by events of the seven prior years: seventeen cases of cougars behaving aggressively toward park visitors and two cougar attacks, one resulting in the death of a hiker. The park was pressed to provide for human safety as well as conserve its resident cougar population. The park's small size (~100 km²), heavy human visitation (~500,000 per year and increasing), accessibility (about 2 km of trail per square km of park), and location adjacent to rapidly developing private lands suggested that cougar-human interactions would probably increase. The research addressed questions on human safety by examining cougar and human activity patterns as well as each species' use of the park in space and time.

From March 2001 through October 2003, researchers captured ten adult cougars (five males, five females, one raising kittens) within or near the CRSP boundary. All were fitted with GPS collars and subsequently tracked for

Figure 13.4 During 16 of 172 close approaches by New Mexico researchers, cougars showed threat responses. One threat was given by a solitary male (like the one shown); however, the majority of threats were given by females with cubs at their sides. Photo © Ken Logan.

an average of 9.4 months (range = 1–20 months). Most GPS collars were programmed to attempt to get a position fix once each hour on Saturdays and Wednesdays and four fixes (0000, 0600, 1200, and 1900 hours) on the other days. These ten cougars were within the park during an average of 31 percent (range = 1–92 percent) of the days they were located by GPS. Researchers also monitored visitor use of trails on a seasonal basis with the aid of TrailMaster monitors and volunteers. ArcView 3.2 and digitized data layers, including roads and trails, were used in cougar location analyses with respect to human activity areas within the park. Trails, roads, and park facilities were initially digitized from 1992 aerial photographs (1:24,000); trails and roads were then updated with ground surveys using a GeoExplorer 3 (Trimble Navigation Limited, Sunnyvale, California). Random points were generated for CRSP and compared with cougar locations to examine whether cougars might be attracted to or avoiding human activity areas.

Based on 4,661 hourly distance measurements obtained for ten adults, cougars traveled the greatest distances during the night and crepuscular (morning and evening) periods, and the shortest distances during the day (Figure 13.6). Cougars were most often stationary during the day (i.e., period 1.5 hours after sunrise to 1.5 hours before sunset), moving less than 30 m during 56 percent of hourly intervals. During the night (i.e., 1.5 hours after sunset to 1.5 hours before sunrise), cougars were typically active, moving greater than 30 m during 77 percent and 67 percent of hourly intervals

for males and females, respectively. Because cougars were more likely to be inactive during the day, there was a lower but not negligible probability of a daytime cougar-human encounter. Assuming cougars that moved greater than 30 m during an hourly interval were "active," overall daytime activity for CRSP cougars was higher than reported in areas of low human presence (43–44 percent versus <30 percent; Ackerman 1982; Sweanor 1990) but lower than reported for a larger California park with an annual visitation of over 300,000 people (>50 percent; Ellingson 2003). The combined research indicates that some degree of daytime activity is normal for cougars and may not be in response to human activity.

In CRSP, human activity on trails was greatest during the day, when an average of 85 percent of people passed by trailhead TrailMaster monitors. On average, only 3 percent, 11 percent, and 1 percent of visitors passed by trailhead monitors during the morning, evening, and night periods, respectively. Visitor use of trails was also 370 percent greater on weekends than during weekdays. When comparing cougar and human activity during the four diel periods (morning, day, evening, and night), researchers found that cougars showed opposite activity patterns to human visitors on CRSP trails. This contributed to a low probability of a cougar-human encounter. The greatest chance for an encounter in CRSP was probably during crepuscular periods, most notably the time around sunset, when activity was waning for one species while waxing for the other (Figure 13.6).

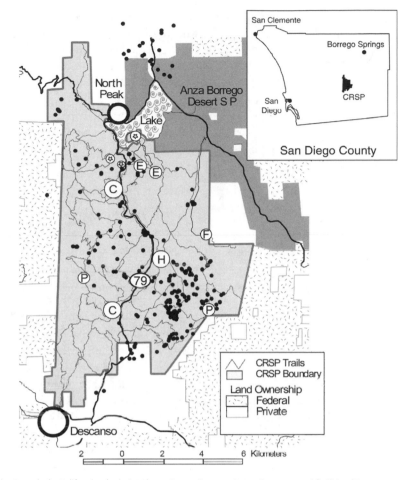

Figure 13.5 Cuyamaca Rancho State Park, California, depicting the major roads, extensive trail system, and facilities (C = car campgrounds, H = horse camps, P = primitive camps, F = fire camp, H = park headquarters and school camp, stars = park residences). Adjacent communities are shown as large circles. Locations (black dots, n = 327) are depicted for one female cougar (F15) over the four months she was GPS-collared.

We examined the juxtaposition of thirty-three prey caches (thirty deer, two domestic animals, and one coyote, *Canis latrans*) to trails, roads and facilities within CRSP. Although eight caches were within 100 m of a trail, two caches were within 100 m of a state highway, and three were within 150 m of private facilities north of CRSP, the cache sites were located randomly in relation to distance from trails, major roads, and facilities. Cougars were neither avoiding nor choosing cache sites that were close to area trails, roads, or facilities. Even though cougar caches were not farther from human activity areas than expected by chance, researchers speculated that most prey animals were killed and fed on during periods when diel human activity was relatively low. Of 196 GPS locations on cougars at twenty-seven prey caches, 81 percent were obtained during crepuscular or night periods. GPS data indicating night activity at cougar kills and caches were consistent with the results of other cougar studies in California (Beier et al. 1995) and Wyoming (Anderson and Lindzey 2003).

Examination of the GPS locations of eight adult cougars in relation to human activity areas (i.e., trails, roads, and park facilities) indicated variation in individual cougar behavior in response to changes in human activity. Two cougars used the park randomly in relation to distance from human activity areas, regardless of the time of day. However, five cougars were farther from trails during the day (when human activity peaked), or closer to trails at night (when human activity was minimal). Four cougars were closer to trails on Wednesday than on Saturday (when more people visited the park), suggesting that the cougars were using those areas more often as resting sites when there were fewer people on the trails. The female raising kittens showed the most consistent behavior; she used areas farthest from trails and facilities more than expected, and areas closest to trails and facilities less than expected, when people were most active in the park.

Cougars were not near trails more often than expected during the crepuscular period; nevertheless, typically one-third of each cougar's crepuscular locations in the park were within

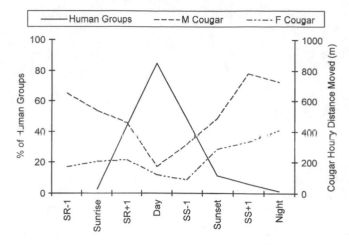

Figure 13.6 Daily human and cougar activity patterns in Cuyamaca Rancho State Park, California, based on human group counts on park trails and cougar distances moved per hour. The figure indicates opposite activity patterns with greatest probability of cougar-human encounters during the early morning and late evening (sunrise and the hour period either side of sunrise, SR-1 and SR+1; sunset and the hour period either side of sunset, SS-1 and SS+1).

100 m of a trail. The frequent tracks we observed indicated cougars were using the trails as travel routes. Beier (1995) found that cougars favored dirt roads and hiking trails as routes through dense chaparral. Considering that 11 percent of human traffic on area trails occurred during the evening, the chance of a cougar-human encounter would probably be highest during evening hours.

Some of the adult cougars in CRSP were most frequently located in human activity areas when human activity was minimal. This finding could be explained two ways: either cougars were avoiding areas with high human activity, or their natural behavior simply put them in certain habitats at times when people were not active there. Deer, the cougar's main prey, may have further confused the relationship. If deer avoided certain areas when there was increased human activity, this may have resulted in a corresponding "avoidance" of those areas by cougars. Regardless of the cause, the research demonstrated a space-time separation between some of these adult cougars and people visiting the park.

The temporal separation we observed between people and some of the resident cougars in CRSP allowed cougars to access food and other resources with a reduced chance of a cougar-human encounter. To avoid all probability of an encounter, cougars or people would have had to avoid CRSP completely. Although we did not document aggressive or defensive encounters between our study animals and park visitors, cougars of different age classes, social status, or temperament may behave differently. The New Mexico research (Sweanor et al. 2005) showed differences in

behavioral responses between and within individual cougars when encountering people. Cougars in CRSP that did not show detectible responses to people may have been exhibiting some level of habituation; if so, this level of habituation did not correlate to an increase in cougar-human conflicts. Regardless of whether the cougar showed some degree of avoidance or habituation, all CRSP cougars were in proximity to humans some of the time.

Modeling Cougar Activity as an Aid to Managing Exposure

A number of human-cougar encounters near Flagstaff, Arizona, during 2000–2001 stimulated local debates over the potential risks cougars posed to human safety as well as the impacts of people and human infrastructure (e.g., facilities, roads) on cougar populations and movements. Consequently, in 2003, researchers began studying cougar behavior and movements in the Flagstaff Uplands to obtain information that could be used to increase human safety, decrease human impacts, and provide insights on cougar ecology in the region (Mattson 2007b). Researchers focused on collecting data that could be used in explanatory models. One desired outcome was that the models could identify hotspots of cougar activity and thus zones with increased potential for cougar-human encounters. Although research is ongoing (probably through 2009), preliminary results provided in the 2003–2006 interim report are promising; they are summarized here from Mattson (2007b).

The study area was centered on Flagstaff, with capture efforts focused on cougars using National Park Service lands and important features of the Flagstaff Uplands (e.g., highways and urban-wildland interfaces). There are obvious human impacts on the area: Flagstaff's present population of 60,000 has increased approximately sixfold since the 1950s; two interstate highways carry over 20,000 vehicles per day through Flagstaff and the study area; and a railroad paralleling one of the interstates is traveled by about a hundred trains each day.

During the three-year period, eleven cougars were captured, fitted with GPS collars programmed to attempt up to six position fixes per day, and tracked via Argos satellites. During that interval, researchers obtained more than 9,000 GPS locations and investigated almost 400 sites where GPS locations were clustered. For analysis of movements and habitat selection, researchers calculated straight-line distance and turn angle between consecutive points, estimated home range size, and developed seasonal models of habitat selection using several explanatory variables (i.e., terrain roughness; elevation; distance to nearest paved highway, human residential area, and known source of free water; habitat type; and land jurisdiction).

Both natural and human features had apparent effects on cougar movements and habitat use. The animals strongly selected for rough terrain and woodland or forest cover. They typically avoided flat, open private lands and residential areas, and also avoided crossing area roads. When habitat selection models were used to map seasonal probability of cougar activity, hotspots of potential activity appeared. Such maps may help wildland and wildlife managers educate people on the relative risks of cougar encounters in space and time, identify areas where more intensive management or planning may be beneficial, and consequently reduce overall risk of a negative cougar-human encounter.

Assessing Risk of Attack

Studies in New Mexico and California (Sweanor et al. 2004, 2005, 2007) suggest that most cougars avoid confrontations with people and, even in areas where the two species share habitats, cougars show a proclivity for using those habitats during times when human use is minimal. Nevertheless, threatening encounters and attacks by cougars occur. Reasons for such encounters and attacks are not well understood. Our understanding of the behaviors and actions that lead to attack and death are mostly limited to documentation of events during actual cougar attacks (e.g., Beier 1991; Etling 2001; Fitzhugh et al. 2003).

Our ability to assess human risk may be improved by the development of models that map the probability of cougar activity (Arundel 2007). Mattson (2007a) attempted to clarify the risk of cougar attack by partitioning the probability of a person being killed by a cougar into four transitional stages: probability of a cougar encounter; probability of an aggressive response by the cougar; probability the cougar will attack; and probability that the attack will result in a human death. The numbers and activity patterns of people and cougars in cougar habitat affect encounter probability; the progression from simple encounter to aggressive encounter and attack are affected by the behaviors and actions of people and cougars once an encounter has occurred. We examine these factors in the sections that follow; a more thorough discussion can be found in Mattson (2007b).

Probability of a Cougar-Human Encounter

Increased encounter probability is likely due in large part to a growing number of people in cougar habitat and to possible increases in cougar numbers (Cougar Management Guidelines Working Group 2005). Hence, more and more people are exposed to cougars. In the last thirty years there have been substantial increases in human populations throughout the western United States and Canada.

California, British Columbia, and Colorado have documented the greatest numbers of cougar attacks on humans (Mattson 2007a). The Californian human population has more than doubled in thirty years, to 33.9 million in 2000. As a result of population growth, human developments encroach on remaining cougar habitat and an increasing number of people probably use it for recreation (Figure 13.7). For example, during 1998–2003, average annual visitation to Cuyamaca Rancho State Park was 17 percent greater than for the preceding five-year period. Visitation to Rocky Mountain National Park in Colorado, where a boy died in a cougar attack in 1997, more than doubled between 1960 and 1995 (Stohlgren et al. 1995). In the next twenty years, the human population in the western United States (Arizona, California, Colorado, Idaho, Montana, Nevada, New Mexico, Oregon, Utah, Washington, and Wyoming) is projected to increase by 28 percent or by over 18 million (U.S. Census Bureau projections 2005); as more people live in and visit cougar country, the probability of a cougar-human encounter of some kind (benign or threatening) will likely increase.

Probability of encounter may also be affected by changes in cougar numbers or localized densities; however, these are more difficult to assess. Most western states and Canadian provinces probably realized increases in cougar numbers in the latter half of the twentieth century, likely as a result of management protections starting in the mid-1960s in conjunction with increasing numbers of deer and elk (Cougar Management Guidelines Working Group 2005). Although dramatic increases in cougar removals in the 1990s (through sport hunting and depredation control; see Figure 4.1) suggest that western cougar populations might be robust and increasing, high harvests in combination with unprecedented habitat loss and fragmentation might result in some cougar populations stabilizing or declining. However, in Florida, where fairly accurate estimates have been made on the intensively studied, and recently augmented, remnant panther population, panther numbers have probably doubled in the past twenty years. Although there are no recorded attacks on humans by panthers, confirmed attacks on pets and livestock have increased from none ten years ago to seven in 2006 (D. Land, Florida Fish and Wildlife Conservation Commission, pers. comm., 2007).

Encounter probability has both spatial and temporal components; basically cougars and people have to be present at the same place at the same time. Regarding the spatial component, Halfpenny and colleagues (1991) used reports of cougar sightings and incidents along the Front Range of Colorado's Boulder County to examine properties of cougar-human interactions. They speculated that increases in interactions were a product of rapidly expanding deer,

Figure 13.7 Radio-tracking cougars from the air reveals southern California's expanding human population and the potential for more cougar-human encounters. The town of Ramona (pictured) is expanding eastward toward Cuyamaca Rancho State Park and overlaps the western edge of an adult male cougar's territory (to the right of the dotted line). Photo © Linda L. Sweanor.

cougar, and human populations in that area. Cougars use features of the habitat selectively; specific habitat features confer advantages to cougars in hunting prey and provide security for cougars and their offspring (Logan and Irwin 1985; Belden et al. 1988; Laing and Lindzey 1991). In Colorado (and likely other western states) human settlement patterns have recently shifted toward habitats that may favor cougars: rural settings often bordering public lands, more rugged terrain with expansive views adjacent to valley floors, and denser vegetation (Riebsame et al. 1996; Theobald et al. 1996). Consequently, more people are choosing to live in habitats that are also preferred by cougars—areas that supply hiding and stalking cover and substantial numbers of vulnerable deer and elk.

Temporal components of encounter probability include daily and seasonal activity patterns of cougars and humans. Cougar attacks on people have historically peaked during daylight and dusk hours (Mattson 2007a), times when human activity is typically high (daylight) and cougar activity is waxing (dusk). Researchers in California (Sweanor et al. 2004, 2007) found that people were most active on park trails during daylight hours and dusk, and that cougar activity increased dramatically between those two time periods. Predominantly crepuscular and night activity at cougar kills and caches has also been observed (Beier et al. 1995; Anderson and Lindzey 2003; Mattson 2007a; Sweanor et al. 2007). More than 40 percent of recorded cougar attacks have occurred during summer (Mattson 2007a). Increased human activity is the most likely cause of the high incidence of cougar attacks during that time of year (Beier 1991). However, the relationship between human activity and the probability of attack is not completely clear. Consider the low number of attacks during fall (~12 percent; Mattson 2007a), when human use of cougar habitat is probably still fairly high, in part because of the popularity of big game hunting.

Why Cougars May Respond Aggressively

Whether a cougar-human encounter leads to an aggressive response by the cougar may be affected by an individual cougar's genetics, behavioral inclinations, life stage, health, and possibly degree of habituation (i.e., a waning response to a repeated neutral stimulus) toward people. Aggressive tendencies (or prey image) may have a genetic component that is passed down from generation to generation; if so, the probability of an aggressive attack may depend partly on the proportion of a cougar population that possesses aggressive genes and the probability that those aggressive cougars encounter people. Consequently, we speculate that some directional selection (in the form of humans killing aggressive cougars and minimizing their occurrence in the gene pool) may occur and may be partly responsible for the low historical incidence of cougar attacks on people. Additionally, cougars may not naturally perceive humans as prey. Their evolutionary history with modern, relatively technologically advanced humans who could defend themselves spans only perhaps 14,000 years, in contrast to the African and Asian big cats that prey on humans much more frequently (Løe 2002; Packer et al. 2005). Those felids coevolved with a variety of pre-hominid and human forms for hundreds of thousands, if not millions, of years (Wells 2003).

A cougar's behavioral inclinations can vary. Individuals may respond differently to the same type of encounter, and the same cougar may behave differently from one encounter to the next (Sweanor et al. 2005). Although it is impossible to know how many benign cougar-human encounters occur each year, research in New Mexico suggests that most cougars will not show aggressive tendencies toward humans and that people may often not even be aware of the cougar's presence (Sweanor et al. 2005). In fact, 24 percent of cougars closely approached by researchers remained in place without exhibiting a threat response, and they sometimes even appeared to relax after tense moments associated with initial human-to-cougar eye contact (Sweanor et al. 2005).

Cougars may exhibit defensive or aggressive responses; it may sometimes be difficult for human observers to discern the difference. A defensive cougar may be displaying threats to protect itself, its food source, or its kittens. The threats observed in the New Mexico research (Sweanor et al. 2005) were likely defensive in nature. A defensive response may simply include vocalizations (e.g., hissing, growling) or it may escalate to an approach, charge, or even contact (attack). Females with kittens may be most likely to exhibit defensive responses (see Sweanor et al. 2005). An aggressive (versus defensive) response most likely results from a cougar including humans in its prey image. Some aggressive behaviors may be similar to defensive ones, including crouching, approaching, and charging. Because a cougar relies on stealth during predatory attacks, many of these behaviors may not be observed. When a cougar is observed, risk of attack can still be assessed, regardless of whether the cougar is behaving defensively or aggressively. Observed cougar behaviors that may indicate an increasing risk of attack are described in the Cougar Management Guidelines (2005, 89; Table 13.2).

Limited data suggest that underweight and young cougars are more likely to be involved in attacks or come into close contact with people (Beier 1991; Ruth 1991; Mattson 2007a). Young cougars are not as skilled at hunting, they may have a more malleable prey image, and their tendency to disperse long distances provides them with increased opportunities for encounters. Based on an examination of previous data (Beier 1991; Etling 2001; Lewis 2006), Mattson (2007a) also suggests that female cougars may have a greater propensity to attack. This observation may be related to the structure of the cougar population; females are typically more abundant due to the species' social structure and higher female survival rates (Logan and Sweanor 2001). Females may also be involved in more defensive attacks because they are protecting kittens.

It has been suggested that cougars can become habituated to human developments and activities (Aune 1991; Halfpenny et al. 1991; Ruth 1991). Habituated cougars would likely be close to areas where humans are active, potentially increasing the probability of a cougar-human encounter. However, research from California suggests that habituation may not readily occur, since most cougars appeared to avoid park trails, facilities, and roads when people were most active there (Sweanor et al. 2004; Sweanor et al. 2007). But it is possible that habituation is adopted by only a small percentage of a local cougar population, and it is manifested in progressive stages. A habituated cougar may progress to more undesirable behaviors, possibly to the point of including humans in its prey image.

Habituation has been argued for cougars in the case of a jogger killed in Colorado in 1991 (Baron 2004); nevertheless, there is no scientific evidence that habituation increases risk of a negative encounter or attack (Cougar Management Guidelines Working Group 2005). People may be able to approach habituated cougars more closely and thus increase the chance of a defensive attack. Habituated cougars may also be more likely to engage in risky behavior such as killing poorly protected domestic livestock and pets and crossing busy roads. More research needs to be done to determine whether cougars are more tolerant, or conversely more dangerous, toward humans when they become habituated. Research on the brown bear (*Ursus arctos*) suggests that habituated bears are more tolerant of hikers or bear viewers and thus less likely to threaten or attack (Aumiller and Matt 1994; Smith et al. 2005). However, bears that

Table 13.2 Interpretation of cougar behaviors, arranged in order of increasing risk to a human interacting with the cougar.

Observation	Interpretation	Human Risk
Opportunistically viewed at a distance	Secretive	Low
Flight, hiding	Avoidance	Low
Lack of attention, various movements not directed toward person	Indifference, or actively avoiding inducing aggression	Low
Various body positions, ears up; may be shifting positions; intent attention; following behavior	Curiosity	Low, provided human response is appropriate
Intense staring; following and hiding behavior	Assessing success of attack	Moderate
Hissing, snarling, vocalization	Defensive behaviors; attack may be imminent	Moderate or high, depending on distance to animal
Crouching; tail twitching; intense staring; ears flattened like wings; body low to ground; head may be up	Pre-attack	High
Ears flat, fur out; tail twitching; body and head low to ground; rear legs "pumping"	Imminent attack	Very high and immediate

Because a cougar's behavior usually is not observed prior to an attack, these behaviors should not be the only criteria used for assessing risk (from Cougar Management Guidelines Working Group 2005, 89).

are rewarded for aggressive, food-seeking behavior around people may treat humans as prey or cause human injury (Gunther 1994; Gniadek and Kendall 1998).

Human Attributes and Behaviors That Affect Outcome

Because of the growing number of people in cougar habitat, people are increasingly sampling the variation in cougar responses to humans. Most aggressive (versus defensive) attacks may occur because the cougar has decided the person fits its prey image, or it is at least willing to experiment with the idea. Consequently, human physical attributes, vulnerability, and behavior will affect the probability of an encounter escalating into an attack. Children (singly or in groups) appear to be at much greater risk of being attacked than adults (Beier 1991; Etling 2001; Fitzhugh et al. 2003), and solitary people (juveniles or adults) are three times as likely to be attacked as groups of two or more (Fitzhugh et al. 2003). Only groups of five or more appear fairly secure against attack (Fitzhugh et al. 2003). This contrasts with New Mexico research where people deliberately initiated the encounter; under those conditions, threat responses were probably defensive and were most often given when there were at least two (and up to five) observers present (Sweanor et al. 2005). This may result in part from greater efforts to capture or see cubs when researchers had more personnel. Solitary observers tended to be more cautious and did not usually attempt to get as close to nursery or bed sites as did multiple observers.

Rapid movement (e.g., running, skiing, bicycling) and position may also influence a cougar's response to humans. Mattson (2007a) analyzed previously published data (Beier 1991; Etling 2001; Fitzhugh et al. 2003) and information compiled by C. Lewis and available online at www.cougarinfo.com (Lewis 2006) and found that people moving rapidly away from or laterally to cougars were generally more likely to trigger a cougar attack. As suggested by the approach behavior of researchers in New Mexico (Sweanor et al. 2005), people who remain still and appear large or are positioned above a cougar may appear dominant and discourage an attack. Supporting research by Leyhausen (1979) shows that cats appear frightened when approached from above. Consequently, a person positioned below a cougar or in a squatting or prostrate position might appear more subordinate, preylike, and vulnerable to attack.

Aggressive, dominant human behavior (e.g., standing ground, looking large, obtaining an elevated position, swinging a stick, throwing objects) may reduce the probability that an aggressive encounter will turn into an attack (Beier 1991; Sanders and Halfpenny 1993; Sweanor et al. 2005). In a few cases, charging or lunging toward a cougar has been shown to cause the animal to back down or leave the area (Fitzhugh et al. 2003). In cases where a cougar is surprised, subsequent identification of the intruder as human also may deter an attack (Sweanor et al. 2005). In New Mexico, cougars confronted, advanced on, or charged observer(s) on six occasions (Sweanor et al. 2005). In all six cases, the observers helped avert a possible attack by making themselves more identifiable, standing erect and beating sticks or other objects against bushes or rocks, making continuous noise, banding together, or backing away to a higher (possibly more dominant) position.

In one of those instances (similar to accounts provided by Fitzhugh et al. 2003), sporadic gunfire did not deter a cougar's advance. Beier (1991) reported that shouting loudly seemed to avert or repel attacks. Fitzhugh and colleagues (2003) emphasized that noise may be effective, but it probably needs to be loud and continuous to interrupt the cougar's activity. Whether maintaining eye contact will affect the probability of attack has been debated (Etling 2001; Mattson 2007a); it may be dependent on whether the interaction is aggressive or defensive. Regardless, maintaining visual contact with the animal will help a person assess the cougar's responses and take appropriate defensive measures (Fitzhugh et al. 2003).

It is still unclear whether the presence of a dog may attract a cougar and subsequently increase the threat that the cougar will attack or kill an accompanying human. Mattson's (2007a) analysis of attack data revealed similar attack probabilities for encounters with and without dogs present (70 and 72 percent, respectively). However, dogs appeared to reduce the odds of a human death during an attack, probably because dogs often harassed or chased off the offending cougar.

Death is most common when cougar attacks involve children who are by themselves (Mattson 2007a). Death rates in children decline dramatically when adults are present, primarily because adults are able to intervene (Etling 2001). In attacks that involved human groups of two or more adults (no children), no human fatalities were reported. This is probably because there is at least one non-target, capable adult available to fight off the attack. In most cases victims have been capable of repelling an attack by fighting back with bare hands, a stick, a knife, a jacket, or a rock (Beier 1991). Most recently, a sixty-five-year-old California woman fought off a cougar that was attacking her seventy-year-old husband by beating the cougar repeatedly with a four-inch-diameter stick (R. Konrad, Associated Press, 1/28/07). Lone individuals, especially those involved in fast-moving activities (e.g., running, skiing) are typically caught by surprise in a predatory attack and have no time to react or fight back.

Reducing Risk

Being informed and using common sense can help minimize a person's risk of a threatening cougar encounter or attack. Anyone living or recreating in cougar habitat should be aware of the risks and act accordingly. Lists of appropriate actions to minimize risk are provided by most western state and provincial wildlife management agencies as well as in several recent books (e.g., Etling 2001; Torres 2004; Cougar Management Guidelines Working Group 2005, 93; Table 13.3).

Individual Behaviors and Actions

Although not all listed responses are proven effective, they do involve common sense. Most are based on the premise that the cougar is determining whether the object of its interest is prey and, if so, that it is vulnerable. Consequently, it is in the person's best interest not to look or act like prey (Mattson 2007b). Logical behaviors include hiking in groups, keeping children close, looking dominant (and not vulnerable) during a confrontation by gaining elevation, grouping together, and making continual noise.

Because continual noise may dissuade a cougar, it may be advantageous to carry an air horn (Etling 2001; Mattson 2007b). Although pepper spray may have helped prevent a possible attack in Cuyamaca Rancho State Park in 1998 (Etling 2001), its effectiveness is mostly untested and would only apply at close range. Avoiding activities during dawn and dusk (when cougars show increased activity) should further reduce the risk of encounter.

Eliminating attractants may reduce the probability that a cougar will venture close to people or residential areas and subsequently engage in undesirable behavior. People may attract cougars unintentionally; substantial increases in deer populations in residential areas that provide shelter and food (in the form of lawns, hedges, and gardens) have been suggested causes for increased sightings and possible habituation of cougars (Halfpenny et al. 1991). Consequently, residents in cougar country should refrain from feeding wildlife (or leaving pet food outside), avoid planting vegetation that attracts deer, and deprive cougars of stalking cover near their homes. Cougars are opportunistic predators (Logan and Sweanor 2001) and scavengers (Bauer et al. 2005) and may also be attracted to domestic prey when it is vulnerable. In California,

Table 13.3 Some of the measures, with supporting information, humans can take during an encounter to prevent injury.

Recommendation	Supporting Information
Keep children under close control, and in view. Pick up small children immediately if you encounter a cougar. Do not hike alone.	66% of victims have been unsupervised children or lone adults (Beier 1991).
Do not run.	Running and quick movements may stimulate chasing and catching response.
Stand. Wave your arms. Raise a jacket over your head. Appear as large as possible. Move to higher ground if nearby. Throw sticks, rocks, or other objects if within reach and accessible without bending too low.	Prey size, vulnerability, and "positioning" influence cougar response.
Avoid dead animals and never approach kittens. Talk calmly. Back away.	Nonprey may be attacked if viewed as a threat.
Maintain eye contact. Do not look away. If cougar appears agitated use peripheral vision to keep track of its location.	Eye-to-eye contact often restrains large cats. Direct eye contact from prey may inhibit predatory action.
Be alert to your surroundings.	Cats exploit all vantage points/cover when investigating prey.
If attacked, fight back. Humans have successfully deterred attacks by becoming aggressive.	A cat grasps with its teeth only if it meets with no resistance. Violently struggling prey may be released (Leyhausen 1979).
Secure pets and hobby animals in predator-proof enclosures between dusk and dawn. Keep pets on leashes and off trails in the backcountry (see text).	Cougars may be attracted to vulnerable prey (Aune 1991, Torres et al. 1996, Sweanor et al. submitted).
Keep garbage under control to avoid attracting raccoons, skunks, etc. Do not feed pets outside and remove extra feed from domestic animal pens. Do not feed deer and wild turkeys.	Cougars may be attracted to concentrations of potential prey.
Keep pets under control. Cougars entering yards or campsites to kill pets may be candidates for removal.	Once a learned behavior develops it may not be possible to modify this behavior (Leyhausen 1979).

SOURCE: Cougar Management Guidelines Working Group 2005, 93

Sweanor and colleagues (2004) found that simply plac-ing domestic animals in cougar-proof enclosures at night could reduce vulnerability. In one notable case, an adult female cougar was captured at the site of an alpaca she had killed. The cougar was GPS-collared and released, and the alpaca's owner immediately began placing the rest of her animals (~12 alpacas and 2 llamas) in pro-tective enclosures at night. Although the alpaca property remained within the home range of two adult cougars (the female and a GPS-collared male) during the following year, no other cougar attacks occurred. People who live in cougar habitat and practice appropriate husbandry on their domestic animals can protect both domestic animals and cougars, reduce the chance of "positive conditioning" of the cougar, and probably decrease the potential for a cougar-human interaction (Cougar Management Guide-lines Working Group 2005).

Management Actions

Reducing the risk of cougar-human conflicts requires invest-ment by people and the agencies that serve them. A pro-active management approach involving education, land use planning, and standardized protocols for dealing with cougars that exhibit unacceptable behaviors toward people or property may help reduce the risk of conflicts and aid cougar conservation.

Education. Wildlife and wildland managers should provide educational materials on cougar presence, cougar behavior (e.g., that cougars on their cache sites can be close to human activity areas), ways to minimize conflicts, and appropriate human responses during an encounter (Figure 13.8; Torres 2004; Cougar Management Guidelines Working Group 2005). This allows for more informed choices by wildland users. Because there is likely to be a continued influx of new human residents into cougar habitat, local communities or landowners' associations have the opportunity to develop their own outreach programs whereby newcomers can quickly be informed of the presence of cougars and how to avoid conflicts.

Habitat management. Because there appears to be a positive correlation between human use of cougar habitats and numbers of cougar attacks (Torres et al. 1996), habitat management practices have the potential to influence risk. Consequently, wildland managers would benefit from assessing habitat use patterns of cougars and their prey (i.e., deer and elk) when planning trails, roads, and facilities. In areas frequented by cougars, more active management could include limitations on type of human activity (e.g., no jogging), on number and composition of users (e.g., no

children or solo hikers), and on time of day (e.g., trails close between dusk and dawn; see Mattson 2007b). Shuey (2005) concluded that city and county developers might also be able to reduce the risk of a cougar-human encounter by restricting the development of land-covers or land-cover patterns that produce the most risk. Models similar to the one being developed by Arundel and colleagues (2007) to map probability of cougar activity can be used as aids to reducing cougar-human interactions.

Research findings highlight the need for land use plan-ning on a larger scale. Cougars can threaten human safety, but people may also affect cougar populations negatively. Human developments (e.g., roads, houses) in cougar habitat impede cougar movements, affect the distribution of cou-gar prey, place cougars proximate to humans, and result in greater cougar mortality, often due to vehicle strikes or depredation control. In California, each of ten GPS-collared adult cougars used a protected state park; however, pri-vate lands in various stages of development also made up a large proportion of each cougar's home range, increasing the probability of further habitat fragmentation and cougar-human encounters (Figure 13.9; Sweanor et al. 2004).

Roads have become barriers to movement (Logan and Sweanor 2001; Beier et al. 2005; Mattson 2007b), some to the degree that viability of a local cougar population is threatened (e.g., Beier and Barrett 1993). Vehicle strikes have also been a leading cause of death in some cougar popula-tions (Beier and Barrett 1993; Maehr 1997b; see Mortality Factors, Chapter 5, and Fine-Scale Movement, Chapter 12). Human-caused deaths were the greatest known cause of mortality in a study population in southern California (Sweanor et al. 2004). During the study (2001–3) and within a 25-km radius of the study area's boundary, thirteen

Figure 13.8 Signs are posted at all trailheads in Cuyamaca Rancho State Park, California, to inform visitors of the presence of cougars and to suggest appropriate safety precautions for hikers. Photo © Linda L. Sweanor.

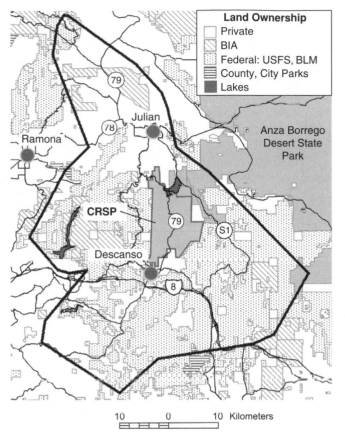

Figure 13.9 Cuyamaca Rancho State Park (CRSP), California, and surrounding land ownership. The heavy black line depicts the conjoined, cumulative home ranges (100 percent minimum convex polygon) for adult pumas with home ranges that overlapped CRSP, 2001–3.

including defining what constitutes a "dangerous" cougar and actions for its removal (Cougar Management Guidelines Working Group 2005; Table 13.4). What constitutes a certain level of risk (low, moderate, high) should be determined in advance, as should the corresponding level of response (e.g., increased educational efforts, posting warning signs, area closure, increased patrol, hazing or scaring the animal away, removal; see Cougar Management Guidelines Working Group 2005, 100). There may also be different levels of response, depending on the environment where the observation or incident occurred (e.g., remote versus highly developed areas). Removal may entail placing cougars in captivity, translocation, or euthanasia. Translocation of cougars has typically been shown to be ineffectual, with cougars suffering low survival rates and exhibiting unpredictable movements or homing behavior. Translocation may be more successful for subadult cougars (i.e., animals that are not yet of breeding age or established an adult home range; Ruth et al. 1998) and may sometimes be a useful tool for handling young individuals that have not exhibited undesirable behavior toward people (e.g., a cougar that has wandered into a residential area during a dispersal move). Nevertheless, translocation entails significant planning, cooperation between agencies, identification of an appropriate release site, and risks of future liability. Removal protocols should be determined in advance and should ensure that there is substantiating evidence of the correct animal being captured or killed.

Further Research

Our understanding of cougar-human interactions comes, to a large degree, from information recorded on cougar attacks and, to a much lesser degree, from reports of cougar sightings and encounters. There have been commendable efforts at analyzing the data on cougar attacks (Beier 1991; Fitzhugh et al. 2003; Mattson 2007b) and sightings (Halfpenny et al. 1991; Shuey 2005), as well as attempts to create or maintain a general data base on the details of individual attacks (Etling 2001; Lewis 2006). However, anyone involved with analyses on cougar attack or sightings data will undoubtedly lament the incompleteness, inaccessibility, or unreliability of the data (e.g., Fitzhugh 1988). Improvements can be made by better collaboration, standardizing data forms (including standardizing definitions), and developing a detailed, active data base that can be easily updated or accessed.

We encourage researchers involved in intensive studies of wild cougars in habitats with variation in human development to record and quantify cougar behaviors in relation to human activities. Managers can thus gain greater insights into cougar behaviors that may or may not con-

cougars (including four study animals) were killed for taking pets and hobby animals, and two cougars (including one study animal) were killed by vehicles (Sweanor et al. 2008). Although the immediate effect of these deaths on the local cougar population is unknown, the deaths may be indicative of a developing crisis, given the rapid human development in the region. Acquisition of key habitat patches and connecting corridors can offset fragmentation, ensure the continued presence of cougars in the area, and perhaps reduce cougar-human conflicts. Fine-scale cougar habitat models can assist with habitat planning by helping identify cougar movement corridors (Dickson et al. 2005) and the characteristics of important road-crossing points (see Chapter 12; Clevenger and Waltho 2005).

Protocols for responding to conflicts. Individuals and responsible management agencies should make every effort to minimize the probability of a negative cougar-human interaction. Nevertheless, dangerous situations will sometimes occur. Responsible agencies would benefit from policies and attendant protocols for responding to cougar-human conflicts, stitute threats toward people. Better yet, we recommend research focused on cougar-human interactions that identify the relative strength of factors contributing to dangerous encounters between cougars and people. Data from the New Mexico research (Sweanor et al. 2005) identified some influential factors that can be used in a conceptual model to be tested through experimental research: cougar sex, cougar reproductive status, and observer approach distance.

More research at the landscape level (Arundel et al. 2007) can help identify hotspots for potential cougar-human interactions and assist with land use and habitat management practices that will help reduce encounter probability. Since cougar behavior may be strongly affected by the activity patterns and distribution of their major prey, further research on prey behavior and vulnerability to cougar predation might also provide insight into reducing the potential for encounters where people and cougars are using the same habitats. Models of habitat selection should include prey distribution and density as explanatory variables.

More controlled experiments on cougar responses to humans and human activity, including the effects of aversive conditioning or other deterrents, are needed. An effective method of aversive conditioning would be useful in cases where the removal of nuisance but nondangerous cougars is not the preferred option (e.g., removal of an endangered Florida panther). Current information about aversive conditioning on cougars is minimal and sketchy. It is still unclear whether a research cougar's experience during the capture and handling process may instill in it some level of aversion toward humans or human activity. In New Mexico (Sweanor et al. 2005), the number of times a cougar was captured did not appear to affect the probability of a threat response during an encounter; cougars that gave threats were captured and handled a similar number of times as those that did not give threats. However, the New Mexico research did not test whether cougars that had never been captured and handled might be more likely to give threats than those that had, or whether intentional aversive conditioning during

Table 13.4 Suggested protocol in decision-making process as related to single or multiple occurrences of cougar behaviors in the risk categories defined in Table 13.2.

Risk Category: Specific Behavior, Number of Occurrences	Recommended Type of Response	Recommended Management Actions
Low risk: single occurrence	Wait and see	Continue or initiate public education.
Low risk: multiple occurrences	Take appropriate action Evaluate circumstances of observations	Post warning signs. Consider use of hazing. Consider data base for observations.
Moderate risk: deliberate approach (curiosity), single occurrence	Take appropriate action Evaluate conditions leading to approach	Post warning signs. Mark and monitor animal if possible. Consider use of hazing. Map observations and document management actions in data base.
Moderate to high risk: multiple occurrences	Take appropriate action Evaluate conditions leading to approaches Evaluate removal	Post warning signs, or close area. Increase education effort. Patrol area with loaded firearm to kill cougar if perceived as dangerous, or haze if perceived as curious.
High risk: near attack, single occurrence	Take immediate action Evaluate if behavior was predatory or defensive Evaluate removal	Post warning signs or close area. Increase education effort. If decision is to remove, patrol area with loaded firearm to kill cougar.
High risk: nonfatal attack, single occurrence	Take immediate action Evaluate if attack was defensive	Secure victim. Post warning signs and close area. Secure incident scene, contact Wildlife Services and kill the cougar. Contact the media.
High risk: fatal attack, single occurrence	Take immediate action	Close area. Secure incident scene, contact Wildlife Services and kill the cougar. Assist and support victim's family. Contact the media.

captures would result in a reduction in subsequent threat responses. It is possible that the general avoidance reaction of cougars toward people during approaches was partly a result of prior captures.

In Florida, four panthers involved in public safety complaints were subjected to a three-stage aversive conditioning program. The initial stage involved treeing an offending panther with hounds with subsequent tranquilizing, handling, and radio-collaring. Repeat offenders were subjected to harassment with hounds or taped hound vocalizations. These techniques resulted in varying responses from the panthers, but the researchers concluded that some degree of avoidance and fear of humans might have been instilled (McBride et al. 2005). In contrast, a cougar in Big Bend National Park attacked a person four months after a capture and handling experience similar to the one imposed on the Florida panthers. A second cougar was shot with rock salt after a near-attack; it returned to aggressive behavior two weeks later (Beier 1991). Consequently, aversive conditioning should be attempted for nuisance cougars on an experimental basis, and data should be gathered for analysis of effectiveness.

Conclusion

Cougar attacks on humans are rare but appear to be increasing. This trend has sharpened apprehension among wildlife personnel and the public about the dangers cougars may pose to human safety. A clearer understanding of cougar behavior and other factors that lead to cougar attacks is needed to help reduce the probability of a negative encounter, to keep fears realistic, and to ensure the continued viability of cougar populations.

Three recent studies have provided some important information on cougar behavior at three scales by quantifying cougar behavior during encounters when cougars do not attack; examining the timing of human and cougar uses of shared environments; and using explanatory models to obtain landscape-level predictions of cougar activity. The first two studies (Sweanor et al. 2005, 2007) bolster other research and observations suggesting a tendency for cougars to avoid people and human activity. Nevertheless, in habitats that support cougar populations and are used by people, it should be expected that cougars and people will sometimes be in proximity. The third study (Arundel et al. 2007) provides ways to determine relative risks of cougar encounters and attacks in relation to space, time, and human characteristics. The information from these studies, in conjunction with recent works that have analyzed reported cougar attacks on humans, can be used to help reduce risk of a cougar attack.

The likelihood of a cougar attack is affected first by the probability of a cougar-human encounter, and subsequently by the behavioral traits of the cougar and the behavior and appearance of the human(s). Encounter probability will undoubtedly increase with increases in human populations and activities in cougar habitats. Local, state, and federal government agencies play a vital role in reducing risk—agencies are getting better at posting signage, distributing brochures, conducting programs alerting people to the presence of cougars, and offering guidance on how to avoid cougar encounters and conflicts. Further analyses of attacks, field research on cougar-human interactions and cougar-prey interactions, and experimental studies on the effectiveness of deterrents can provide the basis for better management and for reducing risk.

Cougars are stalking predators capable of killing animals that are up to six times their own size (Logan and Sweanor 2000) and that are armed with formidable defensive weaponry in the form of sharp hooves and antlers. Cougars can easily kill humans. The fact that they rarely do indicates cougars and people can coexist, albeit this needs to be a *managed coexistence*. Attempting to change cougar behavior so as to eliminate all risk is beyond the capabilities of any management agency. Much of the success of reducing risk of attack is thus in the hands of people who share the habitat. Our actions, our choices, our willingness to educate ourselves, and our acceptance of some level of risk will ultimately affect the survival of cougars. A commonsense approach can save their lives and ours.

Chapter 14 People, Politics, and Cougar Management

David J. Mattson and Susan G. Clark

COUGAR (*Puma concolor*) MANAGEMENT PIVOTS upon two questions: "How will animals be used?" and "Who gets to decide?" At present, it is conflict-laden and controversial as people vie for influence in answering these two questions (e.g., Bates 1988; Mansfield and Weaver 1988; Rieck 1988, Baron 2004; Love 2005; Perry and DeVos 2005). Management can be understood as an ongoing process of people making decisions—not about cougar ecology, predation, or prey populations, but about our own actions. The questions multiply. Should we limit our killing of cougars? How should we go about setting goals? Who should be involved in deciding, and when, and how? In short, decision making in management is about who will be empowered, who will receive material benefits, and whose losses will be minimized. Another word for this is, of course, politics.

The processes we create for making decisions determine how we interact with one another and whether we serve the special interests of a few or the common interests of many. By most indications, participants in cougar management are having difficulty sorting through special interests to find, secure, and sustain common interests. In broadest terms, the widespread corrosive politics in cougar management today can be understood as a result of colliding participant demands and public processes insufficient to resolve differences democratically and fairly.

This observation forces us to look at our institutions of wildlife management and how they operate. Unresolved conflict in cougar management is evidence that the decision-making processes are not functioning as well as they might in the common interest (see Brunner 2002). In this chapter,

we describe the present participants in cougar management, describe and explain how the institutional and decision-making system functions, and review options for moving cougar management forward democratically and in the common interest. These are human-centered dynamics. We focus on people, not cougars, and on how we engage with one another over cougars, thus leaving matters such as cougar population trends to other authors of this book.

Learning about People and Decision Making

We took an interdisciplinary approach in examining people, decision making, and institutions. The humanities and social sciences are little used in wildlife management; the integrative policy sciences even less. Yet these human-focused sciences offer the best prospects for understanding current conflict so as to help build and sustain durable institutions of cougar management to undergird durable cougar conservation. We used the policy sciences as our primary integrating tool. The policy sciences employ a set of propositions, concepts, and analytic categories, using multiple methods, to clarify *context* as a practical means of solving problems (Lasswell 1971; Clark 2002). We organized our chapter around the concepts of "good governance" and "common interest," which have been described in practical terms by McDougal and colleagues (1980, 1981).

We define successful cougar management as enduring public support for sustainable cougar populations and the habitat they need. Enduring support depends, in turn, on good governance that serves common, over special, interests.

The core challenge in our democratic society is to allow citizens freedom to pursue their own interests but with respect for rights of others (Dahl 1982; Shils 1997). Balancing of egoism (self-interest) and altruism (shared interest) depends on institutions that help citizens internalize democratic norms and show "democratic character" (Lasswell 1951). There is no single public or common interest (Ascher 2004), but interests qualifying as common fall within a range that arises from active involvement of everyone with a stake, that are supported by virtually all people who make nonexclusionary claims, that are evidence based, and that produce the desired outcomes (Brunner 2002). Common interests tend to coalesce around ideas about social and decision processes, including fairness, inclusiveness, transparency, civility, factuality, respectfulness, practicality, and amelioration (Lasswell 1971; Clark 2002; Mattson et al. 2006). By these standards, any official decision-making process that breeds conflict, incivility, and disrespect through clear service of a special interest does not qualify as good governance.

Our examination of cougar management included characteristics of participants. We identified generic participant groups from technical literature, management plans, online Web page searches, newspaper and popular journal articles, discussions, and our own observations. For each group we gathered information on perspectives (identities, nature-views, beliefs, and prioritized values), material and symbolic stakes, situations or arenas typifying interactions, characteristic strategies (including knowledge claims), and demands (Lasswell 1971). We employed the well known "nature-views" system of Kellert (1985, 1989, 1996) and two well-established classifications of values (Lasswell and Kaplan 1950; Schwartz 1994). Nature-views are shared narratives about what relations between people and nature are and should be (Table 14.1). We found the earliest version of Kellert's evolving schematic (e.g., Kellert 1989) to be most descriptive of views expressed in cougar management, ranging from the negativistic, dominionistic, and utilitarian at one extreme to the humanistic and moralistic at the other. Naturalistic and ecologistic/scientific views fell nearer the midrange. Schwartz's (1994) value schematic is rooted in psychodynamics, grouping values under self-transcendance (universalism, benevolence), openness to change (self-direction, stimulation, hedonism), self-enhancement (achievement, power), and conservation or conservatism (tradition, conformity, and security). Lasswell and Kaplan's (1950) value schematic for individuals (i.e., respect, affection, enlightenment, well-being, skill, power, wealth, and rectitude) can be directly linked to corresponding societal institutions.

Institution is a complex term that is often confused with *organization*. An institution is a set of rules in the game of society that comes about through decisions that constitute, or "institutionalize," norms of behavior and interactions among people. Institutions thus reflect people's perspectives and demands and set patterns of acceptable behavior (Lasswell and McDougal 1992; see citations in Clark and Rutherford 2005). By contrast, organizations are simply collections of people with a shared goal and working for desired outcomes. For example, the Wyoming Game and Fish Department, Fund for Animals, and Mule Deer Foundation are all

Table 14.1 Characteristics of participants in cougar management and policy.

Nature-View[a]	Associated Narrative and Attitude (Related Values)
Negativistic	Nature is a threatening and fearful place to be avoided or even eliminated (*security*).
Dominionistic	Nature is to be dominated or controlled as an expression of human will and centrality (power, *achievement*, skill, *stimulation*, rectitude).
Utilitarian	Nature is a source of commodities and material goods (wealth, skill, *achievement*).
Ecologistic/Scientific	Nature, as ecosystems, is a source of knowledge and services (enlightenment, well-being).
Naturalistic	Nature is a source of solitude, communion, and naturalness (well-being, *universalism*, *stimulation*).
Aesthetic	Nature is a fount of and venue for experiencing beauty (*hedonism*, well-being, *stimulation*).
Humanistic	Animals, in particular, are a source of individual relations constructed around attributions of human characteristics (affection, *benevolence*).
Moralistic	Nature is to be protected for ethical reasons (*universalism*, respect, rectitude).
Symbolic	Nature is a source of metaphors, allegories, and other symbols for human communication.

[a]Nature-views and associated narratives and attitudes used in this chapter to describe participant perspectives. Narratives and attitudes are based on our interpretation of Kellert (1985, 1989, 1996). We also relate nature-views to values, italicized in parentheses, defined by Schwartz (1994) and Lasswell and Kaplan (1950). Values are described more fully in the text.

organizations within the institution of wildlife management. The institution of wildlife management sets and enforces the rules for how we decide about the use of cougars and who gets to participate. Whether or not people engage in civil negotiation focused on outcomes of common interest is determined largely by institutions. At root, conflict in cougar management is better understood as disagreement over the adequacy and nature of institutions rather than as disagreement over disposition of cougars.

Our involvement in the institution of wildlife management has included ecological field work plus experience with and analysis of organizations. Our ecological work collectively spans twenty-one years in Idaho, Utah, and Arizona (e.g., Laundré and Clark 2003; Laundré et al. 2006; Mattson 2007b). Ameliorating corrosive conflict matters to us. Our goal is to provide practical insights that can help those involved in cougar management discover and secure their common interests in the form of durable, widely supported, and effective policies.

Participants

To understand human interactions in cougar management, we need a sense of who is actively involved in cougar politics. In Appendix 3, we describe eight participant groups in brief profiles intended to cut through otherwise overwhelming complexity, at the admitted cost of obscuring sometimes important diversity of perspectives. For each of these generic "stakeholders," we present nature-views, favored sources of knowledge, stakes in cougar management, and demands for management outcomes, summarized in Table 14.2.

A handful of overarching patterns stand out among participants in cougar management in the United States and Canada (Table 14.2; Appendix 3). Perhaps with the exception of some communities in the Southwest, participants are overwhelmingly Caucasians. Otherwise, the broad spectrum of expressed nature-views is closely identified not only with sometimes antithetical beliefs about cougars and cougar-human relations but also with aspects of identity, such as sex, residence, employment, education, and support for hunting—factors that allow ample opportunity for consolidation of self-identified groups. The groups we describe as participants have narratives asserting their demands regarding cougar management; their respective demands conflict, and participants are often derogatory of groups with opposing demands.

Another remarkable feature of cougar-centered human social dynamics is the comparatively minor material stakes of just about everyone involved. Few livestock producers lose enough stock to cougars to make or break them economically; cougars are not a key revenue generator for any wildlife management agency; effects of cougar predation on big game populations are genuinely a matter of debate (Chapters 9, 10); the outfitters who make a living from hunting cougars probably only number in the hundreds; and a comparative handful of people have ever been attacked by cougars (Chapter 13). Yet a considerable number of people clearly feel strongly about cougars, cougar management, and others who are involved. This force of feeling is most plausibly understood as arising from symbolic attachments to cougars; beliefs about relations between people and nature; about people with potentially threatening nature-views; about access to power; and, in the case of hunters, about opportunities for enjoyment and exercise of skill through hunting and killing big game. Conflict is largely, although not wholly, about *symbolic* rather than material outcomes.

The Changing Climate of Decision Making

Decision-making processes tend to be relatively stable once they are institutionalized. Even so, the history of cougar management reveals significant changes, which led us to divide this section into discussions of long-term trends in wildlife decision making, current decision-making arrangements and their implications, and recent perturbations generating new kinds of arrangements for cougars and for wildlife management in general. We focus on how participants typically assemble to advance their interests, depending on their strategies, shared interests and worldviews, and access to power.

Overview from 1870 to the Present

Official goals for what to do about cougars have varied since overt management by European settlers began in the 1870s. Figure 14.1 shows the proportion of states and provinces in the United States and Canada managing cougars under different classifications from 1872 to 2005. Some interesting patterns emerge. Offerings of bounties increased in the wake of both world wars, a result of expressed concern about the waning of predator control efforts during the considerable distractions of both conflicts (Baron 2004). This was followed by a rapid series of transitions that occurred primarily between 1965 and 1980, entailing first the termination of bounties; then, designation as game animals; and then, implementation of regulations to offer some measure of protection to kittens (see Chapter 4). With the exception of Texas, which still classifies cougars as unprotected predators, and California, where cougars are a specially protected species, almost all states currently manage cougars as game animals, regulating harvest to protect kittens

Table 14.2 Synopsis of characteristics typifying participants in cougar management and policy in the western United States. All of these characteristics vary within the identified groups. (See Appendix 3 for more detailed discussion.)

Participant	Identities	Nature-views	Material Stakes	Value Stakes	Claims and Beliefs	Demands and Preferences
			Characteristics			
Livestock producers	Rural Politically conservative	Dominion-istic Utilitarian Negativistic	Depredation losses	Skill, achievement, security, tradition, wealth	Local knowledge and life-ways should have primacy. Depredation losses are unacceptable.	Compensate and prevent depredation. Reduce or eliminate cougars.
Ungulate hunters	Caucasian males Non-metropolitan Outdoor active Attracted to wildlife Knowledgeable of wildlife Politically conservative	Dominion-istic Utilitarian Naturalistic Ecologistic	Huntable ungulates	Skill, achievement, power	North American conservation ethic. Scientific management. Hunting is necessary and ethical. Hunting instills fear, reduces conflict, and is good for cougars.	Hunt cougars with hounds. Resolve conflicts lethally. Reduce cougar populations to benefit ungulates and increase hunting opportunities.
Cougar hunters	Caucasian males Non-metropolitan Outdoor active Attracted to wildlife Knowledgeable of wildlife Politically conservative	Dominion-istic Utilitarian Naturalistic Ecologistic	Huntable cougars	Skill, achievement, power, wealth	Hunting cougars with hounds is logical and ethical. Hound hunting instills fear, reduces conflicts, and removes large males to benefit other cougars.	Hunt cougars with hounds. Resolve conflicts lethally. Maintain cougar hunting opportunities.
Wildlife agency commissioners	Caucasian males Non-metropolitan Outdoor active Attracted to wildlife Knowledgeable of wildlife	Dominion-istic Utilitarian Naturalistic Ecologistic	Huntable ungulates Agency budgets	Power, skill, achievement, wealth	North American conservation ethic. Scientific management. Total quality management. Hunting is necessary and ethical. Hunting instills fear and reduces conflicts. Conflict can be resolved through education.	Hunt cougars with hounds. Resolve conflicts lethally. Reduce cougar populations to benefit ungulates, increase hunting opportunities, and reduce depredation.
Wildlife agency personnel	Caucasian males Non-metropolitan Outdoor active Attracted to wildlife Knowledgeable of wildlife	Naturalistic Ecologistic Dominion-istic Utilitarian	Huntable wildlife Agency budgets	Skill, achievement, power, wealth, enlightenment	North American conservation ethic. Scientific management. Total quality management. Hunting is necessary and ethical. Hunting instills fear and reduces conflicts. Conflict can be resolved through education.	Hunt cougars with hounds. Resolve conflicts lethally. Reduce cougar populations to benefit ungulates, increase hunting opportunities, and reduce depredation.
"The public"	Caucasian male heads of households	Ecologistic	—	—	Cougars have important ecological role. Cougars are not a major threat. Endangering kittens and hunting with hounds is unethical.	Kill cougars that have injured or killed humans, to protect endangered and threatened species, and to protect children. Do not kill cougars to increase hunting opportunities. Prohibit hunting with hounds.

(Continued)

Table 14.2 (Continued)

Participant	Identities	Nature-views	Material Stakes	Value Stakes	Claims and Beliefs	Demands and Preferences
				Characteristics		
Animal focused activists	Caucasian females Urban Well educated Politically liberal	Humanistic Moralistic Naturalistic Ecologistic	Live cougars Ecologically functional cougar populations	Rectitude	Hunting is unethical Endangering kittens and hunting with hounds is unethical. Cougars have important ecological role. Hunting does not reduce conflicts. Humans responsible for living with cougars.	Prohibit cougar hunting Prohibit hunting females and hunting with hounds. Maintain ecologically functional cougar populations.
Environmentalists	Caucasian Politically liberal	Ecologistic Naturalistic Moralistic Humanistic	Wilderness Ecologically functional cougar populations	Rectitude	Cougars have important ecological role. Endangering kittens and hunting with hounds is unethical. Hunting does not reduce conflicts. Humans responsible for living with cougars.	Prohibit hunting females and hunting with hounds. Maintain ecologically functional cougar populations.

(see Table 4.1). These comparatively abrupt changes followed a societal decline in utilitarian perspectives (Kellert 1996), coinciding with the broad-scale emergence of environmental awareness and reforming environmental movements (see Chapter 1; Dunlap 1992; Brulle 2000).

These rapid transitions in the status of cougars beg for an explanation beyond coincidence with broader societal shifts in nature-views. Numerous observers of cougar management have noted that the transitions of 1965–80 came at the same time as, the emergence of participants who promoted "nonconsumptive" wildlife interests and cougar protection (e.g., Brown 1984; Murphy 1984; Tsukamoto 1984; Herbert 1988). By contrast, the peak in the bounty-based approach during the 1940s and 1950s coincided with emergence of the "scientific era" of wildlife management (Graham 1997) and many current civic wildlife management organizations (Brulle 2000). Beyond these broader patterns, there are only a handful of historical observations that suggest exact mechanisms and agents of change. Most notable among these is David Brown's account (Brown 1984) of how a small group of "dedicated" and "erudite" women were instrumental in changing the status of Arizona's cougar from bountied predator to game animal. Morrison (1984) and Shaw (1994) make similar reference to the role of activist women and, more generally, to "public concern." Baron (2004) suggested not only public pressure, for example in the form of editorials in the *Denver Post*, but

also wildlife agency support arising from interest in control to promote "rational" management. Taken together, these scant items of evidence do not provide a compellingly detailed explanation for how change happened from 1965 to 1980. But they do suggest that those with the primary role were nonagency activists, motivated by emerging ecologistic, humanistic, and moralistic nature-views, and that hunters, agency personnel, and scientific motivations were secondary (for more discussion on the role of nonagency activists see Chapter 15).

Decision Making before 1965

Wildlife management institutions constituted in the first half of the 1900s were deliberately designed to minimize politically expedient interventions by elected officials and to maximize the effects of hunters, then deemed the primary standard-bearers of wildlife conservation (Reiger 2001). This institutionalization of conservation was in reaction to a legacy of unsustainable commercial and other meat hunting, and it gave voice to an ethic organized around sport hunting (Shaw 1994; Reiger 2001; Dizard 2003). Important for cougar management, these ancestral institutions of wildlife management were built on deep positive ties to large ungulates and on beliefs that demonized predators as threats to the very survival of game populations (Reiger 2001; Baron 2004). More specific to the West, sportsmen

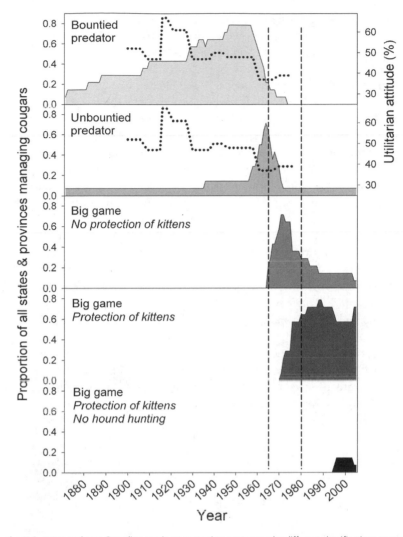

Figure 14.1 Proportions of twelve U.S. states and two Canadian provinces managing cougars under different classifications or management provisions, from 1872 to 2005. Dotted trend lines denote the proportional commonness of utilitarian statements in journalistic media, from Kellert (1996). Dashed vertical lines bound the period of rapid change in management provisions between 1965 and 1980.

expressed widespread antipathy toward cougars, rooted partly in beliefs that the killing or even complete eradication of cougars was necessary to increase mule deer herds (Brown 1984; Shaw 1994; Baron 2004). These antipredator beliefs were naturally aligned with those of ranchers, who had a shared interest in exterminating predators of all sorts. It should be no surprise that incentives to kill cougars were not only formally institutionalized but also acculturated early on in the history of wildlife management.

In practice, the commission structure limits authority of elected officials in wildlife management to the passage of laws that set broad-scale policy and to the appointment of commissioners, who are more or less autonomous thereafter (Nie 2004a, 2004b). Control by elected officials was and still is limited because wildlife agency revenues in almost all states are sequestered from normal budgetary processes,

with most revenue coming directly or indirectly from hunters and anglers, either through license fees and other charges for "service" or from taxes on sales of sporting goods through federal grants, such as under the Pittman-Robertson Act (e.g., Gill 1996a; Hagood 1997; Nie 2004b). We summarized revenues of ten state wildlife management agencies involved in cougar management and found that, on average, hunters and anglers continue to provide 80 percent of all revenues, directly or indirectly (49 percent licenses/related sales, 31 percent from sales tax via federal grants) and that appropriations from general state funds average only 6 percent (range 0–33 percent).

It is clear that authority and control over cougar management in virtually all states prior to 1960 was almost wholly governed by participants closely identified with hunting as a tradition and a management tool. Under these arrangements, hunters, trappers, and fishermen were the almost exclusive constituency of wildlife management (Decker et al. 1996). The only proviso pertained to influences of livestock interests, which were and continue to be substantial. A notable number of wildlife commissioners were identified with livestock interests (Hagood 1997). Moreover, there is a long and often-observed history of livestock interests exercising controlling influences, either directly on wildlife agency commissioners and personnel or, more definitively, indirectly through elected officials with shared interests and worldviews (e.g., Robertson 1984; Thompson 1984; Weeks and Packard 1997; Nie 2004a; Clark and Munno 2005).

It remains unclear exactly what the mechanisms of agricultural influences were and are, but they are manifest in policies mandating measures to address depredation through control and compensation (e.g., Morrison 1984; Pall 1984; Anderson and Tully 1988; Robertson and Bell 1988; Sharma 1988; DeSimone and Jaffe 2003; Woolstenhulme 2003; Winslow 2005). Underlying these patterns before 1960, though, is the probable adherence of virtually all the people with control over cougar management to dominionistic and utilitarian nature-views unfriendly or ambivalent to conservation of large carnivores (Hook and Robinson 1982; Kellert 1985; Bjerke et al. 1998; Pate et al. 1996; Vittersø et al. 1988).

Polarization after 1970

Even with the widespread emergence of stakeholders in cougar management who adhere to ecocentric nature-views, patterns of decision making after 1970 have remained much like those before 1960. Numerous commentators have concluded that state-level cougar as well as other wildlife management primarily serves the special interests of hunters, anglers, trappers, and livestock producers, with little consideration of nonconsumptive stakeholders (Decker et al. 1996; Gill 1996a; Nie 2004a, 2004b; Clark and Munno 2005; Jacobson and Decker 2006). Rutberg (2001, 35) has even gone so far as to liken wildlife management agencies to the "private regulatory bodies that govern professional sports." Reasons for preferential serving of hunting and ranching interests are largely the same as before: the pro-hunting perspectives of commissioners and agency personnel, a hunting-focused culture shared with "customers," control of most revenues by hunters, the limited authority of elected officials, and models of management that appropriate power to agency technical experts and cause inattention to governance (Decker et al. 1996; Gill 1996a, 2001b; Byrd 2002; Nie 2004a; Clark and Rutherford 2005).

As one outcome, animal-focused activists and their allies perceive themselves to be disenfranchised in the normal process of cougar management decision making (e.g., Hagood 1997; Pacelle 1998; Papouchis et al. 2005; Blessley-Lowe 2006). Moreover, the latent constituency for "mutualist" (humanistic and moralistic) management outcomes is large and likely growing (Teel et al. 2005), at the same time that already comparatively small numbers of hunters are proportionally declining (U.S. Fish and Wildlife Service and U.S. Bureau of the Census 1993, 1997, 2003, 2008). All this sets the stage for conflict, primarily between those holding dominionistic and utilitarian nature-views, avowing primacy for hunting, and those holding humanistic and moralistic nature-views, extolling intrinsic or nonconsumptive "values."

Conflict indeed surrounds cougar management. What to do about cougars is one of the most controversial of wildlife management issues (e.g., Mansfield and Weaver 1988; Rieck 1988; Baron 2004; Perry and DeVos 2005). Conflict and controversy are expressed not only in litigation and ballot initiatives, which we cover later (and see Chapter 15), but also as public incivility among participants. Baron (2004), Clark and Munno (2005), and Perry and DeVos (2005) describe meetings in which accusations were freely exchanged and participants otherwise treated with disrespect. In print, animal-focused activists have claimed that agency personnel are "manufacturing paper lions," "purposefully altering numbers," and playing "games of statistical chicanery" to justify policies (Schubert 2002, 2, 11); conversely, hunters have claimed that agency personnel are "conspiring to shut down local guides" (Lermayer 2006, 6) and that animal-focused activists are "emotional," "nuts," "warped," and "cast an almost worthless image [sic] on human life" (Howard 1991, 96; Einwohner 1999a, 66; Arizona Game and Fish Department 2004, B-2).

Although some of this incivility can be explained simply by conflicting demands arising from divergent nature-views, we and Kellert (1996) suspect that identities and personalities of participants have an inflaming effect. Animal activists often pursue their ends with righteous fervor (Appendix 3). We speculate that in response, hunters and agency personnel often show limited empathy and openness because of their orientation toward self-enhancing power and achievement and political or other conservatism, as described in our profiles of participants (Appendix 3). Self-enhancing and conservative values have been closely identified with prejudice, unwillingness to engage constructively with unlike others, and a preference for power arrangements that perpetuate inequity (e.g., Heaven and Bucci 2001; Jost et al. 2003; Sidanius et al. 2004). Moreover, the fact that almost all hunters and agency personnel are men, often with nonmetropolitan upbringings, and that a large majority of animal-focused activists are well-educated urban women, probably aggravates conflict by providing a ready basis for stereotyping and intensification of group boundaries (e.g., Einwohner 1999a; Nie 2003; Skogan and Krange 2003). The ingredients for conflict in cougar management seem to run a full gamut, with the agency personnel who hold primary authority feeling beset in the middle.

Much of the fabric of this conflict is woven around claims and counterclaims about science and the authority of scientific managers. Consistent with their identities and with allocations of power prescribed by scientific management, hunters and agency personnel often invoke technical expertise grounded in scientific knowledge as the authoritative basis for identifying and solving the physical problems they claim typify cougar management (e.g., Howard 1988; Nie 2004a; Beausoleil et al. 2005). This perspective is found in the following quotes: "Let [the agency] manage wildlife, using scientific and statistical methods, rather than having to be concerned about public, political, and media reaction"; and "No commissioner should ever bow to the wishes of the emotional masses when science proves them wrong . . . politics should never interfere in their decisions" (Arizona Game and Fish Department 2004 , A-1, B-4). The second quote highlights a common way that hunters and agency personnel delegitimize animal-focused activists—by ascribing their perspectives and demands to illogical emotion, rooted in "Disney-driven" media and other urban-fostered romantic notions of nature (e.g., Ingram 1984; Howard 1991; Einwohner 1999a; Miniter 2004; Perry and DeVos 2005). Such ascription of emotional motivations to mostly female activists by mostly male hunters and agency personnel is highly suggestive of sex-based stereotyping (Einwohner 1999a). Whatever the basis, animal activists are deprived of legitimate standing in the eyes of those making such claims (Einwohner 1999a).

Assertions of science-based authority by hunters and agency personnel are intriguing given broader patterns of behavior in management of cougars and other wildlife. In other contexts, typically involving management of mule deer, hunters *contest* the validity of agency perspectives and information, usually when agencies are forwarding policies counter to their demands (Zumbo 2002; Freddy et al. 2004; Lermayer 2006). This opposition is suggestive of situational rather than principled support by hunters for scientific management. More to the point here, agency biologists have indeed invoked "feelings" or "beliefs" (e.g., Austin 2003; Apker 2005; Whittaker 2005) and have employed subjective assessments, uncertain area-extrapolated population estimates, and ambiguous harvest and depredation trend data as a common basis for cougar management. As we describe in Appendix 3, agencies also at times base management on assumptions about benefits of hunting that have little or no grounding in scientific studies.

At a more nuanced level, case studies reveal a pattern of agencies allocating burden of proof (Clark and Munno

2005) or overstating certainty of information (Shaw 1994) in ways consistent with agency interests. One example of the former can be found in a report from South Dakota (Gigliotti 2005) in which agency specialists invoked uncertainty about efficacy of nonlethal control methods to justify nonadoption, while in the same document asserting a direct beneficial link between hunting, reduction of human-cougar conflicts, and acceptance of cougars by humans, without any supporting scientific evidence. An example of overstated certainty can be found in the testimony of an agency specialist before a state legislature (Montana Senate, 58th Legislature, Committee on Fish and Game 2003), where, in response to questions about impacts of recreational hound pursuit on cougars, the specialist stated that "mountain lions and bobcats evolved in the presence of wolves, so running from dogs was not much different for the lions" and that a chase season "would not" have unintended consequences on cougar reproduction.

We are not implying here that there is or has been no place for anecdote and subjective judgment in cougar management. Logistics alone preclude certain kinds of scientific or other real-time information about cougars, as is evident from almost every chapter in this book. What we have observed, instead, is a pattern of behavior suggesting that science and expert standing are being used for power purposes and for advancing special interests, rather than for enlightenment or for fostering common ground; this is consistent with "blurring science and values" (Decker et al. 1996; Pacelle 1998; Jones 2002). Animal-focused activists give evidence of the same motivations in their use of information and invocations of science (e.g., Perry and DeVos 2005). However, the focus here is legitimately on agency personnel and their use of authority invested partly on the basis of technical expertise, because this investiture is contingent on fulfilling the public trust, which requires that agency personnel be truthful, just, equitable, and ameliorative (Gill 2001b; Clark and Rutherford 2005; Jacobson and Decker 2006).

State-Level Variation since 1970

Levels of conflict and incivility surrounding cougar management have not been uniform among states. We developed an index for potential conflict based on state-level information given in Teel et al. (2005) by correlating percentages of respondents who were "mutualists," who felt their interests were served by state wildlife management agencies, who trusted the agencies, and who saw discrepancies between current and ideal management arrangements—that is, between funding and constituencies. By this index, potential for conflict was highest in California, Washington, Oregon, Arizona, Colorado, and New Mexico, and lowest in South and North Dakota, Montana, and Wyoming (Figure 14.2). All of the eight successful ballot initiatives or lawsuits we identified, brought by animal-focused activists (see later discussion), occurred in the six states with the highest potential for conflict, suggestive that the index does capture some of the drivers of widespread discontent or ambivalence about state-level cougar management. Successful litigation is relevant here because of its documented dependence on favorable public opinion (Ingram and Mann 1989). In this context, two recent cases of controversy warrant mention, being in states with ostensibly low potential for conflict: in Wyoming, centered on Jackson Hole (Clark and Munno 2005), and in South Dakota, centered on the Black Hills (Love 2005). We ascribe this discrepancy to cultural heterogeneity. Inhabitants of the Jackson Hole region, as a group, are quite different from other inhabitants of Wyoming in their greater adherence to ecocentric nature-views (Clark and Munno 2005); possibly the same applies for enclaves of Black Hills residents in contrast to other South Dakotans.

In examining conflict, we also found that states differed in how much attention they gave nature-views and process questions in their state-level goals for cougar management. To gauge this, we summarized goals for 1984 to 2005 from reports in the *Proceedings* of Mountain Lion Workshops and, more recently, from state-level cougar management plans. We categorized one hundred recorded goals according to whether they advanced interests that were (1) utilitarian or dominionistic, (2) ecologic or scientific, (3) naturalistic or moralistic, or (4) focused on process itself. We found utilitarian-dominionistic emphasis for 38–50 percent of goals, clearly the dominant category; ecologistic-scientific emphasis for 19–31 percent; naturalistic or moralistic ("natural heritage") emphasis for 17–31 percent; and process emphasis for only 4–6 percent of goals. The range depended on the time period, but we found no major trends in frequencies among types of goal statements for the 1980s, 1990s, and 2000s.

We developed a score for each state based on weighting the respective categories with values of 1, 2, 3, or 4—not ascribing moral valuation but rather attempting to capture the degrees of expressed attentiveness to values and to process considerations that have emerged during the last thirty-five years. Washington, Oregon, and South Dakota expressed high levels of attentiveness, and British Columbia, Idaho, Nevada, New Mexico, and Texas expressed low levels (Figure 14.3). For the United States, we found a remarkably high positive correlation (r = 0.69) between our score for expressed goal-attentiveness to emerging nature-views and Putnam's state-level index of social capital or public trust (Putnam 2000). By contrast, we did not find noteworthy correlations with measures such as percentage of "utilitarians" or "mutualists" in any state (Teel et al. 2005).

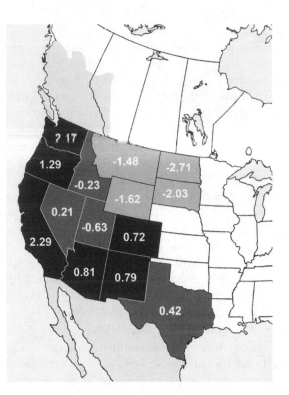

Figure 14.2 Index of potential for conflict associated with state-level wildlife management. Higher positive values denote higher potential for conflict based on percentages of respondents in Teel et al. (2005) who expressed a "mutualist" perspective, distrust in state wildlife agencies, lack of confidence that their interests were considered in wildlife management, and consonant perceptions of ideal and existing funding and constituencies. (The index was the reversed-sign weights for each state derived from the first principal component of correlations among percentages of respondents; mutualists, PC1 loading = − 0.61; interests served, loading = 0.56; trusted agencies, loading = 0.49; discrepancies, loading = 0.29.)

This intriguing correlation with social capital suggests that broad-scale levels of public trust and empathy are perhaps key drivers of attention by wildlife management agencies to state-level diversity of interests and nature-views.

Perturbations

In this final section on institutional arrangements, we examine events that have perturbed the decision making in cougar management since roughly 1970 in such a way as to challenge longstanding norms. Three types of perturbations stand out: litigation, ballot initiatives, and "incidents." Each is distinct because of the social process entailed and the aspects of decision making highlighted. In almost all instances animal-focused activists have used these perturbations to advance their interests, although in the face of their considerable successes, hunters and others with related interests are increasingly using similar strategies to try to reinstate traditional norms favoring utilitarian

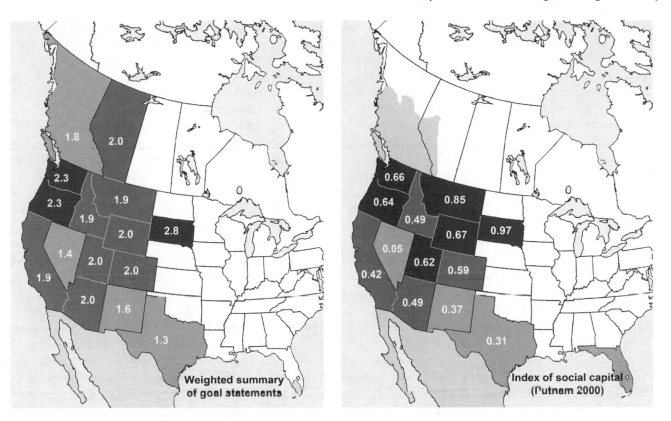

Figure 14.3 Weighted summary of goal statements for cougar management expressed by state wildlife agencies between 1984 and 2005 (left), with progressively higher weights attached to goals expressing greater attention to process or emerging nature-views (see text). The darkest shading denotes a high weighted summary, medium shading indicates medium, and pale shading reflects a low weighted summary. For social capital (right), darker and paler shading similarly correspond to levels of social capital indexed for each U.S. state by Putnam (2000).

and dominionistic preferences. In many instances, animal activists have resorted to perturbations only after repeated unsuccessful efforts to work through routine wildlife commission and legislative processes (DeVos et al. 1998; Pacelle 1998; Einwohner 1999a; Peña 2002). One of the few successful legislative strategies by animal-focused activists occurred in California, in the form of renewed moratoria on cougar hunting lasting from 1972 to 1986 (see Chapter 15; Mansfield and Weaver 1988; Wolch et al. 1997).

Litigation

Litigation is a new arena in cougar management, a ritualized process where formal policy is challenged and a judge evaluates wildlife managers' implementation of the policy. The judge then issues a ruling, applying authority over the policy at issue (cf. Lasswell and McDougal 1992). Although litigation is often accompanied by flurries of media and promotional activity, the real focal point is the formalized communications among judges, lawyers, and clients, usually involving a wildlife agency and groups with an animal or environmentalist focus. Several cases in which litigants were successful, most notably in Oregon (2003 and 2006;

Findholt and Johnson 2005) and New Mexico (1998), have involved claims under the National Environmental Policy Act (NEPA) applied to either the U.S. Fish and Wildlife Service or U.S. Department of Agriculture, Wildlife Services, regarding the administrative processes for decisions setting the stage for killing substantial numbers of cougars as part of either research or depredation control. Animal activists and environmentalists have made other claims under similar state-level policies, such as the California Environmental Quality Act, most of them successful and, again, pertaining to planned or ongoing killing of cougars for research or depredation control. Such claims have been made in New Mexico (see list of litigation summaries, Appendix 4 1986; Robertson and Bell 1988) and California (1986–1989; Mansfield and Weaver 1988; Wolch et al. 1997).

Under current arrangements, litigation affords animal-focused activists one of very few opportunities to employ authoritative control, but it is entirely contingent on persuading a judge and is constrained by how laws read and thus what can be challenged. Not all litigants are successful at persuading judges, and almost all legal claims are made under policies that merely govern *procedure* rather than substantive outcomes (e.g., NEPA), unlike options available to litigants who invoke the U.S. Endangered Species Act. All in all, litigation provides limited opportunities for animal activists to revise management norms by intervening in cougar-related decision processes, and at the risk of turning disregard from agency personnel and commissioners into hostility (as evidenced by the tone of Findholt and Johnson 2005). Successful litigation can nevertheless set the stage for prescriptive measures such as ballot initiatives, as in California during the late 1980s (Wolch et al. 1997), and can publicly elevate issues or create space for discussion of alternatives (Ingram and Mann 1989; Parker 1995).

Ballot Initiatives

Ballot initiatives place policy prescription in the hands of registered voters. However, a prescriptive verdict is the culmination of a long initiation period, involving the collection of signatures sufficient to place the initiative on a regularly scheduled ballot and a long and intense period of promotional activity by all who are motivated and have a perceived stake in the outcome. Ballot initiatives are unique among options available to animal-focused activists in affording them the possibility of substantively reconfiguring the authority framework of cougar management. However, the outcome of ballot initiatives is contingent upon a successful promotional campaign, which predictably depends on a favorably disposed electorate, successful control of media framing, and monetary resources (Pacelle 1998). For this reason, animal-focused activists have been strategic in their employment of ballot initiatives. Even though some animal rights groups advocate the end of all sport hunting of cougars (e.g., Schubert 2002), they forwarded such a measure only in California (in 1990), where a long history of legislative and other public support signaled likely success (see Chapter 15; Wolch et al. 1997), which was indeed the outcome. In Oregon and Washington, ballot initiatives sponsored by animal-focused groups called only for prohibiting the use of hounds (Appendix 5). This successful focus on banning hounds rather than banning cougar hunting outright was pragmatic because most surveys have shown much greater public disapproval of hound-assisted hunting than of other hunting methods. Hunters have framed these and similar initiatives as an assault on hunting, but surveys of voters have shown that the issue for most is not hunting but, rather, "fair chase" and sportsmanship (e.g., Loker and Decker 1995; Kellert 1996).

Hunter reactions to ballot initiatives that limit hunting have included attempts at reversal through state legislatures, efforts to mount countering ballot initiatives, and dismissive publicity claiming failed democracy and corrupted motives among voters and animal activists. To date, almost all of the many legislative attempts and the ballot initiatives in California and Oregon designed to overturn prior ballot

victories have failed (Pozzanghera 1996; Wolch et al. 1997; Kertson 2005). To our knowledge, the only exceptions have been in Washington, where six rural counties got legislative authorization for hound-assisted hunts (Kertson 2005; see later discussion), and in Oregon, where the state fish and wildlife department was given authority in 2007 to employ nonagency persons for management-related hunts.

California's ban on cougar sport hunting has been invoked as an instructive parable for hunters and agency personnel in which "irrational" and "urban" voters have banished "responsible" and "scientific" management (e.g., Miniter 2004; Perry and DeVos 2005). A "rising tide of conflicts" between humans and cougars is attributed directly to the ending of hunting or banning of hounds (e.g., Howard 1988; Portland Chapter of Safari Club International 2003; Hoffman 2004; Miniter 2004). Implicit in this narrative are the ideas that hunting with hounds prevents conflict and is required to keep cougar populations in check; hunting, with or without hounds, reduces conflict; and hunting, especially with hounds, prevents cougar attacks. Evidence for this argument is scant or contradictory. For example, with liberalized regulations, cougar harvest in Oregon and Washington soon recovered to levels higher than before bans on hound-assisted hunting, and at the same time as reported cougar-human conflicts were increasing (Beausoleil et al. 2003; Whittaker 2005). Similarly, on a per capita or unit area basis, people in California have experienced fewer or similar numbers of conflicts with cougars compared to people in states with cougar hunting (Papouchis 2006a).

Incidents

Incidents include more than just cougar attacks on humans. When routine agency implementation of policy triggers short-term and localized intensification of focus to the point of becoming inflammatory, and provokes appraisal of norms and promotion of alternatives, we call it an incident. During incidents the public becomes more attentive and engaged, largely as a function of a perceived rise in the stakes and often under circumstances involving urban dwellers. Most incidents have been triggered by public safety concerns, although others have arisen with agency plans to kill greater numbers of cougars, usually to increase production of huntable mule deer (*Odocoileus hernionus*) and elk (*Cervus elaphus*) or to protect vulnerable bighorn (*Cervis canadensis*) sheep populations (e.g., Clark and Munno 2005).

Perry and DeVos (2005) and Mattson and Clark (in prep.) describe incidents in Arizona in which agency personnel killed cougars to resolve perceived threats to human safety, in both cases on the borders of urban areas. These moves triggered animal-focused activists, who critiqued the lethal approach that was adopted, promoted nonlethal resolution, and framed "the problem" as urban encroachment, lack of responsible human behavior, and lack of inclusive process. By contrast, agency personnel framed the problem as one of dangerous cougars, one best solved by killing them. A key feature of these cases was how differently participants described the problem and hence the solution (cf. Mattson et al. 2006). Civil venues for finding common ground were consistently missing (Baron 2004; Clark and Munno 2005; Perry and DeVos 2005). Different ways of perceiving the problem, lack of ameliorative processes, and heightened incivility are consistent features of incidents.

Incidents can serve multiple purposes for animal activists otherwise faced with exclusion from decision making. First, incidents offer a way to engage not only sympathetic publics and political elites but also naturally allied environmentalists, all of whom are often latent rather than active supporters of reforming cougar management (cf. Birkland 1998). Incidents allow much greater media access, a key to reaching the public and elected officials. Journalists and the media are important gatekeepers during incidents, framing much of the public discourse and regulating the overall level of intensity.

During incidents, journalistic references to cougars can increase to levels five to ten times higher than the rest of the time (Mattson and Clark, in prep). Moreover, content shifts from informational reports on natural history and benign human-cougar encounters to pointed appraisals of agency policies and discussion of participants' motivation, responsibility, and blame (Baron 2004; Clark and Munno 2005; Mattson and Clark, in prep). As in many other natural resources cases (Schlechtweg 1996), media framing has commonly been described as adversarial and inflammatory, which predictably heightens overall incivility and complicates efforts to find common ground (e.g., Nie 2003; Baron 2004; Perry and DeVos 2005). Elected officials sometimes further inflame incidents by issuing public statements condemning various participants (Morton 2003; Baron 2004; Perry and DeVos 2005). In one case, county commissioners, with the apparent intent of promoting a political agenda, declared a state of public emergency after a young boy was attacked (Kertson 2005). Overall, incidents become public phenomena that participants can manipulate to critique routine cougar management and to use as a means for airing alternatives.

Effects of Perturbations

Wildlife management agency personnel and commissioners have responded in various ways to the focusing and perturbing effects of litigation, ballot initiatives, and incidents. Consistent with institutionalized norms, by far the most common response has been a call for more education of the public (e.g., Shroufe 1988; Clark and Munno 2005; Perry and DeVos 2005; South Dakota Department of Game, Fish and Parks 2005). The voiced or implicit rationale is that if only people were better informed, then the public would fall in line behind agency policy and support agency solutions to physical problems by virtue of technical merit. This approach fails to recognize that there are multiple definitions of "the problem" arising from participants' different value-based aspirations for cougars and the world (e.g., Healey and Ascher 1995; Forsyth 2003; Brunner and Steelman 2005).

As a means of public involvement, Pimbert and Pretty (1995) and Decker and Chase (1997) rank education of stakeholders among the lowest in terms of realism, efficacy, long-term sustainability, and democratic character. Some agency personnel and commissioners have recognized that a broader cross section of stakeholder values needs to be accounted for, at least in process, if not in outcomes (Graham 1997; Clark and Munno 2005; South Dakota Department of Game, Fish and Parks 2005). For example, consultative workshops or advisory groups have been convened in Arizona, South Dakota, and Utah to ensure a broad spectrum of stakeholder input on management plans or protocols (Utah Division of Wildlife Resources 1999; Arizona Game and Fish Department 2004; Gigliotti 2005). This kind of midrange public involvement (Pimbert and Pretty 1995; Decker and Chase 1997) constitutes a comfortable fit for most agency personnel, consistent with preferences expressed by employees of the Utah Division of Wildlife Resources (Mortenson and Krannich 2001). Overall, public involvement has increased with the proliferation of cougar-specific management plans since the early and mid-1990s, which at the very least offered stakeholders greater insight into agency goals, strategies, and justifications (Clark and Munno 2005).

Public process and a broader spectrum of nature-views are beginning to appear in wildlife agency plans for cougar management. Idaho and Arizona both recognize "recreational, ecological, intrinsic, scientific, and educational values" in introductory material of their predator management policies (Idaho Fish and Game Department 2000; Arizona Game and Fish Department 2000). Arizona's strategic plan for the years 2001–2006 (Arizona Game and Fish Department 2001) states the need to "work with partners to find common ground" and be "collaborative." However, such language consistently contrasts with the more specific goals that guide cougar management, either within the same documents or as an outcome of policies—goals that invariably emphasize hunting and the primacy of hunting opportunities to the exclusion of other matters. For example, the three primary objectives for Arizona's 2001–2006 cougar management were to "maintain annual harvest at 250 to

300 mountain lions," "provide recreational opportunity for 3000 to 6000 hunters per year," and "maintain existing occupied habitat and maintain the present range" (Arizona Game and Fish Department 2001, 37). Likewise, under terms of its predator management policy, plans by Arizona Game and Fish Department to reduce cougar populations in eleven game management units offered no indication that values other than hunting were considered or that a collaborative process was employed to foster common ground (Arizona Game and Fish Department 2000, 2006). Such divergence between actual practice and verbalized commitments to conciliatory processes may arise from lag effects in cultural change within agencies, but agencies may also be more deliberately restricting the diffusion of nontraditional considerations by incorporating them only partially, at a symbolic rather than substantive level (Brunner and Steelman 2005).

Beyond the content of planning documents, agencies have taken tentative steps toward accounting for a broader spectrum of constituents in decision making by giving serious consideration to structural changes such as diversification of funding and overt embrace of nonconsumptive stakeholders. This agenda was evidenced in a public survey sponsored by the Western Association of Fish and Wildlife Agencies (Teel et al. 2005) in which solicited respondents were asked their preferred structure for wildlife management funding and clientele. Of relevance here, by far the most popular selected alternative called for major funding from nonhunter sources and a clientele that included nonconsumptive users. However, in virtually every written instance of agency personnel describing efforts to broaden constituencies and diversify funding, the ultimate professed goal was to *increase agency legitimacy* or public support for its policies, not to identify and promote common interest solutions (e.g., Shroufe 1988; DeVos et al. 1998; Freddy et al. 2004; Lafon et al. 2004; Perry and DeVos 2005; South Dakota Department of Game, Fish and Parks 2005).

Common Interest Remains Elusive

The conflicts that typify cougar management are frustrating to nearly all participants. Ideally, decision-making systems should allow all valid participants to engage spontaneously, efficiently, fairly, constructively, and civilly, in order to discover, secure, and sustain their common interests (Lasswell 1971, 86–93; Clark 2002, 60). Apparently, much of cougar management does not work that way. The common interest is a powerful democratic concept central to the ideal functioning of our society (McDougal et al. 1980; Dahl 2006) and subject to evaluative standards (Lasswell and McDougal 1992). We have emphasized the role of social

processes—interactions among people making decisions within the institution of wildlife management—in determining whether cougar management proceeds in the common interest. A key question now is whether recent developments move us closer to or farther away from this goal.

Based on evidence presented in this chapter, and referenced to standards of democratic process and human dignity outcomes (Lasswell 1971; McDougal et al. 1980), the institutional arrangements of cougar management have often not functioned well during the last thirty years. As others have noted before, the causes of this suboptimal performance are not difficult to identify and are largely attributable to the structure and function of official decision-making processes (Decker et al. 1996; Gill 1996a, 2001b; Nie 2004a, 2004b; Clark and Munno 2005; Jacobson and Decker 2006). Wildlife management agencies favor the special interests of hunters while too often discounting the interests of virtually all others. The identity and culture of agency commissioners and personnel foster deference to hunting and reflect values of power and achievement rather than values of universalism or benevolence. Agency reliance on hunter funding and the comparative isolation of normal decision making from intervention by elected officials exacerbates these trends. This focus on agencies follows from the fact that, under the commission structure, and as trustees for the public interest, wildlife managers hold not only primary authority but also primary responsibility and accountability (Nie 2004a, 2004b; Clark and Munno 2005).

Scientific and business models of management have compounded difficulties, ultimately justifying concentration of power in the hands of agency personnel and commissioners on the basis of expertise (cf. Brunner and Steelman 2005). Moreover, neither science nor Total Quality Management offers a language to help agencies engage fruitfully with issues related to common ground and common interests (Brunner and Steelman 2005). Governance is not particularly relevant if "problems" are seen not as human but as objectively biophysical, people are seen not as diverse interests but as "customers" for technical services, and the "product" is not inclusive decision making but hunting opportunity. Cougar management today thus suffers from a suite of problems rooted in institutional arrangements that fail to acknowledge the shifts in public perspective away from traditional wildlife management. Current patterns erode trust, diminish social capital, and harden political positions—usually without these effects being intended. Common ground in cougar management remains an elusive and sometimes neglected goal.

What is to be done? One option is to maintain the status quo, perhaps with small changes primarily in the form of symbolic uses of new language. "Dynamic conservatism" is a common strategy for organizations to create the appearance of modernization while preserving basic institutional and decision-making arrangements (Schön 1973). Here, changes occur at the margin, if they occur at all, and consist largely of symbolic adjustments to deflect criticism and enhance public relations. Many recent changes in the institution of cougar management appear to be of this nature. Continuation of this approach will likely perpetuate conflict in yet more ballot initiatives, lawsuits, incidents, and negative press.

Another option is to employ techniques that can influence institutionalized patterns to improve both process and outcomes (Brunner 2002; Brunner and Steelman 2005). Among these are integrative interdisciplinary appraisal (Clark et al. 2001; this chapter), institutional analysis (Clark and Rutherford 2005), social methods such as Q-assessments for clarifying perspectives (Mattson et al. 2006), problem solving and skill-building workshops (Clark et al. 2002), leadership improvements (Clark 2007), participatory or community-based projects (McLaughlin et al. 2005), and prototyping (Wilson and Clark 2007). All these methods can provide insight for upgrading the full spectrum of policy-related decision making.

More fundamentally, institutional change is grounded in a paradigmatic shift from focusing on expert-based authority to focusing on common ground. Current stakeholders do share considerable common ground. In principle, most would probably support civil, open, fair, and participatory decision making (Mattson et al. 2006). There is even common ground regarding physical outcomes (Figure 14.4), most clearly among three participant groups: agency personnel, those belonging to environmental organizations, and those focused on the conservation of predators. Participants currently in conflict could potentially agree on aspects of process, constructive roles for biological and social science, habitat protection, the value of outdoor experiences, and the merits of long-term conservation. If participants choose to focus on their common interests, they have five areas of opportunity to change institutions: (1) changing who participates, (2) affecting the perspectives of those who do, (3) changing strategies that are used, (4) altering the situations within which participants interact, and (5) reconfiguring structural incentives such as revenue streams (Clark 2002, 2007).

Changing Participants, Perspectives, and Strategies

There are probably limited prospects for improvement by focusing overtly on participants, especially without changing situations and revenue incentives. In fact, changed situations are often a prime reason people modify perspectives and adopt different strategies (Lasswell and McDougal 1992). In principle, those who make demands

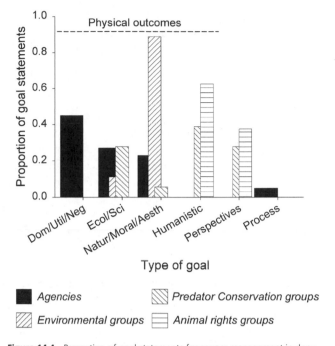

Figure 14.4 Proportion of goal statements for cougar management in documents created by state wildlife management agencies, environmental groups, predator conservation groups, and animal rights groups. Goals are differentiated by which of four kinds of worldviews they reflect—Dom/Util/Neg = dominionistic, utilitarian, or negativistic; Ecol/Sci = ecologistic or scientific; Natur/Moral/Aesth = naturalistic, moralistic, or aesthetic; or Humanistic—or whether they focus on changing perspectives of participants in cougar management, or on the quality of decision-making processes. Goals pertaining to Kellert's (1996) worldview categories focus on achieving physical outcomes.

that categorically exclude other participants' interests do not have legitimate standing (Brunner 2002). Examples of such demands include advocating extirpation of cougars or complete bans on hunting. However, in practice, participants involve themselves on the basis of existing access to power or the media, regardless of the legitimacy of their claims, and with the risk of marginalizing themselves if their demands are deemed patently unreasonable by politicians or the public.

Once established, and especially when rooted in fundamental worldviews, perspectives tend to resist change, even in the face of new information (Kellert 1994, 1996). In the longer term, changes in perspectives, especially among agency personnel and commissioners and about the structure of decision-making processes, could dovetail with changes in incentives to upgrade the quality of governance in cougar management significantly. How can those interested in change catalyze such transformations in perspective? Educational institutions are one logical source of leadership. "Human dimensions" have received increasing attention in wildlife management curricula during the last decade, providing grounds for optimism. But effectively promoting the kind of changes described here would require curricula to move beyond topics such as social surveys and education methods and to deal meaningfully with policy- and governance-oriented professionalism (Clark 2001).

Changing Situations and Incentives

The greatest near-term prospects for improving governance in cougar management lie in changing situations and incentives. In recent years most who have examined cougar or other wildlife management have advocated the creation of processes that engage stakeholders meaningfully in gathering information and in selecting, appraising, and even implementing policy (Gill 2001b; Nie 2002, 2004b; Durant et al. 2004; Clark and Munno 2005; Mattson et al. 2006). Ideally, such processes would be designed to foster common interest outcomes (Clark and Rutherford 2005), employing principles of adaptive governance (Brunner and Steelman 2005) and operating at local scales that capitalize on a place-based sense of community (McLaughlin et al. 2005; Cherney et al. 2008). Local and regional watershed councils are perhaps the best example of this kind of innovation in natural resources management (Wondolleck and Yaffee 2000). To be meaningful, reformation requires that agency personnel and commissioners divest some power. In instances where those in power are reluctant to participate, how can participants innovate with authoritative, collaborative, localized processes? Well-placed agency innovators and legislative interventions may be necessary.

Superseding all, it is difficult to imagine substantive change in cougar management without changes in revenue sources and the commission structure of wildlife management agencies. Numerous observers have concluded that for equitable decision making to happen, revenues must be diversified to reflect the standing of all stakeholders in wildlife management (Decker et al. 1996; Gill 1996a; Hagood 1997; Beck 1998; Pacelle 1998; Rutberg 2001; Nie 2004b; Jacobson and Decker 2006). Moreover, the commission structure is likely to perpetuate divisive and inequitable decision processes. Created specifically to "minimize the intrusion of politics" (Reiger 2001), the commission structure is counterproductive when politics are central and diverse legitimate interests need venues to work out differences and achieve common interest policies (Brunner 2002; Brunner and Steelman 2005)—and doubly so when current arrangements merely institutionalize bias in favor of one set of special interests while marginalizing all others. Invoking the public trust, one alternative would be to retain the commission structure but distribute commission memberships among stakeholder groups proportional to their public numbers. Invoking democratic principles, another alternative would be to elect commissioners directly or make wildlife agencies directly accountable to elected officials. Whatever the alternative adopted, equitable and ameliorative decision-making processes have been recommended by many as a means of achieving good governance (McDougal et al. 1981; Dahl 1982, 2006; Lasswell and McDougal 1992; Clark 2002).

Conclusion

Cougar management today is problematic for numerous reasons. Officials and citizens alike are commonly thwarted by unappreciated, unacknowledged, and unaddressed problems in processes and outcomes. Current institutional arrangements often do not address the valid interests of a diverse public and at times do not produce decision making that is respectful, factual, and fair. Conflict and incivility deplete social capital as participants turn to the media, courts, and other divisive venues to advance their interests. The common interest remains elusive. We have attempted to provide a realistic problem definition focused on patterns of social interaction and decision making to highlight opportunities for constructive change. Our analysis suggests that institutions affecting cougar management need to be changed if participants are to find common ground and clarify shared interests. Opportunities for meaningful change exist, both in how participants interact and in how decisions are made. Of central importance, citizens who are knowledgeable and civic-minded need to be meaningfully involved in all aspects of making decisions if cougar management is to be adaptive, democratic, and in the common interest.

Chapter 15 Cougar Conservation: The Growing Role of Citizens and Government

Sharon Negri and Howard Quigley

I F COUGARS (*Puma concolor*) ARE to continue filling their essential ecological role far into the future, we must ask ourselves, "What can we do today to ensure that one hundred years from now cougars are still living on the landscape?" To answer this question adequately, and to formulate strategies that can ensure the cougar's future, we must first understand something about their past.

The historical record of carnivores shows us devastating extinction rates, with 352 genera going extinct relative to the 129 currently living (McKenna and Bell 1997; Purves et al. 2001, 11; Steneck 2005). Gill reminds us in Chapter 1 that "no group of species faces a more insecure future than large carnivorous mammals." At the same time, cats, overall, are known to be some of the most adaptable and successful species on earth. The cougar, for example, has lived in almost every major habitat type in the Americas (Nowell and Jackson 1996, 134), roaming from Patagonia to northern British Columbia, and from the Atlantic to the Pacific. But severe reduction of its native prey through hunting and extensive loss of their habitat through human development, combined with direct persecution, have eliminated the cougar from roughly two-thirds of its North American range (Culver 2000; see the Range Map).

Today, the cougar's adaptability has allowed it to thrive in some of its native habitat, and even move into areas where it was once extinct. But the exploding human population and the resulting impacts to cougars and other wildlife cannot be ignored. Add human-induced predator control and habitat loss to the evolutionary forces at play, and the situation looks tenuous. More people usually means less room for wildlife, and large carnivores are especially vulnerable to persecution and habitat loss because they live at relatively low densities and need large connected lands to survive (Nowell and Jackson 1996; Gittleman et al. 2001).

Given the complex and increasing threats to cougar populations throughout their range, ensuring a secure future for these large cats in many areas will be no easy task. Fortunately, despite tensions that exist among the diverse interests (see Chapter 14) and major challenges to the conservation of this large and controversial carnivore, important steps are being taken that are benefiting both cougars and humans. Diverse government and citizen groups, both individually and collectively, have launched a wide variety of programs that are helping improve the cougar's future on the landscape. Citizen advocates have demanded an increasing say in cougar management (Hornocker 1991), calling for synergy between "the democratic values of truth, justice, and community with the spiritual values of stewardship and compassion" (Chapter 1; see also Chapter 14). Wildlife agencies are contributing to cougar conservation by launching local and statewide education programs to reduce human-cougar interactions and are working with others to protect critical habitat. Also, some of these agencies are adopting meaningful strategies that represent a more cautionary, science-based approach to cougar management.

In an environment of shrinking budgets, and growing threats to many of the natural systems on the planet, the enormity of the task of cougar conservation clearly should not be left to any one interest alone. It will take a collaborative effort of both government and outside interests to secure the cougar's future. For these reasons, in this chapter we provide real examples of diverse interests working together to conserve cougars and their habitats. Our goals are to define what conservation must entail, to encourage collaborations, to describe the current state of cougar conservation, and to illustrate the evolving roles of citizen advocates and government agencies. Finally, we provide a compass to guide us forward by identifying principles and motivations that are essential to conserving the cougar long into the future.

Defining Conservation

The term *conservation* came into use in the late nineteenth century to describe two related activities, namely, the economically driven management of timber, fish, game, soil, water, pasture, and minerals and the aesthetically or ethically driven preservation of forests, wildlife, parkland, wilderness, and ecosystems. In the United States, this schism was embodied by the public dispute between John Muir (famed writer, naturalist, and founder of the Sierra Club) and Gifford Pinchot (first chief of the U.S. Forest Service and governor of Pennsylvania) in 1897. Muir was deeply opposed to commercializing nature and argued for the preservation of land for its spiritual values. Pinchot saw conservation as a means of managing for the sustainable, wise use of natural resources. This philosophical divide between preservationists led by Muir and Pinchot's camp, who co-opted the term *conservation*, persists today (see Chapter 1; Nash 1967).

Aldo Leopold (1949, 262) defined conservation as "man living in harmony with the land." He promoted stewardship, emphasizing that "Conservation is a state of health in the land-organism. When any part lives by depleting another, the state of health is gone" (Leopold 1942, 265). Leopold embraced both utilitarian and intrinsic-aesthetic values (see Table 14.1), promoting a vision of people caring for life, benefiting from its use, and valuing it in its natural state. We embrace Leopold's more expansive vision of conservation in this chapter. However, important questions remain. How do we preserve the integrity and stability of cougar populations? What conservation measures need to be in place to assure a future for cougars, and for how long? Science can show the way, but only in part: it can tell us what to do but it cannot motivate us or prioritize the actions necessary to ensure the cougar's long-term survival.

The modern field of conservation biology was founded in the early 1990s on the same premise as Aldo's land ethic; emphasizing an interdisciplinary science-based approach to protecting, maintaining, and restoring species, as well as fostering ecological processes (Society for Conservation Biology 2008). This modern concept is already being applied. Studies of cheetahs and tigers show that successful carnivore conservation includes scientists from a variety of disciplines who examine population fragmentation, genetic integrity, and habitat loss. These elements are interrelated to such a degree that adopting only one approach is bound to miss important elements (Gittleman et al. 2001).

Protecting and maintaining critical habitat are essential components in the conservation of wide-ranging species, such as the cougar. In *Principles of Wildlife Management* (1984), James A. Bailey defines conservation as the dynamic social process that seeks to attain wise use of wildlife resources while maintaining the productivity of wildlife habitats. Thus, cougar conservation must be ecologically based, interdisciplinary, and proactive.

Wildlife management has an important part to play in conservation. Ira N. Gabrielson in his book *Wildlife Conservation* argues that "management should be built upon a firm foundation of a constantly increasing body of facts about the complex relationships among living things." He concludes that the job (of wildlife management) is to repair the damage we have done to the natural world as best we can and put natural constructive processes back to work (Gabrielson 1941, 111).

Combining the ideas of Leopold, Bailey, and Gabrielson produces a mission statement for cougar conservation that can guide decision makers, wildlife managers, conservation planners, and stewards of our natural resources.

A healthy and effective approach to cougar conservation:

- is based on an interdisciplinary scientific approach;
- involves and includes public participation;
- anticipates and acts upon threats to cougars and their habitats;
- acknowledges the aesthetic and intrinsic value of nature.

These elements closely describe the components of the discipline of conservation biology, reflecting both "scientific questions and environmental concerns, seeking to reshape how humans understand living nature, and assembling diverse concerns" (Bocking 2006, 59). Ultimately, cougar conservation will be more effective if it is an inclusive and collaborative process. It requires broad respect for the natural world and mutual respect among stakeholders whose standpoints range from utilitarian to aesthetic. Finally, "conservation is not just about saving rare species, but also prudently managing abundant ones" (MacDonald 2001, 527).

Government Mandates and Jurisdictions

Understanding the laws governing cougars in the United States provides a roadmap of the management and conservation responsibilities of various governmental agencies. The legal context differs from Canada and varies across Latin America countries (see Chapters 4, 6, and 7), but the laws are complex and questions of jurisdictional authority are complicated.

States have primary jurisdiction over the wildlife within their borders. In the United States, state legislatures pass wildlife laws, state game commissions interpret the laws, and state wildlife agencies implement and enforce them. Under U.S. common law, the people own the wildlife, and that ownership is undertaken through the people's representative—the government. The U.S. Supreme Court has held that states hold wildlife in trust for citizens for conservation and protection (*Hughes v. Oklahoma,* 441 U.S. 322 1979). State wildlife commissioners are usually appointed by the governor. Wildlife commissions establish hunting seasons, harvest quotas, and management actions, purportedly advised by the biological information and cultural responses provided by agencies and citizens (for further discussion on commissions see Chapter 14). Commission authority is not open-ended, as it is constrained by federal laws such as those governing the protection and management of migratory waterfowl, federally listed endangered species, and legislated mandates, such as state laws related to endangered species.

The missions and mandates of state wildlife commissions are broad, such as "to protect and enhance Oregon's fish and wildlife and their habitats for use and enjoyment for present and future generations" (Oregon Department of Fish and Wildlife 2007) or "to serve the people of Utah as trustee and guardian of the state's wildlife and to ensure its future and values through management, protection, conservation and education" (Utah Division of Wildlife Resources 2005b). Such statements allow commissions to establish specific regulations, yet they are broad enough to accommodate decisions seemingly contradictory to conservation mandates. Hence, for example, a commission can choose to ignore scientific input and/or ecological considerations, and direct the wildlife agency to reduce cougar numbers to enhance deer populations for recreational hunting.

Federal jurisdiction also affects cougars. The U.S. Fish and Wildlife Service (USFWS) manages 548 national wildlife refuges and has authority over endangered species (including subspecies and distinct population segments). Thus, the USFWS is responsible for recovery of the endangered Florida panther, although ceding much of the recovery activity to the state of Florida. Hunting is permitted on many wildlife refuges and is usually, but not always, managed consistent with state hunting laws.

The U.S. Forest Service (USFS) manages wildlife habitat on 191 million acres of forest and grassland, and the Bureau of Land Management (BLM) manages 264 million acres of public land, mostly shrub and grassland. In the western states, most of these lands provide cougar habitat. Both agencies have habitat management plans alongside resource extraction strategies. However, having a higher priority on extracting resources, rather than on the long-term health of ecosystems, has not benefited wildlife on these lands (O'Gara and McCabe). The BLM and USFS generally defer wildlife management authority to state wildlife agencies, although the federal agencies remain responsible for implementing certain federal laws, such as the ESA. The USFS provides special use permits for outfitters using Forest Service land for hunting camps.

Wildlife Services, a program within the U.S. Department of Agriculture, has primary authority over predator control. From 1937 to 1983 federally funded wildlife agents killed approximately 8,000 cougars (O'Gara and McCabe 2004, 66). In some cases, Wildlife Services' authority may not be consistent with state authority, or with that of other federal agencies. For example Wildlife Services is unlikely to be allowed pursuit permits into National Park Service (NPS) lands following depredation complaints on the edge of a park.

The NPS oversees 84 million acres of wildlife habitat, conducts extensive wildlife research (including on cougars), and manages threatened and endangered species within park boundaries. The NPS protects cougars under a statutory mandate to provide protected areas in which the natural resources are subject to minimal human influence or utilization and are managed in a natural state much as they existed when settlers first landed in America. The Department of Defense also has extensive and excellent habitat for cougars. These lands often act as refugia, as many are closed to cougar hunting.

State game laws do not apply on sovereign land of Native American nations, state game laws do not apply. Tribal game laws and annual regulations are set by tribal councils and committees, some of which employ wildlife management professionals. Cougar hunting on tribal lands generally follows regulations established by the states in which those lands lie. However, management can swing further toward cougar protection or toward control, depending on the leaders and the particular sentiments and interests of tribal members at a given time.

In the United States, the courts and Congress have determined that the federal government has a role in protecting and conserving native wildlife. The Property Clause of the U.S. Constitution gives Congress "complete power" over federal lands (see *Kleppe v. New Mexico,* 426 U.S. 529, 535 1976). Under the Property Clause, "Congress may preempt traditional state trustee and police powers over wild animals, giving the federal government authority to regulate and protect wildlife on federal land" (*Kleppe,* 426 U.S. at 535; *Wyoming v. United States,* 279 F.3d 1214 [10th Cir. 2002]; 43 C.F.R. § 24.3). This federal control option is generally delegated to the states on federal lands managed by the BLM and the USFS but is maintained on lands managed by USFWS and the Park Service.

Overlapping jurisdictional authority complicates cougar conservation. Most national parks can prohibit cougar hunting, but rarely will a cougar home range lie completely within the park. Similarly, adjoining states may have different policies for cougars. Improved coordination across jurisdictional boundaries is increasingly important for landscape conservation plans that attempt to promote connectivity across large areas. In such cases, lack of coordination can become a major obstacle. Cougar conservation is also complicated by overlapping and sometimes opposing agency mandates and the growing conservation needs of threatened and endangered species. These factors coincide with pressure on agencies to widen their traditional narrow focus on sport hunting and protection of property and embrace a broader mission, namely, the conserving of all wildlife. Yet, because cougars generate little agency revenue, their conservation needs tend to fall to the bottom of the priority list of wildlife and land agencies, leaving citizen groups to fill an important niche in conserving cougars.

The New Cougar Advocates

The Wilderness Act of 1964, the Endangered Species Act of 1973, and the Federal Land Policy and Management Act of 1976 all signaled an increased public commitment to resource protection. During the same period, western state legislatures shifted the cougar's status from injurious predator to game animal. Nevertheless, the principal focus was on cougar management, providing hunting opportunities and protecting livestock.

This limited view was at odds with the Leopold ethic that emphasized the value of untamed, wild nature, as well as with the public's appreciation of large predators and wild places (see Leopold, 1966). Advocates for this broader view have emphasized the importance of accountability, science, and transparency in management decisions and have highlighted the need to preserve large connected wild lands.

The new cougar advocates of modern times share much in common with sportsmen at the beginning of the twentieth century. Sportsmen challenged the commercial exploitation of wildlife in response to unregulated market hunting, and subsequently influenced the development of wildlife management, becoming a force for conservation. To protect their hunting rights and ensure that game would remain available, hunters became citizen activists. They advocated for game laws, helped set aside hunting preserves, and funded conservation of wildlife habitat (Reiger 2001).

Like sportsmen of a century ago, the new cougar advocates are shaping public opinion and the culture of management agencies. Their methods include promoting and funding sound research, educating the public on the ecological role of the cougar, designing predator-safe management for livestock owners, and educating residents, hikers, and campers on how to stay safe in cougar country. Some advocates have used the courts and ballot box to protect cougars from overharvest and to protect their habitats from development and fragmentation (see litigation and initiative discussion, this chapter; Appendixes 4, 5).

The term *cougar advocates* describes individuals and organization working to improve cougar research, management, and conservation. Some of the wildlife and animal groups that have devoted time to these efforts include Sinapu (now known as WildEarth Guardians), the Mountain Lion Foundation (MLF; formerly the Mountain Lion Preservation Foundation), the Cougar Fund, Animal Protection of New Mexico, Animal Defense League of Arizona, Animal Welfare Institute, Ontario Puma Foundation, and the Humane Society of the United States. Other groups, such as the Cougar Network, focus primarily on disseminating information.

Today, advocacy groups are focusing much of their time reviewing and criticizing state management plans for their lack of sound science, transparency, and stakeholder input. For example, the Cougar Fund (2006) argued that Wyoming's plan lacked goals related to perpetuation of cougars, genetic diversity, and the ecological health of the systems in which they lived. The new conservationists are increasingly asking for better documentation that cougar populations can sustain hunting (Lindzey 1991). In turn, they are putting pressure on agencies to terminate cougar hunting or to significantly reduce the number of cougars that hunters are allowed to kill, until the states can better assess the impact of hunting and how other mortality factors contribute to the long-term health and viability of populations.

To exert more influence on agency decisions, and thereby achieve improvements in the way cougars are managed and conserved, these groups have had to immerse themselves in the science of managing wildlife. At times, they have enlisted biologists to support their positions or help them critique cougar management plans. Many biologists, however, are unwilling to criticize colleagues in state wildlife agencies because they are employed by those agencies, and/ or depend on them for necessary permits to carry out their field research (Bocking 2006). Conservation organizations have also added legal professionals, business people, academics, and former agency employees to their staff or advisory committees. This more integrated approach has allowed scientific and legal experts to critique complex hunting strategies, and has made advocates more effective in influencing wildlife management decisions.

One of the more important roles of these citizen advocates is to broaden the debate and raise awareness of the importance of, and threats to, cougars. For example, they have advocated for cougar habitat and population connectivity, and are urging agencies to manage the total impact of all threats, such as high hunting quotas, overhunting of females, habitat fragmentation and loss, road kills, disease, genetic concerns, and poaching.

These cumulative impacts, gone unchecked, can reduce populations to unsustainable or ecologically insignificant levels (see Figure 15.1). Addressing all impacts was central to recovery of the grizzly bear (*Ursus arctos*) in the Greater Yellowstone Ecosystem, where incidental hunter kill outside the park, depredation removals inside and outside the park, and road kills were significant factors hindering population growth (Schwartz et al. 2007). A cumulative impacts model was developed to analyze and monitor all the factors that negatively affected the desired growth of the grizzly bear population in the ecosystem (Mattson et al. 1992; Mace et al. 1999).

Cougar advocacy groups, as well as conservation biologists and others, have emphasized the important role the cougar plays as a keystone or flagship species (Gittleman et al. 2001). Like grizzly bears, cougars occur at low densities on the landscape, and their sensitivity to human impacts makes them useful for the development of landscape conservation plans (see Chapter 12).

Because of the known and unknown impacts of threats to cougar populations, advocates are asking agencies to use the precautionary principle when developing cougar management plans. The precautionary principle states that an action should be avoided if it has the potential to cause serious or irreversible harm, despite the lack of scientific certainty as to the likelihood or magnitude of that harm (The Precautionary Principle Project 2007). Essentially, the principle encourages decisions that err on the side of caution. When applied to cougar management, it means actions preventing habitat destruction or reducing female cougar harvest, for example, should be presumed appropriate until proven otherwise. Protecting females and habitat may provide the cushion, or "biological savings account," to protect cougars from impacts we do not understand or know little about. As Ray and colleagues (2005, 423) argue, there is ample evidence that top predators can have broad effects over natural systems, and that the precautionary principle should be "front and center" in dealing with large carnivorous animals.

To ensure that their message is heard by a broad audience, cougar advocacy groups have also become skilled at running effective information and lobbying campaigns, including conducting public opinion polls, hiring media consultants, and taking out newspaper advertisements. Through their public outreach efforts, they have recruited thousands of supporters to write letters and/or show up *en masse* at commission meetings and legislative hearings. These actions appear to have impact. At least partly due to public pressure generated by these groups, hunter quotas limiting the take of female cougars have been established or lowered in Colorado, Washington, and Arizona.

Taken as a whole, the new cougar advocates have become knowledgeable and formidable spokespersons for this important carnivore and are filling a crucial conservation niche. They also play another important role: independent advocacy groups can say and do things that government personnel often cannot, providing essential outside support for conservation innovations. Wildlife agency personnel and independent biologists have at times turned clandestinely to cougar advocacy groups to ensure that key information is distributed to the public. These groups also act as conduits between scientists and the media, ensuring the broad distribution of information and again reflecting the willingness of some environmental groups to take on tasks governments are unwilling or unable to perform (Bocking 2006).

Litigation and Initiatives

During the 1980s and 1990s, cougar advocates grew increasingly frustrated with state wildlife commissions and agencies who, they argued, adhered to antiquated policies and practices of cougar management, focusing primarily on hunting and livestock protection without acknowledging and applying sound scientific evidence. When advocacy groups felt their concerns were not being addressed or that existing laws were not enforced by state and federal wildlife and land agencies, they sometimes turned to the courts or the ballot box (see Chapter 14; Appendixes 4, 5).

Citizen groups have used the courts to stop or alter hunting seasons, protect cougar habitat, and challenge the legality of experimentally hunting cougars to protect bighorn sheep or livestock. Groups have filed lawsuits to force agencies to consider the impacts of their actions, to require inclusion of available science in management decisions, and to block development in Florida panther habitat. Judges have sometimes ruled in favor of citizens and at other times have sided with government agencies.

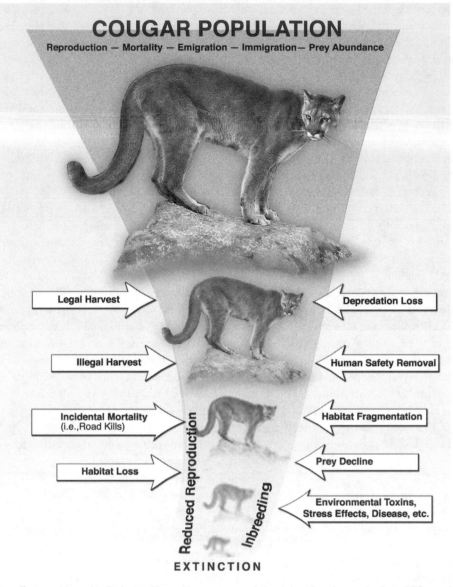

Figure 15.1 The cumulative effects model, graphically depicted here with a cougar population, is based on the premise that wildlife population viability decreases progressively with additional human influences. Beginning with naturally structured and functioning demographic characteristics (births, deaths, immigration, emigration) and environmental components (food, cover, etc.), factors that can decrease population viability include direct killing, loss of habitat, habitat fragmentation, and the like. Increased human impacts can lead to increased stress, susceptibility to disease, reduced reproduction, and inbreeding. Thus, effective, long-term conservation should account for the cumulative impacts of all human-caused influences.

Though legitimate parts of our judicial and democratic system, litigation and ballot initiatives can produce unintended consequences and are expensive for all parties. They are, therefore, frequently used as a last resort strategy to force an agency to alter or reconsider a decision when other efforts to compel such change have failed. Lawsuits filed in California in the late 1980s may best illustrate how groups have used litigation to gain protections for cougars when other efforts failed. Between 1907 and 1972, more than 12,500 cougars were killed for bounty and sport in California. The bounty was removed in 1963 and a

moratorium on hunting was put in place in 1972. Twice citizens persuaded the legislature to renew the moratorium. Then, in 1986, the governor of California vetoed a bill that would have continued the ban, and the California Department of Fish and Game (CDFG) authorized a hunting season on cougars. The Mountain Lion Preservation Foundation (MLPF), along with a host of other organizations, filed suit to stop the first hunting of cougars in two decades.

Thousands of letters and more than fifty newspaper editorials condemned the hunt on scientific, humane, and ethical grounds. Scientific experts, conservationists, and citizen groups raised concerns over the methods used for developing the cougar population assessment that became the basis for calculating the lethal take of cougars. They argued in public hearings and in written comments that too little was known about the impact of a sport hunt and other cumulative impacts (such as loss of habitat) to ensure sustainable cougar populations.

Equally important to many was the belief that the large cat should be protected simply because it had meaning in their lives, despite the fact that they would never see one in the wild. Some also opposed the hunt on religious grounds. The hunt was ultimately thwarted by two lawsuits in 1987 and 1988, on the basis that the California Fish and Game Commission (CFGC) had failed to address the environmental impacts of the hunt, as required by the California Environmental Quality Act. The court's analysis set a new legal standard for how the state's public agencies must draft environmental impact reports. In setting aside the CFGC hunting plan, the court argued that the agency "must provide a cumulative impact analysis to the public *that encourages rather than impedes meaningful public discussion*" (*Mountain Lion Coalition v. Calif. Fish and Game Commission,* 1989, 214 Cal.App.3d 1043; emphasis ours).

Later, unrelated cases (such as *Laurel Heights Improvement Association v. Regents of the Univ. of California,* 1993, 9 Cal. 4th 1112) used the MLPF case as a precedent when environmental impact reports were found to be "meaningless" or public comment was stifled (B. Yeates, pers. comm., 2007).

Besides revealing inadequacies in the CFGC procedures for public input and in its assessment of the impacts of a hunting season, the lawsuits were significant in other ways. They shone a public spotlight on the ecological importance of cougars and on the questionable ethics of shooting a large carnivore for sport. The groundswell of public support included some ranchers and hunters, elected officials, major newspapers, and celebrities. It reflected the social importance of cougars in a state where 95 percent of people did not hunt. This litigation rapidly changed the influence of citizens: cognizant of costly lawsuits in California, some wildlife agencies in other states became more responsive to citizen concerns.

The success or failure of cougar lawsuits depends on the quality of the legal argument, the factual evidence, and the values or predisposition of the judge regarding cougars. The overall success of cougar litigation has been decidedly mixed. In the mid-1990s, efforts led by The Fund for Animals to challenge a decision to permit the construction of a landfill in Florida panther habitat were unsuccessful, as were efforts by Forest Guardians in Arizona to challenge lethal cougar control activities in congressionally designated wilderness areas.

Beginning in 2000, however, conservation and animal protection organizations prevailed in several federal lawsuits related to cougar management and conservation. In New Mexico, Defenders of Wildlife succeeded in stopping the lethal control of cougars by the U.S. Department of Agriculture's Wildlife Services due to its failure to comply adequately with the National Environmental Policy Act (NEPA). NEPA also was used successfully to challenge controversial "scientific studies" involving the lethal removal of cougars in Arizona and Oregon and to stop a mining project within Florida panther habitat. Since NEPA is a procedural statute requiring agencies to consider the environmental impacts of their actions, many NEPA victories are short-lived if the agency eventually prepares a legally sufficient analysis.

While the federal courts have, on occasion, ruled in favor of conservation interests, lawsuits filed in state courts have not been as successful. Challenges to state wildlife agency decisions in both New Mexico and South Dakota, alleging excessive and unwarranted cougar hunting, failed. The courts ruled that the agencies had complied with relevant statutory authority when establishing their hunting regulations (see Appendix 4 for a summary of cougar-related litigation).

In addition to lawsuits, conservation and animal protection groups have also used state initiatives or referenda. In California, the same organizations involved in the litigation challenging the legality of the state's cougar hunt formed the California Wildlife Protection Committee. The Committee, capitalizing on statewide support and polls showing a public favoring cougar protection, launched a statewide initiative drive and placed the Wildlife Protection Act, known as Proposition 117, on the June 1990 ballot. They hired a top media firm and conducted one of the most successful grassroots efforts in the country up to that time. The proposition passed, banning the sport hunting of California cougars, allowing an individual cougar to be killed if it was deemed a threat to humans or livestock, and allocating $30 million dollars annually for thirty years to the protection of critical wildlife habitat (with specific attention to deer and cougar habitat and endangered and threatened species). Proposition 117 also tightened depredation regulations by banning the use of snares (see Appendix 5 for a summary of cougar-related ballot initiatives).

Passage of the Wildlife Protection Act made California the first state to ban cougar hunting by popular vote and dramatically increased the visibility of the cougar conservation issue nationwide. As of 2007, more than a million acres of California habitat and hundreds of miles of waterways have been protected for cougars and other wildlife (Mountain Lion Foundation 2007). Advocates took the issue of hunting cougars to voters, but they also tied it to a practical, land-based conservation effort that would counter the degradation and fragmentation of the California landscape. This broader approach resonated with the people of California.

When agencies are not receptive to public input, the people excluded from the process try to pass laws promoting their agenda (Chapter 14; Decker 2001). Steve Torres, a scientist with the California Department of Fish and Game, noted: "If you don't involve the public in wildlife management, they will involve themselves" (S. Torres, pers. comm., 2007). However, some wildlife agency personnel view initiatives and lawsuits as an encroachment into professional wildlife management decisions. In 1996, Washington voters passed Initiative 655 with 52 percent support. The initiative banned the use of hounds to pursue cougars, initially reducing the number of cougars that could be killed. The Washington Department of Fish and Wildlife responded by replacing a permit-only season with a general season in an effort to mitigate the expected decrease in cougar harvest; the cougar season was extended from six weeks to seven and a half months, the annual bag limit per hunter went from one to two, and the cost of a cougar tag dropped from $24 to $5. The number of cougar tags sold soared from 1,000 to 59,000 and, as a result, more females and young cougars were killed (Beausoleil 2003). The effects of this expanded harvest may be broadly felt in the state for some time (Lambert et al. 2007; Gross 2008).

Thus, some ballot initiatives have had unintended, negative consequences. Other initiatives that have focused on one issue (hunting or banning the use of dogs; see Appendix 5) have neglected important issues such as protection of the cougar habitat (Gabrielson 1941; Crichton et al. 1998; Hopkins 2003). They also have not provided funding to agencies for enforcement of these new laws, exacerbating tensions between agencies and cougar advocates.

Actions Advancing Cougar Conservation

Bocking (2006) wrote that crises can be opportunities for innovation. Many of these new cougar advocates have developed innovative programs that are less costly than initiatives and are improving conditions for both cougars and people. For example, Sinapu, now WildEarth Guardians, an organization dedicated to the restoration and protection of carnivores and their habitat in the southern Rocky Mountains, argued that females with dependent young deserve the same protections as females of other game species, such as deer. Some hunters find it difficult to distinguish a treed male cougar from a female, so Sinapu launched an aggressive campaign for mandatory hunter education on distinguishing female from male cougars.

Sinapu submitted extensive comments to the state wildlife commission, placed ads in local newspapers, and conducted outreach to hunter organizations. As a result, the Colorado Game and Fish Commission passed a mandatory hunter training requirement in 2007 which is now being replicated in several other states (Colorado Department of Natural Resources 2007).

WildFutures, a nonprofit group based in Washington state and directed by author Sharon Negri, identified a need for a comprehensive summary of cougar science for professionals. The group—whose mission it is to bridge the gap between the scientific and conservation communities—led a two-year effort to find a consensus of expert opinion in cougar management, and compiled it in one book. Thirteen experienced cougar biologists and wildlife managers were brought together to produce the first edition of *Cougar Management Guidelines* (Cougar Management Guidelines Working Group 2005), a compendium designed for wildlife managers that outlines cougar biology, describes management principles, and explains large-landscape approaches and adaptive frameworks based on the best available science. While the book was controversial among some agency personnel, the guidelines have been well received and are being utilized by many federal, state, and local land and wildlife agencies, as well as those interested in the research, management, and conservation of cougars (Shaw and Negri 2005).

In 2007, another group, Animal Protection of New Mexico (APNM), provided funds for the development of a Habitat Suitability Map when the state wildlife agency did not have the resources for such an effort. APNM raised the needed funds to hire an independent electronic mapping expert to conduct a habitat analysis that resulted in the identification of suitable habitat for cougars. This effort provided a state-of-the-art Habitat Suitability Map based on prey availability, terrain, roads, human population, and riparian access, which allowed APNM to make a contribution even when there is no revenue mechanism in place for agency support from the nonhunting public. As a result of this effort, New Mexico's cougar population estimates are now based on cougar-specific habitat maps.

Based in Sacramento, California, the MLF has a mission to protect mountain lions and their habitat for future generations. In 2001, in an effort to reduce human-cougar conflicts, MLF initiated a collaborative effort with 4-H and Future Farmers of America groups in several California counties where cougars had been killed due to livestock depredation. The program involved reviewing California Department of Fish and Game depredation records and meeting with agency personnel to identify areas where proactive conservation measures might be most successful. Subsequently, MLF staff worked with community leaders and particularly 4-H groups to educate livestock owners on cougar ecology and conflict-reduction strategies and to construct cougar-proof pens to house livestock at sites where depredations had occurred or were likely to occur. MLF has since expanded the program into Oregon, South Dakota, and Florida (home of the endangered Florida panther).

Individuals are also finding ways to contribute to cougar conservation. Vicki Long of Riverside County, California, discouraged that little was being done to protect cougars and their habitat from explosive human population growth in the county, hired a wildlife biologist. To simulate how future land development would impact the animals, she paid the biologist to develop a landscape suitability model for cougars, incorporating topography, vegetation, population density, location of high speed roads, and core habitat areas. The biologist presented the findings to county officials for integration into regional planning and to educate local communities about cougars. Long's hope is that the maps will help mitigate the damage of development by steering it away from core wildlife areas, ultimately reducing human-cougar conflicts.

In a new era of increasing responsibilities and public involvement, agencies are recognizing the importance of gathering comprehensive information for management decisions and working collaboratively with more diverse stakeholders. Terry Cleveland, former director of the Wyoming Game and Fish Department, recently described the state of Wyoming's wildlife and offered this perspective: "Active engagement by the Wyoming citizenry is key to answering the question of where do we go from here and to charting that course for the future" (Cleveland 2007, 2). "Active engagement" is perhaps the key. During the last decade, agencies have developed innovative programs that require and encourage citizen involvement.

From 2001 to 2008, Project CAT (Cougars and Teachers) a citizens program in the state of Washington (Koehler and Nelson 2003) was an exemplary model of innovative large carnivore conservation. Researchers captured and radio-collared dozens of cougars in a study area around the rural community of Cle Elem, approximately sixty miles east of Seattle. The research was integrated into a variety of class subjects and programs, and students worked directly with the cats. The most innovative aspect of this effort was that students became "teachers" in the community. They had a greater understanding of where cougars live, why they are ecologically important, and how people can coexist with them without conflict. Involving the students was a most effective way to disseminated cougar information to the wider community (G. Koehler, pers. comm., 2008).

Living with Cougars and Carnivores is a collaborative effort between conservation groups and the U.S. Fish and Wildlife Service. It provides information on cougar behavior, safety tips for living and hiking in cougar habitat, and results from current research. Evening presentations have been hosted in selected communities in Idaho, Washington, and Oregon, with a wide variety of sponsors. For example, sponsors in Idaho included Defenders of Wildlife, the Idaho Cattle Association, U.S. Department of Agriculture Wildlife Services, the Wolf Education and Research Center, and the Counter Assault company. The program was expanded to include CDs, Web site outreach, educational handouts, and children's materials to keep the public informed about how to minimize conflict with cougars and other carnivores.

Individual agency personnel can also make a difference. Rocky Spencer, biologist and carnivore specialist with the Washington Department of Game and Fish, worked with communities for twenty-eight years to help them coexist safely with cougars and bears. Until his accidental death on the job in 2007, Spencer was often a first responder for problem animals. Everyone respected his expertise, passion for wildlife, and efforts to educate communities about the important role of large carnivores. There are numerous other people employed by government agencies who, like Spencer, have put in untold hours of outreach into communities to assist in cougar conservation, but are not mentioned here. All these individuals' efforts and all the organizational progress produced small steps in the much larger framework of how cougars are managed and conserved on the landscape.

Changing Cougar Management Policy

In the long run, maintaining the viability of cougars, their prey, and their habitats depends on good policy underpinned by good science. Wyoming is a case in point. The state's management plan for cougars was informed by a long-term field study conducted through the University of Wyoming and funded by the Wyoming Game and Fish Department. In 2006, the plan was the most progressive state plan in the West, utilizing the source-sink concept (described in Chapter 5), adaptive management approaches, and an open display of all data used. The research might have come about without the encouragement of citizen groups, but it became an obvious priority for the state, which committed substantial financial and personnel support. One advocacy group, the Cougar Fund, applauded the plan as a "marked improvement [over previous plans] . . . written with the science the Fund and others had been advocating for years as it acknowledges the best available science and aims to establish a process for setting cougar hunting season and management objectives" (Cougar Fund, August 14, 2006).

The Wyoming Game and Fish Commission set another precedent in 2007 when it directed that hunting quotas include "any human caused mountain lion death documented by the Department" (Wyoming Game and Fish Department 2007a), and further directed that "any mountain lion mortality documented by Department personnel after the close of the season shall be subtracted from the mortality quota for the next biological year" (Wyoming Game and Fish Department 2007b).

These are some of the most progressive steps taken by any state agency to date. For the first time, a state plan included an estimate of the number of kittens orphaned by hunting. Fully considering all sources of mortality in setting hunting quotas is good management practice and a good model for conserving cougars in the long term.

Cougar management plans, harvest quotas, and hunting restrictions have also been changing in other states. For example, after years of no protections, the Arizona Game and Fish Commission (AGFC) adopted hunting regulations in 2004 to protect cougar females with spotted kittens. In 2007, hunters were required to report their lion kills within forty-eight hours rather than the previous ten-day deadline. In a state once known for its year-round hunting season, the AGFC took another progressive step in 2008 and limited hunting to a nine-month season in many parts of the state.

In 2006, the New Mexico Game Commission adopted changes to its evaluation of cougar hunting to incorporate population estimates based on habitat mapping specific to cougars (as earlier discussed); they also required the sexing and aging of cougars killed in the state (via analysis of a tooth extracted by New Mexico Game and Fish personnel). The resulting age and sex data will be used for more precise management recommendations. In Colorado, a similar situation exists. In recent years, as a result of more refined population estimates and indicators, the state has initiated a 30 percent reduction in its lion hunting quota statewide.

In 2004, there was considerable public outcry over removal of two cougars and closure of the Sabino Canyon Recreation Area near Tucson due to concern that cougars posed a risk to humans. As a result of this incident, the Arizona Game and Fish Department involved the public in developing a more transparent process for responding to interactions between cougars and humans. Similarly, the California Department of Fish and Game worked with the MLF and others to develop a statewide range map for cougars and to produce educational materials on living in cougar country.

These agency programs—to improve the hunting quota determination process, protect kittens, and involve the public in human safety issues—are important steps toward improving cougar management. Some steps may seem small, others more significant, but they are all advances in improving the prospects for cougars on the landscape.

Managing Threats to Cougars

Even the most well-defined, most specific list of threats to cougars boils down to two categories: habitat loss and direct killing (Nowell and Jackson 1996; Sunquist and Sunquist 2002). We know from historical records that cougar populations that are heavily exploited can be reduced to ecologically insignificant levels or eliminated. The MLF found that from 1997 to 2004, approximately 30,000 cougars were reported killed in eleven western states (Arizona, California, Colorado, Idaho, Montana, New Mexico, Nevada, Oregon, Utah, Washington, and Wyoming; Papouchis 2006b; see also Figure 4.1). The number killed averaged more than 3,600 each year (Papouchis 2006b). These annual removals have been subjected to varying levels of biological scrutiny. Wildlife managers should not assume that these removals have no effect on populations, especially when combined with habitat losses (Sweanor 2001; Cougar Management Guidelines Working Group 2005).

Yet as Anderson and Lindzey (Chapter 4) point out, although sport hunting may be the threat most visible and easily fixed, ". . . it may be the least important in the long term. Alteration and fragmentation of habitats for cougars and the ungulate prey supporting them are ongoing and insidious and much more difficult to control." *The Cougar Management Guidelines* (2005) likewise cite habitat loss as the greatest long-term threat to cougars in North America. Laundré and Hernández (Chapter 6) concur that this observation can be directly applied to Latin America, and Beier (Chapter 12) reminds us that limited habitat and isolation can doom populations, as in the case of the Florida panther.

Thus, agencies and citizen advocacy groups may be more effective in ensuring the long-term survival of cougar populations by focusing on what ultimately will be the animal's saving grace—the protection of their habitat. As Hopkins (2003, 145) said, "While many of the citizen groups have spent their limited time and resources on critiquing agency management plans, and agencies have spent endless hours in defending them and providing hunting opportunities, lost in this debate is the preservation of those elements that will perhaps truly lead to [cougar] conservation."

Human population growth throughout the cougars' range will present enormous challenges. The U.S. population is projected to increase to nearly 400 million by 2050—more than a 50 percent increase from 1990. Latin America has one of the highest rates of annual human population increase in the world. More people means less room for wildlife, with fragmented habitats separated by substantial barriers to dispersal. Roads are a major cause of mortality for small populations of cats in such fragmented habitat (Nowell and Jackson 1996, 167). Unfortunately, it is estimated that only 6 to 9 percent of extant cougar range lies within fully protected areas (Papouchis 2006b), and the size of these protected areas is insufficient to protect cougar subpopulations (Nowack 1986). This could spell disaster for cougars since they need a minimum habitat area of 1,000–2,200 km^2 to sustain a nonmigratory cougar population with a greater than 98 percent probability of persisting for one hundred years (Beier 1993; Logan and Sweanor 2001). Cougars that live in smaller habitat patches without immigration are more likely to be severely impacted or become extinct (Logan and Sweanor 2001).

More people within and adjacent to cougar habitat means more cougar-human conflicts (Chapter 13) as a consequence of both an increase in people living within cougar habitat and an increase in outdoor recreation, such as fishing, hunting, hiking, and camping. Nonconsumptive use of the outdoors—bicycling, camping, and hiking—is projected to increase 142 percent over the next fifty years (Flather and Cordell 1995). More people living and participating in outdoor activities within cougar habitat will diminish the quality of the habitat, result in habitat abandonment, and/or lead to greater conflicts which generally do not end well for any cougar.

Effective conservation for sustainable cougar populations requires identifying the cumulative impacts of all threats. Such threats are known to include predator control, disease, loss of habitat, sport hunting, isolation of small populations, decreases in prey, removal of depredating animals, and fragmentation of habitat (Figure 15.1). Global climate change is an emerging threat. Ecologist Reed Noss points out: "The fossil record tells us that most species have adjusted to past climate change not by *in situ* evolution but by changes in distribution through dispersal. It is reasonable to speculate that the current rate of global warming would not be so catastrophic to biodiversity were it not for habitat fragmentation at multiple spatial scales. The combination of habitat fragmentation and rapid climate change is devastating because fragmentation makes it difficult for many species to respond to changing climatic conditions through dispersal" (Noss 2007, 4–5). Birds and butterflies can alter their flight patterns, but any range shift by cougars requires running a gauntlet of suburbs and freeways in search of alternative habitat. To complicate matters, recent findings show when ecosystems lose a keystone predator they may be less resilient to the impact of climate change (University of California, Berkeley 2005; Vucetich et al. 2005). Clearly, we must consider all threats—extant and emerging—and develop reasonable management responses to ensure the survival and security of cougars and other species.

As an important first step in addressing threats, inventory and mapping of cougar habitat can provide a geographic base for the identification of critical habitat. Current and accurate habitat suitability maps can help wildlife planners identify blocks of habitat, potential habitat, and critical wildlife corridors (Cougar Management Guidelines Working Group 2005; see Chapter 8).

Reserve designs are another means for identifying habitat for cougars. These designs are pivotal tools in the effort to conserve biodiversity (Terborgh et al. 1999; Ray et al. 2005), and carnivores like the cougar are useful in developing them (Chapter 12). The Wildlands Network engages scientists, conservationists, and citizens in developing reserve designs driven by a long-term vision of connected wild habitat on a continental scale. Its efforts are based on the premise that protected areas in Mexico and North America are too small and isolated and represent too few ecosystem types to protect, sustain, and facilitate adaptation of wildlife populations over the long term. Designs are therefore focused on mapping critical linkages between parks, wilderness areas, and other protected lands. These maps can become essential tools for identifying, prioritizing, and protecting a continental system of pathways for wildlife. Many state and federal agencies have programs to identify and map species distributions. The Wildlands Network's tools can help them identify "conservation gaps," where wildlife habitat protection may be inadequate or entirely absent.

One case study is the South Coast Missing Linkages Project (SCMLP), spearheaded by South Coast Wildlands, which is a collaborative effort to conserve wildlife linkages in California at statewide and ecoregional scales. The SCMLP invited a large and inclusive group of more than 250 people, mostly scientists and land managers, to identify 232 habitat linkage connections across the state of California (Penrod et al. 2001). Because of its collaborative nature it gained almost immediate acceptance by government and private stakeholders and received positive media coverage (Beier 2007). South Coast Wildlands also catalyzes a collaboration of more than two dozen government and nongovernment partners to develop and implement detailed conservation plans for the fifteen most important linkages in southern California. The linkages are being analyzed and implemented by military bases, county planners, state and local transportation agencies, federal land managers, and private conservation groups. The linkage designs have also been incorporated into other regional conservation efforts, including the state wildlife plan (P. Beier, L. Sadler, pers. comm., 2008).

The conservation and restoration of the Coal Canyon biological corridor in southern California illustrates how a wildlife linkage plan can promote coordinated actions. Coal Canyon links the Santa Ana Mountains to the smaller wildlands of the Puente-Chino Hills (sixty square miles). In the mid-1990s the busy ten-lane State Highway 91 (Riverside Freeway) crossed Coal Canyon. A freeway interchange had been built decades earlier to accommodate anticipated urbanization of two private parcels (of 32 and 653 acres) that straddled the freeway in the canyon. A 1993 study (Beier 1995) documented that three cougars used the corridor, and one used it to move between the Chino Hills and Santa Ana Mountains. Without Coal Canyon, the Puente-Chino Hills area would become a wild island in an urban sea. Loss of the corridor would obstruct gene flow, guarantee the loss of cougars from the Puente-Chino Hills, and endanger populations of at least ten other species in the area (Noss et al. 1998). However, in November 2000, governmental agencies, conservation and land organizations, developers, and private citizens came together to help purchase the private land for $40 million and dedicated it as part of Chino Hills State Park, which agreed to fund the habitat restoration work to protect and improve the canyon as a wildlife corridor. To assist the effort, for the first time in U.S. history, a state highway agency removed an existing freeway interchange and replaced it with one that incorporated a wildlife underpass. The scientific evidence regarding cougar use of the corridor was a critical motivating factor. The linkage conservation plan was essential to allow the partners to achieve collectively what none could have done alone (P. Beier, pers. comm., 2007).

Network designs are also being implemented in parts of the Southwest and the southern Rocky Mountains. The Arizona Wildlife Linkage Workgroup is a joint endeavor of the Arizona Game and Fish Department, Arizona Department of Transportation, conservation groups, federal land management agencies, and university scientists. In late 2006, the work group published a statewide assessment of wildlife corridors, or linkages, that were at risk. The publication used maps to document the wildlife corridors and linkages of concern. Beyond merely providing a visual depiction of such corridors, the publication effectively and importantly placed the "natural, native linkages for wildlife" on the agenda of many agencies, most notably transportation planners and county land-use planners. The workgroup is developing conservation plans—specific focal actions recommended to improve and protect areas as biological pathways—for sixteen potential corridors that have high biological importance and are at risk of loss to urbanization or highway projects. By reducing wildlife-vehicle collisions, such planning also promotes the safety of both humans and wildlife.

Cougars can provide an important biological tool that helps organizations plan for the connections that will protect the biota of human-dominated areas; and they are best used in concert with other focal species (see Chapter 12). Connectivity is increasingly recognized as important. For example, the Western Governors' Association is encouraging all western states to develop connectivity maps and adopt strategies to protect important wildlife (or habitat) corridors from energy extraction, urbanization, and transportation projects (Western Governors' Association 2008).

These collaborative efforts by agencies, citizens, biologists, and conservation groups are providing vital solutions to the loss and fragmentation of wildlife habitat. While there are likely other examples of collaborative conservation efforts occurring within the cougar's range, those mentioned are exemplary models of what can be done when multiple stakeholders with diverse backgrounds and expertise work together to address some of the most serious threats facing cougar populations today. These projects require vision, commitment, and resources, from all concerned. Our hope is for more such efforts to occur with even greater success in critical areas across the cougar's range.

A final point bears mentioning: the lack of information about cougars, especially for those who live and recreate in cougar country. This may be one of the most serious deficiencies in cougar conservation efforts. In so many ways, education paves the way for on-the-ground success in protecting both cougars and people. As Laundré and Hernández (Chapter 6) point out, for cougar conservation to succeed in Latin America, it must include education of local people, an arena in which agencies and citizen groups have worked together to great benefit. If applied throughout cougar range, in an inclusive manner and with diverse stakeholders, progressive outreach and education can multiply the potential for success and build a strong base of public support for cougar conservation.

Conclusion

The history of large cats may tell us enough to answer the question of what we should be doing today to ensure that cougars are living on the landscape one hundred years from now. Cougars are resilient, but we know that the cumulative effects of human impacts can reduce them to ecological insignificance, or even extinction.

Science has documented the critical roles the cougar plays in maintaining healthy ecosystems (Ray et al. 2005). We know that by protecting this important predator, we also secure habitat for a wide range of other species within its range. We also have evidence that when we eliminate the cougar, it can have devastating impacts on other species (Terborgh et al. 1999).

These are reasons enough to secure the cougars' future. And, putting adequate conservation measures in place will help make it so. There is no time to lose. As human population continues to escalate throughout many parts of the cougar's range, the prognosis is particularly troubling for large carnivores. The fact that these carnivores occupy large home ranges, need to disperse long distances, and are vulnerable to the cumulative effects of hunting, habitat fragmentation, road kills, disease, and other impacts, makes this species more vulnerable to extinction than smaller carnivores, regardless of population size (Gittleman et al. 2001, 417).

Still, we see signs of hope. The spectrum of opinion about cougars has probably never been wider than it is today. Yet the programs described in this chapter demonstrate that citizen-government collaborations can begin to address the myriad threats facing this large carnivore. Armed with a thorough understanding of the problem and a determination to identify and implement the most effective strategies, government agencies, scientists, and citizen advocates are devising conservation measures to ensure that cougars maintain their ecological significance on the landscape and their socially important role in our culture.

The task is far from finished. To realize fully the vision of cougars surviving far into the future in healthy, functioning populations, securely established in their native habitat, much more work is needed. For the agencies and citizens with a growing role in cougar conservation, we offer the following six principles, providing additional insights and essential mechanisms for success.

Adopt a proactive, long-term approach to conservation to avoid perpetual crisis management. A redefinition of conservation is a crucial step in achieving our goals. We need to shift to a long-range, broad-based ecosystem approach that ensures the cougar's long-term survival. The enormous and costly effort to save the Florida panther is a sobering reminder that it is wiser to implement conservation measures before a species is seriously threatened (Nowell and Jackson 1996). The shift from a policy of eradication to one of maximum sustained yield was a major reorientation in cougar management; so is the current shift toward viewing the cougar within the larger context of carnivore conservation as a key player in biological connectivity, biodiversity conservation, and ecological resiliency in the face of global warming. This more mature mindset can predict and prevent future crises and thus maximize the benefits achieved with limited financial and staff resources.

Use an interdisciplinary and multi-stakeholder approach. Effective conservation is a dynamic and inclusive process that promotes communication among diverse stakeholders, fosters common goals in natural resource management, and counters the stagnation that sometimes grips management priorities. Combinations of expertise can bring about creative solutions. Interdisciplinary work is prevalent in urban design, with inclusion of social scientists and psychologists as well as planners and transportation experts; it is prevalent in medicine, with inclusion of various specialists to help cure disease. For conservation to be successful, urban and landscape planners, habitat specialists, ecologists, social

scientists, citizens, and wildlife managers must all be seen as key allies. Just as cougars need diverse and intact ecosystems to survive, they also need diverse stakeholders working together toward a common purpose. Citizen and advocate groups need to cooperate with one another and with government agencies to increase research funds, map habitat, and advance the overall goal of protecting cougars and their prey. Agencies need to embrace broad mandates, anticipate threats, diversify the funding base by accepting nongovernmental funds from conservation interest groups, work with new conservation constituents, and provide leadership that allows for diverse public opinions. In short, a multitude of interests need to come together and develop a shared vision for ensuring that cougars are on the landscape a hundred years from now.

Maintain the ecological role of the cougar. The ultimate goal in wildlife conservation is to maintain all components of an ecosystem; the loss of any component decreases the health of the whole. Cougars embody one of the most important driving forces in nature: predation. In many ecosystems, they are the only extant large carnivore. Cougars influence ecological communities as top-down regulators and serve as a focal species for habitat protection (Laundré and Clark 2003). Our work must be guided by a clear understanding of this scientific basis for conservation. However, science cannot secure a species; people must accomplish that.

Incorporate cumulative impacts into management strategies. Cougars are not biological islands. As top predators, they are integral parts of the ecosystems they inhabit. Effective conservation requires addressing the *cumulative impacts* of anthropogenic changes. It is futile to study the impacts of hunting while ignoring the impacts of habitat loss. Understanding the ecology of cougars is important, but so is understanding the impacts of hunting, highway design, livestock husbandry, human population growth, deer ecology, genetics, biopolitics, and human attitudes toward predators.

Acknowledge the intrinsic value of cougars. Management of large carnivores has been driven by the need to reduce threats to human property and safety, and more recently by regulating the sport harvest and reducing predation on species of concern (e.g., desert bighorn sheep). Modern conservationists and resource managers have a more comprehensive appreciation of natural systems and value the cougar for its intrinsic and aesthetic values. Teddy Roosevelt was a hunter with utilitarian views of wildlife, yet more than a century ago he advocated aesthetic reasons for conserving species. In an 1899 letter, he wrote: "How immensely it would add to our forests if the great Logcock [ivory-billed woodpecker] were still found among them! The destruction of the Wild Pigeon [passenger pigeon] and the Carolina Paroquest [Carolina parakeet] has meant a loss as severe as if the Catskills or the Palisades were taken away. When I hear of the destruction of a species, I feel just as if the works of some great writer had perished" (Reiger 2001, 69). Perceptive naturalists like Barry Lopez recognize that by cutting ourselves off from nature, by turning nature into scenery and commodities, we lose something vital. The idea that animals can hold meaning and have intrinsic value is a common belief around the world. Only in recent times has the modern world, with its focus on science as the "arbitrator of truth," weakened this notion (Lopez 1998). One of our greatest challenges in today's world is to embrace the notion that animals have value in their own right, and that their future is interwoven with our own.

Respect and build on the successful work of others. A broad array of perspectives and approaches can best advance the conservation of the cougar. In the last forty years, conservationists of every stripe have expanded the debate and drawn attention to pressing issues (Bocking 2005). Resource managers practicing sustainable management and applying sound science are practicing conserva-

tion; and advocates of resource preservation are also making their contribution to the species. Wildlife and land management agencies should recognize and encourage these different contributions just as all stakeholders should recognize the progress that people with different values contribute. For conservation to succeed, civility and respect must prevail among affected parties. We need to build on each one another's successes and applaud each step forward, despite our philosophical differences.

There is wide public support for retaining cougars in most of North America (Nowell and Jackson 1996, 135) as important components—even symbols—of our natural heritage and wild landscapes (Kruuk 2002, 226). This support, added to our understanding of the ecological role of the cougar, should motivate us to move beyond our differences and work together to assure that the cougar will be part of the landscape one hundred years from now. It will not be one agency, organization, or citizen who will ensure this is so, but our collective efforts that will best provide the foundation for success.

Chapter 16 Pressing Business

Maurice Hornocker

I WAS AN undergraduate at the University of Montana when I saw my first cougar. The slender, beautiful female was very dead, hanging head down from the eaves of the Montana Fish and Game headquarters building in Missoula. She had been killed and delivered to Fish and Game for the bounty the state awarded. Cougars were scarce in those days and her picture made the front page of the local newspaper. Fresh from an Iowa farm, I marveled at the magnificent cat and wondered about its life in the mountains west of Missoula. At the time, I had no inkling that I would spend much of the next four decades learning all I could about this remarkable carnivore.

The year was 1956. Cougars (*Puma concolor*) were regarded as vermin in those days in Montana, and bounties were paid to kill them. This bounty system, along with unlimited hunting, was very successful—cougar numbers were reduced drastically. Montana Fish and Game Commission files suggest that from 1932 to 1950, fewer than five cougars were taken annually statewide under the bounty system (see Riley et al. 2004). This pattern was typical across the United States. In the East, decades of persecution, along with habitat destruction, resulted in the loss of cougars. The Florida panther (*Puma concolor coryi*) was the exception only because of the inaccessible swamps of the Everglades. Despite those decades of persecution, cougars hung on in remote areas of the West and, when bounties ended and restrictions were placed on killing the cats, they increased in numbers. Montana has managed cougars as game animals since 1971; in 1998, 776 were legally killed by hunters, illustrating the dramatic increase statewide.

As we have seen in this book, cougar numbers have increased markedly throughout their range in western North America in the last forty years. Our knowledge of cougars has also increased. Back in 1956, about all we had in the scientific literature was Young and Goldman's classic book *The Puma: Mysterious American Cat* (1946a). A fund of knowledge has since accumulated from different regions of cougar range and scientific publications have flourished. But in no one work had all this new knowledge generated by professional biologists been brought together in detail.

This fact was the impetus for proposing *Cougar: Ecology and Conservation*. Sharon Negri and I envisioned a collected work that would be the "go to" source for information on this species, from evolution to biology and from ecology to cougar-human interactions, and we believe this has been achieved by the contributing authors. Several have also thoroughly outlined suggested research directions. As a result, there is little for me to add in wrapping up, but I do want to speculate and encourage research that I see as relevant and necessary in the future. I may repeat what has already been said to underscore a point, but some of what I propose has not been addressed. I discuss the issue of cougar-human conflict and propose research for a possible solution; I offer opinions on cougar genetics and behavior, and big game hunting. These are my opinions, and I recognize that they may differ from the views of some of the other authors expressed in this book.

The Early Work

My cougar research began in 1964 in the central Idaho wilderness and spanned almost ten years, involving both students and colleagues. It was the first intensive life-history study of cougars anywhere. We marked cougars individually, which had not been done before, and relied on capture-recapture methodology to gather data. Radiotelemetry, still in its infancy, was utilized for the latter half of the project.

We developed much new information and learned a great deal. As it turns out, we got some things right but also some things wrong. Subsequent research by others has supported our conclusions concerning social order—manifested in territorialism—in cougar populations. Territorialism, especially in males, is accepted as fact; its function in population dynamics is questioned by some. The concepts of mutual avoidance and transients, now commonly accepted, were first recognized by us. Our conclusions concerning the effects of predation on prey populations have generally been supported, but there are some situations where these conclusions no longer hold.

Our work focused on a well-established unhunted population. Territorial males did not aggressively defend their territories. We concluded that males peacefully maintained their territories because injuries sustained fighting conspecifics would be detrimental to survival. Work in different regions of the western United States and Canada has shown this is not the case in most populations—males fight to gain and hold territories. I further concluded that our self-regulated unexploited population was at maximum density for cougars—that cougars could not live in denser numbers. Evidence now shows that cougar densities may be altered by removal of territorial individuals. Removal disrupts the social system and may, in the short run, allow for more cougars in a given area. There are no "prior rights" established, and thus there is space for incoming transients. Until a new social order is established, more individuals can occupy a given area provided there are adequate food resources.

These findings highlight the fact that there are different "kinds" of cougar populations—those made up of individuals of long standing, which all know and recognize one another and one another's territories. Other populations, because of removal, exploitation, or recent reestablishment, are socially or behaviorally unstable because individuals are of recent origin and are still sorting things out. These populations can also contain more transients or part-time residents. As a result, densities of cougars can be temporarily higher. From a management and conservation perspective, it is of great importance to know what kind of population you are dealing with.

What Lies Ahead?

Following the bounty era, protective measures adopted in the 1960s and 1970s are most responsible for increased numbers of cougars throughout their range. Prey numbers—principally deer and elk—have likewise grown and have contributed not only to cougar population increases but also to the recolonization of many historic cougar ranges. Cougar populations may be at historic highs in parts of their range in North America. But the burgeoning human population threatens to diminish cougar habitat and fragment and isolate populations. Clearly more refined management will be required. Wildlife managers must begin to think "outside the box" in designing research and committing resources to long-term work. Where do we want to be by 2050?

The future for cougar populations in North America appears contradictory. On one front, cougars are continuing to repopulate long-vacated ranges. And they are moving eastward, attempting to occupy suitable habitat in states like the Dakotas, Nebraska, Iowa, and the prairie provinces in Canada (see Chapter 5). On the other hand, some long-standing cougar populations are threatened by human occupation and development (see Chapters 6, 12). Likewise, wildlife managers are challenged by sometimes conflicting objectives—conserving cougars and at the same time seeking to satisfy human needs and desires. Public attitudes enter the picture, making some former management practices (bounties, lethal control) no longer possible in many situations. New management approaches are required and research must provide the information necessary for success.

Today, a largely urban human population questions all killing of cougars (with occasional exceptions when a jogger on the urban periphery is killed by a cougar); but hunters still want to be able to kill more cougars. "Adaptive management" has been proposed to meet any situation (Cougar Management Guidelines Working Group 2005). In the real world, where human safety and interests are critical, this may mean the removal of a cougar population. At the same time, everything possible could be undertaken to conserve cougar populations in other environments, such as the vast wildlands of the West. In the future, all the concepts and scenarios critical to cougar conservation can be addressed by an adaptive landscape program: areas that serve as refuges and reserves, source populations and population sinks, linkages, corridors, and travel ways. Ultimately, human values will determine management objectives and the means used to achieve them (Cougar Management Guidelines Working Group 2005). And it is here that a sharp focus on public information can make the difference. Including an informed public from the start in management proposals can only enhance the success of any program. Cougar biology and behavior are engaging to many people, and information properly shared is crucial to acceptance of management programs.

The wildlife profession does not have a history of effective communication of scientific findings. It used to be enough to communicate just with ranchers and hunters in game department decision-making processes, but that is not enough any more. Now the whole nation is watching. Good communication is by its very nature informative rather than misleading. Making complicated issues personally meaningful will activate public support much more effectively than blinding people with science (Nisbet and Mooney 2007). The U.S. Congress, recognizing that this weakness pertains across the sciences, is considering a bill that would allocate funding to the National Science Foundation to train scientists to become better communicators. If wildlife scientists do not learn how to communicate effectively and win political arguments, change government policies, and influence public opinion, they will be ceding their ability to contribute to the future (Nisbet and Mooney, 2007).

Scientists have traditionally avoided entering the political arena. But politics are integral to our governing system, and policy affecting all wildlife is created within this system. I think it is imperative that scientists become involved in the process. This does not mean becoming a "politician." It means interacting with people, communicating research results, pointing out options, and working to influence policy as well as implement it. It means selling the program by interacting with all interested parties, presenting a good solid plan positively and aggressively. The late Les Pengelly put it this way: "One test of a professional is to act like one—this involves knowledge and integrity; it requires the courage to speak out on vital issues with no fear of consequences; it requires active participation in meaningful activities. And it requires communication" (L. Pengelly, pers. comm., 1980).

Cougars and Big Game

Predation on big game animals has been a principal reason for controlling cougar numbers. Hunters have traditionally believed fewer cougars mean more deer or elk. We now know this is not always the case and that other factors in the environment are important in these predator-prey interactions (Chapters 10, 11). Cougar-ungulate management strategies must consider all factors operating in individual situations, whatever the management objectives. Theory developed over time by rigorous research in different situations and different regions must be applied in management programs. And this management should be flexible so that it can be adjusted in response to changed environmental conditions and subsequent shifts in ungulate and cougar numbers.

I first called for concurrent long-term research on cougars and their prey in 1971 (Hornocker 1971). The need still exists. Experiments, including manipulating both predator and prey populations, are required to define each situation

more clearly. Sterilization of cougars and subsequent lowered reproduction (discussed below) should be considered where predation on small or valuable prey populations is a problem. This research must be long term—ten years at a minimum—but ideally encompassing generational cycles of both cougar and prey.

Understanding the relationship of cougars to economically valuable big game populations is of great importance. This is where wildlife managers receive much social and political pressure to "do something" about cougar predation. That is why it is critical to resolve any issues on what limits cougar numbers. If food alone limits, then the public can rightfully assume that more deer in a given area will mean more cougars. If the objective is more deer for hunters, the call for cougar reduction is valid. If cougars are shown to be capable of limiting their own numbers in a given area, then calls for cougar reduction can be validly countered.

Control of cougar numbers is still a major issue where livestock are threatened. Targeting offending individuals for removal has proven more effective than blanket control (Cougar Management Guidelines Working Group 2005). But we have also seen that removal can sometimes exacerbate the problem. The arguments for sterilization apply here, as well—stabilizing a cougar population behaviorally and reducing reproduction should lessen the likelihood of predation on livestock.

Self-Regulation through Territoriality

Basically there are two theories about what limits cougar numbers. One holds that cougars are capable of limiting their numbers through a system of strict territorialism. In other words, cougars are capable of self-limitation. The other theory holds that food alone ultimately limits population size; territorialism is recognized but is not considered important in population regulation (see Pierce 2000b; Logan and Sweanor 2001). Population limitation is discussed by Quigley and Hornocker in Chapter 5, and later in this chapter.

I believe space is important in self-regulation of cougar numbers. Males simply require a certain amount of space regardless of the food supply. Support for this and the competition theory—that territorialism acts in a self-regulating manner and automatically assumes a self-limiting function—is provided by Quigley et al. (1989; Hornocker 1996). An Idaho population originally studied by Hornocker (1969, 1970) and Seidensticker et al. (1973) was investigated again fifteen years later. The prey population—primarily elk—had increased by 100 percent. The resident cougar population had increased also, from six females in the original study to ten. But the resident male population remained at three on precisely the same territories as in the original study, despite the major increase in food availability. These data further suggest that females also require a certain amount of space. Six new females occupied the same home areas occupied by females fifteen years earlier. All four additional new females were on ranges upstream and apart from the original six home areas.

Recent work further supports the competition theory. De Azevedo and Murray (2007, 1) found that territoriality was the limiting factor in a jaguar population in Brazil. They concluded that "spacing patterns in the local jaguar population were likely based on exclusion through territoriality rather than food limitation." Laundré and Hernández (2003) reject the reproductive strategies hypothesis utilizing data from different research efforts. They contend that all the data presented by Logan and Sweanor (2001) concur with Sweanor's (1990) original conclusion to reject the reproductive strategies hypothesis and provide strong support for the competition model (see Logan et al. 1996). Laundre and Hernandez (2003, 158) further state that "the avoidance of inbreeding is very likely not the driving force behind dispersal in young male pumas, but rather, likely is a serendipitous consequence of dispersal driven by competitive interactions."

Ruth documented philopatry in two males in Yellowstone National Park (T. Ruth, pers. comm., 2007). As adults, both established territories in their natal areas. Such observations of male philopatry support the competition hypothesis and reject the reproductive strategies hypothesis which holds that young males must leave the natal area in order to avoid inbreeding.

Further, current and ongoing research in Washington State supports the contention that tenured territorial males literally hold local populations in check (G.Koehler, pers. comm. 2008)

Summing up, all species seek to maximize reproduction. But all species must first be established in a situation favorable to reproductive effort. Margaret Nice, credited with first recognizing territoriality and its function, stated that all animal life needs nutrition, protection (shelter), and reproduction (Nice 1937). I might add, "in that order." I believe evidence is strong that male cougars establish territories first for food and shelter and that reproductive opportunity is a bonus for a successful territory holder; or, as Laundré and Hernández (2003,158) put it, reproductive opportunity can be a "serendipitous consequence" of competitive interactions. I have also speculated that female cougars must first establish site fidelity before breeding. I am unaware of any evidence that transient females will breed prior to establishing a home area. Familiarity with a home area would be a big advantage to a female in providing for and

rearing offspring. This is further support for the theory that nutrition and shelter must precede reproductive effort.

From a conservation standpoint, it is tremendously important to recognize that cougars are capable of limiting their own numbers. It is crucial to wildlife managers faced with decisions concerning either enhancement or reduction of a cougar population. If food alone limits cougar numbers, then it is fair to assume cougar numbers would increase unchecked in a situation where food was unlimited. Cougars would presumably pack themselves into smaller and smaller home ranges, with resultant overcrowding consequences, but they do not. To my knowledge, this has not been documented anywhere with cougars.

On the contrary, territorialism, with its exclusivity and inherent limiting function, has been documented in every credible study reported. Only in populations that have been heavily exploited or disturbed in some major environmental manner (fires, floods, etc.) are males nonterritorial; they simply have not had time to establish a territory. Further, no study yet supports the contention that an endless supply of food will result in an unlimited rise in cougar numbers in a given area. Certainly food limits some cougar populations, but not all. It seems clear that if food does not limit cougar numbers—in situations where food is unlimited—then something else must do the limiting. I believe numerous studies have shown behavior in the form of territorialism to be that self-limiting mechanism.

Conflicts with People

Human-cougar conflict, real or imagined, will always be with us. The same is true of any big carnivore anywhere in the world. How to meet this challenge is the issue. I believe this is the single biggest issue in cougar conservation in the near future. Loss of habitat looms large, but if we do not resolve human-cougar conflicts it loses significance—cougars could be relegated to parks, reserves, and the "inhospitable" habitats where they survived persecution during the first half of the twentieth century (see Kruuk 2002). A good example involves the Florida panther. Much work has been done to develop a linkage between suitable habitat in southern Florida and north Florida/south Georgia. But until attitudes are changed and panther-human conflicts addressed, the plan will never be realized.

Another example of what is coming happened in 2004 in the southern Iowa county where I grew up. A young male cougar, on a fresh deer kill, was killed by hunters (cougars are not legally protected in Iowa). This sparked a spirited debate in the local media. Approximately half of local citizens supported the destruction of this "dangerous" predator; the other half decried its killing as barbaric and senseless. This scenario will be repeated many times as cougars repopulate former midwestern habitats now teeming with white-tailed deer. And midwestern states will soon need to address whether and where they want cougars and how reestablished populations will be managed.

As the human population increases—projections are for one hundred million more people in the United States by midcentury—and human-cougar interactions become more frequent, there will be calls to remove or at least reduce cougar numbers. In these areas of conflict, the challenge is to accomplish two important objectives: public acceptance and reducing—or at least holding stable—the cougar numbers in local populations. Most of the human population increase will be in urban areas. Traditionally, urban dwellers have been more tolerant of big carnivores than rural people, perhaps understandably, and will call for measures other than lethal control in any management program. And most urban people will support the conservation—even enhancement—of cougar populations in nonconflict areas. The challenge is finding a culturally acceptable and biologically sound "don't kill the cougars" approach but at the same time achieving a "we don't want too many cougars" solution to maintaining behavioral and numerical stability.

Stable cougar populations are made up of mature males, mature females, offspring dependent on their mothers, and occasional transients. Stability in the population is brought about by strict territoriality exhibited by the males. And those males and females that live the longest exhibit the greatest reproductive output. Transients—usually younger individuals—are discouraged by these territorial males and, if not driven out or sometimes killed, they move on. Unexploited cougar populations of long standing all appear similar in this regard, regardless of locale or different habitat types.

Maintenance of these behaviorally stable populations appears to be key to solving different problems. Human-cougar encounters most often involve young, inexperienced cougars (Aune 1991; Beier 1991). Managing cougar populations to discourage the influx of young individuals should reduce conflict. Therefore, any measure that could achieve this should be considered. Removing or killing resident adult individuals may have just the opposite of the desired effect. This action opens up territories to newcomers. The newcomers most often are those young, inexperienced cougars that are more likely to clash with humans.

Evidence for this is provided by two projects in Washington state. The Washington Department of Fish and Wildlife and Washington State University have collaborated since 2002 on research on two separate cougar populations subject to different management strategies. Their objective is to compare cougar demographics, spatial patterns, and habitat use for populations subjected to very different hunting regulations and harvest rates. Regulations permit hunting with hounds in northeastern Washington but not in central Washington. This results in light hunting pressure in the central area (no hound hunting) and much heavier pressure in the northeast (hound hunting). Comparison of results shows lower densities of cougars, older average age, greater survival rates for resident animals, and higher kitten survival for the lightly hunted central population versus the more liberally hunted northeast populations. The central area also has far fewer complaints of cougar activity than does the northeast area (G. Koehler, pers. comm., 2008).

These comparisons suggest that light hunting pressure in the central area has resulted in fewer complaints and cougar-human conflicts. The heavy harvest rate in the northeast area may result in population instability and a population made up of inexperienced young cougars that may be more prone to conflicts with humans. The data suggest that a more liberal cougar hunting strategy may only exacerbate cougar-human conflicts (G. Koehler, pers. comm., 2008).

This trend has already occurred in Oregon. In the last thirteen years, the Oregon Department of Fish and Wildlife has greatly expanded the hunting of lions. In 1993, the year before voters banned the hunting of cougars with hounds, hunters killed 160 cougars, an additional 27 were killed as problem animals, and the state received 276 cougar-related complaints from residents. In 2006, hunters shot 284 cougars, whereas another 128 were killed for their potential threat, and officials heard 443 complaints (Morell 2007). Ironically, instead of the apparently intuitive conclusion that more hunting should reduce problem animal kills and complaints, the reverse applies.

Ruth and Murphy address this issue: "It should be abundantly clear that killing cougars is often a simplistic approach at best. Whether the impulse is experimental design or answering to public pressure, removing cougars is a tool that has severe limitations unless we have a much fuller understanding of the other interacting factors and forces" (T. Ruth and K. Murphy, pers. comm., 2008).

Sterilization—A Solution?

Surgical sterilization holds promise as a culturally acceptable and biologically sound solution. Adult males instill order in a given area through strict territorialism. They keep the lid on the population. If individuals remained territorial after being surgically sterilized, the benefits from a human conflict standpoint would be twofold: behavioral and numerical.

1. Behavioral stability in the population would be maintained—inexperienced newcomers would be excluded.
2. Since females usually breed with the resident territorial male, reproduction would be diminished; the populations would be numerically stabilized.

In my experience, there is anecdotal evidence that territorial behavior persists in sterilized cats. I hand reared a castrated bobcat male (*Lynx rufus*) that remained fiercely territorial; three different domestic Manx males (*Felis cattus*), living in three different locales, were highly territorial after being surgically sterilized. Ewer (1973) and Liberg (1980) report territoriality in feral male house cats, and Walker and colleagues (1983, 1068) state: "There is no reason to think that its behavior and ecology under noncaptive conditions differ greatly from what has been found for *F. silvestris* [the native wildcat]." Leyhausen (1979) stated that behavior in all species of wild and domestic cats is remarkably similar.

Belden and McCown (1996) released nineteen mountain lions from Texas into Florida as surrogates for evaluating the feasibility of reintroducing Florida panthers into unoccupied areas of their historic range. These nineteen mountain lions included eight vasectomized males. "The lions established home ranges, killed large prey at a predicted frequency, and settled into routine movement and feeding patterns" (Belden and Hagedorn 1993b, 388).

Orford et al. (1988) treated thirteen free-ranging lionesses (*Panthera leo*) with a chemical contraceptive in Etosha park, Nambia. The treated lionesses were observed for changes in behavior, birth rate, and mortality. The contraceptives prevented pregnancy, were reversible when removed, and did not alter lion behavior significantly, except that sexual behavior was not recorded in treated lionesses.

Sterilization of canids—wolves (*Canis lupus*) and coyotes (*Canis latrans*)—offers some interesting perspectives. Alaska Department of Fish and Game biologists surgically sterilized the alpha pairs in fifteen wolf packs from 1997 through 2000. The packs continued to function as packs and continued to defend and maintain their territories for as long as seven years (Boertje and Gardner 2000). Surgical procedures were safe and did not change wolf social behavior (Spence et al. 1999). Hayes et al. (2003, 30) state: "We recommend that fertility control be considered whenever there is a management plan to hold wolf densities below natural levels for sustained periods."

Utah State University biologists surgically sterilized fifteen different coyote packs over a three-year period from 1998 through 2000. They found that, behaviorally, sterile packs appeared to be no different than intact packs. They concluded that surgical sterilization did not modify the territorial or affiliative behavior of free-ranging coyotes, and

therefore sterile coyotes could be used as a management tool to exclude other potential sheep-killing coyotes (Bromley and Gese 2001). Notable points in both the wolf and coyote studies were that longevity was increased in the sterilized individuals and that the programs were cost effective in meeting objectives of the research—reducing depredation on caribou and domestic sheep.

The results of these experiments confirm the advantages of a sterilization effort proposed by Balser for coyotes in 1964:

1. It may be more practical to prevent animals from being born than to reduce their numbers after they are partially or fully grown and established in a secure environment.
2. Increasing one or more mortality factors often results in a compensating increase in reproduction or survival or both. This reduces the effectiveness of any control program. By suppressing reproduction, the compensating increase in reproduction may be overcome, while survival may be increased in the remnant population.
3. Movement or ingress which occurs when animals are removed from a population may be lessened by occupation of territories by sterilized adult coyotes.
4. Nontoxic antifertility agents are safer to use than existing lethal agents and devices and likely would be more readily accepted by the public. This could result in more effective population control in areas where the use of lethal techniques is now restricted.

If surgical sterilization in cougars has the same effect as that in wolves and coyotes, the practical benefits are obvious. Behaviorally stable cougar populations made up of experienced, nontroublesome individuals would mean fewer cougar-human encounters. Increased longevity of individuals would mean population stability was maintained and extended, and young cougars were kept out. Reduced reproduction would mean fewer young, inexperienced individuals in the population.

As a species, the cougar lends itself ideally to a sterilization program, and meets all of the criteria Caughley et al. (1992) list as essential for a successful program. Cougars occur in relatively low numbers—all individuals in a local population can be captured by experienced personnel.

Logistics are not a problem. Researchers in every intensive cougar study I am aware of know every individual in a local population. These individuals, wearing radio transmitters, are routinely recaptured for re-instrumenting, for physical inspections, etc. Finally, costs of a sterilization program would be minimal.

Cougars are solitary, with a well-defined social structure. Likewise, the breeding system is simplified, involving only individuals; there is no pack or colony structure to complicate results. The surgery is straightforward, easily performed in the field, and carries no ill side-effects, as can some forms of chemical contraception. The treatment is permanent—no expensive and time-consuming retreatment is required. Cougars may be closely monitored with time-proven radiotelemetry techniques. Global positioning systems (GPS) technology further refines the monitoring effort. Finally, results should be apparent in a relatively short time.

There are, however, emotional and political issues concerning contraception of any kind in wildlife populations (see Kirkpatrick 1997, 2005). Some people will enthusiastically support contraception in humans but oppose application to wild species for emotional reasons. Public and political pressure blocked a proposal to sterilize nuisance dominant male rhesus monkeys (*Macaca mulatta*) in Delhi, India (Malik 2001). Hunters have opposed the technique, fearing widespread application and a reduction in hunting opportunity. The need for a vigorous positive education program is clear.

Some progress has been made in convincing the public of the benefits of contraception. Rowan (2005) believes we now stand at the brink of a revolution in wildlife management where we no longer need to focus on lethal techniques to address situations in which humans and wildlife come into conflict. He was referring to immunocontraception as a technique to control wildlife fertility and population growth, but it also applies to surgical sterilization. Gill (1996b, 265) identifies the areas of conflict and differing ideologies. He points out that currently wildlife agencies respond to many policy challenges reactively and defensively in an attempt to preserve their past. He further states: "If we are to reach a productive compromise over the issue of if, how, when, and where to use wildlife contraception, the wildlife policy decision process must be visionary, wise, bold, accessible, adaptable, and most of all, fair."

The Wildlife Society, the most prominent organization representing professional wildlife biologists, has prepared a formal position statement on wildlife fertility control. This statement urges consideration of different fertility control techniques in different problem situations (The Wildlife Society 2008). The Humane Society of the United States, National Park Service, Bureau of Land Management, and U.S. Fish and Wildlife Service have endorsed the concept of wildlife contraception in some form. Colorado State University is involved in contraceptive research, and the Science and Conservation Center in Billings, Montana, is a leader in the field.

Emotional and political issues will persist with any sterilization program, but as positive results emerge from emotion-evoking species, like wild horses and elephants, where objectives have been met with no behavioral change, the public will become more accepting (Delsink et al. 2004; Kirkpatrick 2005). An example is the Alaska experiment with wolves, where sterilization was more publicly acceptable than killing wolves (Hayes et al. 2003). This is especially applicable since the alternative is killing or removing individuals. And we have seen that with cougars, removal by any means may create rather than solve human-cougar conflicts.

I emphasize that this technique should be applied only to local, defined populations, and only after an assessment of the surrounding regional population—the source of new replacement individuals as tenured cougars die. Examples of hypothetical situations could be cougar populations living in close proximity to urban subdivisions or a population occupying an isolated mountain range that also harbors an endangered prey population.

The objective of a sterilization program is not to compromise the wildness inherent in the species and manifested in wildland populations. But there are, as we have seen, different kinds of cougar populations. The objective is to intensify management of local populations in special problem situations, some of which involve cougars being subjected to the sights and sounds of human activity on a daily basis. Solving these problems means that management of truly wild-living populations may be minimized. Further, limited application of the program would not alter or endanger the genetic fitness of the regional population.

Theoretical Implications of Sterilization

Sterilization has much promise from the theoretical, as well as the practical, standpoint. The fact that male cougars are territorial is undisputed. Territorialism, by whatever definition, involves a limiting function. Other functions, such as reproduction, may be enhanced, but the fact remains that territorialism (and its exclusivity) is an automatic, built-in population-limiting mechanism. It may not be the ultimate limiting factor in all situations, but the very fact that other individuals are excluded qualifies territorialism as a bona fide self-limiting mechanism.

Why and for what purpose territories are established is questioned. One theory holds that male territories are, first and foremost, the result of competition for food and space. Males defend an area that provides food and cover to meet basic requirements for survival and secondarily supports reproduction. A second theory holds to the belief that sexual opportunity and reproduction are the primary underlying reasons. This theory also discounts the self-limiting theory, placing population regulation on food; that is, as prey populations continue to increase, cougar populations respond and cougar density is limited by food (see Chapter 8). As noted, the competition theory holds that territorialism, established for whatever reason, acts in a self-regulating manner; territorialism automatically assumes a self-limiting function. Only so many males may inhabit a given area, even if there is an overabundance of food. Space becomes limiting. Castrated males could shed light on this issue—if food and space are the inherent driving forces responsible for territorialism, then castrated males should remain territorial.

Females do not exhibit territorialism in a strict sense but do appear to require a minimum amount of space. Mutual avoidance behavior is utilized to limit home range crowding (Logan and Sweanor 1994).

Male Sterilization

Males could be either castrated or vasectomized, depending on the research objectives. Castration would end all male sex hormone production. Vasectomizing a male would render him sterile, but hormonal activity would continue, that is, the individual would behave "normally." If males no longer exhibit territorial behavior after being castrated, then the reproductive strategies hypothesis is supported; if males remain territorial after castration, this indicates the territorial imperative is stronger than the reproductive imperative. It lends support to the theory that competition for food and space is first and foremost.

Female Sterilization

From a practical and management standpoint, sterilization of females would contribute to numerical stability—reproduction would cease. From a theoretical standpoint, there are several considerations:

- Would males establish territories in the absence of reproductively viable females? *If the reproductive strategies hypothesis is valid, the answer is no.*
- Would males already resident and territorial abandon their territories if reproductively viable females were sterilized? *If the reproductive strategies hypothesis is valid, the answer is yes.*
- Carrying this further, if females were physically removed from an area, would males establish territories? *No.*
- Would already resident and territorial males abandon their territories? *Yes.*
- Do dispersing young males establish only where females occur? *Yes.*
- Must males await the usually later-dispersing females before they establish in new areas? *Yes.*

Definitive answers to all these questions could be obtained in a rigorously designed research effort utilizing surgical sterilization.

Emerging Perspectives

Some of the advances in our knowledge arising from recent work and presented in this book give rise to new questions about cougars, new options for their management, and potential new alternatives for addressing conflicts between people and cougars. The following observations indicate the expanding range of possibilities.

Genetics and Cougar Behavior

Are regional populations different genetically? Certainly the phenotype differs, for example, from Panama to Vancouver Island. Do regional populations exhibit different behavior? Are some populations more likely to become habituated to humans? Are some populations more likely to become aggressive toward humans? Vancouver Island cougars do look different from central Rocky Mountain cougars, and records indicate they also are far more aggressive toward humans. Some 60 percent of all actual physical encounters between cougars and humans in North America have occurred on Vancouver Island (Beier 1991).

If the Vancouver Islands cougars are, indeed, more aggressive due to genetics, would it not be possible to introduce individuals from kinder, gentler populations to hot spots like Vancouver Island? Removing territorial males from a population with an aggressive history and replacing them with males from a documented less aggressive population should bring about favorable behavior change.

There is evidence such behavioral change can occur and is genetically linked. Russian geneticist Dmitry Belyaev, addressing domestication in species, believed that the patterns of changes observed in domesticated animals resulted from genetic changes that occurred in the course of selection (Trut 1999). He believed that the key factor selected for was not size or reproduction but behavior, specifically, amenability to domestication or "tamability." Belyaev designed a selective-breeding program to reproduce a single major factor, a strong selection pressure for tamability in the silver fox (*Vulpes vulpes*). Today, through genetic selection alone, the research group has created a population of tame foxes fundamentally different in temperament from their wild forebears (Trut 1999).

Iowa State University scientists, working with domestic cattle, have shown that disposition of calves can clearly impact feedlot performance with an even greater impact noted in carcass quality grade (Busby et al. 2006). They assigned a disposition score to each of 13,315 calves, ranking them from very docile to very aggressive, at eight Iowa feedlots. Their conclusion pertinent to our discussion is that disposition is a heritable trait that can be improved by proper culling strategies.

Other species are receiving attention because of obvious behavioral differences. U.S. Agricultural Research Service biologists are investigating Africanized honey bees and European honeybees in an attempt to determine if genetic differences are responsible for differences in their behavior. Africanized bees are more aggressive and defensive of their nests (Kaplan 2007).

Morphological problems have been resolved through genetic manipulation. Negative morphological features in the small endangered Florida cougar population were corrected by importing individuals of different genetic makeup from Texas. In a short period, after interbreeding, the negative features practically disappeared from the Florida population (see Chapter 12; Laud et al. 2002).

As we have seen in Chapter 3, the field of genetics is practically unlimited. The familial relationship of individuals in local and regional populations and its significance is just one of many new avenues of investigation and promises insights useful for conservation strategies.

Public Safety and Feline Learning

Public safety is an important concern (Torres 1997). Encounters and attacks have increased in the last thirty years (see Chapter 13; Aune 1991; Beier 1991; Torres et al. 1996) as both cougar and human numbers have increased. Young, inexperienced cougars are most often involved in attacks (Aune 1991; Beier 1991). Wildlife managers must respond not only to address the problem but also to maintain credibility with the public. Targeting problem individuals and taking swift, decisive action strengthens both public confidence and management strategy. Communicating that strategy coupled with a vigorous education program has proven successful (Torres 1997) as well as useful in gaining public support for population management.

The development of quick-response teams to deal with potentially dangerous nuisance individuals shows promise. Such a program has been very successful with Siberian tigers in Far Eastern Russia. Problem individuals are captured and tranquilized, removed promptly, and either released elsewhere or placed in an appropriate facility. This approach is also very favorable from a public relations standpoint (Hornocker 2007). Siberian tigers are endangered, and every effort is made to keep each individual alive. This is not an issue with cougars. Releasing problem cougars elsewhere often is just shifting the problem elsewhere. If no release area is suitable, then the individual should be humanely euthanized.

We need to design research to enable us to condition wild cougars to avoid humans and human activities. As cougars and humans more frequently come into contact, we need to learn more about how cougars think. We need to learn more about innate and learned behaviors and how we can utilize this knowledge to our advantage. Cougars are very adaptable and intelligent and should be susceptible to aversive conditioning. New and innovative techniques and methodology for studying animal intelligence and learning capacity need to be implemented. An example is the research carried out on dog deductive learning at the University of Vienna (Range et al. 2007). Results of these experiments suggest strongly that canines may be thinkers, not just imitators, and capable of making conscious decisions. The research has sparked renewed interest in research on dogs and dog behavior.

We need to apply the same innovative approach to cat behavior and genetics. While dog behavior has been modified in countless breeds to suit human desires and needs, domestic cat behavior has remained practically constant throughout feline breeds. "Cats are cats," and we should seek greater understanding of all cats' behavior in order to address conservation issues in the future. Knowledge of all species of cats' innate and learned behavior could possibly be the most important and powerful tool in the conservation of wild cats.

An example is killing technique in young cougars. Most biologists concur that the female, through demonstration, teaches her offspring stalking and killing technique. Is this simple imitation or is deduction—thinking—involved? Would it be possible to emulate the mother's teaching, whatever the process, to achieve our desired results?

There is an "orderliness" inherent in populations of all cat species. Certainly all those my colleagues and I have studied intensively—cougars, bobcats, leopards, jaguars, lynx, tigers, ocelots—have shown this social orderliness. All these species can adapt to different conditions in their respective environments, and all show some differences in their behavior. But all have developed social systems that are extremely important in their population dynamics. Knowledge of the behavior responsible for these social systems can be useful in the application of practical conservation measures (see Chapter 8).

Hunting—the Great Experiment

Sport or recreational hunting can meet different objectives: it can aid in control programs in some instances and possibly lessen effects of cougar depredations on livestock and pets; it can possibly reduce human-cougar encounters; it can help lower the numbers in local and regional cougar populations (Anderson and Lindzey 2005). Hunting can, however, have detrimental effects in certain situations. Removing tenured individuals from a population can disrupt the social structure and literally open the door to inexperienced individuals. These cougars have been shown to be more likely to be involved in encounters with humans. Population size can also increase, at least in the short term, because the territorial individuals have been removed, with resultant greater predation pressure on prey species. Removal of mature, territorial males may also have long term effects in an evolutionary sense. These males are presumably the fittest individuals in the population, and the loss of their genetic input may affect the fitness of future members of the population.

Hunting can severely reduce local and regional populations if not carefully monitored and controlled. In the 1990s, Montana responded to public pressure to reduce growing cougar numbers. The legal kill peaked in 1998 when 776 cougars were taken statewide. It dropped steadily after that; in 2006 only 287 were taken. The evidence indicates that hunting achieved the goal of depressing the population (J. Williams, Montana Fish, Wildlife and Parks, pers. comm., 2008). Currently, under stricter hunting regulations widely supported by the public, the population is rebuilding.

Wolves and Cougars

Ruth (2001, 2004a) has shown that wolves dominate cougars in Glacier and Yellowstone national parks. Wolves kill cougars, steal their kills, and have forced cougars to change their ranges. While cougars did not influence elk behavior to any extent or depress elk numbers prior to wolf reestablishment in Yellowstone (Murphy 1998), wolves have brought about change. Elk behavior and habitat use have changed, obviously in response to wolf predation (Creel and Winnie 2005). Evidence suggests elk numbers have declined (Smith et al. 2003; Smith and Ferguson 2006), but climate and hunter harvest may also be involved (Vucetich at al. 2005).

Wolves have become an important component of the ecosystem in the northern Rockies and their presence and influence must be considered in cougar management (see Chapter 11). Theory developed from research on both wolves and cougars is important in structuring management programs where both species occur. This is important regardless of the management objective—wolf or cougar conservation, prey population increase or decrease, or overall biodiversity conservation.

Cougars as Keystone or Umbrella Species

In Chapter 12, Beier thoroughly discussed the role cougars can play in shaping and influencing biodiversity. He points out that cougar conservation can assist in achieving other long-term conservation goals. The ecological role of cougars, including their ability to help dampen oscillations in prey populations, structure biological communities, and direct the evolution of their prey are all reasons why cougars should be conserved (Hornocker 1970; Logan et al. 1996). Moreover, they can be used to define minimum areas required to preserve ecologically intact ecosystems (Clark et al. 1996; Noss et al. 1996; Beier, Chapter 12, this volume). Ray and colleagues (2005) discuss in detail the role all large carnivores play in the conservation of diversity.

Miquelle and colleagues (2005) state that although large carnivore conservation may not be synonymous with biodiversity conservation, the charisma of the big carnivores, their large area requirements (related to prey requirements), and their plasticity in habitat requirements provide a mechanism for achieving other conservation objectives.

Urgency in Latin America

The vast wildlands in western North America will support cougars in the foreseeable future if we make the necessary commitments. But what about cougars in Latin America? It has been pointed out in this volume that five of the six genetically defined cougar subspecies occur in Central and South America (see Figure 3.4). We have learned that all countries in Latin America have many unknowns (see Chapter 6). The puma's status in Bolivia is listed by the Convention on International Trade in Endangered Species (CITES) as an Appendix II species (may become threatened without protection); it is considered endangered in Costa Rica, vulnerable in much of Brazil outside the Pantanal, and vulnerable in Colombia; Mexico lists the puma as a game species but classifies it as requiring special protection. With vulnerable, threatened, and endangered status declared in so many places, Laundré and Hernández (Chapter 6) recognize what should be done: "A tremendous amount of work still needs to be done to fill the gaps in understanding and geography. For several Latin American countries, we lack even the basic information on pumas. What is needed is a massive international effort to raise the state of knowledge about the species."

I could not agree more. Latin America has one of the fastest growing human populations—puma habitat is being altered and destroyed at an alarming rate. In the best interests of *Puma concolor* as a species and looking fifty years ahead, we need to take action now to plug some of the large knowledge gaps. In the 1940s, 1950s, and 1960s, Mexico was viewed as a place where wolves and even grizzlies might persist. They did not. This knowledge should lend urgency to our efforts.

We need to launch a massive international effort throughout Latin America, patterned, perhaps, after the Save the Tiger Fund (STF), a joint project of the National Fish and Wildlife Foundation and the ExxonMobil Corporation. Established in 1995 as a special initiative to support diverse tiger conservation projects, the STF had contributed more than $10 million by 2003 in support of 196 projects in thirteen of the fourteen Asian countries where tigers occur. The program is widely recognized as being extremely successful—regional populations in Far Eastern Russia and India have stabilized, poaching has been reduced in most tiger ranges, and worldwide awareness of the plight of the tiger has increased greatly (STF 2003).

Such an approach is possible in Latin America with pumas. The international scientific community will surely give its blessing and the support of major nongovernmental organizations is a certainty. An international corporate presence is evident in many countries, and some corporations have shown much interest in environmental matters.

An overall program should be developed recognizing that immediate needs are different in different countries. This program should borrow from STF's belief that solid research, planning, implementation of results on the ground, and education are the best hope for long-term conservation. The case can be made that unless such action is taken, it is only a matter of time before many regional populations of pumas are lost. While perhaps lacking the urgency of the tiger situation, the case of the puma is similar—there is no question that the burgeoning human population, development, and habitat loss will result in the demise of pumas in some countries in the not-too-distant future. The clock is ticking.

A Save the Puma initiative must recognize that conservation is a seamless fabric comprising biology, ecology, economics, politics, culture, and even religion. For conservation to truly work in any form, all factors should ideally be addressed positively. Such a program will require leadership and the efforts of people with vision and a "program" approach. Those involved will need people skills, political skills, and, because funding will necessarily come from the private sector, fund-raising expertise. All the above will also be under the umbrella of a passionate desire to save the species throughout its range.

This program should not exclude North America. While it is perhaps more urgent in Latin America from the standpoint of a basic knowledge base, we are at a crossroads in North America. Where do we want cougars? How many? How do we want to manage them? Are we willing to provide for them in midwestern and eastern North America? There are no answers—yet—to some of these questions. But we should seek answers, again through sound research, education, and communication.

Creativity in Conservation

The Cougar Management Guidelines (2005) and authors in this book cite habitat loss as the greatest long-term threat to cougars. This certainly is true in areas like California and Florida and some growing urban areas, but if short-term cougar-human conflicts are not resolved, habitat loss will not matter. Public pressure in some locales may dictate cougar removal, just as it did in the early to mid-twentieth century. Cougars survived only in remote, inhospitable areas. Certainly, we will maintain cougars in protected parks and preserves, but to maintain viable populations as they now exist will require some innovative thinking—and action—outside the box.

If cougars are to survive, the public must direct its management agencies to address all conservation issues. Those agencies should elevate cougars from a problem or "second tier" species in their overall wildlife management plan. Resources and personnel should be committed to meaningful long-term research and management efforts. If they are to be successful, those programs must address the human elements in addition to the biological and ecological concerns. All the economic, political, and cultural realities involved should be considered so that credibility can be established locally and regionally for any management strategy.

The future requires solutions, not conferences (see Bearzi 2007). Those solutions will come from bold and innovative research that seeks insight and understanding, not the support of lofty hypotheses that often narrow rather than broaden the investigator's curiosity (see Chitty 1996). Research should be designed to answer straightforward, meaningful questions with rigor but with a minimum of complexity. Nobel Prize winner Lord Rutherford's advice applies: "If your experiment needs statistics, you ought to have done a better experiment" (http://thinkexists.com).

The issue of hunting and control should be resolved. We have seen that hunting can exacerbate human-cougar problems. Light hunting or no hunting at all has been shown to be more effective in solving the problem. At the same time, there is evidence that regional populations can be severely depleted by overhunting. There is, after all, no argument that overhunting and bounties wiped out regional populations in the first half of the twentieth century. It is clear that any management strategy involving hunting or removal should involve a comprehensive monitoring effort to assess the plan's effectiveness and to assure that objectives are being met.

Cougars have made a remarkable comeback in the past forty years in North America. Compared to other big carnivores worldwide, cougars are a towering success story. And it is worth noting that the comeback took place while cougars were under state or provincial management, which generally included regulated hunting. Many species of big, potentially dangerous carnivores have been exterminated in much of their historic range worldwide. Those surviving in remnant populations have been relegated to postage-stamp-sized reserves, comparatively speaking. And as the human population grows, many of those protected areas will be overwhelmed. In order for these species, including cougars, to survive in meaningful numbers, an informed public must make its wishes known. It is here that science must take an active leadership role. Innovative, visionary, rigorous scientific research, involving not only biology and ecology but all of the cultural, political and economic factors as well, can provide the knowledge essential to the conservation of cougars. But to be constructive, this information must be communicated effectively to the public. An informed and motivated public can change government policies, influence political arguments, and literally alter the course of conservation. This is our charge.

Conclusion

The authors contributing to this book have done their part to provide that knowledge essential to cougar conservation. From the philosophical to the purely biological, from the evolutionary to the cultural, the knowledge these authors have shared reflects their collective wisdom garnered over many years, and Chapters 14 and 15 address the interest and investment of new players and the broader public. Sharon Negri and Howard Quigley have articulated what must be done to ensure that cougars continue to enhance our landscapes indefinitely. They paint a picture of hope.

And there is, indeed, hope for cougars. The peregrine falcon (*Falco peregrinus*) and bald eagle (*Haliaeetus leucocephalus*), once facing extinction in the contiguous forty-eight states, have come back in force. Recovery programs for grizzly bears (*Ursus arctos*) and wolves have received overwhelming support from the general public. This support, both economic and political, has been driven by the cultural concerns of an informed public.

In contrast, the cougar's success in recent years has been largely on its own. With the exception of the endangered Florida panther program and California's Proposition 117, no special legislation, no special commissions, no special recovery programs have been initiated on the cougar's behalf. But as the human population increases and habitats become fragmented and lost, self-sustaining populations of cougars will become difficult to maintain. It will then become necessary to make the cultural, economic, and political commitments to cougars that we have made for grizzlies and wolves.

We should learn to appreciate cougars for the beautiful creatures they are, for the structuring role they play in nature, and we should strive to learn to live harmoniously with them. We must be willing to think and act outside the traditional box to meet sometimes contradictory objectives—promoting the welfare and safety of people while maintaining viable populations of cougars in our wildlands. In achieving this, we will do well to draw from older cultures that revered rather than feared big carnivores (see Arseniev 1941; Boomgaard 2001; Wessing 1986).

It will be worth the commitment, just as it has been with peregrines and eagles, and just as it has been with the successful ongoing grizzly and wolf programs. Thoreau's dictum "In wildness is the salvation of the world" is matched by Leopold's lament at the killing of the last grizzly in the White Mountains of Arizona: "Escudilla still hangs on the horizon, but when you see it you no longer think of bear. It's only a mountain now" (Leopold 1966, 145).

Wallace Stegner shared Leopold's empathy for the natural systems that sustain us as a cultured species. His broadside *Memo to the Mountain Lion* (1984) can help guide us in the future:

Once in every corner of the continent, your passing could prickle the stillness and bring every living thing to the alert. But even then you were more felt than seen. You were an imminence, a presence, a crying in the night, pug tracks in the dust of a trail. Solitary and shy, you lived beyond, always beyond. Your comings and goings defined the boundaries of the unpeopled. If seen at all, you were only a tawny glimpse flowing toward disappearance among the trees or along the ridges and ledges of your wilderness . . .
Controls we may need, what is called game management we may need, for we have engrossed the earth and must now play God to the other species. But deliberate war on any species, especially species of such evolved beauty and precise function, diminishes, endangers, and brutalizes us. If we cannot live in harmony with other forms of life, if we cannot control our hostility toward the earth and its creatures, how shall we ever learn to control our hostility toward each other?

Unlike my first experience of seeing a cougar fifty years ago, if a contemporary midwestern farm boy or girl and University of Montana student paid a visit in 2009 to the Missoula headquarters of the Montana Department of Fish, Wildlife and Parks, he or she wouldn't find a dead cougar hanging outside but, instead, well-prepared brochures on how to behave in cougar country. The student would find maps and hunting regulations defining different management units, and descriptive materials presented positively about this highly prized game animal. By inquiring, he or she could learn the logic behind the customized hunting regulations set for different regions of the state, including the model limited-entry hunts in the Kalispell area some one hundred miles to the north. Here, biologist Jim Williams and his crew have crafted regulations benefiting cougar conservation and at the same time providing limited hunting opportunity. Working closely with hunting groups and environmentally oriented organizations, he has shown that innovative approaches can be successful.

Outside the offices in Missoula and all across western Montana, the cougar is no longer viewed as the evil game-destroying predator it was once thought to be. Now common throughout the region, cougars are accepted by most people as an integral and even valuable component of a natural wild system. And most Montanans realize that such wild natural systems—with their native big carnivore populations intact—are being recognized worldwide for their ecological and cultural value (see Stolzenburg 2008).

Thoreau is smiling somewhere. Commitment and innovative workable management programs like that in Williams's area can ensure that in another fifty years cougars will still be a viable part of the Montana landscape.

Over his career, our hypothetical young biologist will see cougars become common in his native Midwest, and hear more and more reports of their occurrence in Appalachia and New England—the cougar's own great experiment. Cougars can survive and thrive in the future if we will it so. I think we will.

Appendix 1 Genetics Techniques Primer

Genetic diversity is measured using a variety of molecular genetic tools that allow a wealth of valuable high-resolution information to be obtained from individuals, populations, and species. This information has been employed in cougar studies for two decades now, beginning with the first genetic study of cougars in the mid-1980s. Molecular genetic tools examine DNA molecules, or the proteins coded for by DNA. The techniques profiled here have already been used in cougar studies; some definitions and procedures are given in the notes below.

Molecular Genetic Approaches

Molecular markers can be used to describe subdivisions (or taxonomy) within a species, which can be more reliable than taxonomy based on morphology alone. DNA molecules can be used to examine population boundaries, or provide evidence of migration/dispersal among populations, by assessing levels of gene flow among populations. In addition, genetic studies can be used to explain behavioral differences due to gene variation or relatedness/kinship among individuals. Genetic techniques have also been used as a forensic tool to estimate population features for unknown populations, and genetics of a virus infecting populations can be an effective tool in estimating ecological parameters for those populations.

Molecular genetic methods provide several advantages over morphological methods, although both are important tools. However, if an incorrect genetic marker or morpho-logical trait is selected, both can lead to anomalous results. Morphological methods measure the combination of genetics plus environmental effects, whereas genetic methods have an advantage because they tease apart the genetic information from the environmental effects. Another advantage of molecular markers is the ability to choose markers that are undergoing selection or not.

DNA sequence changes that are not under selection are said to be *neutral* (having no effect on physical characteristics and thus not under selection pressure) and rely on the assumption that nondeleterious genetic changes accumulate in a predictable manner over time. If populations do not interbreed for a sufficient length of time, these genetic changes will be able to differentiate those populations. Molecular markers that are neutral are important because they can estimate time elapsed since populations were founded or isolated, or they can supply an indication of the length of time that noninterbreeding populations have been separated.

Molecular genetics also allows us to select markers that are *independent* (not on the same chromosome). Independent markers are essential because they allow us easily to acquire several separate lines of evidence toward our conclusion; if most of the markers provide corroboration of results, then we can be confident that our conclusion is not biased by a single anomalous marker.

The molecular genetic techniques described here are limited to those relevant to studies of cougars. The techniques described involve electrophoresis of either a protein or a DNA fragment, a technique to separate the variants according to net electrical charge, molecular weight, pH, and/or structure.[1] Protein (allozyme) electrophoresis can detect half or fewer of the amino acid changes and does not detect any DNA sequence changes that do not alter the protein (so-called silent changes). DNA-based genetic markers—such as Restriction Fragment Length Polymorphism (RFLP), DNA fingerprint, or DNA sequencing—can potentially reveal changes that do not alter a protein and can examine DNA sequence that does not even code for a protein. The RFLP and DNA fingerprinting techniques detect much less than half the total sequence changes, whereas DNA sequencing detects all sequence changes.

Some techniques are designed to analyze a single region, or locus, of DNA (RFLP, DNA sequencing), whereas other techniques examine many loci at the same time (DNA fingerprinting). Polymerase Chain Reaction (PCR) is a technological advance in molecular genetics that allows easy, quick, and inexpensive amplification of a specific region of DNA across many individuals.[2] This technique thus makes it possible to analyze samples with poor quality and/or very low quantity of DNA (such as museum, forensic, or scat samples). All of the most recent genetic studies in cougars use PCR-based methods.

Phylogenetic reconstruction is a method used to estimate the evolutionary relationships among taxa (individuals or populations) being studied, whereby clustering of taxa is based on sharing a recent common ancestor. Neutral markers have another advantage in that resulting phylogenetic reconstruction is based solely on shared characters not confounded by natural selection. Often, a phylogenetic tree is generated from the molecular data collected, and all types of data discussed here can be used for phylogenetic reconstruction. Several classes of molecular genetic markers have been used in cougar studies, including proteins, mitochondrial DNA (mtDNA), nuclear DNA, and viral DNA. The next section provides details of each marker class.

Protein (Allozyme) Polymorphism

The first molecular genetic studies in cougars included electrophoresis of proteins (enzymes); and *allozyme* refers to all the different forms of the enzyme, due to a single locus, that can be resolved by starch gel electrophoresis (see Table 3.1). Polymorphism (having two or more different forms) is detected by comparing the mobility of the same enzyme among individuals. Compared to other felid species, cougars have moderate to high levels of allozyme polymorphism, with the exception of the Florida panther, which has low allozyme polymorphism (Roelke et al. 1993). Allozymes are useful for comparisons among species but are not always informative for studies of genetic variation within a species.

DNA Polymorphism

As mentioned, DNA-based genetic markers offer more powerful techniques, with the capability of revealing changes that do not alter proteins.

mtDNA. Because many regions of the mtDNA molecule have more rapid evolutionary rates than allozymes and, therefore, have higher levels of polymorphism, mtDNA has become more commonly used. This makes mtDNA more useful for examining genetic variation within a species while still useful for comparisons among species (Avise 1994). With very few exceptions, the mammalian mtDNA is composed of thirty-seven genes and a noncoding region called the control region. There are also hundreds of mtDNA molecules per cell, unlike nuclear DNA with one copy per cell, which greatly increases the ease of genotyping mtDNA.

The first studies of mtDNA used RFLP to identify polymorphisms. This involves using restriction enzymes, which cut the mtDNA molecule at specified bases, generating a set of DNA fragments. An example would be an enzyme such as MspI, which cleaves the DNA at every CCGG sequence of bases, or nucleotides. Even a single base pair change at the restriction site (the bases recognized by the enzyme) will prevent the enzyme from cleaving the DNA. Polymorphism occurs when these fragments differ in size among individuals, indicating sequence changes at the restriction enzyme site. This method allows detection of much less than half of the potential polymorphic sites in the DNA sequence.

Another way to identify polymorphism in mtDNA is by direct sequencing of all or part of the molecule, thereby detecting all of the changes in the sequenced region. After the PCR technique was developed, DNA sequencing of specific regions of the mtDNA became feasible for large data sets, and it quickly became the most popular method to obtain genetic variation information from mtDNA. The mtDNA sequence variation is widely used as a genetic

1. Electrophoresis is the movement of charged molecules through an electric field. As used in this primer, it refers to the separation of proteins or DNA fragments by charge and size of the molecule. Proteins can be positively or negatively charged, whereas DNA fragments are always negatively charged. The electric field is run through a gel medium, which for protein molecule separation is made from starch, and for DNA molecule separation made from agarose or polyacrylamide.

2. Amplification serves to improve accuracy. The DNA region selected for amplification depends on the question being asked and the level of resolution required (e.g., for low-level of polymorphism a nuclear gene may be selected, for moderate-level of polymorphism an mtDNA gene may be selected, and for high-level polymorphism a nuclear microsatellite locus may be the best choice). The reason amplification is used is simply to increase the detectability of the DNA molecule. A single copy of DNA is very difficult to detect, whereas a DNA region that has been amplified by PCR will have millions of copies of the same molecule. This high number of molecules makes it possible to collect accurate data for this DNA molecule.

marker to resolve species or subspecies-level taxonomy and for population subdivision, gene flow, and maternal lineage assessment. In cougars, mtDNA RFLP studies showed that cougars contained low to moderate levels of polymorphism, compared to other felids (Roelke et al. 1993), and mtDNA sequence variation was used to detect subspecies-level taxonomy and population subdivision (Culver et al. 2000a).

Nuclear DNA. Since mtDNA is inherited strictly from mothers, the resulting phylogeny only represents the female lineages for the individuals being studied—sometimes called a "matrilineal phylogeny." Because matrilineal phylogeny constitutes only a portion of the total genealogical information, phylogenetic reconstructions utilizing multiple unlinked nuclear loci may be even more informative because they offer independent support, or opposition, to mtDNA inferences (including some paternal contribution). As with mtDNA, there are several ways to detect polymorphism in nuclear DNA. As used in cougars, these include DNA sequencing, DNA fingerprinting, and microsatellite fragment length polymorphism.

DNA sequencing. DNA sequencing for regions of the nuclear genome is similar to sequencing for mtDNA in that PCR greatly boosted its use, and all base pair changes are revealed. However, PCR requires prior knowledge of the DNA sequence in the region of interest. Because of the larger size of the nuclear genome, only small amounts of the nuclear genomes of most species have been sequenced. In most species nuclear DNA sequence data sets are fairly limited. Other species—human, mouse, rat, cow, domestic dog, and domestic cat—either have complete genome sequences or are very close to having their genome completely sequenced. In these cases, there is plenty of valuable information to be found in the nuclear DNA. And, while most of the nuclear DNA has a slower evolutionary rate relative to mtDNA, the nuclear DNA contains regions of noncoding DNA, which does not code for any functional protein or nucleic acid. Thus, it tends to evolve neutrally (not constrained by natural selection) and allows an abundance of information to be gained. Nuclear DNA sequencing has not been used in cougar-specific studies but has been used for felid evolutionary studies among the thirty-seven felids that include the cougar (Johnson et al. 2006).

DNA fingerprint. The first highly polymorphic nuclear DNA markers used to assess genetic variation were named minisatellites and were defined as a repeated sequence of nucleotides, or bases, usually eight to eighty base pairs in length. Minisatellites were used as single-locus markers (called VNTRs, variable number of tandem repeats), and as multilocus markers (called DNA fingerprints) where many loci throughout the genome are examined simultaneously, giving a sort of "bar code" or "fingerprint" for each individual. DNA fingerprinting was widely used in wildlife population genetic studies as the first highly polymorphic marker able to detect individual uniqueness as well as population subdivision. DNA fingerprinting was used in cougars as a method to characterize the low levels of genetic variation in Florida panthers (Roelke et al. 1993). Western cougars, and other noninbred felid populations, exhibit moderate to high average heterozygosity compared to Florida panthers.

Microsatellite DNA. Another class of highly polymorphic genetic marker is microsatellites (a repeated sequence of nucleotides, or bases, of one to six bases in length). Nuclear microsatellite markers can distinguish among all individuals of a species (similar to other repetitive sequence markers), and have the advantage of detecting one locus at a time (similar to VNTRs), making them more informative than multilocus DNA fingerprinting. The development of microsatellite DNA markers, smaller in size than minisatellite DNA markers, made it possible to amplify a class of highly polymorphic markers using the newly developed PCR technology. The single-locus nature, combined with the smaller size (amenable to PCR technology), makes microsatellites the single most widely used, highly polymorphic genetic marker in wildlife studies today. Microsatellites, like DNA fingerprints, are used to estimate existing population subdivisions, levels of gene flow, and subspecies-level taxonomy; additionally, due to their single-locus nature, microsatellites can examine paternity and relatedness among individuals as well as direct Mendelian inheritance.[3] Microsatellites are randomly distributed through the genome and are abundant in all eukaryotic organisms (animals, plants, fungi, and protists).

The high levels of polymorphism of microsatellites have made them ideal for examining many population genetics questions in cougars, from estimates of population genetic diversity, phylogeographic patterns, and subspecific taxonomy (Culver et al. 2000a) to defining relationships among closely related populations (Walker et al. 2000; Loxterman 2001; Sinclair et al. 2001; Anderson 2003; Ernest et al. 2003; McRae et al. 2005; Biek et al. 2006a), identifying the degree of relatedness among individuals (Culver et al.,

---

3. Inheritance is where alleles for a particular gene, or region of DNA, are transmitted directly from parent to offspring. The offspring receives one allele from its mother and one allele from its father, at random; and, alleles at one gene are inherited independently from alleles at another gene.

in prep.), and determining paternities (Loxterman et al., in prep.; Culver et al. 2008).

Y-chromosome. Paternal lineage assessment that is comparable to the mtDNA maternal lineages can be gained through examining Y-chromosomal DNA, as this chromosome reflects the paternal lineage analogue to the maternal lineage represented in mitochondria. An intron of the ZFY gene offers a potentially variable genetic region for population studies.[4] Studies of ZFY genetic variation indicate extensive variation between felid species (Pecon-Slattery and O'Brien 1998). The level of variation observed in felid species, combined with strict paternal inheritance, makes this intron potentially useful in the analysis of population subdivision and evolution; however, in cougars no variation was found (Culver 1999).

Viral DNA sequencing. The final type of DNA that has been employed in cougar studies is the DNA of external pathogens that infect the species of study. By examining viral DNA sequences, insights can be gained on temporal and spatial characteristics for the species of concern. Viral sequences have been utilized in cougars and provide excellent examples of how the study of a pathogen can provide substantial information about the temporal and spatial characteristics of the host (Carpenter et al. 1996; Biek et al. 2006b).

Each of the markers described in the preceding list is characterized by distinct rates of evolution and thereby possesses different levels of resolution to analyze distant or recent divergences. The assumption that the distribution of genetic variation in natural populations should reflect the influences of historical, geographical, and ecological factors is what makes these markers useful in cougar conservation genetics.

---

4. Introns are segments of a gene that are not involved in coding for the protein. A single gene will have between one and thirty intron segments. Each intron is between 100 and 10,000 bases long.

Appendix 2 Cougar Harvest in the United States

Number of cougars killed from sport harvest in the United States, 1980 through spring 2008.

	State											
Year[a]	ND[b]	SD[c]	WY	NV	OR	NM	WA	AZ	CO	UT	MT	ID
1980	0	0	19	38	47	86	70	204	81	205	62	31
1981	0	0	27	60	38	97	130	191	106	185	114	97
1982	0	0	38	64	60	122	100	316	137	205	107	198
1983	0	0	44	78	61	99	80	221	125	172	136	189
1984	0	0	54	108	78	132	123	184	107	211	165	126
1985	0	0	54	106	62	79	122	246	155	182	143	275
1986	0	0	80	83	117	104	162	191	166	200	141	216
1987	0	0	84	87	167	101	60	205	180	197	171	219
1988	0	0	59	78	135	78	89	183	173	247	159	306
1989	0	0	63	118	145	91	85	130	183	231	170	276
1990	0	0	66	152	154	112	102	188	235	217	227	340
1991	0	0	50	88	157	108	120	179	228	265	236	293
1992	0	0	74	125	184	119	140	201	295	241	357	335
1993	0	0	79	150	163	105	121	188	299	372	424	359
1994	0	0	94	173	145	127	177	215	330	352	566	437
1995	0	0	110	161	34	148	283	234	314	431	535	441
1996	0	0	142	134	47	119	178	225	391	452	567	612
1997	0	0	144	143	60	176	132	269	419	576	728	638
1998	0	0	172	210	153	166	184	289	407	492	776	797
1999	0	0	204	140	161	153	273	247	337	373	654	776
2000	0	0	182	126	135	156	208	276	318	435	584	614
2001	0	0	213	194	219	236	220	326	439	449	509	715

(Continued)

(Continued)

Year[a]	ND[b]	SD[c]	WY	NV	OR	NM	WA	AZ	CO	UT	MT	ID
						State						
2002	0	0	201	167	232	207	136	264	372	406	407	628
2003	0	0	201	128	248	262	147	218	371	427	346	509
2004	0	0	185	203	265	242	191	247	336	448	335	573
2005	5	14	178	104	224	194	187	204	238	321	322	329
2006	5	15	189	118	289	131	184	221	217	339	287	474
2007	5	20	205	134	308	204	172	250	264	291	309	464
2008	NA	NA	208	146	NA	202	204	NA	295	274	NA	441

From Dawn 2002, state management agency contacts, and records reported on state management agency Web sites.

NA = data not available.

[a]Calendar year reported from North Dakota, South Dakota, Oregon, Arizona and Montana. Harvest year (fall of the previous year through spring of reported year) reported from the remaining states.

[b]There were no protections for cougars in North Dakota until 1991, when the species was classified as a "furbearer." The first hunting season did not go into effect until the 2005–6 hunting season (D. Feske, pers. comm., 2008).

[c]The cougar in South Dakota was designated as a state threatened species for about twenty-five years, until it was delisted in 2003 and reclassified as a game species (L.Gigliotti, pers. comm., 2008).

Appendix 3 Groups Participating in Cougar Management

The profiles that follow pertain to the western United States and Canada.

Ranchers

Cougar management has long been a sharp focus for livestock producers (Kellert 1989), who voice perspectives unique among participants in cougar management. Few ranchers lose significant numbers of livestock to cougar predation (e.g., Phelps 1988; Nadeau 2003). Statewide, annual losses range from a several dozen animals in states such as Idaho or New Mexico to between five hundred and one thousand each year in Utah, where losses are consistently the highest, though still amounting to only about 4 percent of losses from all causes, and constituting less than 0.1 percent of total livestock value (Robertson 1984). Ranchers who are victims of cougar depredation view economic losses as unacceptable and typically seek full monetary compensation (e.g., McIvor and Conover 1994; Logan et al. 2003; Morton 2003; Clark and Munno 2005). Even so, several observers have noted that losses are not readily mollified by money, in part because depredation also constitutes an affront to ranchers' skill (Scarce 1988; Naughton-Treves et al. 2003). Livestock producers consistently demand aggressive prevention of depredation, at times including demands for local eradication of cougars (e.g., Hook and Robinson 1982; Murphy 1984; P. Wilson 1984; Kellert 1985; Russ 1988; Bath and Buchanan 1989; McIvor and Conover 1994; Kellert et al. 1996; Peña 2002).

Significant numbers of livestock producers not materially affected by cougars are nonetheless invested in cougar politics, apparently as a subset of overarching predator management issues, as an affront to their nature-views, and by symbolic identification with perceived assaults by other participants on threatened agrarian lifeways (e.g., Scarce 1998; Nie 2002; Clark and Rutherford 2005). In social surveys, livestock producers consistently score highest on metrics that place humans, rather than "nature," at the center of all considerations (e.g., Hills 1993; Reading and Kellert 1993; Reading et al. 1994; Kellert 1996; Kaltenborn et al. 1998; Bjerke and Kaltenborn 1999), and they tend to score low on empathy and high on values of security and tradition (Hills 1993; Kaltenborn and Bjerke 2002). When faced with disputing claims by other stakeholders—not only about depredation losses but also about predator behavior and populations—livestock producers and allied rural residents often claim primacy of authority for their local knowledge (e.g., Morrison 1984; Weeks and Packard 1997; Mattson et al. 2006).

Deer, Elk, and Sheep Hunters

Hunters of deer, elk and bighorn sheep involve themselves in cougar management based primarily on perceptions of how cougars affect game populations. Mule deer hunters have long believed that cougar predation preempts hunting opportunities by limiting deer populations (e.g., Brown 1984; Wolch et al. 1997; Perry and DeVos

2005), and some still aver "kill a cougar and save 50 deer" (Brown 1984; Zumbo 2002). Even so, since the 1970s, ungulate hunters have rarely promoted eradication of cougars. More commonly, they have supported a scientific approach to management, although with primacy for hunting and for maintenance or increase of ungulate populations (e.g., Brown 1984; Kellert et al. 1996; Reiger 2001). Virtually all hunters not only support hunting as necessary and ethical (Howard 1988; Einwohner 1999a), but also disproportionately advocate hunting large predators, use of hounds, use of lethal methods to deal with problem cougars, and reduction of cougar populations to increase recreational hunting opportunities (e.g., Loker and Decker 1995; Teel et al. 2002; Tsukamoto 2002; McKinstry and Anderson 2003; Dayer et al. 2005; Gigliotti 2005). In general, hunters believe cougar hunting instills fear of, and reduces conflicts with, humans, and is otherwise "good" for cougars (e.g., Howard 1988, 1991; Zumbo 2002; Portland Chapter of the Safari Club International 2003; Miniter 2004).

Collectively, hunters have a distinctive identity. Almost all—75–95 percent, depending on locale—are male and Caucasian (Dizard 2003; Teel et al. 2005). They tend to be political conservatives (Dizard 2003), involved in other types of outdoors activities (Zinn 2003), and residents of nonmetropolitan areas (Manfredo et al. 1998; Zinn 2003). Their orientation toward nature and the outdoors is manifest in considerable interest in natural history, knowledge of and attraction to wildlife, having little fear of predators, and expressing elements of naturalistic and ecologistic/scientific nature-views (Kellert 1989, 1996; Kellert et al. 1996; Floyd 1997; Ericsson and Heberlein 2003; Zinn 2003; Teel et al. 2005). Complicating this picture of engagement and potential empathy, hunters voice dominionistic and utilitarian nature-views with strength second only to that of livestock producers (e.g., Butler et al. 2003; Manfredo et al. 2003; Ruther 2005; Gigliotti 2006), and they identify with self-enhancing values of power, skill, and achievement (Shaw 1994; Kellert 1996; Floyd 1997; Einwohner 1999a; McKinstry and Anderson 2003). Faced with opposition to their demands, hunters often claim that opponents are driven by emotion and thus not worthy of participating in wildlife management (e.g., Howard 1988, 1991; Einwohner 1999a,b; Miniter 2004). Perspectives of ungulate hunters on cougar management are plausibly shaped not only by the centrality of hunting to their identity but also by the perceived endangerment of hunting. Hunters have declined in relative numbers with increasing urbanization, education, and mobility of human populations throughout most of the cougar's range (e.g., Zinn et al. 2002; Manfredo et al. 2003).

Cougar Hunters

Cougar hunters are those who are highly motivated specifically to hunt cougars. As a group, cougar hunters share most of the identity and perspectives of other big game hunters, but there are key differences. For example, cougar hunters have a direct, real, and symbolic stake in killing cougars, which makes cougar management an intense interest for them. Houndsmen who guide professionally are central to hunting cougars and have a potentially major wealth stake, which has increased over time from roughly $1,000 in the 1980s to $3,500 per hunt in recent years (e.g., Clark and Munno 2005). With the recent exceptions of Oregon, Washington, and South Dakota, most killing of cougars—67–95 percent—is assisted by the use of hounds (e.g., DeSimone and Jaffe 2003; Wakeling 2003). Although few in number, cougar hunters have long been a voice in cougar management and were among the first concerned with conserving cougar populations, primarily to sustain harvests (Brown 1984; Murphy 1984; Tsukamoto 1984). Even so, there is a minority of predator hunters who support eradication of cougars. Hook and Robinson (1982) and Kellert and colleagues (1996) describe this subset of hunters as older, less educated, fearful of wildlife, and also distrustful of government.

Cougar hunters are a socially embattled group because of their practices and implied motivations. Recreation and the procurement of meat are the primary professed motivations of most other hunters (Dizard 2003), but most cougar hunters seek trophies (Russ 1988; Zumbo 2002; Clark and Munno 2005), an impulse that has very little public support (Kellert 1996; Kellert and Smith 2000; Campbell and Mackay 2003). Use of hounds is widely viewed by the public as unethical, a perspective also shared by some hunters (Peyton 1989). Compounding this, houndsmen are sometimes singled out by wildlife managers for violating hunting ethics and regulations (Robertson 1984; Rieck 1988; Shaw 1994). Much like other big game hunters, cougar hunters respond to these criticisms by invoking logic and science to assert the specific benefits of hunting cougars with hounds, including eliminating surplus cougars, inculcating fear, increasing prey populations, and reducing suffering, especially among young cougars, which might otherwise starve (Howard 1988, 1991; Einwohner 1999a; Portland Chapter of Safari Club International 2003; Miniter 2004).

Wildlife Agency Commissioners

Wildlife management throughout cougar range in the western United States is governed by a commission or board system. Commission members are typically appointed by state governors, approved by state legislatures, and subject to various guidelines (Hagood 1997). These commissioners are responsible for executing legislated wildlife policies and hold ultimate authority over routine budgetary, rule-making, and policy-evaluating processes for wildlife management agencies, in addition to hiring and firing their directors (Hagood 1997; Nie 2004a,b). The system is explicitly designed to minimize opportunities for intervention in routine wildlife management by elected officials (Reiger 2001). One of the few detailed investigations of commission function, in Oregon, concluded that commissioners served largely to justify agency policies publicly and to provide a buffer between agency personnel and the public and elected officials (Price 1963).

Observers of state-level wildlife management have noted consistent behavioral patterns among commissioners, including strong commitment to hunting as a primary management tool, deference to the interests of hunters, preemptive use of power to settle differences, and evidence of dominionistic, utilitarian, and naturalistic nature-views (e.g., Decker et al. 1996; Gill 1996; Hagood 1997; Beck 1998; Pacelle 1998; Einwohner 1999a; Rutberg 2001; Nie 2004a,b; Clark and Munno 2005; Hatch 2006; Jacobson and Decker 2006). To investigate commissioners and their identities, we assembled and summarized information from official commission Web sites and other Web-accessible journalistic media. In the thirteen states that actively manage cougars, we found that 83 percent of eighty-eight commissioners were Caucasian and 92 percent were male—not unlike the overall composition of hunters. We also found that 88 percent of the sixty-one commissioners who gave indications of outdoor interests voluntarily professed strong interest in hunting (e.g., "ardent" or "avid"). Large numbers of commissioners self-identified a business (32 percent), agricultural (24 percent), legal (14 percent), or governmental (10 percent) background. These patterns did not differ substantially from those documented by Hagood (1997) in the 1990s.

Wildlife Agency Personnel

Wildlife agency personnel have unparalleled power over the cougar management arena, including, with commissioners, control over who gets to participate and how (Nie 2004b). For this reason, other participants have a considerable stake in the perspectives and behaviors of agency personnel. Although agency administrators recognize the need to engage with all stakeholders (McMullin 1993; Graham 1997; South Dakota Department of Game, Fish, and Parks 2005), the routes they choose are either to solicit formal input on draft policies or to conduct public education programs on the assumption that lack of public support arises from lack of information (e.g., Shroufe 1988; Sargent-Michaud and Boyle 2002; Freddy et al. 2004; Lafon et al. 2004; Beausoleil et al. 2005; Perry and DeVos 2005; South Dakota Department of Game, Fish, and Parks 2005; Crowl 2006).

Beyond that, agencies have granted only limited standing to nonagency stakeholders (Decker and Chase 1997; Mortenson and Krannich 2001; Clark and Rutherford 2005) or have constructed stakeholders as "customers" to be served under the rubric of Total Quality Management, which postulates that agencies are repositories of all that is needed to solve technical problems related to delivery of "product" (Hunt 1993; McMullen 1993). Such attitudes are rooted in the paradigm of scientific management, which appropriates power to technical experts housed within management agencies, premised on "problems" being objective biophysical phenomena to be defined and solved by agency experts (e.g., Decker et al. 1996; Gill 1996; Pacelle 1998; Nie 2004b; Clark and Munno 2005). This perspective is evidenced in numerous agency statements granting paramountcy to the "needs of the resource" (e.g., Howard 1988, 1991; McMullin 1993; Blackwell 1996; Arizona Game and Fish Department 2001; Nadeau 2003, 2005).

Wildlife agency personnel officially express beliefs that reveal 0few incentives to maintain—much less increase—cougar populations. Agency personnel are mostly (70–90 percent) Caucasian males who overwhelmingly support hunting as a management tool and are more supportive than the general public of killing wildlife to resolve conflicts and to increase hunting opportunities (e.g., Witter and Shaw 1979; Muth et al. 1998; Campbell and Mackey 2003; Gigliotti and Harmoning 2004; Koval and Mertig 2004; Williams 2005; Heydlauff et al. 2006). Agency personnel have expressed beliefs, largely in official documents, that depredation and conflict with humans are tied to cougar population levels and, therefore, amenable to control through increases in sport hunting; that elimination of hound hunting leads to increased cougar populations and resulting increases in conflict with humans; that in the absence of hound hunting, cougars become bolder toward people; that cougar hunting in itself leads to greater public acceptance of cougars; and that cougars exact considerable opportunity costs by limiting deer and bighorn sheep populations (McCarthy 1996; Beausoleil et al. 2003, 2005; Oregon Department of Fish and Wildlife 2003; McLaughlin 2003; Nadeau 2003; Wakeling 2003; DeSimone et al. 2005; Gigliotti 2005).

On top of this, the direct monetary benefits of cougars to agencies are typically modest, given that sport harvest amounts to no more than a few dozen to few hundred animals per year and licenses and tags run roughly $5 to $60 each for state residents (e.g., Bates 1988; Clark and Munno 2005; Woolstenhulme 2005). By contrast, expenditures for cougar-related salaries and programs can run two to five times higher than cougar-related revenues (Sharma 1988; Cox 1996; Clark and Munno 2005), and the sense of shortfall is compounded by depredation compensation payments in the range of $5,000 to $30,000 annually in states such as Idaho, Colorado, and Wyoming (Anderson and Tully 1988; Becker et al. 2003; Nadeau 2003). Added to concerns about liability for human safety (e.g., Mangus 1991; Baron 2004; Perry and DeVos 2005), these beliefs and monetary considerations plausibly explain widespread support by wildlife management agencies for local or even broader-scale cougar population reductions, employing hounds where possible (e.g., Arizona Game and Fish Department 2000, 2006; Idaho Fish and Game Department 2000; Winslow 2005; McLaughlin 2003; Koval and Mertig 2004).

The Public

Surrounding the narrow core of individuals who have pressing professional involvement with cougars is the large mass of people we call the public, who often participate ephemerally via meetings, newspapers, or opinion surveys sponsored by other stakeholders. The public can be differentiated into those who are engaged versus those who are not, with the engaged public consisting of people who care enough about cougar management to participate briefly, typically during ballot initiatives or "incidents" (see Chapter 14), primarily by voting, writing letters to agencies or newspapers, or selectively attending meetings (the "attentive" and "sympathetic" publics of Dunlap 1989). There is no way to know what fraction of the public is engaged, and this probably varies with the nature of issues and intensity of publicity (Dunlap 1989). Because of the methods employed, the public given voice in surveys consists disproportionately of male heads of permanent households, whose expressed views depend on what questions are asked and how. With this proviso, surveys do offer insight into general moods about key cougar management issues.

Respondents in the range of 60–90 percent either "liked," "supported having," or "supported insuring the survival of" cougars (Duda and Young 1998; Gigliotti 2002, 2005; Peña 2002; Decision Research 2004; Ruther and Ostergren 2005). Extreme stances, such as acceptance of eradication or total protection, were comparatively uncommon (10–29 percent; Peña 2002; Gigliotti 2002, 2005). Support for cougars was expressed as belief that cougars played ecologically important roles (50–85 percent; Chintz 2002;

Gigliotti 2002; Peña 2002; Becker et al. 2003; Decision Research 2004; Casey et al. 2005; Ruther and Ostergren 2005). Typically, a minority of respondents worried about their safety (7–27 percent; Duda and Young 1998; Casey et al. 2005; Ruther and Ostergren 2005), although this percentage increased when people were faced with the prospect of confronting a cougar (36 percent; Gigliotti 2005).

Survey respondents have also been asked about their views on management. With the exception of a select group of respondents in South Dakota (72–78 percent; Gigliotti 2005), only a minority or parity (29–50 percent) supported sport hunting of cougars, with support for using hounds lower yet (16–25 percent), and interest in offering some measure of protection for kittens quite high (roughly 80 percent; Peña 2002; Teel et al. 2002; Becker et al. 2003; Decision Research 2004). Where asked, a majority (56–77 percent) favored nonlethal means of self-defense or nonlethal trapping and relocation of problem animals to resolve human-cougar conflicts (Peña 2002; Ruther and Ostergren 2005). Generally, a large majority of respondents favored killing cougars only if they had caused human injury or death (Manfredo et al. 1998; Zinn et al. 1998; Casey et al. 2005; Dayer et al. 2005), although Dayer and colleagues (2005) reported that 58 percent of their respondents favored killing cougars seen near children.

Where questions were expressed in terms of population management, a consistently large majority of respondents favored reducing cougar numbers where human safety was at risk or to protect other species considered endangered, threatened, or declining; a more varied majority supported reductions to mitigate depredation on domesticated animals (Tsukomoto 2002; Beausoleil et al. 2003; Dayer et al. 2005). Unlike hunters and wildlife agency personnel, most of the public (59–68 percent) did not support reduction of cougar populations to increase game populations (Beausoleil et al. 2003; Dayer et al. 2005; Gigliotti 2005). Surprising to us, in all four surveys where the question was asked, a majority (52–71 percent) of respondents supported limiting human development detrimental to cougars (Duda and Young 1998; Peña 2002; Casey et al. 2005; Ruther and Ostergren 2005).

Animal-Focused Activists

Animal-focused activists represent either rights or welfare perspectives in their participation in wildlife politics (Wywialowski 1991; Duda and Young 1998). The stakes for animal-focused activists are mainly of a symbolic rather than material nature, given that physical encounters with cougars are rare for anyone who does not hunt them professionally. The most deeply felt and most commonly expressed issues for animal activists are suffering of cougar kittens caused by hunting and the suffering and fear they attribute to pursuit by hounds, issues that are expressed in demands to end hunting with hounds and hunting that endangers kittens (e.g., Einwohner 1999a; Schubert 2002; Wyoming Game and Fish Department 2006a). Most also view sport hunting of any kind as intrinsically unethical (Shaw 1994; Schubert 2002; Campbell and Mackay 2003), especially when the objective is trophies, as in cougar hunting.

More broadly, animal activists assert the important ecological role of cougars, express concern about impacts of human developments, view sport hunting and depredation control as a threat to cougar populations, assert that sport hunting does not reduce human-cougar conflicts, claim that increasing conflicts are largely a result of growing human populations, and place the responsibility for living peacefully with cougars largely on people (e.g., Wolch et al. 1997; Schubert 2002; Hoffman 2004; Papouchis 2004; Papouchis et al. 2005; Keefover-Ring 2005; Levy 2005; Love 2005). Animal activists tend to emphasize that science is inappropriately used or even manipulated to justify killing of cougars, and that agency management under the commission structure is "hard-wired" to serve special interests of consumptive users while disenfranchising those holding any other values (e.g., Hagood 1997; Pacelle 1998; Rutberg 2001; Schubert 2002; Hopkins 2003; Arizona Game and Fish Department 2004; Blessley-Lowe 2006). The antidote they advocate is expansion of agency constituencies to include all valid interests (e.g., Hagood 1997; Arizona Game and Fish Department 2004), and they call for collaborative, science-based and community-friendly processes to guide cougar management (see Chapter 15; Schubert 2002; Cullens and Papouchis 2003; Papouchis 2004; Wyoming Game and Fish Department 2006a).

Animal-focused activists have a generic identity distinct from those of all others in cougar politics. As with wildlife agency commissions, we used online information to determine sex, ethnicity, and professional background of 132 board members for thirteen animal-focused organizations with a history of involvement in cougar management. Virtually all board members—97 percent—were, yet again, Caucasian, but 44 percent were female. Of the 50 members who offered information on professions, their backgrounds were prominently in business (32 percent) and law (22 percent), much like wildlife commissioners, although some were academics (14 percent) and in the arts (10 percent). Looking beyond organizational boards, animal rights activists are overwhelmingly (70–80 percent) female, Caucasian, urban, and highly educated (Richards and Krannich 1991; Galvin and Herzog 1992; Peek et al. 1997; Einwohner 1999a; Lowe and Ginsberg 2002).

The preponderance of women among activists has been postulated to arise from projecting the experience of structural oppression by men onto programmatic persecution of animals, also largely by men (Peek et al. 1996, 1997). This is consistent with the fact that nonactivist support for animal rights is more often expressed not only by females but also by nonblack minorities, the young, and the less well educated, all of whom may also have experienced disadvantage (Jerolmack 2003). The perspectives of animal activists have been described as empathetic, ethically sophisticated, and morally evolved (e.g., Hills 1993; Block 2003; Swan and McCarthy 2003) and expressed with a religious-style intensity (e.g., Galvin and Herzog 1992; Shaw 1994; Jamison et al. 2000).

Environmentalists

We distinguish environmentalists from animal-focused activists by their greater overt emphasis on place and ecological systems. Environmentalists are typically transient participants, engaging substantively only when cougar management is a focal issue tied to specific places or wilderness, public lands, or other traditional environmental issues. As natural allies, environmentalists tend to be mobilized by animal activists who are more consistently engaged with cougar issues and who often provide direction and cues regarding claims and demands on wildlife managers. Examples of environmental groups that have engaged with cougar management issues on this basis include chapters of the Sierra Club, the Biodiversity Alliance, Jackson Hole Alliance, Klamath-Siskiyou Wildlands Center, and Friends of the Clearwater. Unlike animal-focused activists, who voice stronger humanistic and moralistic nature-views, environmentalists tend to be more closely identified with ecologistic/scientific and naturalistic perspectives (e.g., Kellert 1989, 1996; Kellert and Smith 2000). The gender composition of environmentalists is less expressly female compared to animal activists, as evinced by 68 percent male composition (n = 48) of the five environmental group boards that we surveyed. In common with virtually everyone else involved in cougar management, environmental groups tended to be largely Caucasian, consistent with 93 percent white composition of boards.

Others: Journalists, Lawyers, and Elected Officials

Much like the public, the remaining participants all tend to become involved in cougar management at the behest or prompting of others, primarily animal-focused activists and environmentalists. This residuum of players includes journalists, attorneys, judges, and elected officials. For all, their stake is typically indirect and transient. Nonetheless, they warrant identification because, when linked to professional roles or to potential power, wealth, or ideological stakes, they can at times play a key part in configuring policies affecting cougar management, especially during policy perturbations. The roles of these participants are disproportionately configured by context (see discussion of institutional arrangements in Chapter 14).

Appendix 4 Cougar Litigation Summary, a Partial Listing

Date/State	Plaintiffs/Defendants (Citation If Applicable)	Primary Claims	Opinion/Ruling
1987 CA	*Mountain Lion Preservation Foundation v California Fish and Game Commission et al.* 214 Cal.App.3d 1043 (Cal.App. 1 Dist. 1989).	Plaintiffs alleged violations of the California Environmental Quality Act (CEQA) based on failure of California Fish and Game Commission (CFGC) to prepare required environmental impact analyses prior to approving the resumption of cougar hunting.	Initially, the trial court postponed the hunt until CFGC prepared a cumulative impact report and provided for a thirty-day public comment period required by CEQA. Ultimately, court determined cumulative impact report was deficient and terminated the hunt for failure to comply with CEQA. Court agreed with plaintiffs that the CFGC failed to adequately address key issues raised by National Park Service (NPS), U.S. Forest Service (USFS), local governments, biologists, and the general public. (In 1989, appellate court affirmed this lower court's order because appellants failed to comply with both the peremptory writ and CEQA.)
1988 OR	*Fund for Animals v Oregon Dep't of Fish & Wildlife* 94 Or.App. 211 (Or. App. 1988).	Plaintiffs argued that the state's administrative record failed to show that the Oregon Fish and Wildlife Commission (OFWC) adequately investigated the number and condition of cougars before adopting the rule.	Plaintiffs failed to show the OFWC did not conduct requisite investigation required under state law. Court noted that information used by OFGC in setting season and tag numbers indicated cougars were particularly difficult to observe and that the Oregon Department of Fish and Wildlife (ODFW) had never performed a comprehensive survey of cougar numbers or their condition. Court also found that the Department had kept careful count of the number of cougars taken each year and the number of complaints of cougar damage broken down by management unit, and relied on that information in setting seasons and tag numbers.
1990 OR	*Felis concolor v U.S. Forest Service* No. 89-6428-E, 1990 U.S. Dist. LEXIS 9498 (D. Or. July 12, 1990).	Multiple claims made by plaintiffs whether special use permit requires the National Environmental Policy Act (NEPA) review; whether special use permits were categorically excluded from NEPA review; whether the federal government had a duty to independently evaluate state bear and cougar management practices; whether the National Forest Management Act (NFMA) established an affirmative duty to regulate cougar and bear hunting on USFS lands.	The 9th Circuit Court decision held special use permits, by themselves, do not constitute a "major federal action." The court explained that a memorandum of understanding between the USFS and the ODFW acknowledges the state's primary role of wildlife management on USFS land. Conversely, the court ruled that the USFS improperly relied on categorical exclusions to support its decision not to prepare an environmental assessment (EA) and its decision to allow commercial guides to lead cougar hunting excursions in two national forests. Defendants had prepared an EA on the issuance of special use permits for cougar hunting in the National Forests which the court ruled were legally sufficient. Finally, the court held the federal defendant did not abdicate its responsibility under NEPA by relying upon studies prepared by the ODFW and that NFMA's planning objective of "securing and maintaining desirable populations of wildlife species" does not constitute an affirmative obligation to regulate the hunting of cougar and bear on USFS lands.

(Continued)

(Continued)

Date/State	Plaintiffs/Defendants (Citation If Applicable)	Primary Claims	Opinion/Ruling
1992 OR	*Felis concolor v U.S. Forest Service* No. 90-35593, U.S. App. LEXIS 3579 (9th Cir. Jan. 03, 1992).	In its appeal of the 1990 decision, USFS did not prepare an environmental impact statement prior to issuing special use permits to allow commercial guides to lead cougar and bear hunts in national forests located in Oregon.	Appellate court affirmed judgment of district court, which had held that USFS had complied with NEPA and the NFMA, that the approval of the use of commercial hunting guides was not a major federal action under NEPA and therefore no environmental impact statement (EIS) was required, and that the states play a major role in the management of wildlife on national forests and, therefore, the plan to allow commercial hunting guides did not violate NFMA.
1995 FL	*Fund for Animals Inc. v Rice* No. 94-1913-CIV-T-23E, 1995 U.S. Dist. LEXIS 22389 (Mid. Dist. Fla. Oct. 12, 1995).	Plaintiffs alleged defendants, the U.S. Army Corps of Engineers (Corps), United States Fish and Wildlife Services (USFWS), Environmental Protection Agency (EPA), and a county administrator violated the Clean Water Act (CWA), the Endangered Species Act (ESA), and NEPA in deciding that a landfill could be constructed within Florida panther habitat.	Court ruled that the Corps fully and adequately considered all environmental factors relevant to its decision regarding the permit, and its actions were not arbitrary or capricious. Administrative records established that the Corps' decisions to modify and reinstate the Section 404 permit complied with both the Administrative Procedures Act (APA) and CWA. The court also held that the Corps properly prepared an environmental assessment of the project.
1996 FL	*Fund for Animals, Inc. v Rice* 85 F.3d 535 (11th Cir. 1996).	In appeal of the 1995 decision, plaintiffs argued that the decision to construct the landfill was arbitrary, capricious, or contrary to law violating the APA; that defendants failed to prepare an EIS in violation of the NEPA; and that defendants violated provisions of the CWA and ESA.	Court held that defendants' actions were not an abuse of discretion, arbitrary, capricious, or contrary to law; that they had taken into account all considerations that factored into alternatives analysis, fully considered all pertinent cumulative impacts, and reasonably concluded an additional public hearing was not needed; that they reasonably concluded that preparation of an EIS was unnecessary; that it had not been determined that the site was a critical habitat.
2000 AZ	*Forest Guardians v Animal & Plant Health Inspection Service* No. CV-99-061-TUC-WDB, 2000 U.S. Dist. LEXIS 22260 (D. Ariz. Nov. 14, 2000).	Defendants could not reasonably interpret the Wilderness Act (WA) to allow lethal predator control to protect private livestock grazing.	Federal defendants had reasonably interpreted the WA to include lethal predator control. The WA and congressional grazing guidelines did not mandate the restrictive interpretation of the WA urged by the plaintiffs.
2002 AZ	*Forest Guardians v Animal & Plant Health Inspection Service* 309 F.3d 1141 (9th Cir. 2002).	In their appeal of the 2000 decision, plaintiffs claimed that decision to allow the lethal control of cougars in the Wilderness Area violated the WA and that the defendants failed to comply with NEPA for failing to prepare a new environmental analysis of this action.	The appellate court held that the WA did not expressly prohibit predator control in wilderness areas but did allow for pre-existing grazing operations to continue in areas later designated as wilderness and that the USFS could authorize U.S. Dept. of Agriculture/Animal and Plant Health Inspection Service (APHIS) to perform lethal predator control of cougars in order to protect private livestock. Court found the USFS' EA complied with NEPA.
2000 NM	*Animal Protection of New Mexico et al. v United States et al.* CIV. No. 98-538 JP/LFG (D.N.M. 2000).	Plaintiffs challenged the adequacy of EAs to consider the impacts of lethal lion control by U.S. Dept. of Agriculture Wildlife Services (USDA/WS) and challenged USDA's failure to prepare an EIS as required by NEPA.	Plaintiffs sought to stop USDA/WS, USFS, and Bureau of Land Management (BLM) from continuing to kill cougars on federal lands until they complied with NEPA. Trial Court ruled federal predator control program violated NEPA by failing to consider information about cougars published in a report of a ten-year a study of cougars in the state. Court ruled defendants do not have an accurate estimate of cougar population based on scientific data. Consequently, the assertion in the EA that the cougar population is stable cannot be supported. Because defendants failed to provide a satisfactory explanation of their sustainable harvest level, it is not possible to know whether the cougar population is actually declining in any of three districts. If it is, any impact on the cougar population from the predator damage management program could have a significant impact on the human environment.

(Continued)

Date/State	Plaintiffs/Defendants (Citation If Applicable)	Primary Claims	Opinion/Ruling
2002 NM	*Animal Protection of New Mexico v New Mexico State Game Commission* No. D-101-CV-2002-00187.	Plaintiffs sued claiming that cougar killing regulations issued by defendants were arbitrary and capricious, potentially resulting in cougar kills three times higher than what was recommended in a state-sponsored cougar study.	Trial Court ruled defendant justified in action as it had involved the public in its decision-making process. Court held defendants met the "technical" requirements of the current statute, which states that the Commission "shall give due regard to the zones of temperatures, and to distribution, abundance, economic value and breeding habits of [the cougar]." The court also held the Commission did not act arbitrarily or capriciously by enacting the challenged amendments and that its cougar hunting regulations were supported by substantial evidence and were within the statutory authority of the Commission.
2002 OR	*Sierra Club v U.S. Fish & Wildlife Service* 235 F.Supp.2d 1109 (D. Or. 2002).	Plaintiffs claimed that the USFWS should have prepared an EIS rather than an EA before permitting a controversial study involving the killing of cougars to assess impacts on ungulate populations. Plaintiffs also challenged legality of using Wildlife Restoration Act funds to pay for the study.	Court held that the project was a major federal action under NEPA because of level of federal funding and the USFWS' required role in monitoring the project. The record indicated the FWS failed to address cumulative effects on cougar populations of other causes of mortality in addition to the study; it appeared that the mortality rate in some study areas could reach 100 percent. Wildlife Restoration Act claim failed, because there was sufficient support in the record for the USFWS' decision to fund study.
2004 AZ	*Fund for Animals Inc. v Norton* 322 F.3d 728 (D.C. Cir. 2003).	Plaintiffs alleged violations of the NEPA, the WA, and the Federal Aid in Wildlife Restoration Act.	USFS agreed to prepare an environmental analysis of the impacts of continued use of the Heber-Reno Sheep Driveway for trailing sheep; USFWS agreed to notify plaintiffs' counsel of any request by the AGFD to use Wildlife Restoration Act funds or any decision to fund the project in question, any continuation of the project, or to fund any similar project involving the study of the predator-prey relationship between cougars and bighorn sheep.
2004 DC	*National Wildlife Federation v Norton* 332 F.Supp.2d 170 (D.D.C. 2004).	Plaintiffs alleged violations of the ESA, NEPA, and CWA as a result of defendants issuing a permit to allow mining, which would impact the habitat of the Florida panther.	The court held that the Biological Opinion, and thus also the Corps permit, were invalid since the USFWS failed to articulate a rational connection between the record facts and its no jeopardy decision, and failed to provide proper analysis of cumulative impact of development upon the Florida panther. Finding that the area involved was relatively small in comparison to the habitat acreage was not dispositive by itself of the effect of the mining on panther habitat, and the USFWS improperly disregarded reasonably foreseeable future projects in evaluating the cumulative impact of the mining operation.
2005 DC	*National Wildlife Federation v Brownlee* 402 F.Supp.2d 1 (D.D.C. 2005).	Plaintiffs alleged that the Corps violated the ESA by failing to consult with USFWS over the impacts of its proposed action on the Florida panther before issuing dredge-and-fill permits and violated the ESA by not developing or implementing a program to conserve the Florida panther.	Court found that although the Corps did not deny that some activities authorized under the nationwide permits may have affected panthers, the Corps had not consulted with USFWS on four challenged nationwide permits in violation of the ESA. Since the Corps' finding of no significant impact and minimal impact under CWA were closely intertwined with the Corps' compliance with ESA, both sides' motions for summary judgment were denied on the claims under the CWA and NEPA.
2005 SD	*Black Hills Mountain Lion Foundation v South Dakota Game, Fish and Parks* No. 05-343 (Cir. CT. S.D September 28, 2005).	Plaintiffs claimed that the initiation of a mountain lion hunt in the Black Hills region of South Dakota by the South Dakota Game, Fish and Parks (SDGFP) violated state regulations governing the management of wildlife and that the procedures followed in establishing the hunt violated South Dakota's administrative procedure law.	Court held that plaintiffs failed to prove the regulated South Dakota cougar season will cause irreparable harm to the state's cougar population and the plaintiffs would suffer an irreparable aesthetic harm. Conversely, court ruled that its granting of the temporary restraining order would cause harm to the SDGFP by preventing "harvest" of cougars and the collection of data. The court also held that considerable harm would come to hunters who have planned and traveled to the Hills for the hunt. The court concluded that the SDGFP considered all relevant evidence and objections regarding the proposed cougar season and that its rulemaking process used to establish the hunting season was consistent with state law.

(Continued)

Date/State	Plaintiffs/Defendants (Citation If Applicable)	Primary Claims	Opinion/Ruling
2006 OR	*Big Wildlife et al. v Johanns and U.S.*	Plaintiffs allege that the USDA/WS failed to comply with NEPA prior to agreeing to kill up to 800 cougars a year for up to five years as permitted under the ODFW's Cougar Management Plan published in 2005.	Case was dismissed due to the government's statement it would issue an EA and comply with NEPA.
2008 OR	*Goat Ranchers of Oregon and Big Wildlife et al. v David E. Williams and Dept. of Agriculture/Animal and Plant Health Inspection Service (APHIS)* No.08-97-ST, 2009 U.S. Dist. LEXIS 26472 (D. Or. Mar. 30, 2009).	Plaintiffs allege that the USDA/WS failed to produce an EIS or adequate EA and Finding of No Significant Impact to meet the stated purpose and need of the ODFW's Cougar Management Plan. Plaintiffs also allege that USDA/WS failed to adequately evaluate environmental impacts of the proposed action and failed to consider a reasonable range of alternatives in its pre-decisional environmental assessment.	The court ruled against the plaintiffs, holding that they lacked standing to sue. The decision was based on a determination that the claims raised by the plaintiffs were not redressible, because if Wildlife Services were not involved in implementing Oregon's Cougar Management plan, the state would still implement the plan and engage in a cull of its cougar population as permitted under the plan. This decision has been appealed as of this writing.

(Table prepared by D. J. Schubert)

Appendix 5 Summary of Cougar Ballot Initiatives in the United States

Year	State	Initiative/ Referendum Number	Initiative Summary	Result	Citation
1990	California	Proposition 117	Prohibited the killing of mountain lions unless for protection of life, livestock, or other property. Established the Habitat Conservation Fund and transfers $30 million a year for thirty years for acquisition of deer and mountain lion habitat, and rare and endangered species habitat. Funds are also specifically allocated to the Coastal, Tahoe, and Santa Monica Mountains Conservancies and state and local parks programs. Remaining funds earmarked for wetlands, riparian and aquatic habitat, open space, and other environmental purposes. Confirmation of depredation by a mountain lion shall not be more than forty-eight hours after receiving the report. Also banned the use of poison, leg-hold or metal-jawed traps, and snares.	Approved	California Fish and Game Code (CFGC) 2780–2781, 2785–2799.6, 4800–4809
1994	Oregon	Measure 18	This measure established a new statute that banned the use of bait to attract or kill black bears and banned the use of dogs to hunt or pursue black bears or mountain lions. Measure included exceptions to allow the use of bait or dogs by county, state, and federal employees or agents acting in their official capacity, and/or when a black bear or cougar is damaging a person's land, livestock, farm, or crops.	Approved	Oregon Revised Statutes 12 ORS498.164
1996	California	Proposition 197	Passage of this proposition would have partially repealed Proposition 117, the California Wildlife Protection Act of 1990, which was adopted by California voters. If passed, Proposition 197 would have required the California Department of Fish and Game to prepare, for Commission approval, a mountain lion management plan to promote health and safety protection, and protection for property and other wildlife species, and to implement the general policy of the state to encourage the preservation, conservation, and maintenance of wildlife resources. In addition, the proposition would have authorized the department and/or any appropriate local agency to remove or take any mountain lion that is perceived to be an imminent threat to public health or safety. Landowners would have been authorized to kill mountain lions perceived to be an imminent threat to public health or safety or livestock anywhere in the state, except the state park system.	Rejected	NA

(Continued)

(Continued)

Year	State	Initiative/ Referendum Number	Initiative Summary	Result	Citation
1996	Oregon	Measure 34	Measure, if passed, would have provided exclusive authority to the Oregon Fish and Wildlife Commission to manage wildlife, and repealed all laws other than legislation and Commission rules enacted since 1975 that regulated time, place, and manner of taking wildlife by angling, hunting, or trapping, including Measure 18, the 1994 bear/mountain lion hunting ban initiative.	Rejected	NA
1996	Washington	Initiative 655	This initiative established a new state law banning the use of bait to attract or kill black bears and banning the use of dogs to pursue or hunt black bear, mountain lion, bobcat, or lynx. Exceptions to this ban provide county, state, or federal agency employees or their agents with the authority to use bait to attract and kill bears to protect livestock, domestic animals, private property, or public safety; to establish and operate feeding stations to prevent damage to commercial timber land; and for a public agency, university, or scientific or educational institution from using bait to attract bears for scientific research. State, county, and federal employees and their agents are also authorized to use dogs to aid in the killing of black bear, cougar, bobcat, or lynx for the purpose of protecting livestock, domestic animals, private property, or the public safety. Finally, permits can be issued to allow private property owners or their tenants to use dogs to hunt on their lands or to allow a public agency, university, or scientific or educational institution to use dogs for the pursuit of black bear, mountain lion, bobcat, or lynx for scientific purposes.	Approved	Revised Code of Washington (RCW) 77.16

(Table prepared by D. J. Schubert)

References

Ackerman, B. B. 1982. Cougar predation and ecological energetics in southern Utah. Master's thesis, Utah State University, Logan.

Ackerman, B. B., F. G. Lindzey, and T. P. Hemker. 1984. Cougar food habits in southern Utah. *J Wildl Mgmt* 48:147–55.

———. 1986. Predictive energetics model for cougars. In *Cats of the world: Biology, conservation, and management,* ed. S. D. Miller and D. Everett, 333–352. Washington, D.C.: National Wildlife Federation.

Adams, C. C. 1925. The conservation of predatory mammals. *J Mammal* 6:83–96.

Adams, C. C., J. Dixon, and E. Heller. 1928. Supplementary report of the committee on wild life sanctuaries, including provision for predatory mammals. *J Mammal* 9:357–58.

Adams, D. B. 1979. The cheetah: Native American. *Science* 205:1155–58.

Aebischer, N. J., P. A. Robertson, and R. E. Kenward. 1993. Compositional analysis of habitat use from animal radio-tracking data. *Ecology* 74:1313–25.

Akenson, J., H. Akenson, and H. Quigley. 2005. Effects of wolf reintroduction on a cougar population in the central Idaho wilderness. In *Proceedings of the eighth mountain lion workshop,* ed. R. A. Beausoleil and D. A. Martorello, 177–187. Olympia: Washington Department of Fish and Wildlife.

Alberta Sustainable Resource Development. 2007. *Alberta guide to hunting regulations.* Edmonton, Alberta: Sport Scene Publications.

Alexander, S., T. Logan, and P. Paquet. 2006. Spatio-temporal co-occurrence of cougars (*Felis concolor*), wolves (*Canis lupus*), and their prey during winter: A comparison of two analytical methods. *J Biogeogr* 33:2001–22.

Allee, W. C. 1931. *Animal aggregations: A study in general sociology.* Chicago: University of Chicago Press.

Altendorf, K. B., J. W. Laundré, C.A. López González, and J. S. Brown. 2001. Assessing effects of predation risk on foraging behavior of mule deer. *J Mammal* 82:430–39.

Alvard, M. S., J. G. Robinson, K. H. Redford, and H. Kaplan. 1997. The sustainability of subsistence hunting in the Neotropics. *Conserv Biol* 11:977–82.

Amin, M. 2004. Patrones de alimentación y disponibilidad de presas del jaguar (*Panthera onca*) y del puma (*Puma concolor*) en la Reserva de la Biosfera Calakmul, Campeche, México. Master's thesis, Instituto de Ecología, Universidad Nacional Autónoma de México, Mexico City.

Anderson, A. B. 2006. *Applying nature's design: Corridors as a strategy for biodiversity conservation.* New York: Columbia University Press.

Anderson, A. E. 1983. *A critical review of literature on puma (Felis concolor).* Special Report No. 54. Denver: Colorado Division of Wildlife.

Anderson, A. E., D. C. Bowden, and D. M. Kattner. 1992. *The puma on Uncompahgre Plateau, Colorado.* Technical Publication No. 40. Denver: Colorado Division of Wildlife.

Anderson, A. E., and R. J. Tully. 1988. Status of mountain lion in Colorado. In *Proceedings of the third mountain lion workshop,* ed. R. H. Smith, 19–23. Phoenix: Arizona Game and Fish Department.

Anderson, C. R., Jr. 2003. Cougar ecology, management, and population genetics in Wyoming. Ph.D. diss., University of Wyoming, Laramie.

Anderson, C. R., Jr., and F. G. Lindzey. 2003. Estimating cougar predation rates from GPS location clusters. *J Wildl Mgmt* 67:307–16.

———. 2005. Experimental evaluation of population trend and harvest composition in a Wyoming cougar population. *Wildl Soc Bull* 33:179–88.

Anderson, C. R., Jr., F. G. Lindzey, and D. B. McDonald. 2003. Genetic structure of cougar populations across the Wyoming Basin: Metapopulation or megapopulation (abstract). In *Proceedings of the seventh mountain lion workshop,* ed. S. A. Becker, D. D. Bjornlie, F. G. Lindzey, and D. S. Moody, p. 10. Jackson Hole: Wyoming Game and Fish Department and Wyoming Cooperative Fish and Wildlife Research Unit.

———. 2004. Genetic structure of cougar populations across the Wyoming Basin: Metapopulation or megapopulation. *J Mammal* 85:1207–14.

Anderson, S. 1997. Mammals of Bolivia: Taxonomy and distribution. *Bull Am Mus Nat Hist* 231:1–652.

Andrew, N. G., V. C. Bleich, P. V. August, and S. G. Torres. 1997. Demography of mountain sheep in the East Chocolate Mountains, California. *Calif Fish Game* 83:289–96.

Anonymous. 1989. The eastern puma: Evidence continues to build. *Newsl Intl Soc Cryptozoology* 8:1–8.

Apker, J. A. 2005. Colorado mountain lion status report. In *Proceedings of the eighth mountain lion workshop,* ed. R. A. Beausoleil and D. A. Martorello, 57–65. Olympia: Washington Department of Fish and Wildlife.

Aranda, M., and V. Sánchez-Cordero. 1996. Prey spectra of jaguar (*Panthera onca*) and puma (*Puma concolor*) in tropical forests of Mexico. *Stud Neotropical Fauna Environ* 31:65–67.

Arizona Game and Fish Department. 2000. Predator management policy, at http://www.gf.state.az.us/wildlife_conservation/predator_management.html (accessed October 16, 2006).

———. 2001. *Wildlife 2006: The Arizona Game and Fish Department's wildlife management program strategic plan for the years 2001–2006*. Phoenix: Arizona Game and Fish Department.

———. 2004. *Final report of the mountain lion workshop, May 1, 2004, Tucson, Arizona*. Phoenix: Arizona Game and Fish Department.

———. 2006. *Black Mountain predation management plan*. Phoenix: Arizona Game and Fish Department.

Arjo, W. M. 1998. The effects of wolf colonization on coyote populations, movements, behaviors, and food habits. Ph.D. diss., University of Montana, Missoula.

Arjo, W. M., and D. H. Pletscher. 1999. Behavioral responses of coyotes to wolf recolonization in northwestern Montana. *Can J Zool* 77:1919–27.

Armour, C. L., D. A. Duff, and W. Elmore. 1991. The effects of livestock grazing on riparian and stream ecosystems. *Fisheries* 16:7–11.

Arnstein, S. A. 1969. A ladder of citizen participation. *J Am Inst Planners* 35:216–24.

Arseniev, V. K. 1941. *Dersu the trapper*. Translated by M. Burr. Kingston, N.Y.: McPherson and Co., 1996.

Arundel, T., D. Mattson, and J. Hart. 2007. Movements and habitat selection by mountain lions in the Flagstaff uplands. In *Mountain lions of the Flagstaff uplands, 2003–2006 progress report*, D. J. Mattson, 17–30. Washington, D.C.: U.S. Department of the Interior, U.S. Geological Survey.

Ascher, W. 2004. Scientific information and uncertainty: Challenges for the use of science in policy making. *Sci Eng Ethics* 10:437–55.

Ashman, D., G. C. Christensen, M. L. Hess, G. K. Tsukamoto, and M. S. Wickersham. 1983. *The mountain lion in Nevada*. Nevada Fish and Game Department, Federal Aid in Wildlife Restoration Final Report, Project W-48-15.

Atwood, T. C., E. M. Gese, and K. E. Kunkel. 2007. Comparative patterns of predation by cougars and recolonizing wolves in Montana's Madison Range. *J Wildl Mgmt* 714:1098–106.

Augustine, D., and D. DeCalesta. 2003. Defining deer overabundance and threats to forest communities: From individual plants to landscape structure. *Ecoscience* 10:472–86.

Aumiller, L. D., and C. A. Matt. 1994. Management of McNeil River State Game Sanctuary for viewing of brown bears. *Intl Conf Bear Res Mgmt* 9:51–61.

Aune, K. E. 1991. Increasing mountain lion populations and human–mountain lion interactions in Montana. In *Mountain lion–human interaction symposium and workshop*, ed. C. E. Braun, 86–94. Denver: Colorado Division of Wildlife.

Austin, M. 2003. Mountain lion status report: British Columbia. In *Proceedings of the seventh mountain lion workshop*, ed. S. A. Becker, D. D. Bjornlie, F. Lindzey, and D. S. Moody, p. 87. Lander: Wyoming Game and Fish Department.

———. 2005. British Columbia mountain lion status report. In *Proceedings of the eighth mountain lion workshop*, ed. R. A. Beausoleil and D. A. Martorello, p. 3. Olympia: Washington Department of Fish and Wildlife.

Avise, J. C. 1994. *Molecular markers, natural history and evolution*. New York: Chapman and Hall.

———. 1995. Mitochondrial DNA polymorphism and a connection between genetics and demography of relevance to conservation. *Conserv Biol* 9:686–90.

Avise, J. C., and R. M. Ball, Jr. 1990. Principles of genealogical concordance in species concepts and biological taxonomy. Pp 45–67 in *Oxford surveys in evolutionary biology*, Vol. 7. Oxford: Oxford University Press.

Bailey, J. A. 1984. *Principles of wildlife management*. New York: John Wiley and Sons.

Bailey, V., J. Dixon, E. A. Goldman, E. Heller, and C. C. Adams. 1928. Report of the Committee on Wild Life Sanctuaries, including provision for predatory mammals. *J Mammal* 9:354–57.

Baldi, R., P. D. Carmanchahi, D. De Lamo, M. C. Funes, S. Puig, J. von Thüngen and P. Nugent. 2006. Conservación del Guanaco en la Argentina: Propuesta para un Plan Nacional de Manejo. In *Manejo de Fauna Silvestre en la Argentina*, ed. M. L. Bolkovic and D. E. Ramadori, 137–149. Programas de uso sustentable, Dirección de Fauna Silvestre, Secretaría de Ambiente y Desarrollo Sustentable, Buenos Aires, Argentina.

Ballard, W. B., D. Lutz, T. W. Keegan, L. H. Carpenter, and J. C. deVos, Jr. 2001. Deer—predator relationships: A review of recent North American studies with emphasis on mule and black-tailed deer. *Wildl Soc Bull* 29:99–115.

Balser, D. S. 1964. Antifertility agents in vertebrate pest control. In *Proceedings of the second vertebrate pest control conference*, 133–37. Anaheim, California.

Bank, M. S., and W. L. Franklin. 1998. Puma (*Puma concolor patagonica*) feeding observations and attacks on guanacos (*Lama guanicoe*). *Mammalia* 62:599–605.

Bank, M. S., R. J. Sarno, N. K. Campbell, and W. L. Franklin. 2002. Predation of guanacos (*Lama guanicoe*) by southernmost mountain lions (*Puma concolor*) during a historically severe winter in Torres del Paine National Park, Chile. *J Zool London* 258:215–22.

Barber, S. 2005. Arizona mountain lion status report. In *Proceedings of the eighth mountain lion workshop*, ed. R. A. Beausoleil and D. A. Martorello, 65–69. Olympia: Washington Department of Fish and Wildlife.

Barber-Meyer, S. M. 2006. Elk calf mortality following wolf restoration to Yellowstone National Park. Ph.D. diss., University of Minnesota, St. Paul.

Barnes, C. S. 1928. Letter of request for cougar information. *J Mammal* 9:88.

———. 1960. *The cougar or mountain lion*. Salt Lake City: The Ralton Co.

Barnhurst, D. 1986. Vulnerability of cougars to hunting. Master's thesis, Utah State University, Logan.

Barnhurst, D., and F. G. Lindzey. 1989. Detecting female mountain lions with kittens. *Northwest Sci* 63:35–37.

Baron, D. 2004. *The beast in the garden: A modern parable of man and nature*. New York: Norton and Company.

Barone, M. A., M. E. Roelke, J. Howard, J. L. Brown, A. E. Anderson, and D. E. Wildt. 1994. Reproductive characteristics of male Florida panthers: Comparative studies from Florida, Texas, Colorado, Latin America, and North American zoos. *J Mammal* 75:150–62.

Barrere, P. 1741. Essai sur l'historie naturael France Equinoxiale. In *Account of the "tigre rouge"* (F. c. concolor), p. 166. Quoted in Young and Goldman, *The puma: Mysterious American cat*. Washington, D.C.: The American Wildlife Institute, 1946.

Barrowclough, G. F. 1982. Geographic variation, predictiveness, and subspecies. *Auk* 99:601–3.

Bartel, R. A., and F. F. Knowlton. 2005. Functional feeding responses of coyotes (*Canis latrans*) to fluctuating prey abundance in the Curlew Valley, Utah, 1977–1993. *Can J Zool* 83:569–78.

Bataille, G. 1955. *Lascaux: or, the birth of art: Prehistoric painting*. Lausanne, France: Skira.

Bates, B. 1988. Status of cougar in Utah. In *Proceedings of the third mountain lion workshop*, ed. R. H. Smith, 32–34. Phoenix: Arizona Game and Fish Department.

Bath, A. J., and T. Buchanan. 1989. Attitudes of interest groups in Wyoming toward wolf restoration in Yellowstone National Park. *Wildl Soc Bull* 17:519–25.

Bauer, J. W., K. A. Logan, L. L. Sweanor, and W. M. Boyce. 2005. Scavenging behavior in puma. *Southwest Nat* 50:466–71.

Bearzi, G. 2007. Marine conservation on paper. *Conserv Biol* 21:1–3.

Beausoleil, R. A., D. Dawn, D. A. Martorello, and C. P. Morgan. 2008. Cougar management protocols: A survey of wildlife agencies in North America. In D.E Toweill, S. Nadeau, and D. Smith, eds., *Proceedings of the ninth mountain lion workshop*, 205–41. Boise: Idaho Department of Fish and Game.

Beausoleil, R. A., and D. A. Martorello, eds. 2005. *Proceedings of the eighth mountain lion workshop*. Olympia: Washington Department of Fish and Wildlife.

Beausoleil, R. A., D. A. Martorello, and R. D. Spencer. 2003. Washington cougar status report. In D.E. Toweill, S. Nadeau and D. Smith, eds., *Proceedings of the seventh mountain lion workshop*, ed. S. A. Becker, D. D. Bjornlie, F. G. Lindzey, and D. S. Moody, 60–63. Lander: Wyoming Game and Fish Department.

———. 2005. Washington mountain lion status report. In *Proceedings of the eighth mountain lion workshop*, ed. R. A. Beausoleil and D. A. Martorello, 4–10. Olympia: Washington Department of Fish and Wildlife.

Beausoleil, R. A., K. I. Warheit, and D. A. Martorello. 2005. Using DNA to estimate cougar populations in Washington: A collaborative approach. In *Proceedings of eighth mountain lion workshop*, ed. R. A. Beausoleil and D. A. Martorello. Olympia: Washington Department of Fish and Wildlife.

Beck, T. D. I. 1998. Citizen ballot initiatives: A failure of the wildlife management profession. *Human Dim Wildl* 3:21–28.

Becker, S. A., D. D. Bjornlie, F. G. Lindzey, and D. S. Moody, eds. 2003. *Proceedings of the seventh mountain lion workshop*. Lander: Wyoming Game and Fish Department.

Becker, S. A., D. D. Bjornlie, and D. S. Moody. 2003. Wyoming mountain lion status report, in S. A. Becker, D. D. Bjornlie, F. G. Lindzey, and D. S. Moody, eds., *Proceedings of the seventh mountain lion workshop*, 64–70. Launder: Wyoming Game and Fish Department.

Becker, S. A., D. D. Bjornlie, and D. S. Moody. 2003. Wyoming mountain lion status report. In *Proceedings of the seventh*

mountain lion workshop, ed. S. A. Becker, D. D. Bjornlie, F. G. Lindzey, and D. S. Moody, 64–70. Lander: Wyoming Game and Fish Department.

Beckerman, A. P., M. Uriarte, and O. J. Schmitz. 1997. Experimental evidence for a behavior-mediated trophic cascade in a terrestrial food chain. *Proc Nat Acad Sci* 94:10735–38.

Beecham, J. J. 1983. Population characteristics of black bears in west central Idaho. *J Wildl Mgmt* 47:405–12.

Beier, P. 1991. Cougar attacks on humans in the United States and Canada, 1890–1990. *Wildl Soc Bull* 19:403–12.

———. 1992. Cougar attacks on humans: An update and some further reflections. In *Proceedings of fifteenth vertebrate pest conference*, ed. J. E. Borrecco and R. E. Marsh, 365–367. University of California, Davis.

———. 1993. Determining minimum habitat areas and habitat corridors for cougars. *Conserv Biol* 7:94–108.

———. 1995. Dispersal of juvenile cougars in fragmented habitat. *J Wildl Mgmt* 59:228–37.

———. 1996. Metapopulation modeling, tenacious tracking, and cougar conservation. In *Metapopulations and wildlife conservation*, ed. D. R. McCullough, 293–323. Washington, D.C.: Island Press.

———. 2007. Learning like a mountain: Lessons on conserving habitat corridors. *Wildl Professional* 1:26–29.

Beier, P., and R. H. Barrett. 1993. The cougar in the Santa Ana Mountain Range, California. In Final Report, Orange County Cooperative Mountain Lion Study. Department of Forestry and Resource Management, University of California, Berkeley.

Beier, P., D. Choate, and R. H. Barrett. 1995. Movement patterns of mountain lions during different behaviors. *J Mammal* 76:1056–70.

Beier, P., and S. C. Cunningham. 1996. Power of track surveys to detect changes in puma populations. *Wildl Soc Bull* 24:540–46.

Beier, P., D. R. Majka, and W. D. Spencer. 2008. Forks in the road: Choices in procedures for designing wildlife linkages and corridors. *Conserv Biol* 22:836–51.

Beier, P., D. R. Majka, and S. L. Newell. 2009. Uncertainty analysis of least-cost modeling for designing wildlife linkages. *Ecological Applications*. In press.

Beier, P., K. Penrod, C. Luke, W. Spencer, and C. Cabañero. 2006. South Coast missing linkages: Restoring connectivity to wildlands in the largest metropolitan area in the United States. In *Connectivity conservation*, ed. K. R. Crooks and M. A. Sanjayan, 555–586. Cambridge University Press, Cambridge, UK.

Beier, P., M. R. Vaughan, M. J. Conroy, and H. Quigley. 2003. An analysis of scientific literature related to the Florida panther. Final report by the Florida Fish and Wildlife Conservation Commission, Tallahassee. Available at http://oak.ucc.nau.edu/pb1/publications.htm (accessed June 8, 2007).

———. 2006. Evaluating scientific inferences about the Florida panther. *J Wildl Mgmt* 70:236–45.

Bekoff, M. 2002. *Minding animals: Awareness, emotions and heart*. New York: Oxford University Press.

Belden, R. C. 1986. Florida panther recovery plan implementation. In *Proceedings of the second international cat symposium*, ed. S. D. Miller and D. D. Everett, 159–172. National Wildlife Federation, Washington, D.C.

Belden, R. C., and B. W. Hagedorn. 1993a. Feasibility of translocating panthers into northern Florida. *J Wildl Mgmt* 57:388–97.

———. 1993b. Feasibility of translocating panthers into northern Florida. *J Wildl Mgmt* 57:388–97.

Belden, R. C., and J. W. McCown. 1996. Florida panther reintroduction feasibility study. Final report by the Florida Game and Fresh Water Fish Commission, Tallahassee.

Belden, W. C., W. B. Frankenberger, R. T. McBride, and S. T. Schwikert. 1988. Panther habitat use in southern Florida. *J Wildl Mgmt* 52:660–3.

Bellati, J., and J. von Thüngen. 1990. Lamb predation in Patagonian ranches. In *Proceedings of the fourteenth vertebrate pest conference*, ed. L. R. Davis and R. E. Marsh, 263–268. University of California, Davis.

Berg, J. 2007. The carnivore assemblage of La Payunia Reserve, Patagonia, Argentina: Dietary niche, prey availability, and selection. Master's thesis, University of Montana, Missoula.

Berger, J., P. B. Stacy, L. Bellis, and M. P. Johnson. 2001. A mammalian predator-prey imbalance: Grizzly bear and wolf extinction affect avian neotropical migrants. *Ecol Appl* 11:947–60.

Berger, J., and J. D. Wehausen. 1991. Consequences of a mammalian predator–prey disequilibrium in the Great Basin Desert. *Conserv Biol* 5:244–48.

Berger, K. M. 2006. Carnivore–livestock conflicts: Effects of subsidized predator control and economic correlates on the sheep industry. *Conserv Biol* 20:751–61.

Bergerud, A. T. 1992. Rareness as an antipredator strategy to reduce predation risk for moose and caribou. In *Wildlife 2001: Populations*, ed. D. R. McCullough and R. H. Barrett, 1008–1021. London: Elsevier Applied Science.

Bergmann, C. 1847. Über die Verhältnisse der Wärmeökonomie der Thiere zu ihrer Grösse. *Göttinger Studien* 3:595–708.

Best, A. 2005. How dense can we be? *High Country News*, June issue. Also at http://www.hcn.org/servlets/hcn.Article?article_id=15571 (accessed December 5, 2007).

Beyer, H. L., E. H. Merrill, N. Varley, and M. S. Boyce. 2007. Willow on Yellowstone's northern range: Evidence for a trophic cascade? *Ecol Appl* 17:1563–71.

Biek, R., N. Akamine, M. K. Schwartz, T. K. Ruth, K. M. Murphy, and M. Poss. 2006a. Genetic consequences of sex-biased dispersal in a solitary carnivore: Yellowstone cougars. *Biol Lett* 2:312–15.

Biek, R., A. J. Drummond, and M. Poss. 2006b. A virus reveals population structure and recent demographic history of its carnivore host. *Science* 311:538–41.

Biknevicius, A. R., and B. B. Ruff. 1992. The structure of the mandibular corpus and its relationships to feeding behaviors in exant carnivorans. *J Zool London* 228:479–507.

Biknevicius, A. R., and B. Van Valkenburgh. 1996. Design for killing: Craniodental adaptations of predators. In *Carnivore biology, ecology, and evolution*, ed. J. L. Gittleman, 393–428. Ithaca, NY: Cornell University Press.

Birkland, T. A. 1998. Focusing events, mobilization, and agenda setting. *J Public Policy* 18:53–74.

Bixler, A., and Z. Tang-Martinez. 2006. Reproductive performance as a function of inbreeding in prairie voles. *J Mammal* 87:944–9.

Bjerke, T., and B. P. Kaltenborn. 1999. The relationship of ecocentric and anthropocentric motives to attitudes toward large carnivores. *J EnvironPsych* 19:415–21.

Bjerke, T., O. Reintan, and S. R. Kellert. 1998. Attitudes toward wolves in southeastern Norway. *Soc Natur Resour* 11:169–78.

Blackwell, B. H. 1996. State status report–Utah. In *Proceedings of the fifth mountain lion workshop*, ed. W. D. Padley, 120–121. San Diego: Southern California Chapter of the Wildlife Society.

Bleich, V. C., R. T. Bowyer, and J. D. Wehausen. 1997. Sexual segregation in mountain sheep: Resources or predation? *Wildl Monogr* 134:3–50.

Bleich, V. C., and T. J. Taylor. 1998. Survivorship and cause-specific mortality in five populations of mule deer. *Great Basin Nat* 58:265–72.

Bleich, V. C., J. D. Wehausen, and S. A. Holl. 1990. Desert-dwelling mountain sheep: Conservation implications of a naturally fragmented distribution. *Cons Biol* 4:383–90.

Blessley-Lowe, C. 2006. Why such disregard for our cougars? *Jackson Hole News and Guide*, February 15.

Block, G. 2003. The moral reasoning of believers in animal rights. *Soc Anim* 11:167–80.

Blouin, M. S., M. Parsons, V. Lacaille, and S. Lotz. 1996. Use of microsatellite loci to classify individuals by relatedness. *Mol Ecol* 5:393–401.

Boas, F. 1930. *The religion of the Kwakiutl Indians, Vol 2*. New York: Columbia University Press.

Bocking, S. 2006. *Nature's experts: Science, politics and the environment*. New Brunswick, NJ: Rutgers University Press.

Boertje, R. D., and D. C. L. Gardner. 2000. Reducing mortality on the Fortymile caribou herd. Alaska Dept. of Fish and Game, Federal Aid in Wildlife Restoration, Grant W-27-3.

Boitani, L. 2003. Wolf conservation and recovery. In *Wolves behavior, ecology, and conservation*, ed. L. D. Mech and L. Boitani, 317–340. Chicago: University of Chicago Press.

Bolgiano, C. 1995. *Mountain lion: An unnatural history of pumas and people*. Mechanicsburg, PA: Stackpole Books.

———. 2001. *Mountain lion: An unnatural history of pumas and people*. Mechanicsburg, PA: Stackpole Books.

Bonnell, M. L., and R. K. Selander. 1974. Elephant seals: Genetic variation and near extinction. *Science* 184:908–9.

Bonney, R. C., H. D. M. Moor, and D. M. Jones. 1981. Plasma concentrations of oestradiol-17B and progesterone, and laparoscopic observations of the ovary in the puma (*Felis concolor*) furing oestrus, pseudopregnancy and pregnancy. *J Reprod Fertil* 63:523–31.

Boomgaard, P. 2001. *Frontiers of fear: Tigers and people in the Malay world, 1600–1950*. New Haven, CT: Yale University Press.

Booth, A., and M. J. Harvey. 1990. Ties that bind: Native American beliefs as a foundation for environmental consciousness. *Environ Ethics* 12:27–43.

Bortolus, A., and E. Schwindt. 2006. What would Darwin have written now? *Biodivers Conserv* 16:337–45.

Bowyer, R. T. 1984. Sexual segregation in southern mule deer. *J Mammal* 65:410–17.

Bowyer, R. T., D. K. Person, and B. M. Pierce. 2005. Detecting top-down versus bottom-up regulation of ungulates by large carnivores: Implications for conservation of biodiversity. In *Large carnivores and the conservation of biodiversity*, ed. J. C. Ray, K. H. Redford, R. S. Steneck, and J. Berger, 342–361. Washington, D.C.: Island Press.

Boyd, D., and G. K. Neale. 1992. An adult cougar (*Felis concolor*) killed by gray wolves (*Canis lupus*) in Glacier National Park, Montana. *Can Field Nat* 106:524–25.

Boyd, D., and B. O'Gara. 1985. Cougar predation on coyotes. *Murrelet* 66:17.

Branch, L. C., M. Pessino, and D. Villareal. 1996. Response of pumas to a population decline of the plains vizcachas. *J Mammal* 77:1132–40.

Braun, C. E., ed. 1991. *Mountain Lion-Human Interaction Symposium and Workshop*. Denver: Colorado Division of Wildlife.

Brea, J. A. 2003. Population dynamics in Latin America. *Popul Bull* 58:1–36.

Bright, A. D., and M. J. Manfredo. 1996. A conceptual model of attitudes toward natural resource issues: A case study of wolf reintroduction. *Hum Dim Wildl* 1:1–21.

Brito, B. F. A., C. Bassi, R. Garla, and S. Mendes. 1998. Ecologia alimentar da onça-parda *Puma concolor* (Carnivora, Felidae) na Mata Atlântica de Lindares. ES. XXII Congreso Brasileño de Zoología. Quoted in de Oliveira, T. G., 2002.

Bromley, D., and E. M. Gese. 2001. Surgical sterilization as a method of reducing coyote predation on domestic sheep. *J Wildl Mgmt* 65:510–19.

Brooks, P. 1972. *The house of life: Rachel Carson at work*. Boston: Houghton Mifflin Company.

Brown, D. 1984. A lion for all seasons. In *Proceedings of the second mountain lion workshop*, ed. J. Roberson and F. Lindzey, 13–22. Salt Lake City: Utah Division of Wildlife Resources.

Brown, D. E., and E. A. López González. 2001. *Borderland jaguars: Tigres de la frontera*. Salt Lake City: University of Utah Press.

Brown, J. E. 1997. *Animals of the soul: Sacred animals of the Oglala Sioux*. Rockport, MA: Element.

Brown, J. L., and G. H. Orians. 1970. Spacing patterns in mobile animals. *Annu Rev Ecol Syst* 1:239–62.

Brown, L. R. 2006. *Plan B 2.0: Rescuing a planet under stress and a civilization in trouble*. New York: W. W. Norton.

Brulle, R. J. 2000. *Agency, democracy, and nature: The U.S. environmental movement from a critical theory perspective*. Cambridge, MA: MIT Press.

Brunner, R. D. 2002. Problems of governance. In *Finding common ground: Governance and natural resources in the American West*, ed. R. D. Brunner, C. H. Colburn, C. M. Cromley, R. A. Klein, and E. A. Olson, 1–47. New Haven, CT: Yale University Press.

Brunner, R. D., and T. A. Steelman. 2005. Toward adaptive governance. In *Adaptive governance: Integrating science, policy, and decision making*, ed. R. D. Brunner, T. A. Steelman, L. Coe-Juell, C. M. Cromley, C. M. Edwards, and D. W. Tucker, 268–304. New York: Columbia University Press.

Brunner, R. D., T. A. Steelman, L. Coe-Juell, C. M. Cromley, C. M. Edwards, and D. W. Tucker. 2005. *Adaptive governance: Integrating science, policy, and decision making*. New York: Columbia University Press.

Bueno-Cabrera, A. 2001. Hábitos alimentarios del puma (*Puma concolor*) en la Sierra San Pedro Mártir, Baja California, México. B.Sc. thesis. Benemérita Universidad Autónoma de Puebla, Mexico.

———. 2004. Impacto del puma (*Puma concolor*) en ranchos ganaderos del Área Natural Protegida "Cañón de Santa Elena," Chihuahua. Master's thesis, Instituto de Ecología, A.C., Xalapa, Veracruz, Mexico.

Bueno-Cabrera, A., L. Hernandez-Garcia, J. Laundre, A. Contreras-Hernandez, and H. Shaw. 2005. Cougar impact

on livestock ranches in the Santa Elena Canyon, Chihuahua, Mexico. In *Proceedings of the eighth mountain lion workshop*, ed. R.A. Beausoleil and D.A. Martorello, 141–149. Olympia: Washington Department of Fish and Wildlife.

Buotte, P. C., T. K. Ruth, K. M. Murphy, M. G. Hornocker, and H. B. Quigley. 2005. Spatial distribution of cougars (*Puma concolor*) in Yellowstone National Park before and after wolf (*Canis lupus*) reintroduction. In *Proceedings of the eighth mountain lion workshop*, ed. R. A. Beausoleil and D. A. Martorello, p. 176. Olympia: Washington Department of Fish and Wildlife.

Burkett, V. R. 2001. Climate change in the United States: Implications for fish and wildlife management. *Trans N Am Wildl Nat Res* 66:275–99.

Busby, D., D. Strohbehn, P. Beedle, and M. King. 2006. Effect of disposition on feedlot gain and quality grade. *Natl Cattlemen* 21:32–3.

Butler, J. S., J. Shanahan, and D. J. Decker. 2003. Public attitudes toward wildlife are changing: A trend analysis of New York residents. *Wildl Soc Bull* 31:1027–36.

Byers, J. A. 1997. *American pronghorn: Social adaptations and the ghost of predators past*. Chicago: University of Chicago Press.

Byrd, K. 2002. Mirrors and metaphors: Contemporary narratives of wolf in Minnesota. *Ethics Place Environ* 5:50–65.

Cabrera, A. 1963. Los felidos vivientes de la Republica Argentina. *Ciencias Zoologicas*, 6.

Cajal, J., and N. E. Lopez. 1987. El puma como depredador de camélidos silvestres en la Reserva San Guillermo, San Juan, Argentina. *Revista Chilena de Historia Natural* 60:87–91.

Calicott, J. B. 1982. Traditional American Indian and western European attitudes toward nature: An overview. *Environ Ethics* 4:293–318.

Calva-Mercado, A. 2007. México 2000–2012: Fin e inicio de un sexenio. *Entorno Económica* 17:22–3.

Cameron, J. 1929. *The bureau of biological survey: Its history, activities, and organization*. Baltimore: Johns Hopkins Press.

Campbell, J. M., and K. J. Mackay. 2003. Attitudinal and normative influences on support for hunting as a wildlife management strategy. *Hum Dim Wildl* 8:181–97.

Carbone, C., G. M. Mace, S. C Roberts, and D. W. Macdonald. 1999. Energetic constraints on the diet of terrestrial carnivores. *Nature* 402:286–8.

Carbyn, L. N., S. M. Oosenbrug, and D. W. Anions. 1993. *Wolves, bison and the dynamics related to the Peace Athabaska Delta in Canada's Wood Buffalo National Park*. Circumpolar Research Series, No. 4, Canadian Circumpolar Institute, University of Alberta, Edmonton.

Carbyn, L. N., and T. Trottier. 1987. Responses of bison on their calving grounds to predation by wolves in Wood Buffalo National Park. *Can J Zool* 65:2072–8.

Cardoza, J. B., and S.A Langlois. 2002. The eastern cougar: A management failure? *Wildl Soc Bull* 30:265–73.

Caro, T. M. 1994. *Cheetahs of the Serengeti Plains: Group living in an asocial species*. Chicago: University of Chicago Press.

Carpenter, M. A., E. W. Brown, M. Culver, W. E. Johnson, J. Pecon-Slattery, D. Brousset, and S. J. O'Brien. 1996. Genetic and phylogenetic divergence of feline immunodeficiency virus in the puma (*Puma concolor*). *J Virol* 70:6682–93.

Carrillo, E., G. Wong, and A. D. Cuarón. 2000. Monitoring mammal populations in Costa Rican protected areas under different hunting restriction. *Conserv Biol* 14:1580–91.

Carson, R. 1962. *Silent spring.* Boston: Houghton Mifflin.

Casey, A. L., P. R. Krausman, W. W. Shaw, and H. G. Shaw. 2005. Knowledge of and attitudes toward mountain lions: A public survey of residents adjacent to Saguaro National Park, Arizona. *Hum Dim Wildl* 10:29–38.

Cashman, J. L., M. Pierce, and P. L. Krausman. 1992. Diets of mountain lions in southwestern Arizona. *Southwest Nat* 37:324–26.

Caughley, G. 1977. *Analysis of vertebrate populations.* New York: John Wiley and Sons.

Caughley, G., R. Peck, and D. Grice. 1992. Effect of fertility control on a population's productivity. *Wildl Res.* 19:623–27.

Caughley, G., and A. R. E. Sinclair. 1994. *Wildlife ecology and management.* Oxford, UK: Blackwell Science.

Ceballos, G., C. Chávez, H. Zarza, and C. Manterola. 2005. Ecología y conservación del jaguar en la región de Calakmul. *Biodiversitas* 62:1–7.

CEPAL. 2002. Social Panorama of Latin America. United Nations Publication LC/G.2193-P. Santiago, Chile.

Chase, L. C., T. B. Lauber, and D. J. Decker. 2001. Citizen participation in wildlife management decisions. In *Human dimensions of wildlife management in North America,* ed. D. J. Decker, T. L. Brown, and W. F. Siemer, 153–170. The Wildlife Society, Bethesda, MD.

Chávez Tovar, J. C. 2005. Puma. In *Los mamíferos silvestres de México,* ed. G. Ceballos and G. Oliva, 364–367. Mexico City: Comisión Nacional para el Conocimiento y Uso de la Biodiversidad.

Cherney, D. N., A. Bond, and S. G. Clark. 2008. Understanding patterns of human interactions and decision making: An initial map of Podocarpus National Park, Ecuador. *J Sustain Forest* In Press.

Chesson, J. 1978. Measuring preference in selective predation. *Ecology* 59:211–5.

Chiarello, A. G. 1999. Effects of fragmentation of the Atlantic forest on mammal communities in south eastern Brazil. *Biol Conserv.* 89:71–82.

Chinchilla, F. A. 1997. La dieta del jaguar (*Panthera onca*), el puma (*Felis concolor*) y el manigordo (*Felis pardalis*) (Carnivora, Felidae) en el Parque Nacional Corcovado, Costa Rica. *Revista de Biología Tropical* 45:1223–29.

Chintz, A. E. 2002. Laying the groundwork for public participation in cougar (*Puma concolor*) management: A case study of southwestern Oregon. Master's thesis, University of Oregon, Eugene.

Choate, D. M., M. L. Wolfe, and D. C. Stoner. 2006. Evaluation of cougar population estimators in Utah. *Wildl Soc Bull* 34:782–9.

Christensen, G. C., and R. J. Fischer, eds. 1976. *Proceedings of the first mountain lion workshop.* Nevada Game and Fish Department and U. S. Fish and Wildlife Service, Sparks.

City Population. 2007. http://www.citypopulation.de (accessed May 21, 2007).

Clark, S. G. 2007. *Ensuring Greater Yellowstone's future: Choices for leaders and citizens.* New Haven, CT: Yale University Press.

Clark, T. W. 2001. Developing policy-oriented curricula for conservation biology: Professional leadership education in the public interest. *Conserv Biol* 15:31–9.

———. 2002. *The policy process: A practical guide for natural resource professionals.* New Haven, CT: Yale University Press.

Clark, T. W., R. G. Begg, and K. W. Lowe. 2002. Appendix: Interdisciplinary problem-solving workshops for natural resource professionals. In *The policy process: A practical guide for natural resource professionals,* ed. T. W. Clark, 173–189. New Haven, CT: Yale University Press.

Clark, T. W., A. P. Curlee, and R. P. Reading. 1996. Crafting effective solutions to the large carnivore conservation problem. *Conserv Biol* 10:940–8.

Clark, T. W., D. J. Mattson, R. P. Reading, and B. J. Miller. 2001. Interdisciplinary problem solving in carnivore conservation: An introduction. In *Carnivore conservation,* ed. J. L. Gittleman, S. M. Funk, D. McDonald, and R. K. Wayne, 223–240. Cambridge, UK: Cambridge University Press.

Clark, T. W., and L. Munno. 2005. Mountain lion management: Resolving public conflict. In *Coexisting with large carnivores: Lessons from Greater Yellowstone,* ed. T. W. Clark, M. B. Rutherford, and D. Casey, 71–98. Washington, D.C.: Island Press.

Clark, T. W., P. C. Paquet, and A. P. Curlee. 1996. Introduction: Large carnivore conservation in the Rocky Mountains of the United States and Canada. *Conserv Biol* 10:940–8.

Clark, T. W., and M. B. Rutherford. 2005. The institutional system of wildlife management: Making it more effective. In *Coexisting with large carnivores: Lessons from Greater Yellowstone,* ed. T. W. Clark, M. B. Rutherford, and D. Casey, 211–253. Washington, D.C.: Island Press.

Clark, T. W., M. B. Rutherford, and D. Casey, eds. 2005. *Coexisting with large carnivores: Lessons from Greater Yellowstone.* Washington, D.C.: Island Press.

Cleveland, T. 2007. Director's opinion. *Wyoming Wildl News,* July–August, p. 2.

Clevenger, A. P., B. Chruszcz, and K. E. Gunson. 2001. Drainage culverts as habitat linkages and factors affecting passage by mammals. *J Appl Ecol* 38:1340–9.

Clevenger, A. P., and N. Waltho. 2005. Performance indices to identify attributes of highway crossing structures facilitating movement of large mammals. *Biol Conserv* 121:453–64.

Clevenger, A. P., and J. Wierzchowski. 2006. Maintaining and restoring connectivity in landscapes fragmented by roads. In *Connectivity conservation,* ed. K. Crooks and M. Sanjayan, 502–535. Cambridge, UK: Cambridge University Press.

Clutton-Brock, T. H., and G. R. Iason. 1986. Sex ratio variation in mammals. *Q Rev Biol* 61:339–74.

Clutton-Brock, T. H., M. Major, and F. E. Guinness. 1985. Population regulation in male and female red deer. *J Anim Ecol* 54:831–46.

CMGWG. 2005. Puma guía de manejo, 1ª edición. Translated by L. Hernández. Wildfutures, Washington and Instituto de Ecologia, A.C., Veracruz, Mexico.

Collier, G. E., and S. J. O'Brien. 1985. A molecuar phylogeny of the Felidae: Immunological distance. *Evolution* 39:437–87.

Colorado Department of Natural Resources. 2007. Hunter education. Colorado Division of Wildlife. http://wildlife.state.co.us/Hunting/HunterEducation (accessed September 28, 2007).

Coltman, D. W., P. O'Donoghue, J. T. Jorgenson, J. T. Hogg, C. Strobeck, and M. Festa-Bianchet. 2003. Undesirable evolutionary consequences of trophy hunting. *Nature* 426:655–8.

Colwell, R. K., and D. J. Futumaya. 1971. On the measurement of niche breadth and overlap. *Ecology* 52:567–77.

Conforti, V. A., and F. C. C. de Azevedo. 2003. Local perceptions of jaguars (*Panthera onca*) and pumas (*Puma concolor*) in the Iguaçu National Park area, south Brazil. *Biol Conserv* 111:215–21.

Connolly, E. J. 1949. Food habits and life history of the mountain lion (*Felis concolor hippolestes*). Master's thesis, University of Utah, Salt Lake City.

Cook, R. C., J. G. Cook, and L. D. Mech. 2004. Nutritional condition of northern Yellowstone elk. *J Mammal* 85:714–22.

Cooley, H. S., H. S. Robinson, R. B. Wielgus, C. S. Lambert. 2008. Cougar prey selection in a white-tailed deer and mule deer community. *J Wildl Mgmt* 72:99–106.

COSEWIC. 2007. Canadian species at risk. Committee on the status of endangered wildlife in Canada. http://www.cosewic. gc.ca/eng/scto/rpt/rpt_csar_e.cfm (accessed September 2, 2007).

Costa, L. P., Y. L. R. Leite, S. L. Mendes, and A. D. Ditch-field. 2005. Mammal conservation in Brazil. *Conserv Biol* 19:672–9.

Cougar Management Guidelines Working Group. 2005. *Cougar management guidelines*. WildFutures, Bainbridge Island, WA.

The Cougar Network. 2009. http://easterncougarnet.org (accessed January 13, 2009).

Courchamp, F., R. Woodroffe, and G. Roemer. 2003. Removing protected populations to save endangered species. *Science* 302:1532.

Courtin, S. L., N. V. Pacheco, and W. D. Eldridge.1980. Observaciones de alimentacion, movimientos y preferencias de habitat del puma, en el islote rupanco. *Medio Ambiente* 4:50–55.

Cox, J. J., D. S. Maehr, and J. L. Larkin. 2006. Florida panther habitat use: New approach to an old problem. *J Wildl Mgmt* 70:1778–85.

Cox, M. 1996. Nevada. In *Proceedings of the fifth mountain lion workshop*, ed. W. D. Padley, 112–113. San Diego: Southern California Chapter of the Wildlife Society.

Crabtree, R. L., and J. W. Sheldon. 1999. Coyotes and canid coexistence in Yellowstone. In *Carnivores in ecosystems: The Yellowstone experience*, ed. T. W. Clark, A. P. Curlee, S. C. Minta, and P. M. Kareiva, 127–164. New Haven, CT: Yale University Press.

Craighead, F. C., and J. J. Craighead. 1965. Tracking grizzly bears. *Bioscience* 15:88–92.

Crawshaw, P. G., Jr. 1995. Comparative ecology of ocelot (*Felis pardalis*) and jaguar (*Panthera onca*) in a protected subtropical forest in Brazil and Argentina. Ph.D. dissertation, University of Florida, Gainesville.

———. 2003. A personal view on the depredation of domestic animals by large cats in Brazil. *Natureza and Conservação* 1:71–3.

Crawshaw, P. G., Jr., and H. B. Quigley. 2002. Food habits of jaguars and cougars in the Pantanal, Brazil. In *Jaguars in the new millennium*, ed. R. A. Medellin, C. Chetkiewicz, A. Rabinowitz, K. H. Redford, J. G. Robinson, E. Sanderson, and A. Taber, 223–235. National Autonomous University of Mexico, Mexico City. In Spanish with abstract in English.

Creeden, P. J., and V. K. Graham. 1997. Reproduction, survival, and lion predation in the Black Ridge/Colorado National Monument desert bighorn herd. *Desert Bighorn Council Trans* 41:37–43.

Creel, S., and N. M. Creel. 1996. Limitation of African wild dogs by competition with larger carnivores. *Conserv Biol* 10:526–38.

Creel, S., G. Spong, and N. Creel. 2001. Interspecific competition and the population biology of extinction–prone carnivores. In *Carnivore conservation*, ed. J. L. Gittleman, S. M. Funk, D. MacDonald, and R. K. Wayne, 35–60. Cambridge University Press, Cambridge, UK.

Creel, S., and J. Winnie, Jr. 2005. Responses of elk herd size to fine-scale spatial and temporal variation in the risk of predation by wolves. *Anim Beh* 69:1181–9.

Creel, S., J. Winnie, Jr., B. Maxwell, K. Hamlin, and M. Creel. 2005. Elk alter habitat selection as an antipredator response to wolves. *Ecology* 86:3387–97.

Crichton, V. F. J., W. E. Regelin, A. W. Franzmann, and C. C. Schwartz. 1998. The future of moose management and research. In *Ecology and management of the North American moose*, ed. A. W. Franzmann and C. C. Schwartz, 655–664. Herndon, VA: Smithsonian Institution Press.

Cronemiller, F. P. 1948. Mountain lion preys on bighorn. *J Mammal* 29:68.

Crowl, D. 2006. Familiar neighbors: Mountain lions thriving in areas of county. *The Daily Times-Call*, Longmont, CO, April 23.

Cruickshank, H. S. 2004. Prey selection and kill rates of cougars in northeastern Washington. Master's thesis, Washington State University, Pullman.

Cullens, L. W., and C. M. Papouchis. 2003. Community-based conservation of mountain lions (abstract). In *Proceedings of the seventh mountain lion workshop*, ed. S. A. Becker, D. D. Bjornlie, F. G. Lindzey, and D. S. Moody, p. 147. Lander: Wyoming Game and Fish Department.

Culver, M. 1999. Molecular genetic variation, population structure, and natural history of free-ranging pumas (*Puma concolor*). Ph.D. dissertation, University of Maryland, College Park.

Culver M. Hedrick P. W., K. Murphy, O'Brien S. J. and M. G. Hornocker. 2008. Estimation of the bottleneck size in Florida panthers. *Anim Conserv* 11:104–110.

Culver, M., W. E. Johnson, J. Pecon-Slattery, and S. J. O'Brien. 2000a. Genomic ancestry of the American puma (*Puma concolor*). *J Hered* 91:186–97.

———. 2000b. A phylogeographic study of pumas (*Puma concolor*) using mitochondrial DNA markers and microsatellites. *Proceedings of the sixth mountain lion workshop*. Texas Parks and Wildlife Department, Austin.

Culver, M., J. C. Stephens, M. A. Menotti-Raymond, K. Murphy, J. Laundré, L. Sweanor, K. Logan, M. Roelke-Parker, M. Hornocker, and S. J. O'Brien. In preparation. Use of microsatellite loci to examine kinship levels in free-ranging North American puma (*Puma concolor*) populations.

Cunningham, E. B. 1971. A cougar kills an elk. *Can Field Nat* 85:253–4.

Cunningham, S. C., L. A. Haynes, C. Gustavson, and D. D. Haywood. 1995. Evaluation of the interaction between mountain lions and cattle in the Aravaipa-Klondyke area of southeast Arizona. Research Branch Technical Report No. 17, Phoenix: Arizona Game and Fish Department.

Curio, E. 1976. *Ethology of predation*. Berlin: Springer-Verlag.

Currier, M. J. P. 1983. Felis concolor. *Mamm Species* 200:1–7.

Cushing, F. H. 1883. Zuni fetishes. Annual Report of the Bureau of American Ethnology to the Secretary of the Smithsonian Institution 2:9–43.

Dahl, R. A. 1982. *Dilemmas of pluralist democracy: Autonomy vs control*. New Haven, CT: Yale University Press.

———. 2006. *On political equality*. New Haven, CT: Yale University Press.

Dalrymple, G. H., and O. L. Bass. 1996. The diet of the Florida panther in Everglades National Park, Florida. *Bull Florida Mus Nat Hist* 39:173–93.

Danvir, R. E., and F. G. Lindzey. 1981. Feeding behavior of a captive cougar on mule deer. *Encyclia* 58:50–6.

Danz, H. P. 1999. *Cougar*. Athens, OH: Ohio University Press.

Darlington, P. J. 1975. *Zoogeography*. New York: John Wiley and Sons.

Darwin, C. 1859. *On the origin of species by means of natural selection*. London: J. Murray.

Dary, D. 2004. *The Oregon Trail: An American saga*. New York: Knopf.

Davis, M. B. 1996. *Eastern old-growth forests: Prospects for rediscovery and recovery*. Washington, D.C.: Island Press.

Dawkins, R. 2004. *The ancestor's tale: A pilgrimage to the dawn of evolution*. Boston: Houghton Mifflin.

Dawkins, R., and J. R. Krebs. 1979. Arms races between and within species. *Proc Roy Soc Lond* 205:489–511.

Dawn, D. 2002. Management of cougars (*Puma concolor*) in the western United States. Master's thesis, San Jose State University, San Jose, CA.

Dayer, A. A., M. J. Manfredo, T. L. Teel, and A. D. Bright. 2005. State report for Arizona from the research project entitled "Wildlife values in the West" Project Report No. 59 for Arizona Game and Fish Department. Colorado State University, Human Dimensions in Natural Resources Unit, Fort Collins, CO.

de Boer, W. F., and H. H. T. Prins. 1990. Large herbivores that strive mightily but eat and drink as friends. *Oecologia* 82:264–7.

de Oliveira, T. G. 2002. Ecología comparativa de la alimentación del jaguar y del puma en el Neotrópico. In *El Jaguar en el Nuevo Milenio*, ed. R. A. Medellín, C. Equihua, C. L. B. Chetkiewicz, P. G. Crawshaw Jr., A. Rabinowitz, H. H. Redford, J. G. Robinson, E. W. Sanderson, and A. B. Taber, 265–288. Universidad Nacional Autónoma de México/Wildlife Conservation Society, Mexico.

DeAzevedo, F. C. C., and D. L. Murray. 2007. Spatial organization and food habits of jaguars (*Panthera onca*) in a flood plain forest. *Biol Conserv* 137:391–402.

Decision Research. 2004. *Arizona statewide polling results*. San Diego: Decision Research.

Decker, D. J., T. L. Brown, and W. F. Siemer. 2001. *Human dimensions of wildlife management in North America*. Bethesda, MD: The Wildlife Society.

Decker, D. J., and L. C. Chase. 1997. Human dimensions of living with wildlife: A management challenge for the twenty-first Century. *Wildl Soc Bull* 25:788–95.

Decker, D. J., and G. R. Goff. 1987. *Valuing wildlife: Economic and social perspectives*. Boulder, CO: Westview Press.

Decker, D. J., C. C. Kruger, R. A. Baer, Jr., B. A. Knuth, and M. E. Richmond. 1996. From clients to stakeholders: A philosophical shift for fish and wildlife management. *Hum Dim Wildl* 1:70–82.

Delsink, A. K., H. J. Bertschinger, J. F. Kirkpatrick, D. Grobler, J. J. Van Altena, and R. Slotow. 2004. The preliminary behavioural and population dynamic response of African elephants to immunocontraception. In *Proceedings of the fifteenth symposium on tropical animal health and production*, 19–22. Utrecht University, Utrecht, Netherlands.

Demarais, S., K. V. Miller, and H. A. Jacobson. 2002. White-tailed deer. In *Ecology and management of large mammals in North America*, ed. S. Demarais and P. R. Krausman, 601–628. Upper Saddle River, NJ: Prentice Hall.

DeSimone, R., V. Edwards, and B. Semmens. 2005. Montana mountain lion status report. In *Proceedings of the eighth mountain lion workshop*, ed. R. A. Beausoleil and D. A. Martorello, 22–25. Olympia: Washington Department of Fish and Wildlife.

DeSimone, R., and R. Jaffe. 2003. Montana mountain lion status report. In *Proceedings of the seventh mountain lion workshop*, ed. S. A. Becker, D. D. Bjornlie, F. G. Lindzey, and D. S. Moody, 29–30. Lander: Wyoming Game and Fish Department.

DeSimone, R., and B. Semmens. 2005. Garnet Mountains mountain lion research progress report. Montana Fish, Wildlife, and Parks.

Deurbrock, J., and D. Miller. 2001. *Cat attacks—True stories and hard lessons from cougar country*. Seattle: Sasquatch Books.

Devall, B., and G. Sessions. 1985. *Deep ecology: Living as if nature mattered*. Salt Lake City: Gibbs Smith Publishers.

DeVault, T. L., O. E. Rhodes, Jr., and J. A. Shivik. 2003. Scavenging by vertebrates: Behavioral, ecological, and evolutionary perspectives on an important energy transfer pathway in terrestrial ecosystems. *Oikos* 102:225–34.

DeVos, J. C., Jr., D. L. Shroufe, and V. C. Supplee. 1998. Managing wildlife by ballot initiative: The Arizona experience. *Hum Dim Wildl* 3:60–6.

Dias, P. C. 1996. Sources and sinks in population biology. *Trends Ecol Evol* 11:326–30.

Dice, L. R. 1925. The scientific value of predatory mammals. *J Mammal* 6:25–7.

Dickson, B. G., and P. Beier. 2002. Home range and habitat selection of adult cougars in southern California. *J Wildl Mgmt* 66:1235–45.

———. 2007. Quantifying the influence of topographic position on cougar movement in southern California, USA. *J Zool London* 271:270–7.

Dickson, B. G., J. S. Jenness, and P. Beier. 2005. Influence of vegetation, topography, and roads on cougar movement in southern Califorinia. *J Wildl Mgmt* 69:264–76.

Dixon, J. 1925. Food predilections of predatory and furbearing mammals. *J Mammal* 6:34–46.

Dixon, K. R., and R. J. Boyd. 1967. Evaluation of the effects of mountain lion predation. Job completion report, research project segment. Colorado Game, Fish and Parks Department. Research Center Library, Fort Collins, CO.

Dizard, J. E. 2003. *Mortal stakes: Hunters and hunting in contemporary America*. Amherst, MA: University of Massachusetts Press.

Dobzhansky, T. 1950. Evolution in the tropics. *Am Sci* 38:208–21.

Donadio, E., A. J. Novaro, S. W. Buskirk, A. Wurstten, M. S. Vitali, and M. J. Monteverde. 2008. Evaluating a potentially strong trophic interaction: Pumas and wild camelids in protected areas of Argentina. Unpublished manuscript.

Donadio, E., and S. W. Buskirk. 2006. Diet, morphology, and interspecific killing in carnivora. *Am Nat* 167:524–36.

Driscoll, C. A., M. Menotti-Raymond, G. Nelson, D. Goldstein, and S. J. O'Brien. 2002. Genomic microsatellites as evolutionary chronometers: A test in wild cats. *Genome Res* 12:414–23.

Duda, M. D., and K. C. Young. 1997. Floridians' knowledge, opinions and attitudes toward panther habitat and panther-related issues. The Florida Panther Society, Inc. http://www.panthersociety.org/duda.html (accessed October 19, 2006).

———. 1998. American attitudes toward scientific wildlife management and human use of fish and wildlife: Implications for effective public relations and communications strategies. *Trans N Am Wildl Nat Res* 63:589–603.

Duke, D. L., M. Hebblewhite, P. C. Paquet, C. Callaghan, and M. Percy. 2001. Restoring a large-carnivore corridor in Banff National Park. In *Large mammal restoration: Ecological and sociological challenges in the twenty-first century*, ed. D. S. Maehr, R. F. Noss, and J. L. Larkin, 261–275. Washington, D.C.: Island Press.

Duke, R. R., R. C. Klinger, R. Hopkins, and M. Kutilek. 1987. Yuma puma (*Felis concolor browni*) feasibility report population status survey. Harvey and Stanley Associates Report, File No 277–01.

Dunlap, R. E. 1989. Public opinion and environmental policy. In *Environmental politics and policy: Theories and evidence*, ed. J. P. Lester, 87–134. Durham, NC: Duke University Press.

———. 1992. Trends in public opinion toward environmental issues: 1965–1990. In *American environmentalism: The U.S. enviromental movement 1970–1990*, ed. R. E. Dunlap and A. G. Mertig, 89–116. New York: Taylor and Francis.

Dunlap, T. R. 1988. *Saving America's wildlife*. Princeton, NJ: Princeton University Press.

———. 2006. Environmentalism: A secular faith. *Environ Value* 15:321–30.

Durant, R. F., D. J. Fiorino, and R. O'Leary. 2004. *Environmental governance reconsidered: Challenges, choices and opportunities*. Cambridge, MA: MIT Press.

Dzus, E. 2001. Status of the woodland caribou (*Rangifer tarandus caribou*) in Alberta. Wildlife status report 30. Alberta Environment, Fisheries and Wildlife Management Division, Edmonton, Alberta.

Eaton, R. L. 1976. Why some felids copulate so much. In *The world's cats*, ed. R. L. Eaton, 73–94. Seattle: Carnivore Research Institute.

Eaton, R. L., and K. A. Velander. 1977. Reproduction in the puma: Biology, behavior, and ontogeny. In *The world's cats*, ed. R. L. Eaton, 45–70. Seattle: Carnivore Research Institute.

Echeverria, J., and R. B. Eby. 1995. *Let the people judge: Wise use and the property rights movement*. Washington, D.C.: Island Press.

Edmonds, E. J. 1991. Status of woodland caribou in western North America. *Rangifer Special Issue* 7:91–107.

Einwohner, R. L. 1999a. Gender, class, and social movement outcomes: Identity and effectiveness in two animal rights campaigns. *Gender Soc* 13:56–76.

———. 1999b. Practices, opportunities, and protest effectiveness: Illustrations from four animal rights campaigns. *Soc Probl* 46:169–86.

Eisenberg, J. F., and P. Leyhausen. 1972. The phylogenesis of predatory behavior in mammals. *Z Tierpsychology* 30:59–93.

Eisenberg, J. 1989. *Mammals of the Neotropics. Vol. 1, The Northern Neotropics*. Chicago: University of Chicago Press.

Ellingson, J. J. 2003. Mountain lion habitat use, activity patterns, movements, and human interaction in Redwood National and State Parks, California. Thesis, University of Idaho, Moscow, ID.

El Diario de La Pampa. 2007a. Involucran a La Pampa en casos de tráfico de pumas para sacrificar. *El Diario*, January 17.

———. 2007b. La provincia fue alertado por "tráfico" de pumas. *El Diario*, January 18.

———. 2007c. Sugieron que los pumas han sido traficados desde Cordoba. *El Diario*, January 29.

Emmons, L. H. 1987. Comparative feeding ecology of felids in a neotropical rainforest. *Beh Ecol Soc* 20:271–83.

Erdoes, R. 1976. *Lame Deer: Seeker of visions*. New York: Simon and Schuster.

Ericsson, G., and T. A. Heberlein. 2003. Attitudes of hunters, locals, and the general public in Sweden now that the wolves are back. *Biol Conserv* 111:149–59.

Ernest, H. B., W. M. Boyce, V. C. Bleich, B. May, S. J. Stiver, and S. G. Torres. 2003. Genetic structure of mountain lion (*Puma concolor*) populations in California. *Conserv Genet* 4:353–66.

Ernest, H. B., M. C. T. Penedo, B. P. May, M. S. Syvanen, and W. M. Boyce. 2000. Molecular tracking of mountain lions in the Yosemite Valley region in California: Genetic analysis using microsatellites and faecal DNA. *Mol Ecol* 9:433–41.

Ernest, H. B., E. S. Rubin, and W. M. Boyce. 2002. Fecal DNA analysis and risk assessment of mountain lion predation of bighorn sheep. *J Wildl Mgmt* 66:75–85.

Errington, P. 1946. Predation and vertebrate populations. *Q Rev Biol* 21:144–7.

Escamilla, A., M. Sanvicente, M. Sosa, and C. Galindo-Leal. 2000. Habitat mosaic, wildlife availability, and hunting in the tropical forest of Calakmul, Mexico. *Conserv Biol* 14:1592–601.

Estes, J. A. 1996. Predators and ecosystem management. *Wildl Soc Bull* 24:390–6.

Estes, J. A., D. O. Duggins, and G. B. Rathbun. 1989. The ecology of extinctions in kelp forest communities. *Conserv Biol* 3:252–64.

Etling, K. 2001. *Cougar attacks: Encounters of the worst kind*. Guilford, CT: The Lyons Press.

Evans, S. B., L. D. Mech, P. J. White, and G. A. Sargeant. 2006. Survival of adult female elk in Yellowstone following wolf restoration. *J Wild Mgmnt* 70:1372–8.

Evans, W. 1976. Status report for New Mexico. In *Proceedings of the first mountain lion workshop*, pp. 25–27. Portland: U. S. Fish and Wildlife Service.

Ewer, R. F. 1973. *The carnivores*. Ithaca, NY: Cornell University Press.

Farner, D. S. 1955. Birdbanding in the study of population dynamics. In *Recent studies in avian biology*, ed. A. Wolfson, 397–449. Urbana, IL: University of Illinois Press.

Farrell, L., and M. Sunquist. 1999. La ecología del puma y el jaguar en los llanos Venezolanos. In *Manejo y Conservación de Fauna Silvestre en America Latina*, ed. T. G. Fany, O. L. Montenegro, and R. E. Bodmer, 391–396. Instituto de Ecología, NDP/GEF/UNAP and the University of Florida.

Fascione, N., A. Delach, and M. E. Smith. 2004. *People and predators: From conflict to coexistence*. Washington, D.C.: Island Press.

Fecske, D. M. 2003. Distribution and abundance of American martens and cougars in the Black Hills of South Dakota and Wyoming. Ph.D. dissertation, South Dakota State University, Brookings.

Fecske, D. M. 2006. Mountain lions in North Dakota. *Wild Cat News*, April. Also at http://www.cougarnet.org/wcn2-1/North%20Dakota%5b1%5d.pdf

Felicetti, L. A., C. C. Schwartz, R. O. Rye, M. A. Haroldson, K. A. Gunther, and C. T. Robbins. 2003. Use of sulfur and nitrogen stable isotopes to determine the importance of whitebark pine nuts to Yellowstone grizzly bears. *Can J Zool* 81:763–70.

Findholt, S. L., and B. K. Johnson. 2005. Cougars in Oregon: Biopolitics of a recent research project (abstract). In *Proceedings of the eighth mountain lion workshop*, ed. R. A. Beausoleil and D. A. Martorello, p. 210. Olympia: Washington Department of Fish and Wildlife.

Finke, D. L., and R. F. Denno. 2005. Predator diversity and the functioning of ecosystems: The role of intraguild predation in dampening tropic cascades. *Ecol Lett* 8:1299–306.

Fischer, A. G. 1960. Latitudinal variation in organic diversity. *Evolution* 14:64–81.

Fischer, F. 1995. *Greening environmental policy: The politics of a sustainable future*. New York: Saint Martin's Press.

———. 2000. *Citizens, experts, and the environment: The politics of local knowledge*. Durham, NC: Duke University Press.

Fitzhugh, E. L. 1988. Managing with potential for lion attacks against humans. In *Proceedings of the third mountain lion workshop*, ed. R. H. Smith Phoenix: Arizona Game and Fish Department.

Fitzhugh, E. L., S. Schmid-Holmes, M. W. Kenyon, and K. Etling. 2003. Lessening the impact of a puma attack on a human. In *Proceedings of the seventh mountain lion workshop*, ed. S. A. Becker, D. D. Bjornlie, F. G. Lindzey, and D. S. Moody, 89–103. Lander: Wyoming Game and Fish Department.

Fjelline, D. P., and T. M. Mansfield. 1988. Method to standardize the procedure for measuring mountain lion tracks. In *Proceedings of the third mountain lion workshop*, ed. R. H. Smith, p. 49. Phoenix: The Wildlife Society and Arizona Game and Fish Department.

Flather, C. H., and H. K. Cordell. 1995. Outdoor recreation: Historical and anticipated trends. In *Wildlife and recreationists: Coexistence through management and research*, ed. R. Knight and K. Gutzwiller, 3–16. Washington, D.C.: Island Press.

Fleischner, T. L., 1994. Ecological costs of livestock grazing in western North America. *Conserv Biol* 8:629–44.

Floyd, M. F. 1997. Pleasure, arousal, and dominance: Exploring affective determinants of recreation satisfaction. *Leisure Sci* 19:83–96.

Foerster, C. 1996. Researcher attacked by puma in Corcovado National Park, Costa Rica. *Vida Silvestre Neotropical* 5:57–8.

Forsyth, T. 2003. *Critical political ecology: The politics of environmental science*. London: Routledge.

Fortin, D., H. L. Beyer, M. S. Boyce, D. W. Smith, T. Duchesne, and J. S. Mao. 2005. Wolves influence elk movements: Behavior shapes a trophic cascade in Yellowstone National Park. *Ecology* 86:1320–30.

Foster, D., G. Motzkin, J. O'Keefe, E. Boose, D. Orwig, J. Fuller, and B. Hall. 2004. The environmental and human history of New England. In *Forests in time: The environmental consequences of 1,000 years of change in New England*, ed. D. R. Foster and J. D. Aber, 43–100. New Haven, CT: Yale University Press.

Foster, M. L., and S. R. Humphrey. 1995. Use of highway underpasses by Florida panthers and other wildlife. *Wildl Soc Bull.* 23:95–100.

Fox, S. 1981. *The American conservation movement: John Muir and his legacy*. Madison: University of Wisconsin Press.

Franklin, W. L., W. E. Johnson, R. J. Sarno, and J. A. Iriarte. 1999. Ecology of the Patagonia puma (*Felis concolor patagonica*) in southern Chile. *Biol Conserv* 90:33–40.

Franzmann, A. W., and P. D. Arneson. 1976. Marrow fat in Alaskan moose femurs in relation to mortality factors. *J Wild Mgmnt* 40:336–9.

Freddy, D. J., G. C. White, M. C. Kneeland, R. H. Kahn, J. W. Unsworth, W. J. deVergie, V. K. Graham, J. H. Ellenberger, and C. H. Wagner. 2004. How many mule deer are there? Challenges of credibility in Colorado. *Wildl Soc Bull* 32:916–27.

Freeman, S., and J. C. Herron. 1998. *Evolutionary analysis*. Upper Saddle River, NJ: Prentice Hall.

Fritts, S. H., R. O. Stephenson, R. D. Hayes, and L. Boitani. 2003. Wolves and humans. In *Wolves: Behavior, ecology, and conservation*, ed. L. D. Mech and L. Boitani, 289–316. Chicago: University of Chicago Press.

Fryxell, J. M., J. Greever, and A. R. E. Sinclair. 1988. Why are migratory ungulates so abundant? *Am Nat* 131:781–98.

Fuller, T. K., L. D. Mech, and J. F. Cochrane. 2003. Wolf population dynamics. In *Wolves: Behavior, ecology, and conservation*, ed. L. D. Mech and L. Boitani, 161–191. Chicago: University of Chicago Press.

Fuller, T. K., and P. R. Sievert. 2001. Carnivore demography and the consequences of changes in prey availability. In *Carnivore conservation*, ed. J. L. Gittleman, S. M. Funk, D. Macdonald, and R. K. Wayne, 163–178. Cambridge, UK: Cambridge University Press.

Funes, M. C., A. J. Novaro, O. B. Monsalvo, O. Pailacura, G. Sanchez Aldao, M. Pessino, R. Dosio, C. Chehébar, E. Ramilo, J. Bellati, S. Puig, F. Videla, N. Oporto, R. González del Solar, E. Castillo, E. García, N. Loekemeyer, F. Bugnest y G. Matcazzi. 2006. El manejo de los zorros en Argentina. Compatibilizando las interacciones entre la ganadería, la caza comercial y la conservación. In *Manejo de Fauna Silvestre en la Argentina*, ed. M. L. Bolkovic and D. E. Ramadori, 151–166. Programas de uso sustentable, Dirección de Fauna Silvestre, Secretaría de Ambiente y Desarrollo Sustentable, Buenos Aires, Argentina.

FVSN. 2006. Programa Nacional para la conservación de los felinos en Colombia. Ministerio de Ambiente, Vivienda y Desarrollo Territorial, Columbia.

Gabrielson, I. N. 1941. *Wildlife conservation*. New York: The Macmillan Company.

Galvin, S. L., and H. A. Herzog, Jr. 1992. Ethical ideology, animal rights activism, and attitudes towards treatment of animals. *Ethics Behav* 2:141–9.

Garreston, M. S. 1938. *The American bison*. New York: New York Zoological Society.

Garshelis, D., and H. Hristienko. 2006. State and provincial estimates of American black bear numbers versus assessments of population trend. *Ursus* 17:1–7.

Gashwiler, J. S., and W. L. Robbinette. 1957. Accidental fatalities of the Utah cougar. *J Mammal* 38:123–64.

Gauthier, M., C. Lanthier, F. J. La Pointe, L. D. Lang, N. Tessier, and V. Stroeher. 2005. Cougar tracking in the Northeast: Years of research finally rewarded. In *Proceedings of the eighth mountain lion workshop*, ed. R. A. Beausoleil and

D. A. Martorello, 86–88. Olympia: Washington Department of Fish and Wildlife.

Gay, S. W., and T. L. Best. 1995. Geographic variation in sexual dimorphism of the puma (*Puma concolor*) in North and South America. *Southwest Nat* 40:148–59.

Geist, V. 1981. Behavior: Adaptive strategies in mule deer. In *Mule and black-tailed deer of North America*, ed. O. C. Wallmo, 157–223. Lincoln, NE: University of Nebraska Press.

———. 1982. Adaptive behavioral strategies. In *Elk of North America: Ecology and management*, ed. J. W. Thomas and D. E. Toweill, 219–277. Harrisburg, PA: Stackpole Books.

Gibson, L. 2006. The role of lethal control in managing the effects of apparent competition on endangered prey species. *Wildl Soc Bull* 34:1220–4.

Gigliotti, L. M. 2002. Report to survey participants: Mountain lions in South Dakota. A public opinion survey. Pierre: South Dakota Department of Game, Fish and Parks.

———. 2005. Analysis of public opinions towards mountain lion management in South Dakota. Pierre: South Dakota Department of Game, Fish and Parks, Division of Wildlife.

———. 2006. Wildlife values and beliefs of Idaho residents – 2004. Report prepared for Idaho Fish and Game. Pierre: Human Dimensions Consulting.

Gigliotti, L. M., and A. K. Harmoning. 2004. Evaluation of the North Dakota Game and Fish Department using a communications assessment model. *Hum Dim Wildl* 9:79–81.

Gilbert, F. F., and D. G. Dodds. 1987. *The philosophy and practice of wildlife management*. Malabar, FL: Krieger Publishing.

Gill, F. B. 1982. Might there be a resurrection of the subspecies? *Auk* 99:598–659.

Gill, R. B. 1996a. The wildlife professional subculture: The case of the crazy aunt. *Hum Dim Wildl* 1:60–9.

———. 1996b. Thunder in the distance: The emerging policy debate over wildlife contraception. In *Contraception in wildlife management*, ed. T. J. Kreeger. Washington, D.C.: U.S. Government Printing Office.

———. 2001a. Declining mule deer populations in Colorado: Reasons and responses. Special Report No. 77. Colorado Division of Wildlife, Denver.

———. 2001b. Professionalism, advocacy, and credibility: A futile cycle? *Hum Dim Wildl* 6:21–32.

———. 2002. Build an experience and they will come. In *Wildlife viewing: A management handbook*, ed. M. J. Manfredo, 218–253. Corvallis, OR: Oregon State University Press.

Gilpin, M. 1996. Metapopulations and wildlife conservation: Approaches to modeling spatial structure. In *Metapopulations and wildlife conservation*, ed. D. R. McCullough, 11–27. Covelo, CA: Island Press.

Gittleman, J. L., S. M. Funk, D. Macdonald, and R. K. Wayne. 2001. *Carnivore conservation*. New York: Cambridge University Press.

Gniadek, S. J., and K. C. Kendall. 1998. A summary of bear management in Glacier National Park, Montana, 1960–1994. *Ursus* 10:155–9.

Golluscio, R., J. Paruelo, J. Mercau and V. Deregibus. 1998. Urea supplementation effects on low-quality forage utilization and lamb production in Patagonian rangelands. *Grass Forage Sci* 53:47–56.

Gonyea, W. J. 1976. Adaptive differences in the body proportions of large felids. *Acta Anat* 96:81–96.

Gonyea, W. J., and R. Ashworth. 1975. The form and function of retractile claws in the Felidae and other representative carnivorans. *J Morph* 145:229–38.

Goodrich, J. M., and S. W. Buskirk. 1995. Control of abundant native vertebrates for conservation of endangered species. *Conserv Biol* 9:1357–64.

Gottlieb, R. 2004. *Forcing the spring: The transformation of the American environmental movement*. Washington, D.C.: Island Press.

Graham, P. J. 1997. The evolution of a fish and wildlife agency. In *Proceedings of the seventy-seventh annual conference of the Western Association of Fish and Wildlife Agencies*, 91–95. Cheyeme, WY: Western Association of Fish and Wildlife Agencies.

Green, G. I., D. J. Mattson, and J. M. Peek. 1997. Spring feeding on ungulate carcasses by grizzly bears in Yellowstone National Park. *J Wildl Mgmt* 61:1040–55.

Groom, M. J., G. K. Meffe, and C. R. Carroll. 2006. *Principles of conservation biology*. Sunderland, MA: Sinauer Associates, Inc.

Gross, L. 2008. No place for predators? *PLoS Biol* 6:e40.

Guerra-Benítez, M. F., J. W. Laundré, and E. Martínez-Meyer. 2007. Modeling spatial puma distribution in the Mexican Chihuahuan Desert through the Mahalanobis Distance method. In Review.

Guggisberg, C. A. W. 1975. *Wild cats of the world*. New York: Taplinger Publishing Co.

Guichon, M., and A. Novaro. 2007. Dinámica poblacional del guanaco (*Lama guanicoe*) en el norte de la provincia de Neuquén: una aproximación desde la ecología de paisaje para su conservación y manejo. Unpublished post-doctoral research.

Gunther, K. A., 1994. Bear management in Yellowstone National Park, 1960–93. In *Proceedings of the ninth international conference on bear research and management*, 549–60. Missoula, MT: International Association for Bear Research and Management.

Gunzburger, M. S., and J. Travis. 2005. Effects of multiple predator species on green treefrog (*Hyla cinerea*) tadpoles. *Can J Zool* 83:996–1002.

Hadingham, E. 1979. *Secrets of the Ice Age: The world of the cave artists*. New York: Walker.

Hagood, S. 1997. *State wildlife management: The pervasive influence of hunters, hunting, culture and money*. Washington, D.C.: The Humane Society of the United States.

Haines, F. 1970. *The buffalo: The story of bison and their hunters from prehistoric times to the present*. New York: Crowell.

Halfpenny, J. C., M. R. Sanders, and K. A. McGrath. 1991. Human-lion interactions in Boulder County, Colorado: Past, present, and future. In *Mountain lion-human interaction symposium and workshop*, ed. C. E. Braun. Denver: Colorado Division of Wildlife.

Hamlin, K. L., S. J. Riley, D. Pyrah, A. R. Dood, and R. J. Mackie. 1984. Relationships among mule deer fawn mortality, coyotes, and alternate prey species during summer. *J Wildl Mgmt* 48:489–99.

Hansen, K. 1992. *Cougar, the American lion*. Flagstaff, AZ: Northland Publishing.

———. 1993. *Cougar, the American lion*. Flagstaff, AZ: Northland Publishing.

———. 2007. *Bobcat: Master of survival*. New York: Oxford University Press.

Hanski, I. A., and M. E. Gilpin. 1997. *Metapopulation biology: Ecology, genetics, and evolution.* San Diego: Academic Press.

Harlow, H. J., F. G. Lindzey, and W. A. Gern. 1992. Stress response of cougars to non-lethal pursuit by hunters. *Can J Zool* 70:136–9.

Harris, R. B., W. A. Wall, and F. W. Allendorf. 2002. Genetic consequences of hunting: What do we know and what should we do? *Wildl Soc Bull* 30:634–43.

Harrison, S. 1990. Cougar predation on bighorn sheep in the Junction Wildlife Management Area, British Columbia. Master's thesis, University of British Columbia, Vancouver.

Harveson, L. A. 1997. Ecology of a mountain lion population in southern Texas. Ph.D. dissertation, Texas A&M University, Kingsville.

Harveson, L. A., P. M. Harveson, and R.W.Adams, eds. 2003. *Proceedings of the sixth mountain lion workshop.* Austin: Texas Parks and Wildlife Department.

Harveson, L. A., M. E. Tewes, N. J. Silvy, and J. Rutledge. 1997. Mountain lion research in Texas: Past, present, and future. In *Proceedings of the fifth mountain lion workshop,* ed. W. D. Padley, 40–43. San Diego: California Department of Fish and Game.

Hass, C. C. 1989. Bighorn lamb mortality: Predation, inbreeding, and population effects. *Can J Zool* 67:699–705.

Hass, C. C., and D. Valenzuela. 2002. Anti-predator benefits of group living in white-nosed coatis (*Nasua narica*). *Behav Ecol Sociobiol* 51:570–8.

Hatch, C. 2006. State holds lion's fate: Conservationists contend that protection for female cats crucial to healthy population. *Jackson Hole News and Guide,* June 19.

Hayes, C. J., E. S. Rubin, M. C. Jorgensen, R. A. Botta, and W. M. Boyce. 2000. Mountain lion predation of bighorn sheep in the peninsular ranges, California. *J Wildl Mgmt* 64:954–9.

Hayes, R. D., R. Farnell, R. M. Ward, J. Carey, M. Dehn, G. Kuzyk, A. Baer, C. Gardner, and M. O'Donoghue. 2003. Experimental reduction of wolves in the Yukon: Ungulate responses and management implications. *Wildl Monogr* 152:1–35.

Haynes, L., D. Swann, and M. Culver. 2003. Monitoring mountain lions in Tucson Mountain District of Saguaro National Park, Arizona, using noninvasive techniques. In *Proceedings of the seventh mountain lion workshop,* ed. S. A. Becker, D. D. Bjornlie, F. G. Lindzey, and D. S. Moody. Lander: Wyoming Game and Fish Department and Wyoming Cooperative Fish and Wildlife Research Unit.

Healey, R. G., and W. Ascher. 1995. Knowledge in the policy process: Incorporating new environmental information in natural resources policy making. *Policy Sci* 28:1–19.

Heaven, P. C. L., and S. Bucci. 2001. Right-wing authoritarianism, social dominance orientation and personality: An analysis using the IPIP measure. *Eur J Personality* 15:49–56.

Hedrick, P. W. 1995. Gene flow and genetic restoration: The Florida panther as a case study. *Conserv Biol* 9:996–1007.

———. 1996. Genetics of metapopulations: Aspects of a comprehensive perspective. In *Metapopulations in wildlife conservation,* ed. D. R. McCullough, 29–51. Washington, D.C: Island Press.

Heffelfinger, J. 2006. *Deer of the southwest: A complete guide to the natural history, management, and biology of southwestern mule deer and white-tailed deer.* College Station, TX: Texas A&M University Press.

Heilprin, A. 1974. *The geographical and geological distribution of animals.* New York: Arno Press.

Heisey, D. M., and T. K. Fuller. 1985. Evaluation of survival and cause-specific mortality rates using telemetry data. *J Wildl Mgmt* 49:668–74.

Helvarg, D. 1997. *The war against the greens: The "wise-use" movement, the new right, and anti-environmental violence.* San Francisco: Sierra Club Books.

Hemker, T. P. 1982. Population characteristics and movement patterns of cougars in southern Utah. Master's thesis, Utah State University, Logan.

Hemker, T. P., F. G. Lindzey, and B. B. Ackerman. 1984. Population characteristics and movement patterns of cougars in southern Utah. *J Wildl Mgmt* 48:1275–84.

Hemming, J. 1970. *The conquest of the Incas.* San Diego: Harcourt Brace and Company.

Herbert, D. 1988. The status and management of cougar in British Columbia. In *Proceedings of the third mountain lion workshop,* ed. R. H. Smith, 11–14. Phoenix: Arizona Game and Fish Department.

Hernández, L., and J. W. Laundré. 2000. Status of the puma in the Mexican Chihuahuan Desert. In *Proceedings of the sixth mountain lion workshop,* ed. L. A. Harveson, P. M. Harveson, and R. W. Adams, p. 60. Austin: Texas Parks and Wildlife Department.

———. 2005. Foraging in the "landscape of fear" and its implications for habitat use and diet quality of elk (*Cervus elaphus*) and bison (*Bison bison*). *Wildl Biol* 11:215–20.

Hess, Jr., K. 1992. *Visions upon the land: Man and nature on the western range.* Washington, D.C.: Island Press.

Heydlauff, A. L., P. R. Krausman, W. W. Shaw, and S. E. Marsh 2006. Perceptions regarding elk in northern Arizona. *Wildl Soc Bull* 34:27–35.

Hibben, F. C. 1937. A preliminary study of the mountain lion (*Felis oregonensis sp*). University of New Mexico Bulletin No. 318. Albuquerque: University of New Mexico Press.

———. 1939. The mountain lion and ecology. *Ecology* 20:584–86.

———. 1948. *Hunting American lions.* New York: Thomas Y. Crowell Company.

Hildebrand, M. 1959. Motions of the running cheetah and horse. *J Mammal* 40:481–95.

Hills, A. M. 1993. The motivational basis of attitudes toward animals. *Soc Anim* 1:111–28.

Hixon, M. A. 1980. Food production and competitor density as the determination of feeding territory size. *Am Nat* 4:510–30.

Hoctor, T. S., M. H. Carr, and P. D. Zwick. 2000. Identifying a linked reserve system using a regional landscape approach: The Florida ecological network. *Conserv Biol.* 14:984–1000.

Hoffman, C. 2004. Welcome to the neighborhood. *Adventure,* October.

Hold, W. V., A. R. Pickard, and R. S. Prather. 2004. Wildlife conservation and reproductive cloning. *Reproduction* 127:317–24.

Holechek, J. L., R. D. Pieper, and C. H. Herbel. 2004. *Range management: Principles and practices.* Upper Saddle River, NJ: Prentice-Hall.

Holling, C. S. 1959. The components of predation as revealed by a study of small-mammal predation of the European pine sawfly. *Can Entomol* 91:293–320.

———. 1965. The functional response of predators to prey density and its role in mimicry and population regulation. *Mem Entomol Soc Can* 45:1–60.

Holmes, B. R., and J. W. Laundré. 2006. Use of open, edge and forest areas by pumas (*Puma concolor*) in winter: Are pumas foraging optimally? *Wildl Biol* 12:201–9.

Holt, R. D. 1977. Predation, apparent competition, and the structure of prey communities. *Theor Popul Biol* 12:197–229.

Holt, R. D., and G. A. Polis. 1997. A theoretical framework for intraguild predation. *Am Nat* 149:745–64.

Holt, S. J., and L. M. Talbot. 1978. New principles for the conservation of wild living resources. *Wildl Monogr* 59:3–33.

Holt, W. V., P. M. Bennett, V. Volobouev, and P. F. Watson. 1996. Genetic resource banks in wildlife conservation. *J Zool London* 238:531–44.

Hone, J., and T. H. Clutton-Brock. 2007. Climate, food, density and wildlife population growth rate. *J Anim Ecol* 76:361–7.

Hoogesteijn, R. 2000. *Manual on the problem of depredation caused by jaguars and pumas on cattle ranches.* New York: Wildlife Conservation Society.

Hoogesteijn, R., A. Hoogesteijn, and E. Mondolfi. 1993. Jaguar predation and conservation: Cattle mortality caused by felines on three ranches in the Venezuelan Llanos. In *Mammals as predators,* ed. N. Dunstone and M. L. Gorman, 391–407. Symposia Zoological Society of London 65. London: Oxford University Press.

Hook, R. A., and W. L. Robinson. 1982. Attitudes of Michigan citizens toward predators. In *Wolves of the world: Perspectives of behavior, ecology, and conservation,* ed. F. H. Harrington and P. C. Paquet, 382–394. Park Ridge, NJ: Noyes Publications.

Hopkins, R. A. 1990. Ecology of the puma in the Diablo Range, California. Ph.D. dissertation, University of California, Berkeley.

———. 2003. Mystery, myth and legend: The politics of cougar management in the new millennium (abstract). In *Proceedings of the seventh mountain lion workshop,* ed. S. A. Becker, D. D. Bjornlie, F. G. Lindzey, and D. S. Moody, 145. Lander: Wyoming Game and Fish Department.

Hopkins, R. A., M. J. Kutilek, and G. L. Shreve. 1986. Density and home range characteristics of mountain lions in the Diablo Range of California. In *Cats of the world: Biology, conservation, and management,* ed. S. D. Miller and D. D. Everett, 223–235. Washington, D.C.: National Wildlife Federation.

Horn, H. S. 1966. Measurement of "overlap" in comparative ecological studies. *Am Natur* 100:419–24.

Hornocker, M. G. 1969. Winter territoriality in mountain lions. *J Wildl Mgmt* 33:457–64.

———. 1970. An analysis of mountain lion predation upon mule deer and elk in the Idaho Primitive Area. *Wildl Monogr* 21:3–39.

———. 1971. Suggestions for the management of mountain lions as trophy species in the intermountain region, in *proceedings of the fifty-first Annual Conferece of Western Association of State Game and Fish Commissioners,* 399–402. Salt Lake City: Utah Game and Fish Department.

———. 1991. A synopsis of the symposium and challenges for the future. In *Mountain lion-human interaction symposium and workshop,* ed. C. E. Braun. Denver: Colorado Division of Wildlife.

———. 1996. The mountain lion in western North America: A modern-day success story. In *The Chiles Award Papers.* Bend, OR: The High Desert Museum.

———. 2007. Conserving the cats: Cougar management as a model: A review. In *Wildlife science: Linking ecological theory and management applications,* ed. T. E. Fulbright and D. G. Hewitt. Boca Raton, FL: CRC Press.

Hornocker, M. G., and H. S. Hash. 1981. Ecology of the wolverine in northwestern Montana. *Can J Zool* 59:1286–1301.

Houston, D. B. 1982. *The northern Yellowstone elk.* New York: MacMillan Publishing.

Houston, D. C. 1988. Digestive efficiency and hunting behavior in cats, dogs, and vultures. *J Zool London* 216:603–5.

Howard, W. E. 1988. Why lions need to be hunted. In *Proceedings of the third mountain lion workshop,* ed. R. H. Smith, 66–68. Phoenix: Arizona Game and Fish Department.

———. 1991. Mountain lion and the Bambi syndrome. In *Mountain lion–human interaction symposium and workshop,* ed. C. E. Braun, 96–97. Denver: Colorado Division of Wildlife.

Hudson, P. J., A. P. Dobson, and D. Newborn. 1992. Do parasites make prey vulnerable to predation? Red grouse and parasites. *J Anim Ecol* 61:681–92.

Hughes, A. 1977. The topography of vision in mammals of contrasting lifestyle: Comparative optics and retinal organization. In *The visual system in vertebrates: Handbook of sensory physiology* 7(5):613–756. Berlin: Springer.

Hughes, D. 1996. *North American Indian ecology.* El Paso: Texas Western Press.

Hultkrantz, A. 1981. *Belief and worship in native North America.* Syracuse, NY: Syracuse University Press.

Hummel, M., and S. Pettigrew. 1991. *Wild hunters: Predators in peril.* Toronto: Roberts Rinehart Publishers.

Hunt, V. D. 1993. Quality management for government. Pp. 21–52 in *A government manager's guide to quality management.* Milwaukee, WI: ASQC Quality Press.

Hunter, R. 1999. *South Coast regional report: California wildlands project vision for wild California.* Davis: California Wilderness Coalition.

Hurley, M. A., J. W. Unsworth, P. Zager, E. O. Garton, D. M. Montgomery, and C. L. Maycock. Mule deer survival and population response to experimental reduction of coyotes and mountain lions. Idaho Department of Fish and Game, unpublished data.

Husseman, J. S. 2002. Prey selection patterns of wolves and cougars in east-central Idaho. Master's thesis, University of Idaho, Moscow.

Husseman, J. S., D. L. Murray, G. Power, C. Mack, C. R. Wenger, and H. Quigley. 2003. Assessing differential prey selection patterns between two sympatric large carnivores. *Oikos* 101:591–601.

Idaho Fish and Game Department. 2000. Policy for avian and mammalian predation management. http://fishandgame .idaho.gov/wildlife/manage_issues/mam_predation.cfm (accessed August 30, 2004).

INDEC. 2002. *Censo Nacional Agropecuario.* Buenos Aires: Instituto Nacional de Estadística y Censos.

Ingram, H. M., and D. E. Mann. 1989. Interest groups and environmental policy. In *Environmental politics and* policy: *Theories and evidence*, ed. J. P. Lester, 135–157. Durham, NC: Duke University Press.

Ingram, R. 1984. Oregon. In *Proceedings of the second mountain lion workshop*, ed. J. Roberson and F. Lindzey, 53–55. Salt Lake City: Utah Division of Wildlife Resources.

Iriarte, J. A., and T. Clark. 2003. Managing puma hunting in the western United States through a metapopulation approach. *Anim Conserv* 6:159–70.

Iriarte, J. A., W. L. Franklin, W. E. Johnson, and K. H. Redford. 1990. Biogeographic variation of food habits and body size of the American puma. *Oecologia* 85:185–90.

Iriarte, J. A., and L. Hernández. 2002. Growth curve models and age estimation of young cougars in the northern Great Basin. *J Wildl Mgmt* 66:849–58.

Iriarte, J. A., W. E. Johnson, and W. L. Franklin. 1991. Feeding ecology of the Patagonia puma in southernmost Chile. *Revista Chilena de Historia Natural* 64:145–56.

Irwin, E. R., and M. C. Freeman. 2002. Proposal for adaptive management to conserve biotic integrity in a regulated segment of the Tallapoosa River, Alabama, USA. *Conserv Biol* 16:1212–22.

IUCN. 2006. *2006 IUCN Red List of Threatened Species.* International Union for Conservation of Nature and Natural Resources. www.iucnredlist.org (accessed January 8, 2007).

Jackson, H. H. T. 1955. The Wisconsin puma. *Proc Bio Soc Wash* 68:149–50.

Jacobson, C. A., and D. J. Decker. 2006. Ensuring the future of state wildlife management: Understanding challenges for institutional change. *Wildl Soc Bull* 34:531–6.

Jaguar Conservation Fund. 2007. http://www.jaguar.org.br/index.htm (accessed May 21, 2007).

Jalkotzy, M. G., P. I. Ross, and J. R. Gunson. 1992. *Management plan for cougars in Alberta.* Province of Alberta Wildlife Management Planning Series Number 5, Alberta Forestry, Lands and Wildlife, Edmonton, Alberta.

Jalkotzy, M. G., P. I. Ross, and J. Wierzchowski. 2000. Regional scale cougar habitat modeling in Southwestern Alberta, Canada (abstract). In *Proceedings of the sixth mountain lion workshop*. Austin: Texas Parks and Wildlife Department.

Jamison, W. V., C. Wenk, and J. V. Parker. 2000. Every sparrow that falls: Understanding animal rights activism as functional religion. *Soc Anim* 8:305–30.

Janczewski, D. N., W. S. Modi, J. C. Stephens, and S. J. O'Brien. 1995. Molecular evolution of 12S and cytochrome b sequences in the pantherine lineage of Felidae. *Molec Biol Evol* 12:690–707.

Janis, M. W., and J. D. Clark. 2002. Responses of Florida panthers to recreational deer and hog hunting. *J Wildl Mgmt* 66:839–48.

Jardine, S. W. 1834. *The naturalist's library: Mammalia Vol. 2: The natural history of the Felinae.* Edinburgh: W. H. Lizars.

Jedrezejewska, B. and W. Jedrezejewska. 2005. Large carnivores and ungulates in European temperate forest ecosystems: Bottom-up and top-down control. In *Large carnivores and the conservation of biodiversity*, ed. J. C. Ray, K. H. Redford, R. S. Steneck, and J. Berger, 230–246. Washington, D.C.: Island Press.

Jerolmack, C. 2003. Tracing the profile of animal rights supporters: A preliminary investigation. *Soc Anim* 11:245–63.

Johnson, D. H. 1980. The comparison of usage and availability measurements for evaluating resource preference. *Ecology* 61:65–71.

Johnson, W. E., M. Culver, J. A. Iriarte, E. Eizirik, K. L. Seymour, and S. J. O'Brien. 1998. Tracking the evolution of the elusive Andean mountain cat (*Oreailurus jacobita*) from mitochondrial DNA. *J Hered* 89:227–32.

Johnson, W. E., P. A. Dratch, J. S. Martenson, and S. J. O'Brien. 1996. Resolution of recent radiations within three evolutionary lineages of *Felidae* using mitochondrial restriction fragment length polymorphism variation. *J Mammal Evol* 3:97–120.

Johnson, W. E., E. Eizirik, J. Pecon-Slattery, W. J. Murphy, A. Antunes, E. Teeling, and S. J. O'Brien. 2006. The late Miocene radiation of modern felidae: A genetic assessment. *Science* 311:73–7.

Johnson, W. E., and S. J. O'Brien. 1997. Phylogenetic reconstruction of the *Felidae* using 16s rRNA and NADH-5 mitochondrial genes. *J Mol Evol* 44:S98–S116.

Jones, K. 2002. A fierce green fire: Passionate pleas and wolf ecology. *Ethics Place Environ* 5:35–43.

Jordan, D. B. 1994. *Final preliminary analysis of some potential Florida panther population reestablishment sites.* Atlanta, GA: U.S. Fish and Wildlife Service.

Jorgenson, J. P., and K. H. Redford. 1993. Humans and big cats as predators in the Neotropics. *Symp Zool Soc Lond* 65:367–90.

Jost, J. T., J. Glaser, A. W. Kruglanski, and F. J. Sulloway. 2003. Political conservatism as motivated social cognition. *Psych Bull* 129:339–75.

Justad, P. M., L. H. McAvoy, and D. McDonald. 1996. Native American land ethics: Implications for natural resource management. *Soc Nat Res* 9:565–81.

Jung, T. S., and P. J. Merchant. 2005. First conformation of cougar (*Puma concolor*) in the Yukon. *Can Field Nat* 119:581.

Kaltenborn, B. P., and T. Bjerke. 2002. The relationship of general life values to attitudes toward large carnivores. *Hum Eco Rev* 9:55–61.

Kaltenborn, B. P., T. Bjerke, and E. Strumse. 1998. Diverging attitudes towards predators: Do environmental beliefs play a part? *Hum Ecol Rev* 5:1–9.

Kamler, J. F., R. M. Lee, J. C. deVos, Jr., W. B. Ballard, and H. A. Whitlaw. 2002. Survival and cougar predation of translocated bighorn sheep in Arizona. *J Wildl Mgmt* 66:1267–72.

Kaplan, J. K. 2007. *Bee researchers: Meanness genetic?* USDA Agriculture Research Service Information Staff. Tucson, AZ: USDA-ARS Carl Hayden Bee Research Center.

Karanth, K. U., and M. E. Sunquist. 1995. Prey selection by tiger, leopard and dhole in tropical forests. *J Anim Ecol* 64:439–50.

Katnik, D. D. 2002. Predation and habitat ecology of mountain lions (*Puma concolor*) in the southern Selkirk Mountains. Ph.D. dissertation, Washington State University, Pullman.

Kauffman, M. J., N. Varley, D. W. Smith, D. R. Stahler, D. R. MacNulty, and M. S. Boyce. 2007. Landscape heterogeneity shapes predation in a newly restored predator-prey system. *Ecol Lett* 10:690–700.

Kautz, R., R. Kawula, T. Hoctor, J. Comiskey, D. Jansen, D. Jennings, J. Kasbohm, F. Mazzotti, R. McBride, L. Richardson, and K. Root. 2006. How much is enough?

Landscape level conservation for the Florida panther. *Biol Conserv* 130:118–33.

Keefover-Ring, W. 2005. State of pumas in the West: Heading towards overkill? (abstract). In *Proceedings of the eighth mountain lion workshop*, ed. R. A. Beausoleil and D. A. Martorello, p. 213. Olympia: Washington Department of Fish and Wildlife.

Keiter, R. B., and H. Locke. 1996. Law and large carnivore conservation in the Rocky Mountains of the U.S. and Canada. *Conserv Biol* 10:1003–12.

Kellert, S. R. 1985. Public perception of predators, particularly the wolf and coyote. *Biol Conserv* 31:167–89.

———. 1989. Perceptions of animals in America. In *Perceptions of animals in American culture*, ed. R. J. Hoage, 5–24. Washington, D.C.: Smithsonian Institution Press.

———. 1994. Public attitudes toward bears and their conservation. In *Proceedings of the ninth international conference on bear research and management*, 43–50. Missoula, MT: International Association for Bear Research and Management.

———. 1996. *The value of life: Biological diversity and human society.* Washington, D.C.: Island Press.

Kellert, S. R., M. Black, C. Reid Rush, and A. J. Bath. 1996. Human culture and large carnivore conservation in North America. *Conserv Biol* 10:977–90.

Kellert, S. R., and C. P. Smith. 2000. Human values toward large mammals. In *Ecology and management of large mammals in North America*, ed. S. Demaris and P. R. Krausman, 38–63. Upper Saddle River, NJ: Prentice-Hall.

Kellert, S. R., and M. O. Westervelt. 1982. Historical trends in American animal use and perception. *Trans N Am Wildl Nat Res* 47:649–64.

Kelly, M. J., A. J. Noss, M.S. DiBitetti, L. Maffei, R. L. Arispe, A Paviolo, C. D. De Angelo, and Y. E. Di Blanco. 2008. Estimating puma densities from camera trapping across their study sites: Bolivia, Argentina, and Belize. *J Mammal* 89:408–18.

Kendall, P. T., D. Holme, P. M. Smith. 1982. Comparative evaluation of net digestive and absorptive efficiency in dogs and cats fed a on a variety of contrasting diet types. *J Small Anim Pract* 23:577–87.

Kenn, S. 2002. *Recovery strategy and management plan.* Beeton, Ontario: Ontario Puma Foundation.

Kerr, R. 1792. In *The Animal Kingdom*, 151.

Kertson, B. N. 2005. Political and socio-economic influences on cougar management legislation in Washington state: Post Initiative 655. In *Proceedings of the eighth mountain lion workshop*, ed. R. A. Beausoleil and D. A. Martorello, 92–103. Olympia: Washington Department of Fish and Wildlife.

Kie, J. G., and B. Czech. 2000. Mule and black-tailed deer. In *Ecology and management of large mammals in North America*, ed. S. Demarais and P. R. Krausman, 629–657. Upper Saddle River, NJ: Prentice Hall.

King, M. M., and H. D. Smith. 1980. Differential habitat utilization by the sexes of mule deer. *Great Basin Nat* 40:273–81.

Kinley, T., and C. Apps. 2001. Mortality patterns in a sub-population of endangered mountain caribou. *Wildl Soc Bull* 29:158–64.

Kintigh, M. 2003. State of South Dakota mountain lion status report. In *Proceedings of the seventh mountain lion workshop*, ed. S. A. Becker, D. D. Bjornlie, F. G. Lindzey, and

D. S. Moody, 43–48. Lander: Wyoming Game and Fish Department.

Kirkpatrick, J. F. 2005. The wild horse fertility control program. In *Humane Wildlife Solutions: The Role of Immunocontraception*, ed. A. T. Rutberg, 63–75. Washington, D.C.: Humane Society of the United States Press.

Kirkpatrick, J. F., and K. M. Frank. 2005. Fertility control in free-ranging wildlife. In *Wildlife contraception: Issues, methods and application*, ed. C. Asa and I. Porton, 195–221. Baltimore: Johns Hopkins University Press.

Kirkpatrick, J. F., and L. Points. 1997. The role of educational controversial research in national parks. *Legacy* 8:6–33.

Kleese, D. 2002. Contested natures: Wolves in late modernity. *Soc Nat Res* 15:313–26.

Kleiman, D. G., and J. F. Eisenberg. 1973. Comparison of canid and felid social systems from an evolutionary perspective. *Anim Behav* 21:637–59.

Kline, B. 1997. *First along the river: A brief history of the U.S. environmental movement.* San Francisco: Acada Books.

Knight, R. L., and K. J. Gutzwiller, eds. 1995. *Wildlife and recreationists: Coexistence through management and research.* Washington, D.C.: Island Press.

Koehler, G. M., and M. G. Hornocker. 1991. Seasonal resource use among mountain lions, bobcats, and coyotes. *J Mammal* 72:391–6.

Koehler, G. M., and E. Nelson. 2003. Project CAT (Cougars and Teachers): Integrating science, schools and community in development planning. In *Proceedings of the seventh mountain lion workshop*, ed. S. A. Becker, D. D. Bjornlie, F. G. Lindzey, and D. S. Moody, p. 140. Lander: Wyoming Game and Fish Department.

Kohn, M. H., W. J. Murphy, E. A. Ostrander, and R. K. Wayne. 2006. Genomics and conservation genetics. *TREE* 21:629–37.

Konrad, R. 2007. Mountain lion victim heads to San Fransisco for more surgery. *Associated Press*, January 28.

Kortello, A. 2005. Interactions between cougars (*Puma concolor*) and wolves (*Canis lupus*) in the Bow Valley, Banff National Park, Alberta. Master's thesis, University of Idaho, Moscow.

Kortello, A. D., T. E. Hurd, and D. L. Murray. 2007. Interactions between cougars and gray wolves in Banff National Park, Alberta. *Ecoscience* 14:214–222.

Kortello, A., and D. L. Murray. 2005. Interactions between wolves (*Canis lupus*) and cougars (*Puma concolor*) in the Bow Valley, Banff National Park, Alberta. In *Proceedings of the eighth mountain lion workshop*, ed. R. A. Beausoleil and D. A. Martorello, p. 175. Olympia: Washington Department of Fish and Wildlife.

Koval, M. H., and A. G. Mertig. 2004. Attitudes of the Michigan public and wildlife agency personnel toward lethal wildlife management. *Wildl Soc Bull* 32:232–43.

Kracht, B. R. 2000. Kiowa religion in historical perspective. In *Native American spirituality: A critical reader*, ed. Lee Irwin, 236–255. Lincoln, NE: University of Nebraska Press.

Krausman, P. R., and D. M. Shackleton. 2000. Bighorn sheep. In *Ecology and management of large mammals in North America*, ed. S. Demarais and P. R. Krausman, 517–44. Upper Saddle River, NJ: Prentice Hall.

Krebs, C. J. 1989. *Ecological methodology.* New York: Harper Collins Publishers.

———. 1994. *Ecology: The experimental analysis of distribution and abundance.* New York: Harper Collins College Publishers.

Krebs, J., E. Lofroth, J. Copeland, V. Banci, D. Cooley, H. Golden, A. Magoun, R. Mulders, and B. Schultz. 2004. Synthesis of survival rates and causes of mortality in North American wolverines. *J Wildl Mgmt* 68:493–502.

Kruuk, H. 1972. *The spotted hyena: A study of predation and social behavior.* Chicago: University of Chicago Press.

———. 1986. Interactions between *Felidae* and their prey species: A review. In *Cats of the world: Biology, conservation, and management,* ed. S. D. Miller and D. Everett, 353–74. Washington, D.C.: National Wildlife Federation.

———. 2002. *Hunters and hunted: Relationships between carnivores and people.* Cambridge, UK: Cambridge University Press.

Kunkel, K. E. 1997. Predation by wolves and other large carnivores in Northwestern Montana and Southeastern British Columbia. Ph.D. dissertation, University of Montana, Missoula.

Kunkel, K. E., T. K. Ruth, D. H. Pletscher, and M. G. Hornocker. 1999. Winter prey selection by wolves and cougars in and near Glacier National Park, Montana. *J Wild Mgmt* 63:901–10.

Kuroiwa, A., and C. Ascorra. 2002. Dieta y densidad de posibles presas de jaguar en las inmediaciones de la zona de Reserva Tambopata-Candamo, Perú. In *El Jaguar en el Nuevo Milenio,* ed. R. A. Medellín, C. Equihua, C. L. B. Chetkiewicz, P. G. Crawshaw Jr., A. Rabinowitz, H. H. Redford, J. G. Robinson, E. W. Sanderson, and A. B. Taber, 199–207. Universidad Nacional Autónoma de México/Wildlife Conservation Society, Mexico.

Kurtén, B. 1976. Fossil puma (*Mammalia: Felidae*) in North Ameirca. *Neth J Zool* 26:502–34.

Kurtén B., and E. Anderson. 1980. *Pleistocene mammals of North America.* New York: Columbia University Press.

Kurushima, J. D., J. A. Collins, J. A. Well, and H. B. Ernest. 2006. Development of 21 microsatellite loci for puma (*Puma concolor*) ecology and forensics. *Mol Ecol* 6:1260–2.

Kurz, W. A., and R. N. Sampson 1991. North American forests and global climate change. *Trans N Am Wildl Nat Res* 56:587–94.

La Arena. 2007a. Las inspecciones eran con aviso previo. *La Arena,* February 3. Santa Rosa, La Pampa, Argentina.

———. 2007b. Suspenden caza de puma en La Pampa. *La Arena,* February 3. Santa Rosa, La Pampa, Argentina.

———. 2007c. El puma y el bosque, futuro incierto. *La Arena,* February 5. Santa Rosa, La Pampa, Argentina.

Lafon, N. W., S. L. McMullin, D. E. Steffen, and R. S. Schulman. 2004. Improving stakeholder knowledge and agency image through collaborative planning. *Wildl Soc Bull* 32:220–31.

Laing, S. P. 1988. Cougar habitat selection and spatial use patterns in southern Utah. Master's thesis, University of Wyoming, Laramie.

Laing, S. P., and F. P. Lindzey. 1991. Cougar habitat selection in south-central Utah. In *Mountain lion–human interaction symposium and workshop,* ed. C. E. Braun, 86–94. Denver: Colorado Division of Wildlife.

Lambeck, R. J. 1997. Focal species: A multi-species umbrella for nature conservation. *Conserv Biol* 11:849–56.

Lambert, C. M. S., R. B. Wielgus, H. S. Robinson, D. D. Katnik, H. S. Cruickshank, R. Clarke, and J. Almack. 2006. Cougar population dynamics and viability in the Pacific Northwest. *J Wildl Mgmt* 70:246–57.

Land, E. D., and R. C. Lacy. 2000. Introgression level achieved through Florida panther genetic restoration. *Endangered Species Update* 17:99–103.

Lasswell, H. D. 1951. *The world revolution of our time: A framework for basic policy research.* Stanford, CA: Stanford University Press.

———. 1971. *A preview of policy sciences.* New York: Elsevier.

Lasswell, H. D., and A. Kaplan. 1950. *Power and society.* New Haven, CT: Yale University Press.

Lasswell, H. D., and M. S. McDougal. 1992. *Jurisprudence for a free society.* New Haven, CT: New Haven Press.

Laud, D., M. Cunningham, M. Lotz, and D. Shindle. 2002. Florida panther genetic restoration and management, July 2001–June 2002. Progress report to Florida Fish and Wildlife Conservation Commission.

Laundré, J. W. 2005. Puma energetics: A recalculation. *J Wild Mgmt* 69:723–32.

———. 2008. Summer predation rates on ungulate prey by a large keystone predator: How many ungulates does a large predator kill? *J Zool* (London). 275:341–8.

Laundré, J. W. 2009. Behavioral response race, ecology of fear, and patch use of a large predator and its ungulate prey. In Review.

Laundré, J. W., and T. W. Clark. 2003. Managing puma hunting in the western United States: Through a metapopulation approach. *Anim Conserv* 6:159–70.

Laundre, J. W., and L. Hernández. 2000. Habitat composition of successful kill sites for lions in southeastern Idaho and northwestern Utah (abstract). In *Proceedings of the sixth mountain lion workshop,* 24–25. Austin: Texas Parks and Wildlife.

———. 2003a. Factors affecting dispersal in young male pumas. In *Proceedings of the seventh mountain lion workshop,* ed. S. A. Becker, D. D. Bjornlie, F. G. Lindzey, and D. S. Moody, 151–160. Lander: Wyoming Game and Fish Department.

———. 2003b. Winter hunting habitat of pumas (*Puma concolor*) in northwestern Utah and southern Idaho, USA. *Wildl Biol* 9:123–9.

———. 2007. Do female pumas (*Puma concolor*) exhibit a birth pulse? *J Mammal* 88:1300–4.

Laundré, J. W., L. Hernández, and K. B. Altendorf. 2001. Wolves, elk, and bison: Reestablishing the "landscape of fear" in Yellowstone National Park, USA. *Can J Zool* 79:1401–9.

Laundré, J. W., L. Hernández, I. Arías, and G. Fowles. 2002. The landscape of fear and its implications to sheep reintroductions. *Bienn Symp N Wild Sheep Goat Council* 13:103–8.

Laundré, J. W., L. Hernández, and S. G. Clark. 2006. Impact of puma predation on the decline and recovery of a mule deer population in southeastern Idaho. *Can J Zool* 84:1555–65.

———. 2007. Numerical and demographic responses of pumas to changes in prey abundance: Testing current predictions. *J Wildl Mgmt* 71:345–55.

Laundré, J. W., and J. Loxterman. 2007. Impact of edge habitat on summer home range size in female pumas. *Am Midl Nat* 157:221–9.

Laundré, J.W., L. Hernández, J. Loredo Salazar, and D Nuñez López. 2009. Evaluating potencial factors affecting puma abundante in the Mexican Chihuahuan Desert. *Wildlife Biology.* In Press.

Laurance, W. F., B. M. Croes, L. Tchignoumba, S. A. Lahm, A. Alonso, M. E. Lee, P. Campbell, and C. Ondzeano. 2005. Impacts of roads and hunting on Central African rainforest mammals. *Conserv Biol* 20:1251–61.

Law, A. M., and W. D. Kelton. 2000. *Simulation modeling and analysis.* New York: McGraw-Hill.

Law, R. 2001. Phenotypic and genetic changes due to selective exploitation. In *Conservation of exploited species,* ed. J. D. Reynolds, G. M. Mace, K. H. Redford, and J. G. Robinson, 323–342. Cambridge, UK: Cambridge University Press.

Lawrence, R. L., S. E. Daniels, and G. H. Stankey. 1997. Procedural justice and public involvement in natural resource decision making. *Soc Nat Res* 10:577–89.

Lee, H. 1960. *To kill a mockingbird.* Philadelphia: Lippincott.

Lee, R. M., J. D., Yoakum, B. W. O'Gara, T. M. Pojar, and R. A Ockenfels. 1998. *Pronghorn management guide.* eighteenth Pronghorn Antelope Workshop, Prescott, AZ.

Leite, M. R. P., and F. Galvão. 2002. El jaguar, el puma, y el hombre en tres áreas protegidas del bosque Atlántico costero de Paraná, Brasil. In *El Jaguar en el Nuevo Milenio,* ed. R. A. Medellín, C. Equihua, C. L. B. Chetkiewicz, P. G. Crawshaw, Jr., A. Rabinowitz, H. H. Redford, J. G. Robinson, E. W. Sanderson, and A. B. Taber, 237–250. Universidad Nacional Autónoma de México/Wildlife Conservation Society, Mexico.

Leite, M. R. P., R. L. P. Boulhosa, F. Galvão, and L Cullen, Jr. 2002. Conservación del jaguar en las áreas protegidas del bosque Atlántico de la costa de Brasil. In *El Jaguar en el Nuevo Milenio,* ed. R. A. Medellín, C. Equihua, C. L. B. Chetkiewicz, P. G. Crawshaw, Jr., A. Rabinowitz, H. H. Redford, J. G. Robinson, E. W. Sanderson, and A. B. Taber, 25–42. Universidad Nacional Autónoma de México/Wildlife Conservation Society, Mexico.

Leonard, H. J., M. Yudelman, J. D. Stryker, J. O. Browder, A. J. DeBoer, T. Campbell, and A. Jolly. 1989. *Environment and the poor: Development strategies for a common agenda.* New Brunswick, NJ: Transaction Publishers.

Leopold, A. 1933. *Game management.* Madison, WI: University of Wisconsin Press.

———. 1942. Land use and democracy. *Audubon* 44:259–65.

———. 1949. *A Sand County almanac and sketches here and there.* New York: Oxford University Press.

———. 1991. *The river of the mother of God and other essays.* Madison, WI: University of Wisconsin Press.

Leopold, A., and C. W. Schwartz. 1966. *A Sand County almanac, with other essays on conservation from Round River.* New York: Oxford University Press.

Leopold, B. D., and P. R. Krausman. 1986. Diets of 3 predators in Big Bend National Park, Texas. *J Wild Mgmt* 50:290–5.

Lermayer, R. M. 2006. Ed's response: Mountain lion conspiracy and mountain lion mismanagement. *Predator Xtreme,* February.

Levy, S. 2005. Can cougars and people live side by side? *Natl Wildl* 43:14–5.

Lewis, L. 2006. Mountain lion attacks introduction. http://www.cougarinfo.com (accessed January 11, 2007).

Leyhausen, P. 1979. *Cat behavior: The predatory and social behavior of domestic and wildcats.* New York: Garland STPM Press.

Liberg, O. 1980. Spacing patterns in a population of rural free roaming domestic cats. *Oikos* 35:336–49.

Lidicker, W., Jr. 1975. The role of dispersal in the demography of small mammals. In *Small mammals: Their Productivity and Population Dynamics,* ed. F. B. Golley, K. Petrusewicz, and L. Ryszkowski, 103–34. New York: Cambridge University Press.

Light, A. 2000. Restoration, the value of participation and the risks of professionalization. In *Restoring nature: Perspectives from the social sciences and humanities,* ed. P. H. Gobster and R. B. Hull, 163–181. Washington, D.C.: Island Press.

Lilly, B. 1998. *Ben Lilly's tales of bears, lions and hounds.* N. B. Carmony, ed. Silver City, NM: High Lonesome Books.

Lindzey, F. G. 1987. Mountain lion. In *Wild furbearer management and conservation in North America,* ed. M. Novak, J. A. Baker, M. E. Obbard, and B. Malloch, 657–68. Toronto, Ontario: Ontario Trappers Association, Ontario Ministry of Natural Resources.

———. 1991. Needs for mountain lion research and special management studies. In *Mountain lion–human interaction symposium and workshop,* ed. C. E. Braun, 52–53. Denver: Colorado Division of Wildlife.

Lindzey, F. G., B. B. Ackerman, D. Barnhurst, and T. P. Hemker. 1988. Survival rates of mountain lions in southern Utah. *J Wildl Mgmt* 52:664–7.

Lindzey, F. G., W. D. Van Sickle, B. B. Ackerman, D. Barnhurst, T. P. Hemker, and S. P. Laing. 1994. Cougar population dynamics in southern Utah. *J Wildl Mgmt* 58:619–24.

Lindzey, F. G., W. D. Van Sickle, S. P. Laing, and C. S. Mecham. 1992. Cougar population response to manipulation in southern Utah. *Wildl Soc Bull* 20:224–7.

Linneaus, C. 1771. *Mantissa Plantarum, 522.* New York: Hafner.

Linnell, J. D. C., and O. Strand. 2000. Interference interactions, co-existence and conservation of mammalian carnivores. *Divers Distrib* 6:169–76.

Linnell, J. D. C., J. Odden, M. E. Smith, R. Aanes, and J. E. Swenson. 1999. Large carnivores that kill livestock: Do "problem individuals" really exist? *Wildl Soc Bull* 27:698–705.

Løe, J. 2002. Large carnivore related deaths: A conservation issue. Master's thesis, Norwegian University of Science and Technology.

Logan, K. A., and L. L. Irwin. 1985. Mountain lion habitats in the Big Horn Mountains, Wyoming. *Wildl Soc Bull* 13:257–62.

Logan, K. A., L. L. Irwin, and R. Skinner. 1986. Characteristics of a hunted mountain lion population in Wyoming. *J Wildl Mgmt* 50:648–54.

Logan, K. A., and L. L.Sweanor. 1994. Mountain lion management recommendations for the San Andres Mountains, New Mexico. Memorandum to New Mexico Department of Game and Fish and White Sands Missile Range.

———. 1998. Cougar management in the West: New Mexico as a template. Pp. 101–110 in *Proceedings of the Western Association of Fish and Wildlife Agencies.* Jackson: Wyoming Game and Fish Department.

———. 2000. Puma. In *Ecology and management of large mammals in North America,* ed. S. Demarais and P. R. Krausman, 347–377. Englewood Cliffs, NJ: Prentice Hall.

———. 2001. *Desert puma: Evolutionary ecology and conservation of an enduring carnivore.* Washington. D.C.: Island Press.

Logan, K. A., L. L. Sweanor, and M. G. Hornocker. 2003. Reconciling science and politics in puma management in the West: New Mexico as an example (abstract). In *Proceedings of the seventh mountain lion workshop,* ed. S. A. Becker, D. D. Bjornlie, F. G. Lindzey, and D. S. Moody, p. 146. Lander: Wyoming Game and Fish Department.

———. 2005. Reconciling science and politics in puma management in the West: New Mexico as a template. *Trans N Am Wildl Nat Res* 69:35–50.

Logan, K. A., L. L. Sweanor, T. K. Ruth, and M. G. Hornocker. 1996. *Cougars of the San Andres Mountains, New Mexico.* Final report. Federal Aid in Wildlife Restoration Project W-128-R. New Mexico Department of Game and Fish, Santa Fe.

Logan, T., A. C. Eller Jr., R. Morrell, D. Ruffner, and J. Sewell. 1993. *Florida panther habitat conservation plan: South Florida population.* Florida Panther Interagency Committee.

Loker, C. A., and D. J. Decker. 1995. Colorado black bear hunting referendum: What was behind the vote? *Wildl Soc Bull* 23:370–6.

Lopez, B. 1998. *The language of animals.* Bainbridge Island, WA: Arbor Fund.

Lopez, J. V., S. Cevario, and S. J. O'Brien. 1996. The complete sequence of the domestic cat (*Felis catus*) mitochondrial genome and a transposed tandem repeat (Numt) in the nuclear genome. *Genomics* 33:229–46.

López-González, C.A. 1999. Implicaciones para la conservacion y el manejo do pumas (*Puma concolor*) utilizando como modelo una poblacion sujeta a caderia deportiva. Ph.D. dissertation, National Autonomous University of Mexico, Mexico City.

López-González, C. A., and S. E. Carrillo-Percastegui. 2003. Ecology of sympatric pumas and jaguars in northwestern Mexico. In *Proceedings of the seventh mountain lion workshop,* ed. S. A. Becker, D. D. Bjornlie, F. J. Lindzey, and D. S. Moody, p. 121. Lander: Wyoming Game and Fish Department.

Loredo-Salazar, J. 2003. Evaluación de los componentes de hábitat que afectan la abundancia de pumas in el desierto Chihuahuense a través de un análisis comparative. Master's thesis, Instituto de Ecología, A. C. Xalapa, Veracruz, Mexico.

Lott, D. F. 1991. *Intraspecific variation in the social systems of wild vertebrates.* New York: Cambridge University Press.

Lotz, M. A. 2005. Florida mountain lion status report. In *Proceedings of the eighth mountain lion workshop,* ed. R. A. Beausoleil and D. A. Martorello, 73–77. Olympia: Washington Department of Fish and Wildlife.

Lotz, M. A., D. Land, M. Cunningham, and B. Ferree. 2005. *Florida panther annual report 2004–2005.* Florida Fish and Wildlife Conservation Commission.

Love, J. 2005. Living with lions. *South Dakota Magazine,* January/February.

Lovejoy, T. E., and L. Hannah 2005. *Climate change and biodiversity.* New Haven, CT: Yale University Press.

Lowe, B. M., and G. F. Ginsberg. 2002. Animal rights as a postcitizenship movement. *Soc Anim* 10:203–15.

Loxterman, J. L. 2001. The impact of habitat fragmentation on the population genetic structure of pumas (*Puma concolor*) in Idaho. Ph. D. dissertation, Idaho State University, Pocatello.

Loxterman, J. L., M. Culver, K. Murphy, S. J. O'Brien, J. Laundre, and M. Ptacek. In preparation. Reproductive success in Idaho pumas.

Lozada, C. 2003. Latin America. *Foreign Policy,* March/April.

Luoma, J. R. 1992. Eco-backlash. *Wildl Conserv* 95:26–36.

Lyra-Jorge, MC, G. Ciocheti, and V. R. Pivello. 2008. Carnivore mammals in a fragmented landscape northeast of São Paulo State, Brazil. *Biodivers Conserv* 17:1573–80.

MacArthur, R. H. 1960. On the relation between reproductive value and optimal predation. *Proc Nat Acad Sci* 46:143–5.

MacArthur, R. H., and E. R. Pianka. 1966. On optimal use of a patchy environment. *Am Nat* 100:603–9.

Mace, R. D., J. S. Waller, T. L. Manley, K. Ake, and W. T. Wittinger. 1999. Landscape evaluation of grizzly bear habitat in Western Montana. *J Conserv Biol* 13:367–77.

Mackie, R. J., D. E. Pac, K. L. Hamlin, and G. L. Dusek. 1998. Ecology and management of mule deer and white-tailed deer in Montana. Montana Fish, Wildlife and Parks, Wildlife Division, Federal Aid to Wildlife Restoration Report, Project W-120-R, Helena.

Maehr, D. S. 1990. The Florida panther and private lands. *Conserv Biol* 4:167–70.

———. 1997a. *The Florida panther: Life and death of a vanishing carnivore.* Washington, D.C.: Island Press.

———. 1997b. The comparative ecology of bobcat, black bear, and Florida panther in South Florida. *Bull Florida Mus Nat Hist* 40:1–176.

Maehr, D. S., R. C. Belden, E. D. Land, and L. Wilkins.1990. Food habits of panthers in southwest Florida. *J Wild Mgmt* 54:420–3.

Maehr, D. S., and C. T. Caddick. 1995. Demographics and genetic introgression in the Florida panther. *Conserv Biol* 9:1295–8.

Maehr, D. S., and J. A. Cox. 1995. Landscape features and panthers in Florida. *Conserv Biol* 9:1008–19.

Maehr, D. S., and J. P. Deason. 2002. Wide ranging carnivores and development permits: Constructing a multi-scale model to evaluate impacts on the Florida panther. *Clean Tech Environ Pol* 3:398–406.

Maehr, D. S., T. S. Hoctor, and L. D. Harris. 2001. The Florida panther: A flagship for regional restoration. In *Large mammal restoration: Ecological and sociological challenges in the twenty-first century,* ed. D. S. Maehr, R. F. Noss, and J. L. Larkin, 293–312. Washington, D.C.: Island Press.

Maehr, D. S., E. D. Land, and J. C. Roof. 1991. Social ecology of Florida panthers. *Natl Geogr Res* 7:414–31.

Maehr, D. S., E. D. Land, J. C. Roof, and J. W. McCown. 1989. Early maternal behavior in the Florida panther (*Felis concolor coryi*). *Amer Midl Nat* 22:34–43.

Maehr, D. S., E. D. Land, D. B. Shindle, O. L. Bass, and T. S. Hoctor. 2002. Florida panther dispersal and conservation. *Biol Conserv* 106:187–97.

Main, M. B., and B. E. Coblentz. 1990. Sexual segregation among ungulates: A critique. *Wildl Soc Bull* 18:204–10.

———. 1996. Sexual segregation in Rocky Mountain mule deer. *J Wild Mgmt* 60:497–507.

Main, M. B., F. M. Roka, and R. F. Noss. 1999. Evaluating costs of conservation. *Conserv Biol* 13:1262–72.

Major, R. H. 1847. *Select letters of Christopher Columbus, with other original documents relating to his four voyages to the new world.* London: Hakluyt Society.

Malik, I. 2001. Monkey menace: Who is responsible? *ENVIS Bulletin: Wildlife and Protected Areas, Non-Human Primates of India* 1:169–171.

Manfredo, M. J., T. L. Teel, and A. D. Bright. 2003. Why are public values toward wildlife changing? *Hum Dim Wildl* 8:287–306.

Manfredo, M. J., H. C. Zinn, L. Sikorowski, and J. Jones. 1998. Public acceptance of mountain lion management: A case study of Denver, Colorado, and nearby foothills areas. *Wildl Soc Bull* 26:964–70.

Mangus, G. 1991. Legal aspects of encounters on federal lands and in state programs. In *Mountain lion–human interaction symposium and workshop*, ed. C. E. Braun, 43–44. Denver: Colorado Division of Wildlife.

Mansfield, K. G., and E. D. Land. 2002. Cryptorchidism in Florida panthers: Prevalence, features, and influence of genetic restoration. *J Wildl Dis* 38:693–8.

Mansfield, T. M., and R. A. Weaver. 1988. The status of the mountain lions in California. In *Proceedings of the third mountain lion workshop*, ed. R. H. Smith, 15–18. Phoenix: Arizona Game and Fish Department.

Mao, J. S., M. S. Boyce, D. W. Smith, F. J. Singer, D. J. Vales, J. M. Vore, and E. H. Merrill. 2005. Habitat selection by elk before and after wolf reintroduction in Yellowstone National Park. *J Wildl Mgmt* 69:1691–1707.

Marcgrave, G. 1648. Historiae reerum naturalium. *Brasiliae*: 234. Quoted in Young, S. P., and E. A. Goldman, *The puma: Mysterious American cat*, ed. Young, S. P., and E. A. Goldman. 1946. Washington, D.C.: The American Wildlife Institute.

Mares, M. A. 1992. Neotropical mammals and the myth of Amazonian biodiversity. *Science* 255:976–9.

Marino A. and R. Baldi. 2008. Vigilance patterns of territorial guanacos (*Lama guanicoe*): The role of reproductive interests and predation risk. *Ethology* 114:413–23.

Marshall, L. G., S. D. Webb, J. J. Sepkoski, and D. M. Raup. 1982. Mammalian evolution and the great American interchange. *Science* 215:1351–7.

Martin, P. S., and R. G. Klein. 1995. *Quaternary extinctions, a prehistoric revolution*. Tucson: University of Arizona Press.

Martorello, D. A., and R. A. Beausoleil. 2003. Cougar harvest characteristics with and without the use of hounds. In *Proceedings of the seventh mountain lion workshop*, ed. S. A. Becker, D. D. Bjornlie, F. G. Lindzey, and D. S. Moody, 129–35. Lander: Wyoming Game and Fish Department.

Massone Mezzano, M. 2001. Los cazadores después del hielo. Master's thesis, University of Chile, Santiago.

Matthes, S. M. 1988. *Brave and other stories*. Carlotta, CA: Vera Orton Matthes.

Matthiessen, P. 1959. *Wildlife in America*. New York: Viking Press.

Mattson, D. J. 2007a. Managing for human safety in mountain lion range. In *Mountain lions of the Flagstaff Uplands*, 2003–2006 progress report, ed. D. J. Mattson, 43–62. Washington, D.C.: U.S. Department of the Interior, U.S. Geological Survey.

———, ed. 2007b. *Mountain lions of the Flagstaff Uplands*, 2003–2006 progress report. Washington, D.C.: U.S. Department of the Interior, U.S. Geological Survey.

Mattson, D. J., B. M. Blanchard, and R. R. Knight. 1992. Yellowstone grizzly bear mortality, human habituation, and whitebark pine seed crops. *J Wildl Mgmt* 56:432–42.

Mattson, D. J., K. L. Byrd, M. B. Rutherford, S. R. Brown, and T. W. Clark. 2006. Finding common ground in large carnivore conservation: Mapping contending perspectives. *Environ Sci Pol* 9:392–405.

Mattson, D. J., and S. G. Clark. Incidents as a policy phenomenon in wildlife management: Cougars in Arizona and South Dakota. In preparation.

Mattson, D. J., J. Hart, M. Miller, and D. Miller. 2007. Predation and other behaviors of mountain lions in the Flagstaff Uplands. In *Mountain lions of the Flagstaff Uplands, 2003–2006 progress report*, ed. D. J. Mattson, 31–42. Washington, D.C.: U.S. Department of the Interior, U.S. Geological Survey.

Maughan, R., and D. Nilson. 1993. *What's old and what's new about the wise use movement?* Corpus Christi, TX: Western Social Science Convention.

Maxit, I. E. 2001. Prey use by sympatric jaguar and puma in the Venezuelan Llanos. Master's thesis, University of Florida, Gainesville.

Mayr, E. 1982a. Of what use are subspecies? *Auk* 99:593–5.

———. 1982b. *The growth of biological thought: Diversity, evolution, and inheritance*. Cambridge, MA: Harvard University Press.

———. 1996. The modern evolutionary theory. *J Mammal* 77:1–7.

———. 2001. *What evolution is*. New York: Basic Books.

Mazzolli, M. 1993. Ocorrência de *Puma concolor* (Linnaeus)(Felidae, Carnivora) em áreas de vegetação remanescente de Santa Catarina, Brasil. *Revista Brasileira de Zoologia* 10:581–7.

———. 2000. A comparison of habitat use by the mountain lion (*Puma concolor*) and kodkod (*Oncifelis guigna*) in the southern Neotropics with implications for the assessment of their vulnerability status. Master's thesis, University of Durham, North Carolina.

Mazzolli, M., M. E. Graipel, and N. Dunstone. 2002. Mountain lion depredation in southern Brazil. *Biol Conserv* 105:43–51.

McBride, R. T. 1976. The status and ecology of the mountain lion (*Felis concolor stanleyana*) on the Texas-Mexico border. Master's thesis, Sul Ross State University, Alpine, Texas.

McBride, R. T., D. K. Jansen, R. McBride, and S. R. Schulze. 2005. Aversive conditioning of Florida panthers by combining painful experiences with instinctively threatening sounds. In *Proceedings of the eighth mountain lion workshop*, ed. R. A. Beausoleil and D. A. Martorello, p. 136. Olympia: Washington Department of Fish and Wildlife.

McCarthy, J. J. 1996. Montana. Pp. 110–111 in W. D. Padley, ed., *Proceedings of the fifth mountain lion workshop*. San Diego: Southern California Chapter of the Wildlife Society.

McClinton, P. L., S. M. McClinton, and G. J. Guzman. 2000. Utilization of fish as a food item by a mountain lion (*Puma concolor*) in the Chihuahuan Desert. *Tex J Sci* 52:261–3.

McCoy, T. D., M. R. Ryan, E. W. Kurzejeski, and L. W. Burger. 1999. Conservation reserve program: Source or sink habitat for grassland birds in Missouri? *J Wildl Mgmt* 63:530–8.

McCrae, B., P. Beier, L. E. DeWald, and P. Keim. 2005. Gene flow among mountain lion populations in the southwestern USA (abstract). In *Proceedings of eighth mountain lion workshop*, ed. R. A. Beausoleil and D. A. Martorello, p. 210. Olympia: Washington Department of Fish and Wildlife.

McCullough, D. R. 1979. *The George Reserve deer herd: Population ecology of a K-selected species*. Ann Arbor, MI: University of Michigan Press.

McDougal, M. S., H. D. Lasswell, and L.-C. Chen. 1980. *Human rights and world public order: The basic policies of an international law of human dignity*. New Haven, CT: Yale University Press.

McDougal, M. S., H. D. Lasswell, and W. M. Reisman. 1981. The world constitutive process of authoritative decision. In *International law essays: A supplement to International Law in Contemporary Perspective*, ed. M. S. McDougal and W. M. Reisman, 191–268. New York: The Foundation Press.

McHugh, T. 1958. Social behavior of the American buffalo (*Bison bison*). *Zoologica* 43:1–40.

McIntyre, R. 1995. *War against the wolf: America's campaign to exterminate the wolf.* Stillwater, MN: Voyageur Press.

McIvor, D. E., J. A. Bisonette, and G. S. Drew. 1995. Taxonomic and conservation status of the Yuma mountain lion. *Conserv Biol* 9:1033–40.

McIvor, D. E., and M. R. Conover. 1994. Perceptions of farmers and non-farmers toward management of problem wildlife. *Wildl Soc Bull* 22:212–9.

McKinney, T., J. C. deVos, W. B. Ballard, and S. R. Boe. 2006. Mountain lion predation of translocated desert bighorn sheep in Arizona. *Wildl Soc Bull* 34:1255–63.

McKinney, T., T. W. Smith, and J. C. deVos, Jr. 2006. Evaluation of factors potentially influencing a desert bighorn sheep population. *Wildl Monogr* 164:1–36.

McKinstry, M. C., and S. H. Anderson. 2003. Trappers in Wyoming: Opinions on trends in mammalian predator populations, motivations for trapping, and methodologies. *Intermountain J Sci* 9:1–11.

McLaughlin, C. R. 2003. Utah mountain lion status report. In *Proceedings of the seventh mountain lion workshop,* ed. S. A. Becker, D. D. Bjornlie, F. G., Lindzey, and D. S. Moody, 51–59. Lander: Wyoming Game and Fish Department.

McLaughlin, G. P., S. Primm, and M. B. Rutherford. 2005. Participatory projects for coexistence: Rebuilding civil society. In *Coexisting with large carnivores: Lessons from Greater Yellowstone*, ed. T. W. Clark, M. B. Rutherford, and D. Casey, 177–210. Washington, D.C.: Island Press.

McMullin, S. L. 1993. Characteristics and strategies of effective state fish and wildlife agencies. *Trans N Am Wildl Nat Res* 58:206–10.

McNay R. S., and J. M. Voller. 1995. Mortality causes and survival estimates for adult female Columbian black-tailed deer. *J Wildl Mgmt* 59:138–46.

McNeely, J. A., K. H. Redford, and A. S. Carter. 2006. *Friends for life: New partners in support of protected areas.* Washington, D.C.: Island Press.

McRae, B. H. 2004. Integrating landscape ecology and population genetics: Conventional tools and a new model. Ph.D. dissertation, Northern Arizona University, Flagstaff.

McRae, B. H., P. Beier, L. E. Dewald, L. Y. Huynh, and P. Keim. 2005. Habitat barriers limit gene flow and illuminate historical events in a wide-ranging carnivore, the American puma. *Mol Ecol* 14:1965–77.

Mech, L. D. 1970. *The wolf: The ecology and behavior of an endangered species.* Garden City, NY: Natural History Press.

———. 1977. Population trend and winter deer consumption in a Minnesota wolf pack. In *Proceedings of the 1975 predator symposium*, ed. R. L. Phillips and C. Jonkel, 55–83. Montana Forest and Conservation Experiment Station, University of Montana, Missoula.

———. 1996. A new era for carnivore conservation. *Wildl Soc Bull* 24:397–401.

Mech, L. D., and L. Boitani. 2003. *Wolves: Behavior, ecology, and conservation.* Chicago: University of Chicago Press.

Mech, L. D., and G. D. Delgiudice. 1985. Limitations of the marrow-fat technique as an indicator of body condition. *Wildl Soc Bull* 13:204–6.

Mech, L. D., T. J. Meier, J. W. Burch, and L. G. Adams. 1995. Patterns of prey selection by wolves in Denali National Park, Alaska. In *Ecology and conservation of wolves in a changing world*, ed. L. N. Carbyn, S. H. Fritts, and D. R. Seip, 231–243. Canadian Circumpolar Institute Occasional Publication 35.

Mech, L. D., and R. O. Peterson. 2003. Wolf-prey relations. In *Wolves: Behavior, ecology, and conservation*, ed. L. D. Mech and L. Boitani, 131–157. Chicago: University of Chicago Press.

Medellín, R. A. 1994. Mammal diversity and conservation in the Selva Lacandona, Chiapas, Mexico. *Conserv Biol* 8:780–99.

Mehrer, C. F. 1975. Some aspects of reproduction in captive mountain lions (*Felis concolor*), bobcats (*Lynx rufus*), and lynx (*Lynx canadensis*), Ph.D. dissertation, University of North Dakota, Grand Forks.

Meinke, C. W., R. T. Golightly, H. B. Quigley, J. Ellingson, and T. Hofstra. 2004. Mountain lion habitat use relative to human activities in the Redwood Basin of northwest California (abstract). Presented at the American Society of Mammalogists Annual Meeting.

Menotti-Raymond, M., V. David, J. C. Stephens, L. A. Lyons, and S. J. O'Brien. 1997. Genetic individualization of domestic cats using feline STR loci for forensic applications. *J Forensic Sci* 42:1037–50.

Menotti-Raymond, M., and S. J. O'Brien. 1995. Evolutionary conservation of ten microsatellite loci in four species of Felidae. *J Hered* 86:319–22.

Merchant, C. 2002. *The Columbia guide to American environmental history.* New York: Columbia University Press.

Merriam, C. H. 1901. Preliminary revision of the pumas (*Felis concolor* group). *Proceedings of the Washington Academy of Sciences* 3:577–600.

Merriam, J. C. 1997. Community wildlife management by Mayangna Indians in the Bosawas Reserve, Nicaragua. Master's thesis, Idaho State University, Pocatello.

Messier, F. 1994. Ungulate population models with predator: A case study with the North American moose. *Ecology* 75:478–88.

———. 1995. On the functional and numerical responses of wolves to changing prey density. In *Ecology and conservation of wolves in a changing world*, ed. L. N. Carbyn, S. H. Fritts, and D. R. Seip, 187–97. Canadian Circumpolar Institute Occasional Publication 35.

Michalski, F., R. L. P. Boulhosa, A. Faria, and C. A. Peres. 2006. Human—wildlife conflicts in a fragmented Amazonian forest landscape: Determinants of large felid depredation on livestock. *Anim Conserv* 9:179–88.

Michalski, F., and C. A. Peres. 2005. Anthropogenic determinants of primate and carnivore local extinctions in a fragmented forest landscape of southern Amazonia. *Biol Conserv* 124:383–96.

Miller, F. L. 1970. Distribution patterns of black-tailed deer (*Odocoileus hemionus columbianus*) in relation to environment. *J Mammal* 51:248–60.

Miller, S. D., and D. D. Everett, eds. 1986. Cats of the world: Biology, conservation, and management. *Proceedings of the second international symposium*, Caesar Kleberg Wildlife Research Insitute College of Agriculture, Texas A&I University. Washington, D.C.: National Wildlife Federation.

Miller, M. W., H. M. Swanson, L. L. Wolfe, F. G. Quartarone, S. L. Huwer, C. H. Southwick, and P. M. Lukacs. 2008. Lions and prions and deer demise. *PLoS ONE* 3:e4019.

Mills, L. S. 2007. *Conservation of wildlife populations: Demography, genetics, and management*. Malden, MA: Blackwell Publishing.

Mills, M. G. L. 2005. Large carnivores and biodiversity in African savanna ecosystems. In *Large carnivores and the conservation of biodiversity*, ed. J. C. Ray, K. H. Redford, R. S. Steneck, and J. Berger, 208–229. Washington, D.C.: Island Press.

Miniter, F. 2004. Cougars coast to coast. *Outdoor Life*, June/July.

Minnis, D. L. 1998. Wildlife policy-making by the electorate: An overview of citizen-sponsored ballot measures on hunting and trapping. *Wildl Soc Bull* 26:75–83.

Miquelle, D., P. Stephens, E. Smirnov, J. Goodrich, O. Zaumyslova, and A. Myslenkov. 2005. Tigers and wolves in the Russian Far East: Competitive exclusion, functional redundancy, and conservation implications. In *Large carnivores and the conservation of biodiversity*, ed. J. C. Ray, K. H. Redford, R. S. Steneck, and J. Berger. Washington, D.C.: Island Press.

Molina, G. I. 1782. In *Sazzio sulla storia naturale del Chilli*, 295–300.

Montana Senate, Fifty-eighth Legislature, Committee on Fish and Game. 2003. *Minutes of executive action on HB 32*. Tape 1, Side A, Time Counter 11.8–16.8 minutes, March 13, 2003.

Mooring, M. S., T. A. Fitzpatrick, T. T. Nishihira, and D. D. Reisig. 2004. Vigilance, predation risk, and the Allee effect in desert bighorn sheep. *J Wildl Mgmt* 68:519–32.

Morell, V. 2007. Oregon cougars to be hounded. *Science Now Daily News*, June 29.

Moreno, F. P. 1997. *Viaje a la Patagonia Austral*. Buenos Aires: El Elefante Blanco.

Moreno, R. S., R. W. Kays, and R. Samudio Jr. 2006. Competitive release in diets of ocelot (*Leopardus pardalis*) and puma (*Puma concolor*) after jaguar (*Panthera onca*) decline. *J Mammal* 87:808–16.

Morrison, B. 1984. New Mexico. Pp. 49–52, in J. Roberson and F. Lindzey, eds., *Proceedings of the second mountain lion workshop*. Salt Lake City: Utah Division of Wildlife Resources.

Mortenson, K. G., and R. S. Krannich. 2001. Wildlife managers and public involvement: Letting the crazy aunt out. *Hum Dim Wildl* 6:277–90.

Morton, B. 2003. Livestock damaged by state's wildlife deserves compensation. Washington State Senate Replican Caucus. http://www.src.wa.gov (accessed December 10, 2006).

Mosher, F. C. 1982. *Democracy and the public service*. New York: Oxford University Press.

Mountain Lion Foundation. 2007. *The habitat conservation fund, since 1990 – Serving the people of California*. Sacramento.

Munoz, R. 1982. Movements and mortalities of desert bighorn of the San Andres Mountains, New Mexico. *Desert Bighorn Sheep Council Trans* 26:107–8.

Muñoz-Pedreros, A., J. R. Rau, M. Valdebenito, V. Quintana, and D. R. Martínez. 1995. Densidad relativa de pumas (*Felis concolor*) en un ecosistema forestal del sur de Chile. *Revista Chilena de Historia Natural* 68:501–7.

Murphy, K. M. 1983. Relationships between a mountain lion population and hunting pressure in western Montana. Master's thesis, University of Montana, Missoula.

———. 1984. Montana. Pp. 39–43, in J. Roberson and F. Lindzey, eds., *Proceedings of the second mountain lion workshop*. Salt Lake City: Utah Division of Wildlife Resources.

———. 1998. Ecology of the cougar (*Puma concolor*) in the northern Yellowstone ecosystem: Interactions with prey, bears, and humans Ph D. dissertation, University of Idaho, Moscow.

Murphy, K. M., G. S. Felzien, M. G. Hornocker, and T. K. Ruth. 1998. Encounter competition between bears and cougars: Some ecological implications. *Ursus* 10:55–60.

Murray, D. 2006. On improving telemetry-based survival estimation. *J Wildl Mgmt* 70:1530–42.

Murray, D., and B. R. Patterson. 2006. Wildlife survival estimation: Recent advances and future directions. *J Wildl Mgmt* 70:1499–1503.

Murray, D. L., S. Boutin, M. O'Donoghue, and V. O. Nams. 1995. Hunting behaviour of a sympatric felid and canid in relation to vegetative cover. *Anim Behav* 50:1203–10.

Murray, D. L., J. R. Cary, and L. B. Keith. 1997. Interactive effects of sublethal nematodes and nutritional status on snowshoe hare vulnerability to predation. *J Anim Ecol* 66:250–64.

Musters, G. C. 1964. *Vida entre los patagones*. Buenos Aires: Edial Solar/Hachette.

Muth, R. M., D. A. Hamilton, J. F. Organ, D. J. Witter, M. E. Mather, and J. J. Daigle. 1998. The future of wildlife and fisheries policy and management: Assessing the attitudes and values of wildlife and fisheries professionals. *Trans N Am Wildl Nat Res* 63:604–27.

Nadeau, S. 2003. Idaho mountain lion status report. In *Proceedings of the seventh mountain lion workshop*, ed. S. A. Becker, D. D. Bjornlie, F. G., Lindzey, and D. S. Moody, 25–28. Lander: Wyoming Game and Fish Department.

———. 2005. Idaho mountain lion status report. In *Proceedings of the eighth mountain lion workshop*, ed. R. A. Beausoleil and D. A. Martorello, 17–21. Olympia: Washington Department of Fish and Wildlife.

Nash, R. 1967. *Wilderness and the American mind*. New Haven, CT: Yale University Press.

Naughton-Treves, L., R. Grossberg, and A. Treves. 2003. Paying for tolerance: Rural citizens' attitudes toward wolf depredation and compensation. *Conserv Biol* 17:1500–11.

Neal, D. L. 1990. The effect of predation on deer in the central Sierra Nevada. In *Predator management in North Costal California: Proceedings of a Workshop held in Ukiah and Hopeland, California, March 10–11, 1990*, ed. G. A. Guisti, R. M. Timm, and R. H. Schmidt, 53–61. University of California, Hopland Field Station Publication 101, University of California.

Neal, D. L., G. N. Steger, and R. C. Bertram. 1987. Mountain lions: Preliminary findings on home–range use and density in the Central Sierra Nevada. Res. Note PSW–392, Pacific Southwest Forest and Range Experiment Station, U.S. Department of Agriculture, Berkeley, CA.

Nebel, B. J., and R. T. Wright. 1996. *Environmental science: The way the world works*. Fifth ed. Upper Saddle River, NJ: Prentice Hall.

Neihardt, J. G. 1961. *Black elk speaks: Being the life story of a holy man of the Oglala Sioux*. Lincoln, NE: University of Nebraska Press.

Newell, S. L. 2006. An evaluation of a science-based approach to habitat linkage design. Master's thesis, Northern Arizona University, Flagstaff.

Nice, M. 1937. *Studies in the life history of the song sparrow.* New York: Linnaean Society of New York.

Nicholson, M. C. 1995. Habitat selection by mule deer: Effects of migration and population density. Ph.D. dissertation, University of Alaska, Fairbanks.

Nie, M. 2002. Wolf recovery and management as value-based political conflict. *Ethics Place Environ* 5:65–71.

———. 2003. Drivers of natural-resource-based political conflict. *Pol Sci* 36:307–41.

———. 2004a. State wildlife governance and carnivore conservation. In *People and predators: From conflict to coexistence,* ed. N. Fascione, A. Delach, and M. E. Smith, 197–218. Washington, D.C.: Island Press.

———. 2004b. State wildlife policy and management: The scope and bias of political conflict. *Public Admin Rev* 64:221–33.

Nisbet, M. C., and C. Mooney. 2007. Science and society: Framing science. *Science* 316:56.

NOAA. 2007. Glossary of terminology: U. http://www8.nos.noaa.gov/coris_glossary/index.aspx?letter=u (accessed January 23, 2007).

Noss, A. J., M. J. Kelly, H. B. Camblos, and D. I. Rumiz. 2006. Pumas y jaguares simpátricos: Datos de trampas-cámara en Bolivia y Belize. In *Procceedings of the fifth Congreso, Manejo de fauna silvestre en Amazonia y Latinoamérica,* 229–237.

Noss, R. F. 2007. Climate change intensifies need for land conservation. *Conservation Northwest Quarterly* Fall:4–5.

Noss, R. F., P. Beier, and W. Shaw. 1998. Evaluation of the Coal Canyon Biological Corridor. Report to the Puente-Chino Wildlife Corridor Conservation Authority.

Noss, R. F., H. B. Quigley, M. G. Hornocker, T. Merrill, and P. C. Paquet. 1996. Conservation biology and carnivore conservation in the Rocky Mountains. *Conserv Biol* 10:949–63.

Novaro, A. J., M. C. Funes, M. B. Bongiorno, O. B. Monsalvo, E. Donadío, R. S. Walker, G. Sanchez, and O. Pailacura. 1999. Proyecto integrado de investigación sobre especies predadoras y perjudiciales de la producción ganadera en la provincia del Neuquén. I Report to Secretaría de Agricultura y Ganadería de la Nación. Centro de Ecología Aplicada del Neuquén.

Novaro, A. J., M. C. Funes, and R. S. Walker. 2000. Ecological extinction of native prey of a carnivore assemblage in Argentine Patagonia. *Biol Conserv* 92:25–33.

———. 2005. An empirical test of source-sink dynamics induced by hunting. *J Appl Ecol* 42:910–20.

Novack, A. J., M. B. Main, M. E. Sunquist, and R. F. Labisky. 2005. Foraging ecology of jaguar (*Panthera onca*) and puma (*Puma concolor*) in hunted and non-hunted sites within the Maya Biosphere Reserve, Guatemala. *J Zool London* 267:167–78.

Novaro, A. J., and R. S. Walker. 2005. Human-induced changes in the effect of top carnivores on biodiversity in the Patagonian Steppe. In *Large carnivores and the conservation of biodiversity,* ed. J. C. Ray, K. H. Redford, R. S. Steneck, and J. Berger, 268–288. Washington, D.C.: Island Press.

Nowack, M. C. 1999. Predation rates and foraging ecology of adult female mountain lions in northeastern Oregon. Master's thesis, Washington State University, Pullman.

Nowak, R. M. 1976. *The cougar in the United States and Canada.* Washington, D.C.: U.S. Department of the Interior, Fish and Wildlife Service and New York: New York Zoological Society.

Nowell, K, and P. Jackson, eds. 1996. Wild cats: Status and conservation action plan. The World Conservation Union, Species Survival Commission, Cat Specialist Group, Gland, Switzerland.

Núñez, R., B. Miller, and F. Lindzey. 2000. Food habits of jaguars and pumas in Jalisco, Mexico. *J Zool London* 252:373–9.

———. 2002. Ecología del jaguar en la Reserva de la Biosfera Chamela-Cuixmala, Jalisco, México. In *El Jaguar en el Nuevo Milenio,* ed. R. A. Medellín, C. Equihua, C. L. B. Chetkiewicz, P. G. Crawshaw, Jr., A. Rabinowitz, H. H. Redford, J. G. Robinson, E. W. Sanderson, and A. B. Taber, 107–126. Universidad Nacional Autónoma de México/Wildlife Conservation Society, Mexico.

O'Brien, S. J. 1994. Genetic and phylogenetic analysis of endangered species. *Annu Rev Genet* 28:467–89.

O'Brien, S. J., and E. Mayr. 1991. Bureaucratic mischief: Recognizing endangered species and subspecies. *Science* 251:1187–8.

O'Brien, S. J., M. E. Roelke, N. Yuhki, K. W. Richards, W. E. Johnson, W. L. Franklin, A. E. Anderson, O. L. Bass, Jr., R. C. Belden, and J. S. Martenson. 1990. Genetic introgression within the Florida panther (*Felis concolor coryi*). *Natl Geogr Res* 6:485–94.

O'Brien, S. J., D. E. Wildt, and M. Bush. 1986. The cheetah in genetic peril. *Sci Am* 254:84–92.

O'Gara, B. W., and R. B. Harris. 1988. Age and condition of deer killed by predators and automobiles. *J Wildl Mgmt* 52:316–20.

Ockenfels, R. A. 1994. Factors affecting adult pronghorn mortality rates in central Arizona. *Arizona Game and Fish Department Wildlife Digest* 16:1–11.

Oelschlaeger, M. 1994. *Caring for creation: An ecumenical approach to the environmental crisis.* New Haven, CT: Yale University Press.

O'Gara, B. W., and R. E. McCabe. 2005. From exploitation to conservation. In *Pronghorn: Ecology and management,* ed. B. W. O'Gara and J. D. Yoakum, 42–73. Boulder, CO: University Press of Colorado.

O'Gara, B. W., and J. D. Yoakum, eds. 2005. *Pronghorn: Ecology and management.* Boulder, CO: University Press of Colorado.

Ontario Ministry of Natural Resources. 2007. Ontario species at risk (module). http://www.rom.on.ca/ontario/risk.php?doc_type=fact&lang=&id=135 (accessed September 2, 2007).

Ontario Puma Foundation. 2007. Sightings map. http://www.ontariopuma.ca/sightings.htm (accessed October 14, 2007).

Ordway, L. L., and P. R. Krausman. 1986. Sexual segregation in mule deer. *J Wildl Mgmt* 50:677–83.

Oregon Department of Fish and Wildlife. 2003. "Cougar conflict challenges Lane County." http://www.dfw.state.or.us/springfield/cougarconflict.html (accessed April 22, 2003).

———. 2007. Agency information. http://www.dfw.state.or.us/agency/ (accessed January 21, 2008).

Orford, H. J. L., M. Perrin, and H. Berry. 1988. Contraception, reproduction, and demography of free-ranging Etosha lions (*Panthera leo*). *J Zool London* 216:717–33.

Orr, P. C. 1969. *Felis trumani,* a new radiocarbon dated cat skull from Crypt Cave, Nevada. *Santa Barbara Mus Nat Hist Bull* 2:1–8.

Orren, G. 1997. Fall from grace: The public's loss of faith in government. In *Why people don't trust government,* ed. J. S. Nye, Jr., P. D. Zelikow, and D. C. King, 77–107. Cambridge, MA: Harvard University Press.

Pace, M. L., J. J. Cole, S. R. Carpenter, and J. F. Kitchell. 1999. Trophic cascades revealed in diverse ecosystems. *Trends Ecol Evol* 14:483–8.

Pacelle, W. 1998. Forging a new wildlife management paradigm: Integrating animal protection values. *Hum Dim Wildl* 3:42–50.

Pacheco, L. F., A. Lucero, and M. Villca. 2004. Dieta del puma (*Puma concolor*) en el Parque Nacional Sajama, Bolivia y su conflicto con la ganadería. *Ecología en Bolivia* 39:75–83.

Pacheco, L. F., and J. A. Salazar. 1996. Bases para la conservación de félidos en Bolivia. *Ecología en Bolivia* 26:71–92.

Packer, C. 1986. The ecology of sociality in felids. In *Ecological aspects of social evolution,* ed. D. Rubenstein and R. Wrangham, 429–451. Princeton, NJ: Princeton University Press.

Packer, C., L. Herbst, A. E. Pusey, J. D. Bygott, J. P. Hanby, S. J. Cairns, and M. B. Mulder. 1988. Reproductive success of lions. In *Reproductive success: Studies of individual variation in contrasting breeding systems,* ed. T. H. Clutton-Brock, 363–383. Chicago: University of Chicago Press.

Packer, C., R. D. Holt, P. J. Hudson, K. D. Lafferty, and A. P. Dobson. 2003. Keeping the herds healthy and alert: Implications of predator control for infectious disease. *Ecol Lett* 6:797–802.

Packer, C., Ikanda, D., Kissui, B., and Kushnir, H. 2005. Lion attacks on humans in Tanzania. *Nature* 436:927–8.

Packer, C., A. E. Pussey, H. Rowley, D. A. Gilbert, J. Martenson, and S. J. O'Brien. 1991. Case study of a population bottleneck: Lions of the Ngorongoro Crater. *Conserv Biol* 5:219–30.

Packer, C., D.A. Gilbert, A.E. Pusey, and S.J. O'Brien. 1991. A molecular genetic analysis of kinship and cooperation in African lions. *Nature* 351:562–5.

Packer, C., D. Scheel, and A. E. Pusey. 1990. Why lions form groups: Food is not enough. *Am Nat* 136:1–19.

Padley, W. D. 1990. Home ranges and social interactions of mountain lions (*Felis concolor*) in the Santa Ana Mountains, California. Master's thesis, California State Polytechnic University, Pomona.

———. 1996. Mountain lion (*Felis concolor*) vocalizations in the Santa Ana Mountains, California. In *Proceedings of the fifth mountain lion workshop,* ed. W. D. Padley. San Diego: Southern California Chapter of the Wildlife Society.

Pall, O. 1984. Alberta. In *Proceedings of the second mountain lion workshop,* ed. J. Roberson and F. Lindzey, 1–8. Salt Lake City: Utah Division of Wildlife Resources.

Palomares, F., and T. M. Caro. 1999. Interspecific killing among mammalian carnivores. *Am Nat* 153:492–508.

Palomares, F., P. Gaona, P. Ferreras, and M. Delibes. 1995. Positive effects on game species of top predators by controlling smaller predator populations: An example with lynx, mongooses, and rabbits. *Conserv Biol* 9:295–305.

Papouchis, C. 2004. Conserving mountain lions in a changing landscape. In *People and predators: From conflict to coexistence,* ed. J. Roberson and F. Lindzey, 219–39. Washington, D.C.: Island Press.

———. 2006a. "Living with lions: Does sport hunting mountain lions reduce attacks on people and livestock?" Mountain Lion Foundation. http://www.pumaconservation.org/html/printable_version.html (accessed October 15, 2006).

———. 2006b. "Human exploitation of mountain lions in the American West: A review of human caused mountain lion mortality." Mountain Lion Foundation. http://www.pumaconservation.org/html/overview1.html (accessed September 6, 2007).

Papouchis, C., R. A. Hopkins, and D. Dawn. 2005. A new paradigm for cougar conservation and "management" in the twenty-first Century (abstract). In *Proceedings of eighth mountain lion workshop,* ed. R. A. Beausoleil and D. A. Martorello, 217. Olympia: Washington Department of Fish and Wildlife.

Parera, A. 2002. *Los mamíferos de la Argentina y de la región austral de Sudamérica.* Buenos Aires: El Ateneo.

Parker, V. 1995. Natural resources management by litigation. In *A new century for natural resources management,* ed. R. L. Knight and S. F. Bates, 209–220. Washington, D.C.: Island Press.

Parkins, J. R., and R. E. Mitchell. 2005. Public participation as public debate: A deliberative turn in natural resource management. *Soc Nat Res* 18:529–40.

Pate, J., M. J. Manfredo, A. D. Bright, and G. Tischbein. 1996. Coloradan's attitudes toward reintroducing the gray wolf into Colorado. *Wildl Soc Bull* 24:421–8.

Patterson, B., and R. Pascual. 1972. The fossil mammal fauna of South America. In *Evolution, mammals and Southern continents,* ed. A. Keast, C. Erk, and B. Glass. Albany, NY: State University of New York Press.

Pecon-Slattery, J., and S. J. O'Brien. 1998. Patterns of Y and X chromosome DNA sequence divergence during the *Felidae* radiation. *Genetics* 148:1245–55.

Peek, C. W., N. J. Bell, and J. Dunham. 1996. Gender, gender ideology, and animal rights advocacy. *Gender Soc* 10:464–78.

Peek, C. W., C. C. Dunham, and B. E. Dietz. 1997. Gender, relational role orientation, and affinity for animal rights. *Sex Roles* 37:905–20.

Peña, I. A. 2002. Assessing public knowledge, attitudes, and beliefs regarding mountain lions in Texas. Master's thesis, Texas A&M University, Kingsville.

Penrod, K. L., R. Hunter, and M. Merrifield. 2001. Missing linkages: Restoring connectivity to the California landscape. California Wilderness Coalition, The Nature Conservancy, US Geological Survey, Center for Reproduction of Endangered Species, and California State Parks.

Peres, C.A. 2000. Effects of subsistence hunting on vertebrate community structure in Amazonian forests. *Conserv Biol* 14:240–53.

———. 2001. Synergistic effects of subsistence hunting and habitat fragmentation on Amazonian forest vertebrates. *Conserv Biol* 15:1490–1505.

Perovic, P. G. 2002. Conservación del jaguar en el noroeste de Argentina. In *El Jaguar en el Nuevo Milenio,* ed. R. A. Medellín, C. Equihua, C. L. B. Chetkiewicz, P. G. Crawshaw Jr., A. Rabinowitz, H. H. Redford, J. G. Robinson, E. W. Sanderson, and A. B. Taber, 465–475. Universidad Nacional Autónoma de México/Wildlife Conservation Society, Mexico.

Perry, G. L., and J. C. DeVos, Jr. 2005. A case study of mountain lion-human interaction in southeastern Arizona.

In *Proceedings of the eighth mountain lion workshop*, ed. R. A. Beausoleil and D. A. Martorello, 104–13 Olympia: Washington Department of Fish and Wildlife.

Perry, R. 1965. *The world of the tiger*. New York: Atheneum.

Pessino, M. E. M., J. H. Sarasola, C. Wander, and N. Besoky. 2001. Respuesta a largo plazo del puma (*Puma concolor*) a una declinación poblacional de la vizcacha (*Lagostomus maximus*) en el desierto del Monte, Argentina. *Ecología Austral* 11:61–7.

Peters, G., and W. C. Wozencraft. 1989. Acoustic communication by fissiped carnivores. In *Carnivore behavior, ecology and evolution*, ed. G. L. Gittleman, 14–56, New York: Cornell University Press.

Peterson, E. A., W. C. Heaton, and S. D. Wruble. 1969. Levels of auditory response in fissiped carnivores. *J Mammal* 50:566–78.

Peterson, R. O. 1977. *Wolf ecology and prey relationships on Isle Royale*. U.S. National Park Service Scientific Monograph Series, no. 11. Bethesda, MD.

Peterson, R. O., and P. Ciucci. 2003. The wolf as a carnivore. In *Wolves: Behavior, ecology, and conservation*, ed. L. D. Mech and L. Boitani, 104–130. Chicago: University of Chicago Press.

Peyton, B. 1989. A profile of Michigan bear hunters and bear hunting issues. *Wildl Soc Bull* 17:463–70.

Phelps, J. S. 1988. Status of mountain lions in Arizona. In *Proceedings of the third mountain lion workshop*, ed. R. H. Smith, 7–9. Phoenix: Arizona Game and Fish Department.

Pia, M., and A. J. Novaro. 2005. Monitoreo de poblaciones de carnívoros y análisis de alternativas para reducir conflictos con la ganadería en el área circundante al Parque Nacional Monte León. Final report presented to the Argentine Administración de Parques Nacionales.

Pianka, E. R. 1973. The structure of lizard communities. *Annu Rev Ecol Syst* 4:53–74.

Pielou, E. C. 1991. *After the Ice Age: The return of life to glaciated North America*. Chicago: University of Chicago Press.

Pierce, B. M. 1987. Ecology and conservation of the jaguar in the Pantanal Region, Mato Grosso do Sul, Brazil. Ph.D. dissertation, University of Idaho, Moscow.

Pierce, B. M., V. C. Bleich, and R. T. Bowyer. 1999. Population dynamics of mountain lions and mule deer: Top-down or bottom-up regulation? Final report. Deer Herd Management Plan Implementation Program, California Department of Fish and Game, Sacramento.

———. 2000a. Selection of mule deer by mountain lions and coyotes: Effects of hunting style, body size, and reproductive status. *J Mammal* 81:462–72.

———. 2000b. Social organization of mountain lions: Does a land-tenure system regulate population size? *Ecology* 81:1533–43.

Pierce, B. M., V. C. Bleich, J. D. Wehausen, and R. T. Bowyer. 1999. Migratory patterns of mountain lions: Implications for social regulation and conservation. *J Mammal* 80:986–92.

Pierce, B. M., R. T. Bowyer, and V. C. Bleich. 2004. Habitat selection by mule deer: Forage benefits or risk of predation? *J Wild Mgmt* 68:533–41.

Pimbert, M. P., and J. N. Pretty. 1995. *Parks, people and professionals: Putting 'participation' into protected area management*. Discussion Paper No. 57. United Nations Research Institute for Social Development, Geneva.

Pimm, S. L., L. Dollar, and O. L. Bass. 2006. Genetic rescue of the Florida panther. *Anim Conserv* 9:115–22.

Pocock, R. I. 1940. Description of a new race of puma (*Puma concolor*), with note on an abnormal tooth in the genus. *Ann Mag Nat Hist Set 11*, 6:307–13.

Polis, G. A. 1994. Food webs, trophic cascades and community structure. *Australian J Ecol* 19:121–36.

Polis, G. A., and R. D. Holt. 1992. Intraguild predation: The dynamics of complex trophic interactions. *Trends Ecol Evol* 7:151–4.

Polis, G. A., C. A. Myers, and R. D. Holt. 1989. The ecology and evolution of intraguild predation: Potential competitors that eat each other. *Annu Rev Ecol Syst* 20:297–330.

Polisar, J. 2000. Jaguars, pumas, their prey base and cattle ranching: Ecological perspectives of a management issue. Ph.D. dissertation, University of Florida, Gainesville.

Polisar, J., I. Maxit, D. Scognamillo, L. Farrell, M. E. Sunquist, and J. F. Eisenberg. 2003. Jaguars, pumas, their prey base, and cattle ranching: Ecological interpretations of a management problem. *Biol Conserv* 109:297–310.

Portland Chapter of the Safari Club International. 2003. Oregonians for responsible wildlife management. http://www.sciportland.org/bear_and_cougar_bill.htm (accessed April 22, 2003).

Pozzanghera, S. 1996. Washington. In *Proceedings of the fifth mountain lion workshop*, ed. W. D. Padley, 122–3. San Diego: Southern California Chapter of the Wildlife Society.

The Precautionary Principle Project. 2007. http://www.pprinciple.net/ (accessed January 12, 2007).

Price, J. L. 1963. The impact of governing boards on organizational effectiveness and morale. *Admin Sci Q* 8:361–78.

Prince, L. B. 1903. *The stone lions of Cochiti*. Whitefish, MT: Kessinger Publishing.

Pritchart, P. C. H., ed. 1976. Proceedings of the Florida Panther Conference. Florida Audubon Society and Florida Game and Freshwater Fish Commission, Tallahassee.

Pronatura Mexico. 2007. http://www.pronatura.org.mx/index_ing.php (accessed May 21, 2007).

Puig, S. 1986. Ecología poblacional del guanaco (*Lama guanicoe, Camelidae, Artiodactyla*) en la Reserva Provincial La Payunia (Mendoza). Ph.D. dissertation, University of Buenos Aires, Buenos Aires, Argentina.

Puig, S., G. Ferraris, M. Superina, and F. Videla. 2003. Distribución de densidades de guanacos (*Lama guanicoe*) en el norte de la reserva La Payunia y su área de influencia (*Mendoza, Argentina*). *Multiequina* 12:37–48.

Pulliam, H. R. 1988. Sources, sinks, and population regulation. *Am Nat* 132:652–61.

Purvis, A., G. M. Mace, and J. L. Gittleman. 2001. Past and future carnivore extinctions: a phylogenetic perspective. In *Carnivore conservation*, ed. J. L. Gittleman, S. M. Funk, D. Macdonald, and R. K. Wayne, 11–34. Cambridge, UK: Cambridge University Press.

Pusey, A. E., and C. Packer. 1987a. The evolution of sex-biased dispersal in lions. *Behaviour* 4:275–310.

———. 1987b. Dispersal and philopatry. In *Primate societies*, ed. D. L. Cheney, R. M. Seyfarth, B. B. Smuts, T. Struhsaker, and R. W. Wrangham, 136–148. Chicago: University of Chicago Press.

Putman, R. J. 1996. *Competition and resource partitioning in temperate ungulate assemblies*. London: Chapman and Hall.

Putnam, R. D. 2000. *Bowling alone: The collapse and revival of American community*. New York: Simon and Schuster.

Quaratiello, A. R. 2004. *Rachel Carson: A biography*. Westport, CT: Greenwood Press.

Quigley, H. B. 1987. Ecology and conservation of the jaguar in the Pantanal Region, Mato Grosso do Sul, Brazil. Ph.D. dissertation, University of Idaho, Moscow.

Quigley, H. B., and D. Craighead. 2004. Teton Cougar Project Annual Report, Craighead Beringia South.

Quigley, H. B., and P. G. Crawshaw, Jr. 1992. A conservation plan for the jaguar (*Panthera onca*) in the Pantanal region of Brazil. *Biol Conserv* 61:149–57.

Quigley, H. B., and M. G. Hornocker. 1992. Large carnivore ecology: From where do we come and to where shall we go? In *Wildlife 2001: Populations*, ed. D. McCullough and R. Barrett, 1089–1097. New York: Elsevier Applied Science.

Quigley, H. B., G. M. Koehler, and M. G. Hornocker. 1989. Dynamics of a mountain lion population in central Idaho over a 20-year period. In R.H. Smith, ed., *Proceedings of the third mountain lion workshop*, 54. Phoenix: Arizona Game and Fish Department.

Rabinowitz, A., and B. G. Nottingham. 1986. Ecology and behaviour of the jaguar (*Panthera onca*) in Belize, Central America. *J Zool London* 210:149–59.

Ralls, K., and P. J. White. 1995. Predation on San Joaquin kit foxes by larger canids. *J Mammal* 76:723–9.

Range, F., L. Huber, and Z. Viranyi. 2007. Selective imitation in dogs. *Current Biology* May 15. Also at http://www.current-biology.com/ (accessed May 6. 2008).

Rau, J. R., and J. Jiménez. 2002. Diet of puma (*Puma concolor*, Carnivora: Felidae) in coastal and Andean ranges of southern Chile. *Stud Neotropical Fauna Environ* 37:201–5.

Rau, J. R., M. S. Telleria, D. R. Martinez, and A. H. Muñoz. 1991. Dieta de *Felis concolor* (Carnivora: Felidae) en áreas silvestres protegidas del sur de Chile. *Revista Chilena de Historia Natural* 64:139–44.

Ray, J. 1693. Shynopsis methodica animalium quadrupedum et serpentini generia. London: 169. Quoted in Young, S. P., and E. A. Goldman. 1946. *The puma: Mysterious American cat*. Washington, D.C.: The American Wildlife Institute.

Ray, J. C., K. Redford, R. Steneck, and J. Berger, eds. 2005. *Large carnivores and the conservation of biodiversity*. Washington, D.C.: Island Press.

Reading, R. P., T. W. Clark, and S. R. Kellert. 1994. Attitudes and knowledge of people living in the Greater Yellowstone ecosystem. *Soc Nat Res* 7:349–65.

Reading, R. P., and S. R. Kellert. 1993. Attitudes toward a proposed reintroduction of black-footed ferrets (*Mustela nigripes*). *Conserv Biol* 7:569–80.

Redford, K. H. 1992. The empty forest. *Bioscience* 42:412–22.

Reid, F. A. 1997. *A field guide to the mammals of Central America and Southeast Mexico*. New York: Oxford University Press.

Reiger, J. F. 2001. *American sportsmen and the origins of conservation*. Corvallis, OR: Oregon State University Press.

Renkonen, O. 1938. Statisch-okologische Untersuchungen uber die terrestiche daferwelt der finnischen bruchmoore. *Ann Zool Soc Bot Fenn Vanamo* 6:1–231.

Richards, R. T., and R. S. Krannich. 1991. The ideology of the animal rights movement. *Trans N Am Wildl Nat Res* 56:363–71.

Ricklefs, R. E. 1990. *Ecology*. Third ed. New York: W. H. Freeman and Co.

Riebsome, W. E., H. Gosnell, and D. M. Theobald. 1996. Land use and landscape change in the Colorado mountains I: Theory, scale, and pattern. *Mt Res Dev* 16:395–405.

Rieck, J. M. 1988. Status of the cougar in Washington. In *Proceedings of the third mountain lion workshop*, ed. R. H. Smith, 35–37. Phoenix: Arizona Game and Fish Department.

Rikiishi, K., E. Hashya, and M. Imai. 2004. Linear trends of the length of snow-cover season in the Northern Hemisphere as observed by satellites in the period 1972–2000. *Ann Glaciol* 38:229–37.

Riley, S. J., and D. Decker. 2000. Wildlife stakeholder acceptance capacity for cougars in Montana. *Wildl Soc Bull* 28:931–9.

Riley, S. J., G. M. Nesslage, and B. A. Maurer. 2004. Dynamics of early wolf and cougar eradication efforts in Montana: Implications for conservation. *Biol Conserv* 119:575–9.

Ripple, W. J., and R. L. Beschta. 2006. Linking a cougar decline, trophic cascade, and catastrophic regime shift in Zion National Park. *Biol Conserv* 133:397–408.

Ripple, W. J., E. J. Larsen, R. A. Renkin, and D. W. Smith. 2001. Trophic cascades among wolves, elk, and aspen on Yellowstone National Park's northern range. *Biol Conserv* 102:227–34.

Robertson, J. 1984. Utah. In *Proceedings of the second mountain lion workshop*, ed. J. Roberson and F. Lindzey, 60–79. Salt Lake City: Utah Division of Wildlife Resources and Utah Cooperative Wildlife Research Unit.

Robertson, J., and F. Lindzey, eds. 1984. *Proceedings of the second mountain lion workshop*. Salt Lake City: Utah Division of Wildlife Resources and Utah Cooperative Wildlife Research Unit.

Robertson, M., and M. Bell. 1988. Status of the mountain lion in New Mexico. In *Proceedings of the third mountain lion workshop*, ed. R. H. Smith, 24–5. Phoenix: Arizona Game and Fish Department.

Robinette, W. L., J. S. Gashwiler, and O. W. Morris. 1959. Food habits of the cougar in Utah and Nevada. *J Wildl Mgmt* 23:261–73.

———. 1961. Notes on cougar productivity and life history. *J Mammal* 42:204–17.

Robinette, W. L., N. V. Hancock, and D. A. Jones. 1977. *The Oak Creek mule deer herd in Utah*. Salt Lake City: Utah Division of Wildlife Resources.

Robinson, H. S., R. B. Wielgus, H. S. Colley, and S. W. Cooley. 2008. Sink populations in carnivore management: Cougar demography and immigration in a hunted population. *Ecol Apps* 18:1028–37.

Robinson, H. S., R. B. Wielgus, and J. C. Gwilliam. 2002. Cougar predation and population growth of sympatric mule deer and white-tailed deer. *Can J Zool* 80:556–68.

Robinson, J. G., and K. H. Redford. 1991. *Neotropical wildlife use and conservation*. Chicago: University of Chicago.

Robinson, M. J. 2005. *Predatory bureaucracy: The extermination of wolves and the transformation of the West*. Boulder, CO: University Press of Colorado Press.

Robinson, W. L., and E. G. Bolen. 1989. *Wildlife ecology and management*. Second ed. New York: Macmillan Publishing Co.

Roelke, M. E., Martenson J. S., and S. J. O'Brien. 1993. The consequences of demographic reduction and genetic depletion in the endangered Florida panther. *Curr Biol* 3:340–50.

Romero, S. 2007. Suspenden la temporada de caza de pumas en los cotos de La Pampa. *La Nación*, February 3.

Rominger, E. M., H. A. Whitlaw, D. L. Weybright, W. C. Dunn, and W. B. Ballard. 2004. The influence of mountain lion predation on bighorn sheep translocations. *J Wildl Mgmt* 68:993–9.

Romo, M. C. 1995. Food habits of the Andean fox (*Pseudalopex culpaeus*) and notes of the mountain cat (*Felis colocolo*) and puma (*Felis concolor*) in the Rio Abiseo National Park, Peru. *Mammalia* 59:335–43.

Roosevelt, T. 1901. With the cougar hounds. *Scribner's Magazine*, October–November.

Root, R. B. 1967. The niche exploitation pattern of the blue-gray gnatcatcher. *Ecol Monogr* 37:317–50.

Root, T. L., and S. H. Schneider. 2005. Conservation and climate change: The challenges ahead. *Conserv Biol* 20:706–8.

Rosas-Rosas, O. C., R. Valdez, L. C. Bender, and D. Daniel. 2003. Food habits of pumas in northwestern Sonora, México. *Wildl Soc Bull* 31:528–35.

Rosenzweig, M. L. 1966. Community structure in sympatric Carnivora. *J Mammal* 47:602–12.

———. 1978. Aspects of biological exploitation. *Q Rev Biol* 52:371–80.

Ross, P. I., and M. G. Jalkotzy. 1992. Characteristics of a hunted population of cougars in southwestern Alberta. *J Wildl Mgmt* 56:417–26.

———. 1996. Cougar predation on moose in southwestern Alberta. *Alces* 32:1–8.

Ross, P. I., M. G. Jalkotzy, and P. Daoust. 1995. Fatal trauma sustained by cougars (*Felis concolor*) while attacking prey in southern Alberta. *Can Field Nat* 109:261–3.

Ross, P. I., M. G. Jalkotzy, and M. Festa-Bianchet. 1997. Cougar predation on bighorn sheep in southwestern Alberta during winter. *Can J Zool* 74:771–5.

Ross, P. I., M. G. Jalkotzy, and J. R. Gunson. 1996. The quota system of cougar harvest management in Alberta. *Wildl Soc Bull* 24:490–4.

Rowan, A. N. 2005. Foreword in *Humane wildlife solutions*, ed. A. T. Rutberg. Washington, D.C.: Humane Society Press.

Russ, W. B. 1988. Status of the mountain lion in Texas. In *Proceedings of the third mountain lion workshop*, ed. R. H. Smith, 30–31. Phoenix: Arizona Game and Fish Department.

Russell, K. R. 1978. Mountain lion. In *Big game of North America*, ed. J. L. Schmidt and D. L. Gilbert, 207–225. Mechanicsburg, Pa.: Stackpole Books.

Rutberg, A. T. 2001. Why agencies should not advocate hunting or trapping. *Hum Dim Wildl* 6:33–7.

Ruth, T. K. 1991. Mountain lion use of an area of high recreational development in Big Bend National Park, Texas. Master's thesis, Texas A&M University, College Station.

———. 2001. Cougar-wolf interactions in Yellowstone National park: Competition, demographics, and spatial relationships. Annual technical report. Wildlife Conservation Society/Hornocker Wildlife Institute, Bozeman, Montana.

———. 2004a. Ghost of the Rockies: the Yellowstone cougar project. *Yellowstone Sci* 12:13–24.

———. 2004b. Patterns of resource use among cougars and wolves in northwestern Montana and southeastern British Columbia. Ph.D. dissertation, University of Idaho, Moscow.

Ruth, T. K., and P. C. Buotte. 2007. Cougar ecology and cougar-carnivore interactions in Yellowstone National Park. Final technical report. Hornocker Wildlife Institute/Wildlife Conservation Society, Bozeman, Montana.

Ruth, T. K., and S. Gniadek. 1996. Winter bear activity in Glacier National Park. *Intl Bear News* 5:15–16.

Ruth, T. K., K. A. Logan, L. L. Sweanor, M. G. Hornocker, and L. J. Temple. 1998. Evaluating cougar translocation in New Mexico. *J Wildl Mgmt* 62:1264–75.

Ruth, T. K., D. W. Smith, P. C. Buotte, D. R. Stahler, V. Asher, and M. G. Hornocker. Interspecific killing between cougars and wolves in and near Yellowstone National Park. Hornocker Wildlife Institute/Wildlife Conservation Society, Yellowstone National Park, and Turner Endangered Species Fund. Unpublished data.

Ruth, T. K., D. W. Smith, M. A. Haroldson, P. C. Buotte, C. C. Schwartz, H. B. Quigley, S. Cherry, K. M. Murphy, D. Tyers, and K. Frey. 2003. Large-carnivore response to recreational big game hunting along the Yellowstone National Park and Absaroka–Beartooth Wilderness boundary. *Wildl Soc Bull* 31:1150–61.

Ruther, E. J. 2005. Cougar policy preferences influenced by the valuation of nature: A survey of northern Arizona residents. Master's thesis, Northern Arizona University, Flagstaff.

Ruther, E. J., and D. M. Ostergren. 2005. Attitudes towards and perceptions of mountain lions: A survey of northern Arizona residents. In *The Colorado Plateau II: Biophysical, socioeconomic, and cultural research*, ed. C. van Riper III and D. J. Mattson, 37–45. Tucson: University of Arizona Press.

Sánchez, O., J. Ramírez-Pulido, U. Aguilera-Reyes, and O. Monroy-Vilchis. 2002. Felid record from the state of Mexico, Mexico. *Mammalia* 66:289–94.

Sandell, M. 1989. The mating tactics and spacing patterns of solitary carnivores. In *Carnivore behavior, ecology and evolution*, ed. G. L. Gittleman, 164–182. Ithaca, NY: Cornell University Press.

Sanders, M. R., and J. C. Halfpenny. 1991. Human-lion interactions in Boulder County, Colorado: Past, present, and future. In *Mountain lion-human interaction symposium and workshop*, ed. C. E. Braun. Denver: Colorado Division of Wildlife.

Sanderson, E. W., K. H. Redford, C. B. Chetkiewicz, R. A. Medellin, A. R. Rabinowitz, J. G. Robinson, and A. B. Taber. 2002. Planning to save a species: The jaguar as a model. *Conserv Biol* 16:58–72.

Sargent, G. A., and R. L. Ruff. 2001. Demographic response of black bears at Cold Lake, Alberta, to the removal of adult males. *Ursus* 12:59–68.

Sargent-Michaud, J., and K. J. Boyle. 2002. Public perceptions of wildlife management in Maine. *Hum Dim Wildl* 7:163–78.

Saskatchewan Department of the Environment. 2007. Saskatchewan cougars. http://www.environment.gov.sk.ca/Default.aspx?DN=0d6c5f37-8b18-4a6a-8f6e-1001067161eb (accessed October 14, 2007).

Saunders, N. J. 1998. *Icons of power: Feline symbolism in the Americas*. London: Routledge.

Sauvajot, R. M., P. Beier, and S. P. D. Riley. 2006. Mountain lions in fragmented, urban landscapes: Experiences from southern California (abstract). 2006 Annual Meeting of the Wildlife Society.

Savage, D. E., and D. E. Russell. 1983. *Mammalian paleofaumas of the world*. Reading, MA: Addison-Wesley.

Save the Tiger Fund. 2003. *Save the tiger fund: A model for success*. Washington, D.C.: National Fish and Wildlife Foundation.

Sawyer, H., and F. Lindzey. 2002. *A review of predation on bighorn sheep* (Ovis canadensis). Laramie: Wyoming Cooperative Fish and Wildlife Research Unit.

Scarce, R. 1998. What do wolves mean? Conflicting social constructions of Canis lupus in "Bordertown." *Hum Dim Wildl* 3:26–45.

Schaefer, J. A., A. M. Veitch, F. H. Harrington, W. K. Brown, J. B. Theberge, and S. N. Luttich. 1999. Demography of decline of the Red Wine Mountains caribou herd. *J Wildl Mgmt* 63:580–7.

Schaefer, R. J., S. G. Torres, and V. C. Bleich. 2000. Survivorship and cause-specific mortality in sympatric populations of mountain sheep and mule deer. *Calif Fish Game* 86:127–35.

Schaller, G. B. 1967. *The deer and the tiger*. Chicago: University of Chicago Press.

———. 1972. *The Serengeti lion: A study of predator prey relations*. Chicago: University of Chicago Press.

Schlechtweg, H. P. 1996. Media frames and environmental discourse: The case of "focus:logjam." In *The symbolic earth: Discourse and our creation of the environment*, ed. J. G. Cantrill and C. L. Oravec, 257–277. Lexington, KY: University Press of Kentucky.

Schlegel, M. 1976. Factors affecting calf elk survival in northcentral Idaho: A progress report. *Western Association of State Game and Fish Commissioners* 56:342–55.

Schmidt, K. P., and Gunson, J. R. 1985. Evaluation of wolf-ungulate predation near Nordegg, Alberta: Second year progress report, 1984–85. Alberta Energy and Natural Resources Fish and Wildlife Division, Alberta.

Scholz, Z. M., Director of programs and operations of the Cougar Fund to T. Cleveland, Director of Wyoming Fish and Game Department, August 14, 2006.

Schön, D. A. 1973. *Beyond the stable state: Public and private learning in a changing society*. London: Penguin Books.

Schubert, D. J. 2002. "Letter from the Fund for Animals (D. J. Schubert) to M. Golightly (chairman) and D. Shroufe (director)." Humane Society Legislative Fund. http://www.fund.org/library/libraryuploads/huntcom.asp (accessed September 5, 2002).

Schubert, D. J. of Animal Welfare Institute to L. Kruckenburg of Wyoming Game and Fish Department, August 15, 2006.

Schullery, P. 1992. *The bears of Yellowstone*. Worland, WY: High Plains Publishing Co.

Schwartz, C. C., M. A. Haroldson, G. C. White, R. B. Harris, S. Cherry, K. A. Keating, D. Moody, and C. Servheen. 2007. Temporal, spatial, and environmental influences on the demographics of grizzly bears in the Greater Yellowstone ecosystem. *Wildlife Monogr* 161:1–68.

Schwartz, C. C., S. D. Miller, and M. A. Haroldson. 2003. Grizzly bear. In *Wild mammals of North America: Biology, management, and conservation*, Second ed., ed. G. A. Feldhamer, B. C. Thompson, and J. A. Chapman, 556–586. Baltimore: The Johns Hopkins University Press.

Schwartz, J. E., II, and G. E. Mitchell. 1945. The Roosevelt elk on the Olympic Peninsula, Washington. *J Wildl Mgmt* 9:295–322.

Schwartz, S. H. 1994. Are there universal aspects in the structure and contents of human values? *J Soc Issues* 50:19–45.

Scognamillo, D. G., I. E. Maxit, M. E. Sunquist, and J. Polisar. 2003. Coexistence of jaguar (*Panthera onca*) and puma (*Puma concolor*) in a mosaic landscape in the Venezuelan llanos, *J Zool London* 259:269–79.

Seger, J. H. 1905. *Tradition of the Cheyenne Indians*. Colony, OK: Arapaho Bee Print.

Seidensticker, J. 1991. Pumas. In *Great cats*, ed. J. Seidensticker, S. Lumpkin, 130–7. Emaus, PA: Rodale Press.

Seidensticker, J. C., IV, M. G. Hornocker, W. V. Wiles, and J. P. Messick. 1973. Mountain lion social organization in the Idaho Primitive Area. *Wildl Monogr* 35:3–60.

Servheen, C., J. S. Walker, and P. Sandstrom. 2001. Identification and management of linkage zones for grizzly bears between the large blocks of public land in the northern Rocky Mountains. In *Proceedings of international conference on ecology and transportation*, 161–79. Raleigh, NC: Center for Transportation and the Environment, North Carolina State University.

Shabecoff, P. 2000. *Earth rising: American environmentalism in the twenty-first century*. Washington, D.C.: Island Press.

Sharma, G. 1988. Status of the mountain lion in Wyoming. In *Proceedings of the third mountain lion workshop*, ed. R. H. Smith, 38–9. Phoenix: Arizona Game and Fish Department.

Shaw, H. G. 1977. Impact of mountain lion on mule deer and cattle in northwestern Arizona. In *Proceedings of the 1975 Predator Symposium*, ed. R. L. Phillips and C. J. Jonkel, 17–32. Montana Forest and Conservation Experiment Station, University of Montana, Missoula.

———. 1980. Ecology of the mountain lion in Arizona. Final report, P-R Project W-78-R, Work Plan 2, Job 13. Phoenix: Arizona Game and Fish Department.

———. 1981. Comparison of mountain lion predation on cattle on two study areas in Arizona. In *Proceedings of the wildlife–livestock relationships symposium*, 306–318. Idaho Forest, Wildlife and Range Experiment Station, University of Idaho, Moscow.

———. 1983. *Mountain lion field guide*. Special Report No. 9. Phoenix: Arizona Game and Fish Department.

———. 1989. *Soul among lions: The cougar as a peaceful adversary*. First ed. Boulder, CO: Johnson Books.

———. 1994. *Soul among lions: The cougar as peaceful adversary*. Second ed. Tuscon, AZ: University of Arizona Press.

———. 2000. *Soul among lions: The cougar as peaceful adversary*. Third ed. Tuscon: University of Arizona Press.

———. 2006a. Puma research: Now that you have it, what do you do with it? *Wild Cat News*, September. Also available at http://www.cougarnet.org/WCN2-2/Puma%20Research.pdf (accessed January 24, 2008).

———. 2006b. Wood plenty, grass good, water none: Vegetation changes in Arizona's upper Verde River watershed from 1850 to 1997. General Technical Report, RMRS-GTR-177, Fort Collins, CO: U. S. Department of Agriculture, Forest Service, Rocky Mountain Research Station.

Shaw, H. G., and S. Negri. 2005. International cougar management guidelines: Processes for collaboration and implementation. *J Intl Wildl Law Policy* 8:367–73.

Shaw, H. G., N. G. Woolsey, J. R. Wegge, and R. L. Day, Jr. 1988. Factors affecting mountain lion densities and cattle depredation in Arizona. Final report. Phoenix: Arizona Game and Fish Department.

Shelford, V. E., ed. 1926. *Naturalist's guide to the Americas.* Baltimore: Williams and Wilkins.

Shils, E. 1997. The virtue of civility. In *The virtue of civility: Selected essays on liberalism, tradition, and civil society,* ed. E. Shils and S. Grosby, 320–355. Indianapolis: Liberty Fund.

Shindle, D., D. Land, M. Cunningham, and M. Lotz. 2001. *Florida panther genetic restoration. Annual Report 2000–2001.* Tallahassee, FL: Florida Fish and Wildlife Conservation Commission.

Shindler, B., and K. A. Cheek. 1999. Integrating citizens in adaptive management: A prepositional analysis. *Conserv Ecol* 3:9–23.

Shroufe, D. L. 1988. Welcoming address. In *Proceedings of the third mountain lion workshop,* ed. R. H. Smith, p. 1. Phoenix: Arizona Game and Fish Department.

Shuey, M. L. 2005. Land-cover characteristics of cougar/human interactions in and around an urban landscape. In *Proceedings of the eighth mountain lion workshop,* ed. R. A. Beausoleil and D. A. Martorello, 117–26. Olympia: Washington Department of Fish and Wildlife.

Shutkin, W. A. 2000. *The land that could be: Environmentalism and democracy in the twenty-first century.* Cambridge, MA: MIT Press.

Sidanius, J., F. Pratto, C. van Laar, and S. Levin. 2004. Social dominance theory: Its agenda and method. *Polit Psychol* 25:845–80.

Silveira, L. 2004. Ecologia comparada e conservação da Onça-pintada (*Panthera onca*) e onça-parda (*Puma concolor*), no Cerrado e Pantanal. Ph.D. diss., Universidade de Brasilia, Brasilia, D.F. Brasil.

Silveira, L., and A. T. A. Jácomo. 2002. Conservación del jaguar en el centro del Cerrado de Brasil. In *El Jaguar en el Nuevo Milenio,* ed. R. A. Medellín, C. Equihua, C. L. B. Chetkiewicz, P. G. Crawshaw, Jr., A. Rabinowitz, H. H. Redford, J. G. Robinson, E. W. Sanderson, and A. B. Taber, 437–450. Universidad Nacional Autónoma de México/Wildlife Conservation Society, Mexico.

Sinclair, A. R. E., S. Mduma, and J. S. Brashares. 2003. Patterns of predation in a diverse predator-prey system. *Nature* 425:288–90.

Sinclair, A. R. E., R. P. Pech, C. R. Dickman, D. Hik, P. Mahon, and A. E. Newsome. 1998. Predicting effects of predation on conservation of endangered prey. *Conserv Biol* 12:564–75.

Sinclair, E. A., E. L. Swenson, M. L. Wolfe, D. C. Choate, B. Gates, and K. A. Crandall. 2001. Gene flow estimates in Utah's cougars imply management beyond Utah. *Anim Conserv* 4:257–64.

Singer, F. J., A. Harting, K. K. Symonds, and M. B. Coughenour. 1997. Density dependence, compensation, and environmental effects on elk calf mortality in Yellowstone National Park. *J Wildl Mgmt* 61:12–25.

Singer, F. J., and J. E. Norland. 1994. Niche relationships within a guild of ungulate species in Yellowstone National Park, Wyoming, following release from artificial controls. *Can J Zool* 72:1383–94.

Singleton, P H., J. F. Lehmkuhl, and W. Gaines. 2001. Large-scale carnivore habitat connectivity in Washington. In *Proceedings of international conference on ecology and transportation,* 583–594. Raleigh, NC: Center for Transportation and the Environment, North Carolina State University.

Sittler, J. 2000. *Evocations of grace: Writings on ecology, theology, and ethics.* Grand Rapids, MI: William B. Eerdmans Publishing Company.

Sitton, L. W., and S. Wallen. 1976. *California mountain lion study.* Sacramanto: California Department of Fish and Game.

Skogan, K., and O. Krange. 2003. A wolf at the gate: The anti-carnivore alliance and the symbolic construction of community. *Sociologia Ruralis* 43:309–25.

Smallwood, K. S., and E. L. Fitzhugh. 1995. A track count for estimating mountain lion (*Felis concolor californica*) population trend. *Biol Conserv* 71:251–9.

Smith, B. L., and S. H. Anderson. 1996. Patterns of neonatal mortality of elk in northwestern Wyoming. *Can J Zool* 74:1229–37.

Smith, B. L., and T. L. McDonald. 2002. Criteria for improving field classification of antlerless elk. *Wildl Soc Bull* 30:200–7.

Smith, B. L., E. S. Williams, K. C. McFarland, T. L. McDonald, G. Wang, and T. D. Moore. 2006. Neonatal mortality of elk in Wyoming: Environmental, population, and predator effects. Biological technical publication, BTP-R6007–2006, Washington, D.C.: U.S. Department of Interior, U.S. Fish and Wildlife Service.

Smith, D. W. 2005. Ten years of Yellowstone wolves, 1995–2005. *Yellowstone Sci* 13:7–33.

Smith, D. W. and K. E. Bangs. 2009. Reintroduction of wolves to Yellowstone National Park: History, values, and ecosystem restoration. In *Reintroduction of top-order predators,* ed. M. Hayward. Hoboken, NJ: Wiley-Blackwell Publishing.

Smith, D. W., and G. Ferguson. 2006. *Decade of the wolf: Returning the wild to Yellowstone.* Guilsford, CT: The Lyons Press.

Smith, D. W., R. O. Peterson, and D. B. Houston. 2003. Yellowstone after wolves. *Bioscience* 53:330–40.

Smith, M. S. 2001. "Silence, Miss Carson!": Science, gender, and the reception of *Silent Spring. Feminist Stud* 27:733–52.

Smith, P. D., and M. H. McDonough. 2001. Beyond public participation: Fairness in natural resource decision making. *Soc Nat Res* 14:239–41.

Smith, R. H., ed. 1988. *Proceedings of the third mountain lion workshop.* Phoenix: Arizona Game and Fish Department and The Arizona Chapter of the Wildlife Society.

Smith, T. E., R. R. Duke, M. J. Kutileck, and H. T. Harvey. 1986. Mountain lions (*Felis concolor*) in the vicinity of Carlsbad Caverns National Park, New Mexico, and Guadalupe Mountains National Park, Texas. Final Report, Santa Fe: U. S. Department of Interior, National Park Service.

Smith, T. S., S. Herrero, and T. D. Debruyn. 2005. Alaskan brown bears, habituation and humans. *Ursus* 16:1–10.

Society for Conservation Biology. 2007. http://www.conbio.org/Resources/Programs/About/faq.cfm#a10 (accessed February 22, 2008)

Soluk, D. A. 1993. Multiple predator effects: Predicting combined functional response of stream fish and invertebrate predators. *Ecology* 74:219–25.

Soulé, M. E., D. T. Bolger, A. C. Alberts, J. Wright, M. Sorice, and S. Hill. 1988. Reconstructed dynamics of raid extinctions of chaparral-requiring birds in urban habitat islands. *Conserv Biol* 2:75–92.

Soulé, M. E., J. A. Estes, J. Berger, and C. Martinez Del Rio. 2003. Ecological effectiveness: Conservation goals for interactive species. *Conserv Biol* 17:1238–50.

South Coast Wildlands. 2006. http://www.scwildlands.org/ (accessed April 13, 2007).

South Dakota Department of Game, Fish and Parks. 2005. *South Dakota mountain lion management plan: 2003–2012. Adaptive Management System, Game Program, South Dakota Department of Game, Fish and Parks, Pierre.

———. 2007. South Dakota mountain lion hunting season. http://www.sdgfp.info/Wildlife/MountainLions/MtLionhuntingseason.htm (accessed June 6, 2007).

Spalding, D. J., and J. Lesowski. 1971. Winter food of the cougar in south-central British Columbia. *J Wild Mgmt* 35:378–81.

Spence, C. E., J. E. Kenyon, D. R. Smith, R. D. Hayes, and A. M. Baer. 1999. Surgical sterilization of free-ranging wolves. *Can Vet J* 40:118–21.

Spencer, R. D., D. J. Pierce, G. A. Schirato, K. R. Dixon, and C. B. Richards. 2001. Mountain lion home range, dispersal, mortality, and survival in the western Cascade Mountains of Washington. Final report. Olympia: Washington Department of Fish and Wildlife.

Spreadbury, B. 1989. Cougar ecology and related management implications and strategies in southeastern British Columbia. Master's thesis, University of Calgary, Calgary.

Spreadbury, B., K. Musil, J. Musil, C. Kaisner, and J. Kovak. 1996. Cougar population characteristics in southeastern British Columbia. *J Wildl Mgmt* 64:962–9.

Stahler, D. R. 2000. Interspecific interactions between the common raven (*Corvus corax*) and the gray wolf (*Canis lupus*) in Yellowstone National Park, Wyoming: Investigations of a predator and scavenger relationship. Master's thesis, University of Vermont, Burlington.

Stehli, F. G., and S. D. Webb. 1985. *The great American biotic interchange*. New York: Plenum Press.

Stelfox, B. 1993. *Hoofed mammals of Alberta*. Edmonton, Alberta: Lone Pine Publishing.

Steneck, R. S. 2005. An ecological context for the role of large carnivores in conserving biodiversity. In *Large carnivores and the conservation of biodiversity*, ed. J. C. Ray, K. H. Redford, R. S. Steneck, and J. Berger, 9–33. Washington, D.C.: Island Press.

Stegner, Wallace. 1984. "Memo to the Mountain Lion." Broadside illusration. Wolfgang Lederer. Nevada City, CA: Harold Berliner.

Stenseth, N. C., and W. C. Lidicker, Jr., eds. 1992. *Animals dispersal: Small mammals as a model*. London: Chapman and Hall.

Stephens, D. W., and J. R. Krebs. 1986. *Foraging theory*. Princeton, NJ: Princeton University Press.

Stoddart, L. C., R. E. Griffiths, and F. F. Knowlton. 2001. Coyote responses to changing jackrabbit abundance affect sheep predation. *J Range Mgmt* 54:15–20.

Stohlgren, T., J. J. Baron, T. G. F. Kittel, and D. Brinkley. 1995. Ecosystem trends in the Colorado Rockies. In *Our living resources: A report to the nation on the distribution, abundance, and health of U.S. plants, animals, and ecosystems*, ed. E. T. LaRoe, G. S. Farris, C. E. Puckett, P. D. Doran, and M. J. Mac, 310–2. Washington, D.C.: U.S. Department of the Interior, National Biological Service.

Stolzenburg, W. 2008. *Where the wild things were: Life, death, and ecological wreckage in a land of vanishing predators*. New York: Bloomsbury.

Stoner, D. C, W. R. Rieth, M. L. Wolfe, M. B. Mecham, and A. Neville. 2007. Long distance dispersal of a female cougar in a basin and range landscape. *J Wildl Mgmt* 72:933–9.

Stoner, D. C, M. L. Wolfe, and D. M. Choate. 2006. Cougar exploitation levels in Utah: Implications for demographic structure, population recovery, and metapopulation dynamics. *J Wildl Mgmt* 70:1588–1600.

Suminski, H. R. 1982. Mountain lion predation on domestic livestock in Nevada. In *Proceedings of the tenth vertebrate pest conference*, ed. R. E. March, 62–6. Universtiy of California, Davis.

Sunquist, M. E., and F. C. Sunquist. 1989. Ecological constraints on predation by large felids. In *Carnivore behavior, ecology, and evolution*, ed. J. L. Gittleman, 283–301. Ithaca, NY: Cornell University Press.

———. 2002. *Wild cats of the world*. Chicago: University of Chicago Press.

Swan, D., and J. C. McCarthy. 2003. Contesting animal rights on the Internet: Discourse analysis of the social construction of argument. *J Lang Soc Psychol* 22:297–320.

Sweanor, L. L. 1990. Mountain lion social organization in a desert environment. Master's thesis, University of Idaho, Moscow.

Sweanor, L. L., K. A. Logan, J. Bauer, and W. Boyce. 2004. Southern California puma project: Final report for interagency agreement No. C0043050 (Southern California Ecosystem Health Project) between California State Parks and the UC Davis Wildlife Health Center. University of California, Davis.

Sweanor, L. L., K. A. Logan, J. W. Bauer, B. Millsap, and W. M. Boyce. 2008. Puma and human spatial and temporal use of a popular California state park. *J Wildl Mgmt* 72:1076–84.

Sweanor, L. L., K. A. Logan, and M. C. Hornocker. 2000. Cougar dispersal patterns, metapopulation dynamics, and conservation. *Conserv Biol* 14:798–808.

———. 2005. Puma responses to close approaches by researchers. *Wildl Soc Bull* 33:905–13.

Sweitzer, R. A., S. H. Jenkins, and J. Berger. 1997. Near-extinction of porcupines by mountain lions and consequences for ecosystem change in the Great Basin Desert. *Conserv Biol* 11:1407–17.

Swenson, J. E. 2003. Implications of sexually selected infanticide for the hunting of large carnivores. In *Animal behavior and wildlife conservation*, ed. M. Festa-Bianchet and M. Apollonio, 171–189. Washington, D.C.: Island Press.

Taber, A. B., A. J. Novaro, N. Neris, and F. H. Colman. 1997. The food habits of sympatric jaguar and puma in the Paraguayan Chaco. *Biotropica* 29:204–13.

Taylor, A. 1995. Wasty ways. *Environ Hist* 3:291–310.

Taylor, B. 2004. A green future for religion? *Futures* 36:991–1008.

Taylor, M. E. 1989. Locomotor adaptations in carnivores. In *Carnivore behavior, ecology, and evolution*, ed. J. L. Gittleman, 382–409. Ithaca, NY: Cornell University Press.

Taylor, P. 2005. *Beyond conservation: A wildland strategy*. Sterling, VA: Earthscan Publications.

Taylor, R. J. 1984. *Predation*. New York: Chapman and Hall.

Teel, T. L., A. A. Dayer, M. J. Manfredo, and A. D. Bright. 2005. Regional results from the research project entitled "Wildlife values in the West." Project report for the Western Association of Fish and Wildlife Agencies. Human Dimensions in Natural Resources Unit, Colorado State University, Fort Collins.

Teel, T. L., R. S. Krannich, and R. H. Schmidt. 2002. Utah stakeholders' attitudes toward selected cougar and black bear management policies. *Wildl Soc Bull* 30:2–15.

Temple, S. A. 1987. Do predators always capture substandard individuals disproportionately from prey populations? *Ecology* 68:669–74.

Tenhumberg, B., A. J. Tyre, A. R. Pople, and J. P. Possingham. 2004. Do harvest refuges buffer kangaroos against evolutionary responses to selective harvesting? *Ecology* 85:2003–17.

Terborgh, J. 1988. The big things that run the world—A sequel to E. O. Wilson. *Conserv Biol* 2:402–3.

Terborgh, J., J. A. Estes, P. Paquet, K. Ralls, D. Boyd-Hager, B. J. Miller, and R. F. Noss. 1999. The role of top carnivores in regulating terrestrial ecosystems. In *Continental conservations*, ed. M. E. Soulé and J. Terborgh, 39–64. Washington, D.C.: Island Press.

Terborgh, J., K. Feeley, M. Silman, P. Nunez, and B. Balukjian. 2006. Vegetation dynamics of predator-free land-bridge islands. *J Ecol* 94:253–63.

Terborgh, J., L. Lopez, P. Nuñez, V., M. Rao, G. Shahabuddin, G. Orihuela, M. Riveros, R. Ascanio, G. H. Adler, T. D. Lambert, and L. Balbas. 2001. Ecological meltdown in predator-free forest fragments. *Science* 294:1923–6.

Thatcher, C. A., F. T. van Manen, and J. D. Clark. 2006. Identifying suitable sites for Florida panther reintroduction. *J Wildl Mgmt* 70:752–63.

Thinkexists.com. 2008. http://thinkexists.com/ (accessed December 10, 2008).

Theobald, D. M., H. Gosnell, and W. E. Riebsame. 1996. Land use and landscape change in the Colorado Mountains II: A case study of the East River Valley. *Mt Res Dev* 16:407–18.

Thomas, J. W., and D. E. Toweill, eds., 1982. *Elk of North America: Ecology and management*. Harrisburg, PA: Stackpole Books.

Thompson, B. C. 1984. Texas. In *Proceedings of the second mountain lion workshop*, ed. J. Roberson and F. Lindzey, 56–9. Salt Lake City: Utah Division of Wildlife Resources,.

Thompson, D. J., and J. A. Jenks. 2005. Long distance dispersal by a subadult male cougar from the Black Hills, South Dakota. *J Wildl Mgmt* 69:818–20.

Thompson, M. J., and W. C. Stewart. 1994. Cougar(s) (*Felis concolor*) with a kill for 27 days. *Can Field Nat* 108:497–8.

Thorne, J. H., D. Cameron, and J. F. Quinn. 2006. A conservation design for the central coast of California and the evaluation of mountain lion as an umbrella species. *Nat Area J* 26:137–48.

Tinsley, J. B. 1987. *The puma: Legendary cat of the Americas*. El Paso: Texas Western Books.

Tokar, B. 1995. The "wise use" backlash: Responding to militant anti-environmentalism. *Ecologist* 25:150–6.

Torres, S. G. 1997. *Mountain lion alert: Safety for pets, landowners, and outdoor adventurers*. Helena, MT: Falcon Publishing Company.

———. 2005. *Lion sense: Traveling and living safely in mountain lion country*. Helena, MT: The Globe Pequot Press.

Torres, S. G., T. M. Mansfield, J. E. Foley, T. Lupo, and A. Brinkhaus. 1996. Mountain lion and human activity in California: Testing speculations. *Wildl Soc Bull* 24:451–60.

Toweill, D., and E. C. Meslow. 1977. Food habits of cougars in Oregon. *J Wild Mgmt* 41:576–8.

Travaini, A., S. C. Zapata, R. Martinez-Peck, and M. Delibes. 2000. Percepción y actitud humanas hacia la predación ganado ovino por el zorro colorado (*Pseudalopex culpaeus*) en Santa Cruz, Patagonia Argentina. *Mastozoología Neotropical* 7:117–20.

Trefethen, J. B. 1975. *An American crusade for wildlife*. Alexandria, VA: Boone and Crockett Club.

Trivers, R. L. 1972. Parental investment and sexual selection. In *Sexual selection and the descent of man*, ed. B. Campbell, 136–79. Chicago: Aldine Press.

Trolle, M. 2003. Mammal survey in the southeastern Pantanal, Brazil. *Biodivers Conserv* 12:823–36.

Trolle, M., A. J. Noss, E. de S. Lima, and J. C. Dalponte. 2007. Camera-trap studies of maned wolf density in the Cerrado and the Pantanal of Brazil. *Biodivers Conserv* 16:1197–204.

Trut, L. N. 1999. Early canid domestication: The farm fox experiment. *Am Sci* 87:160–9.

Tsukamoto, G. 1984. Nevada. In *Proceedings of the second mountain lion workshop*, ed. J. Roberson and F. Lindzey, 44–8. Salt Lake City: Utah Division of Wildlife Resources.

Tsukamoto, G. K. 2002. "Changes in attitudes about predator management in Washington." Washington Department of Fish and Wildlife. http://www.wa.gov/wdfw/wlm/hunter/gametrails/predator_attitudes.htm (accessed April 22, 2003).

Turner, A. 1997. *The big cats and their fossil relatives*. New York: Columbia University Press.

Turner, J. W., M. L. Wolfe, and J. F. Kirkpatrick. 1992. Seasonal mountain lion predation on a feral horse population. *Can J Zool* 70:929–34.

U.S. Census Bureau. 2005. "Interim projections of the total population for the United States and states: April 1, 2000 to July 1, 2030." U. S. Census Bureau, Population Division. http://www.census.gov/population/www/projections/projectionsagesex.html (accessed August 1, 2007).

U.S. Department of Agriculture, 2007. "Wildlife damage." Animal and Plant Health Inspection Service. http://www.aphis.usda.gov/wildlife_damage/prog_data_report.shtml (accessed February 10, 2008).

U.S. Fish and Wildlife Service. 1995. Second revision Florida panther recovery plan. U.S. Fish and Wildlife Service, Atlanta, GA.

———. 1999. South Florida multi-species recovery plan. U.S. Fish and Wildlife Service, Atlanta, GA.

U.S. Fish and Wildlife Service and National Oceanographic and Atmospheric Administration. 1996. Policy regarding the recognition of distinct vertebrate population segments under the Endangered Species Act. Federal Register 61:4722.

U.S. Fish and Wildlife Service and U.S. Bureau of the Census. 1993. *1991 national survey of fishing, hunting and wildlife-related recreation*. U.S. Fish and Wildlife Service FHW/91.

———. 1997. *1996 national survey of fishing, hunting and wildlife-related recreation*. U.S. Fish and Wildlife Service FHW/96.

———. 2003. *2001 national survey of fishing, hunting and wildlife-related recreation*. U.S. Fish and Wildlife Service FHW/01.

———. 2008. *2006 national survey of fishing, hunting and wildlife-related recreation*. U.S. Fish and Wildlife Service FHW/06.

University of California, Berkeley. 2005. Wolves alleviate impact of climate change on food supply, finds new study. *Science Daily* March 22. Also available at http://www.sciencedaily.com/releases/2005/03/050322134346.htm.

Utah Division of Wildlife Resources. 1999. *Utah cougar management plan*. Publication 99–1, Salt Lake City.

———. 2005a. *Utah big game annual report 2005*. Annual Performance Report for Federal Aid Project W-65-M-53. Publication 06–21, Salt Lake City.

———. 2005b. "Mission statement of the division of wildlife resources." http://www.wildlife.utah.gov/about/mission.php (accessed January 21, 2008)

van Andel, J., and J. Aronson. 2006. *Restoration ecology*. Malden, MA: Blackwell.

Van Ballenberghe, V., and W. B. Ballard. 1994. Limitation and regulation of moose populations: The role of predation. *Can J Zool* 72:2071–7.

Van Dyke, F. G., R. H. Brocke, H. G. Shaw, B. B. Ackerman, T. P. Hemker, and F. G. Lindzey. 1986. Reactions of mountain lions to logging and human activity. *J Wildl Mgmt* 50:95–102.

van Hoogstraten, H.D. 2001. *Deep economy: Caring for ecology, humanity, and religion*. Cambridge, UK: James Clarke and Company.

Van Sickle, W. D., and F. G. Lindzey. 1991. Evaluation of a cougar population estimator based on probability sampling. *J Wildl Mgmt* 55:738–43.

Van Valkenburgh, B. 1989. Carnivore dental adaptations and diet: A study of trophic diversity within guilds. In *Carnivore behavior, ecology, and evolution*, ed. J. L. Gittleman, 410–436. Ithaca, NY: Cornell University Press.

VanValkenburgh, B., F. Grady, and B. Kurtén. 1990. The Plio-Pleistocene cheetah-like cat *Miracinonyx inexpectatus* of North America. *J Vert Paleo* 10:434–54.

Van Valkenburgh, B., and C. B. Ruff. 1987. Canine tooth strength and killing behavior in large carnivores. *J Zool London* 212:1–19.

Vaughan, C., and S. Temple. 2002. Conservación del jaguar en Centroamérica. *El Jaguar en el Nuevo Milenio*, ed. R. A. Medellín, C. Equihua, C. L. B. Chetkiewicz, P. G. Crawshaw Jr., A. Rabinowitz, H. H. Redford, J. G. Robinson, E. W. Sanderson, and A. B. Taber, 355–366. Universidad Nacional Autónoma de México/Wildlife Conservation Society, Mexico.

Vining, J., E. Tyler, and B-S Kweon. 2000. Public values, opinions, and emotions in restoration controversies. *Restoring nature: Perspectives from the social sciences and humanities*, ed. P. H. Gobster and R. B. Hull, 143–161. Washington, D.C.: Island Press.

Vittersø, J., B. P. Kaltenborn, and T. Bjerke. 1998. Attachment to livestock and attitudes toward large carnivores among sheep farmers in Norway. *Anthrozoös* 11:210–7.

Voight, Jr., W. 1976. *Public grazing lands: Use and misuse by industry and government*. New Brunswick, NJ: Rutgers University Press.

Vucetich, J. A., D. Smith, and D. Stahler. 2005. Influence of harvest, climate, and wolf predation on Yellowstone elk, 1961–2004. *Oikos* 111:259–70.

Waid, D. 1990. Movements, food habits, and helminth parasites of mountain lions in southwestern Texas. Ph.D. dissertation, Texas Technical University, Lubbock.

Wakeling, B. F. 2003. Status of mountain lion populations in Arizona. In *Proceedings of the seventh mountain lion workshop*, ed. S. A. Becker, D. D. Bjornlie, F. G., Lindzey, and D. S. Moody, 1–5. Lander: Wyoming Game and Fish Department.

Walker, C. W., L. A. Harveson, M. T. Pittman, M. E. Tewes, and R. L. Honeycutt. 2000. Microsatellite variation in two populations of mountain lions (*Puma concolor*) in Texas. *Southwest Nat* 45:196–203.

Walker, E. P., and J. L. Paradiso. 1975. *Mammals of the world*, Baltimore: The Johns Hopkins University Press.

Walker, E. P., R. M. Nowak, and J. L. Paradiso. 1983. *Walker's mammals of the world*. Baltimore: The John Hopkins University Press.

Wallace, A. R. 1878. *Tropical nature and other essays*. New York: MacMillan.

Wallace, S. 2007. Amazonia, nos estamos quedando sin selvas. *National Geographic en Español* 20:2–31.

Walls, G. L. 1942. *The vertebrate eye*. New York: Harper.

Waser, P. M., C. Strobeck, and D. Paetkau. 2001. Estimating interpopulation dispersal rates. In *Carnivore conservation*, ed. J. L. Gittleman, S. M. Funk, D. Macdonald, and R. K. Wayne, 484–497. New York: Cambridge University Press.

Watkins, B. 2005. Cougars confirmed in Manitoba. *Wildcat News* 1:6–8.

Wayne, R. K., R. E. Benveniste, D. N. Janczewski, and S. J. O'Brien. 1989. Molecular and biochemical evolution in the carnivora. In *Carnivore behavior, ecology and evolution*, ed. J. L. Gittleman, 465–94. New York: Cornell University Press.

Webb, S. D. 1976. Mammalian faunal dynamics of the Great American Interchange. *Paleobiology* 2:220–34.

———. 1978. A history of savanna vertebrates in the New World, Part 2, South America and the great interchange. *Annu Rev Ecol Syst* 9:393–426.

Webb, S. D., and L. G. Marshall. 1981. Historical biogeography of recent South American land mammals. In *Mammalian biology in South America*, ed. M. A. Mares, H. H. Genoways, 39–52. Special publication series of the Pymatuning Laboratory of Ecology. Pittsburgh, PA: University of Pittsburgh.

Webb, S. D., and A. Rancy. 1996. Late Cenozoic evolution of the neotropical mammal fauna. In *Evolution and environment in tropical America*, ed. J. B. Jackson, A. F. Budd, and A. G. Coates, 335–358. Chicago: University of Chicago Press.

Weeks, P., and J. M. Packard. 1997. Acceptance of scientific management by natural resources dependent communities. *Conserv Biol* 11:236–45.

Wehausen, J. D. 1996. Effects of mountain lion predation on bighorn sheep in the Sierra Nevada and Granite Mountains of California. *Wildl Soc Bull* 24:471–9.

Wells, S. 2003. *The journey of man: A genetic odyssey*. New York: Random House.

Wemmer, C., and K. Scow. 1977. Communications in the *Felidae* with emphasis on scent marking and contact patterns. In *How animals communicate*, ed. T. A. Sebeok, 749–766. Bloomington: Indiana University Press.

Werdlin, L. 1989. The radiation of felids in South America: When and where did it occur? In *Abstract of papers and posters*, 290–1. Rome: Fifth International Theriological Congress.

Wessing, R. 1986. The *soul of ambiguity: The tiger in Southeast Asia*. Monographs series on Southeast Asia, special report 24, Center for Southeast Asian Studies. Northern Illinois University, DeKalb.

Western Governors' Association. 2008. Wildlife Corridor Initiative Report, Jackson, Wyoming. June.

White, P. A., and D. K. Boyd. 1989. A cougar (*Felis concolor*) kitten killed and eaten by gray wolves (*Canis lupus*) in Glacier National Park, Montana. *Can Field Nat* 103:408–9.

White, P. J., and R. A. Garrott. 2005. Yellowstone's ungulates after wolves: Expectations, realizations, and predictions. *Biol Conserv* 125:141–52.

Whittaker, D. G. 2005. Oregon mountain lion status report. In *Proceedings of the eighth mountain lion workshop*, ed. R. A. Beausoleil and D. A. Martorello, 11–16. Olympia: Washington Department of Fish and Wildlife.

Wiens, J. A. 1989. *The ecology of bird communities. Volume 2: Processes and variations.* Cambridge, UK: Cambridge University Press.

Wildlands Project. 2007. http://www.twp.org/cms/index .cfm?group_id=1000 (accessed February 10, 2008)

The Wildlife Society. 2008. Official policy statement on sterilization and fertility control in wildlife populations.

Wilkins, L., J. M. Arias-Reveron, B. Stith, M. E. Roelke, and R. C. Belden. 1997. The Florida panther (*Puma concolor coryi*): A morphological investigation of the subspecies with a comparison to other North and South American cougars. *Bull Florida Mus Nat Hist* 40:221–69.

Williams, B. K., J. D. Nichols, and M. J. Conroy. 2002. *Analysis and management of animal populations. Modeling, estimation, and decision making.* San Diego: Academic Press.

Williams, J. 2005. A cat race tale . . . of houndsmen, biologists, administrator, committees and lawmakers in northwest Montana—A history of Montana HB 142 (abstract). In *Proceedings of the eighth mountain lion workshop*, ed. R. A. Beausoleil and D. A. Martorello, 91. Olympia: Washington Department of Fish and Wildlife.

Williams, J. S., J. J. McCarthy, and H. D. Picton. 1995. Cougar habitat use and food habits on the Montana Rocky Mountain Front. *Intermountain J Sci* 1:16–28.

Wilmers, C. C., R. L. Crabtree, D. W. Smith, K. M. Murphy, and W. M. Getz. 2003. Trophic facilitation by introduced top predators: Grey wolf subsidies to scavengers in Yellowstone National Park. *J Anim Ecol* 72:909–16.

Wilson, D., and S. Ruff. 1999. *The Smithsonian book of North American mammals.* Washington, D.C.: Smithsonian Institution Press.

Wilson, E. O. 2002. *The future of life.* New York: Alfred A. Knopf.

Wilson, E. O., and W. L. Brown, Jr. 1953. The subspecies concept and its taxonomic application. *Sys Zool* 2:97–111.

Wilson, M. 2002. *Broken promises: The U.S. Government and Native Americans in the nineteenth century.* Bromall, PA: Mason Crest Publishers.

Wilson, M. A. 1997. The wolf in Yellowstone: Science, symbol, or politics? Deconstructing the conflict between environmentalism and wise use. *Soc Nat Res* 10:453–65.

Wilson, P. 1984. Puma predation on guanacos in Torres del Paine National Park, Chile. *Mammalia* 48:515–22.

Wilson, R. 1984. Wyoming. In *Proceedings of the second mountain lion workshop*, ed. J. Roberson and F. Lindzey, 88–92. Salt Lake City: Utah Division of Wildlife Resources.

Wilson, S., and S. G. Clark. 2007. Resolving human-grizzly bear conflicts: An integrated approach in the common interest. In *Fostering integration: Concepts and practice in resource and environmental management*, ed. S. Hanna and D. S. Slocombe. London: Cambridge University Press.

Wilson, S. F., A. Hahn, A. Gladders, K. M. L. Goh, and D. M. Shackleton. 2004. Morphology and population characteristics of Vancouver Island cougars, *Puma concolor vancouverensis*. *Can Field Nat* 118:159–63.

Wilson, S. M., M. J. Madel, D. J. Mattson, J. M. Graham, T. Merrill. 2006. Landscape conditions predisposing grizzly bears to conflicts on private agricultural lands in western USA. *Biol Conserv* 130:47–59.

Winslow, R. 2005. New Mexico mountain lion status report. In *Proceedings of the eighth mountain lion workshop*, ed. R. A. Beausoleil and D. A. Martorello, 70–2. Olympia: Washington Department of Fish and Wildlife.

Witter, D. J., and W. W. Shaw. 1979. Beliefs of birders, hunters, and wildlife professionals about wildlife management. *Trans N Am Wildl Nat Res* 44:298–305.

Wittmer, H. U., B. N. McLellan, D. R. Seip, J. A. Young, T. A. Kinley, G. S. Watts, and D. Hamilton. 2005. Population dynamics of the endangered mountain ecotype of woodland caribou (*Rangifer tarandus caribou*) in British Columbia, Canada. *Can J Zool* 83:407–18.

Wittmer, H. U., A. R. E. Sinclair, B. N. McLellan. 2005. The role of predation in the decline and extirpation of woodland caribou. *Oecologia* 144:257–67.

Wolch, J. R., A. Gullo, and U. Lassiter. 1997. Changing attitudes toward California's cougars. *Soc Anim* 5:95–116.

Wolfe, M. L., and J. A. Chapman. 1987. Principles of furbearer management. In *Wild furbearer management and conservation in North America*, ed. M. Novak. Ontario: Ontario Trappers Association.

Wondolleck, J., and S. Yaffee. 2000. *Making collaboration work: Lessons form innovation in natural resources management.* Washington, D.C.: Island Press.

Woolstenhulme, R. 2003. Nevada mountain lion status report. In *Proceedings of the seventh mountain lion workshop*, ed. S. A. Becker, D. D. Bjornlie, F. Lindzey, and D. S. Moody, 31–8. Lander: Wyoming Game and Fish Department.

———. 2005. Nevada mountain lion status report. In *Proceedings of the eighth mountain lion workshop*, ed. R. A. Beausoleil and D. A. Martorello, 49–56. Olympia: Washington Department of Fish and Wildlife.

Wozencraft, W. C. 1993. Order carnivora. In *Mammal species of the world: A taxonomic and geographic reference*, Second ed., ed. D. E. Wilson and D. M. Reeder, Washington D.C. Washington D.C.: Smithsonian Institution Press.

Wrangham, R., and D. Peterson. 1996. *Demonic males: Apes and the origins of human violence.* New York: Houghton Mifflin C0o.

Wrangham, R. W., and D. I. Rubenstein. 1986. Social evolution in birds and mammals. In *Ecological aspects of social evolution*, ed. D. I. Rubenstein and R. W. Wrangham, 452–470. Princeton: Princeton University Press.

Wright, G. M. 1934. Cougar surprised at a well-stocked larder. *J Mammal* 15:321.

Wyoming Game and Fish Department. 2006a. Department analysis of written comments received on the draft mountain lion management plan. Cheyenne: Wyoming Game and Fish Department.

———. 2006b. Mountain lion management plan. http://gf.state. wy.us/downloads/pdf/MLPlanFinal9-7-06.pdf (accessed October 31, 2007).

———. 2007a. Emergency Rule, Section 2(c). Wyoming Game and Fish Commission. September 2007.

———. 2007b. Emergency Rule, Section 4(i). Wyoming Game and Fish Commission. September 2007.

Wywialowski, A. P. 1991. Implications of the animal rights movement for wildlife damage management. In *Proceedings of the tenth Great Plains wildlife damage conference*, ed. S. E. Hygnstrom, R. M. Chase, and R. J. Johnson, 28–32. Lincoln, NE: University of Nebraska.

Yañez, J. L., J. C. Cárdenas, P. Gezelle, and F. M. Jaksic. 1986. Food habits of the southernmost mountain lions (*Felis concolor*) in South America: Natural versus livestocked range. *J Mammal* 67:604–6.

Yankelovich, D. 1991. *Coming to public judgment: Making democracy work in a complex world*. Syracuse, NY: Syracuse University Press.

———. 1998. *The magic of dialogue: Transforming conflict into cooperation*. New York: Simon and Schuster.

Yáñez, J. L., J. C. Cárdenas, P. Gezelle, and F. M. Jaksić. 1986. Food habits of the southernmost mountain lions (*Felis concolor*) in South America: Natural versus livestocked ranges. *J Mammal* 67:604–6.

Yellowstone to Yukon Conservation Initiative. 2006. http://www.y2y.net/ (accessed September 5, 2007).

Young, I. M. 2000. *Inclusion and democracy*. New York: Oxford University Press.

Young, J. 2003. Mountain lion status report for Texas. *Proceedings of the seventh mountain lion workshop*, ed. S. A. Becker, D. D. Bjornlie, F. G. Lindzey, and D. S. Moody, 49–50. Lander: Wyoming Game and Fish Department.

Young, S. P. 1946a. *The wolf in North American history with fifty-three illustrations*. Caldwell, ID: Caxton Printers.

———. 1946b. History, life habits, economic status, and control, Part 1. In *The puma: Mysterious American cat*, ed. in S. P. Young and E. A. Goldman, 1–173. Washington, D.C.: The American Wildlife Institute.

Young, S. P., and E. A. Goldman. 1944. *The wolves of North America*. Washington, D.C.: The American Wildlife Institute.

———. 1946a. *The puma: Mysterious American cat*. Washington, D.C.: The American Wildlife Institute.

———. 1946b. *The Puma: 99 illustrations*. London: Constable and Company, Ltd.

Zacari, M. A., and L. F. Pacheco. 2005. Depredación vs. problemas sanitarios como causas de mortalidad de ganado camélido en el Parque Nacional Sajama. *Ecología en Bolivia* 40:58–61.

Zapata, S. C. 2005. Reparto de recursos en gremios de carnívoros: una aproximación ecológica y morfológica. Ph.D. dissertation, University of Buenos Aires, Argentina.

Zapata, S. C., A. Travaini, P. Ferreras, and M. Delibes. 2007. Analysis of trophic structure of two carnivore assemblages by means of guild identification. *Eur J Wildl Res* 53:276–86.

Zapata, S. C., A. Travaini, M. Delibes, and R. Matinez-Peck. 2008. Identificación de morfogremios como aproximación al estudio de reparto de recursos en ensambles de carnívoros terrestres. *Mastozoología Neotropical* 15:85–101.

Zanon, J. I. 2006. Puesta a prueba de la extinción ecológica de las presas autóctonas del puma (*Puma concolor*) en Patagonia. Undergraduate thesis. University of La Pampa, Santa Rosa, La Pampa, Argentina.

Zinn, H. C. 2003. Hunting and sociodemographic trends: Older hunters from Pennsylvania and Colorado. *Wildl Soc Bull* 31:1004–14.

Zinn, H. C., and M. J. Manfredo. 1996. *Societal preferences for mountain lion management along Colorado's Front Range*. Human Dimensions in Natural Resources Unit, Colorado State University, Fort Collins.

Zinn, H. C., M. J. Manfredo, and S. C. Barro. 2002. Patterns of wildlife value orientations in hunters' families. *Hum Dim Wildl* 7:147–62.

Zinn, H. C., M. J. Manfredo, J. J. Vaske, and K. Wittmann. 1998. Using normative beliefs to determine the acceptability of wildlife management actions. *Soc Nat Res* 11:649–62.

Zumbo, J. 2002. Predator control: Trapping is the key to saving game populations. *Outdoor Life*, December.

Contributors

Charles R. Anderson Jr. currently works as a research biologist for the Colorado Division of Wildlife, where he is developing a landscape-scale research project to address mitigation methods and development practices that benefit mule deer populations in areas experiencing extensive energy development. Chuck is also an adjunct professor of zoology and physiology at the University of Wyoming. From 1994 to 1997 and 2004 to 2006, he was a large carnivore biologist for the Wyoming Game and Fish Department, where he developed management strategies for black bear and cougar populations and directed research evaluating grizzly bear-cattle interactions and DNA-based bear population estimates. Chuck investigated cougar predation, population monitoring methods, and population genetics during his Ph.D. work in Wyoming. He received his B.S. in wildlife biology from Colorado State University in 1990 and his M.S. and Ph.D. in zoology and physiology from the University of Wyoming in 1994 and 2003, respectively.

Paul Beier conducts research in wildlife ecology and conservation biology with a focus on projects that directly support conservation planning at landscape scales. His 1988–92 study of cougars is best known for documenting that, during juvenile dispersal, cougars find and use habitat corridors in urban southern California to reach areas where they establish home territories as breeding adults. Since then he has worked on "missing linkages" in California and Arizona to promote science-based efforts to maintain wildlife corridors at the regional scale. A professor at Northern Arizona University, Paul has also studied Mexican spotted owls, northern goshawks, and forest bird communities in the United States and West Africa. Since 2000, he has worked with traditional chiefs in West Africa to create and manage community-based wildlife sanctuaries for hippopotami, elephants, and rare forest birds. He serves on the boards of the Society for Conservation Biology and South Coast Wildlands.

Mark S. Boyce holds the Alberta Conservation Association Chair in Fisheries and Wildlife at the University of Alberta. He has supervised numerous graduate students and has authored many academic papers and several books. He has studied a wide variety of species and ecological systems but has focused extensively on North American large mammals, particularly elk, wolves, grizzly bears, cougars, and black bears. Mark completed his B.S. in fish and wildlife biology at Iowa State University in 1972, his M.S. in wildlife management at the University of Alaska in 1974, his M.Phil. in wildlife ecology at Yale University in 1975, and his Ph.D. in wildlife ecology at Yale in 1977. From 1976 to 1993, Mark was

on the faculty at the University of Wyoming, serving as director of the university's National Park Service Research Center from 1989 to 1993. He was the Vallier Chair of Ecology and Wisconsin Distinguished Professor at the University of Wisconsin–Stevens Point from 1993 to 1999.

Susan G. Clark is the Joseph F. Cullman 3rd adjunct professor of Wildlife Ecology and Policy Sciences in the School of Forestry and Environmental Studies at Yale University, as well as a fellow at the Yale Institution for Social and Policy Studies. She is a member of three species-survival commissions of the IUCN–World Conservation Union and currently works on large carnivore conservation in western North America. Her interests include conservation biology, organization theory and management, and natural resources policy. Susan has written more than 390 papers and several books on small mammals, large carnivores, endangered species recovery, and natural resource management and policy. Honors she has received include the Outstanding Contribution Award from the U.S. Fish and Wildlife Service, the Presidential Award from the Chicago Zoological Society, and a Best Teacher designation from her students at the Yale School of Forestry and Environmental Studies.

Melanie Culver is an assistant professor in the Wildlife and Fisheries Conservation and Management Program at the University of Arizona and works as a U.S. Geological Survey geneticist with the Arizona Cooperative Fish and Wildlife Research Unit. Melanie's work encompasses all aspects of conservation genetics. She currently supervises several graduate students studying felids, carnivores, and other wildlife and fish species. Melanie earned her Ph.D. in biology from the University of Maryland in 1999. During graduate school, she worked on many aspects of puma genetics, including population subdivision, phylogeography, and taxonomy, as well as paternity and forensic testing. Following her graduate work, she spent three years as a postdoctoral research associate at Virginia Tech, where she worked on conservation genetics studies in black bears, fish, eagles, salamanders, several fish species, and freshwater mussels. Her published works include more than fifty scientific articles on many aspects of genetics and wildlife conservation, including seven book chapters.

R. Bruce Gill was the leader of mammals research for the Colorado Division of Wildlife for twenty-seven years. In that capacity, Bruce supervised research on several predatory species, including cougars. He was also actively involved in two major conflict-resolution processes that sought to defuse public controversy, one concerning black bear hunting management and damage control, the other furbearer management and trapping. These processes gave him special expertise in the human dimension of wildlife management and public involvement in the policy-making process. His published work includes scientific articles regarding the ecology of deer, elk, pronghorns, and bighorn sheep, in addition to conflict resolution and democratic public involvement. Bruce has retired from public service and now writes articles for popular publications, illustrating them with his own photographs.

Lucina Hernández is currently the director of the Rice Creek Field Station of the State University of New York at Oswego, New York. She was a senior researcher at the Regional Center of the Instituto de Ecología, A.C., in Durango, Mexico, for over twenty years. Luciana has worked with wildlife in Mexico since 1981 and received her doctorate in National Polytechnical Institute of Mexico in Mexico City in 1995. She has authored or coauthored more than twenty-five published works, including six book chapters. During much of her career, she has worked with carnivores, including cougars, in the Chihuahuan Desert. As the former coordinator of the Durango center, Lucina worked closely with local, state, and federal resource agencies on wildlife issues. She is an expert on the many challenges facing wildlife conservation in Mexico and Latin America.

Maurice Hornocker earned his academic degrees at the University of Montana and the University of British Columbia. He began working with cougars in 1964 in the central Idaho wilderness. Since that first project he, his students, and colleagues have studied cougars in environments throughout western North America. During Maurice's career, he has researched numerous species of carnivores—including ground-breaking research on wolverines—but he is best known for his work with cats. These include the cougar, bobcat, lynx, and ocelot in North America; jaguars in Central and South America; leopards in Africa and Asia; and tigers in India and Far Eastern Russia. Leader of the Cooperative Wildlife Research Unit at the University of Idaho from 1968 to 1985, he mentored forty graduate students. In 1985, he founded the Hornocker Wildlife Institute and was its director until 2000, when it merged with the Wildlife Conservation Society (WCS). Maurice was a senior conservationist at WCS until his retirement in 2006. He is widely published in both scientific and popular contexts, and his wildlife photographs have been published throughout the world. In 2006, he founded the Selway Institute, a nonprofit wildlife research and education foundation, where he currently serves as director.

Martin G. Jalkotzy is currently a senior wildlife biologist with Golder Associates in Calgary, Alberta, where he leads the wildlife team, providing direction and advice to Golder wildlife personnel in western Canada. During his thirty years in the field, Martin has worked on diverse issues, primarily addressing the effects of human development on wildlife, particularly ungulates and large carnivores. He has hands-on field experience with a wide variety of species, including river otters, cougars, and grizzly bears in North America, elephants in southern Africa, and pandas in China. From 1980 to 1994, he and his partners, Orval Pall and Ian Ross, pursued cougars in Alberta's Rocky Mountains and authored several pioneering papers on cougar ecology in Canada. Martin holds a B.Sc. (1977) from the University of Guelph and a Master's of Environmental Design (1983) from the University of Calgary.

Kyle H. Knopff is currently a Ph.D. candidate at the University of Alberta and project leader for the Central East Slopes Cougar Study (CESCS), a large-scale study of several aspects of cougar ecology in west-central Alberta, Canada. His dissertation research addresses several applied questions focused on cougar-prey interactions, cougar population estimation, and cougar habitat selection. Other areas of research include investigations of cougar ecology in rural-residential landscapes and the niche separation among top carnivores. Kyle completed his B.Sc. in anthropology at the University of Calgary in 2002 and began his field research career studying the behavioral ecology of howler monkeys in the jungles of Belize. That study became the subject of his master's thesis (University of Calgary, 2004) and also of several academic papers.

John W. Laundré was a senior researcher for the Instituto de Ecología, A.C., in Durango, Mexico. He has thirty years' experience working with wildlife and studied cougars for twenty years, directing a seventeen-year intensive study of cougars in Idaho and Utah, the results of which appear in at least ten current publications, with more forthcoming. In 2000, John worked with cougar conservation in the Chihuahuan Desert of northern Mexico, determining the status and defining the ecological role of cougars in this vast ecoregion. He currently lives in Oswego, New York, and has become active in carnivore research in the New England area.

Frederick Lindzey is an emeritus professor at the University of Wyoming, having retired from the Department of Interior's Cooperative Research Unit program and the university in 2004. Fred also served as a faculty member at the University of New Hampshire and Utah State University. He is currently an appointed member of the Wyoming Game and Fish Commission, a member of the Wyoming Toad Recovery Team, a board member of Idea Wild, a nonprofit conservation organization, and a fellow of The Wildlife Society. Fred and his students have conducted cougar research directed at answering management questions in Utah, Wyoming, and Colorado, and he has consulted on cougar management issues with many western states. In addition to cougars, he has worked on other carnivores including badgers, black-footed ferrets, black and grizzly bears, coyotes, jaguars, and ocelots.

Kenneth A. Logan has been studying cougars since 1981. His research has focused on cougar population dynamics, behavior, social organization, prey use, habitat use, effects of cougar predation on prey, and cougar-human interactions. Ken's work has been conducted in Wyoming, New Mexico, California, and Colorado. Currently, he is a Carnivore Researcher for the Colorado Division of Wildlife and is studying cougar ecology on the Uncompahgre Plateau. Ken has published numerous peer-reviewed scientific writings on cougars, including a book he co-authored with Linda Sweanor titled *Desert Puma: Evolutionary Ecology and Conservation of an Enduring Carnivore*, which received the Outstanding Publication in Wildlife Ecology and Management award from The Wildlife Society in 2002. He has played a key role in developing cougar management strategies. In addition, Ken is a co-author of the *Cougar Management Guidelines*, first edition. He earned a B.S. in range and wildlife management from Texas A&I University, an M.S. in zoology and physiology from the University of Wyoming, and a Ph.D. in wildlife sciences from the University of Idaho.

David J. Mattson is a research wildlife biologist with the U.S. Geological Survey in Flagstaff, Arizona, and has taught at Yale, MIT, and Northern Arizona University. He has been studying large carnivores for twenty-four years, focusing on cougar ecology and human-cougar interactions in Arizona, as well as on the conservation and behavioral ecology of grizzly bears in the Yellowstone Ecosystem of Wyoming, Montana, and Idaho. David spent fourteen years intensively observing the foraging behavior and diet of the grizzly bear. More recently, he has focused on conservation issues and broad-scale evaluations of habitat conditions. These studies have concerned not only the details of human–large carnivore interactions but also the social, political, and organizational dynamics that shape the policies and practices of carnivore conservation programs. His research has been featured in several scientific journals.

Kerry Murphy serves as an endangered species and mid-sized carnivore biologist with the U.S. National Park Service at Yellowstone National Park. He has worked with a variety of state, tribal, and federal agencies as a wildlife biologist, and has been involved in the study and management of numerous large and medium-sized carnivores. Kerry earned his Ph.D. in forestry, wildlife, and range sciences at the University of Idaho in 1998. His dissertation concerned the ecology of cougars in Yellowstone National Park, including their predation on ungulates and interactions with other carnivores and humans.

Sharon Negri has worked with government and nongovernment organizations on natural resource issues since 1980. She directs WildFutures, an Earth Island Institute project she founded in 1994, with the purpose of bridging the gap between science and conservation. Emphasis is on advancing strategies to promote sound carnivore and ecosystem function, working with a host of organizations, biologists, and wildlife agencies. As director of Wild-Futures, Sharon co-produced the award-winning film *On Nature's Terms* (2001) and published the *Cougar Management Guidelines* (2005), a resource of scientific information and management strategies for wildlife professionals. She co-founded the Mountain Lion Foundation in 1986 and served as its director until 1990. She also co-founded the Wild Felid Research and Management Association in 2007 and serves as an interim board member. Sharon has a B.S. in Environmental Policy and Planning at the University of California, Davis, and a consulting business providing a wide range of organizational and program planning services.

Andrés Novaro has studied carnivores in Patagonia since 1987. He is now a researcher with the Argentine National Research Council, as well as director for the Wildlife Conservation Society's Patagonian and Southern Andean Steppe program. His current research addresses the effects of puma predation on the recovery of native Patagonian herbivore populations, and the effects of abundant exotic prey on the interaction between pumas and native prey. Andrés also works with livestock producers in the vicinity of Patagonian protected areas to reduce predation on livestock by pumas and other carnivores. His Ph.D. from the University of Florida in 1997 was based on a study of source-sink dynamics in a population of hunted culpeo foxes.

Howard Quigley is director of Western Hemisphere programs for Panthera and serves as senior ecologist at Craighead Beringia South. With Maurice Hornocker, Howard initiated and co-directed the Siberian Tiger Project in Far Eastern Russia for nearly ten years, and he currently directs the Teton Cougar Project in northwestern Wyoming. He has conducted black bear research in Tennessee, helped initiate the giant panda project in China in 1980, and conducted a three-year intensive examination of jaguar ecology in the

Brazilian Pantanal. Howard served as president of the Hornocker Wildlife Institute for nearly eight years, after which, from 2000 to 2002, he was director of the Global Carnivore Program at the Wildlife Conservation Society. He obtained his Ph.D. from the University of Idaho in 1988 and subsequently joined the faculty of the University of Maryland system. He has authored more than thirty professional papers and chapters.

Toni K. Ruth has been involved in cougar research in Florida, Texas, New Mexico, Montana, and Wyoming since 1987, and is currently a research scientist with Idaho's Selway Institute. She serves on the Board of Directors of Salmon Valley Stewardship, a nonprofit organization working to promote a sustainable economy and a healthy environment in the Salmon River region. After eight years of conducting research in Yellowstone National Park with the Hornocker Wildlife Institute/Wildlife Conservation Society, Toni is now analyzing data and writing a book about the effects of wolf reestablishment on cougar population and predation characteristics in the park. Toni has published numerous scientific and popular articles, and her research has been highlighted in various films, including programs for National Geographic and the BBC. She received her Ph.D. in wildlife ecology from the University of Idaho.

Harley Shaw has worked in wildlife management for fifty-four years, twenty-seven of which were spent as a research biologist studying mule deer, wild turkey, bighorn, and cougar for the Arizona Game and Fish Department. Harley retired from that agency in 1990 and continues to write, consult, and work with citizen groups on wildlife monitoring. He has published two books, *Soul among Lions: The Cougar as Peaceful Adversary* (University of Arizona Press, 1989, 1994, 2000) and *Stalking the Big Bird: A Tale of Turkeys, Biologists, and Bureaucrats*, through University of Arizona Press. He is also the author of *The Mountain Lion Field Guide* (Arizona Game and Fish Department, 1983) which has undergone multiple printings, and a bulletin *WOOD PLENTY, GRASS GOOD, WATER NONE* (2006), on historic vegetation change in northern Arizona. Harley organized and hosted the Third Mountain Lion Workshop (1988) in Prescott, Arizona, and is co-author of the *Cougar Management Guidelines*. His current projects include a book about early biologist J. Stokely Ligon, as well as research on the history of the upper Canadian River in New Mexico.

Linda L. Sweanor has been involved in cougar research since 1985. She studied cougar population ecology, cougar-prey relationships, and cougar social organization in New Mexico from 1985 to 1995. From 2000 to 2004, Linda's research focused on cougar-human interactions in California. She has published scientific papers on cougar metapopulation dynamics and behavior and co-authored with her husband, Ken Logan, *Desert Puma: Evolutionary Ecology and Conservation of an Enduring Carnivore*, a book about cougar ecology and conservation. She received her M.S. in wildlife sciences from the University of Idaho in 1990. Linda is currently a research associate, assisting with a felid disease transmission study in western Colorado for Colorado State University, where she also volunteers on a cougar population study. Linda is also interim president of the Wild Felid Research and Management Association.

Susan Walker has worked on carnivore issues on three continents for more than twenty years. She has spent thirteen years in Patagonia, where she works for the Wildlife Conservation Society's

Patagonian and Southern Andean Steppe program. Susan's work addresses the evaluation and recovery of landscape connectivity for wildlife and, at a regional scale, the "re-wilding" of Patagonia through improvement of protected areas and connectivity for wildlife on private lands. Her current research uses genetic markers to analyze the recolonization of northern Patagonia by pumas. Susan received her M.S. and Ph.D. in wildlife ecology from the University of Florida.

Index

Activists for cougars, 14, 138, 178, 212–217, 224–228, 257–258

Adaptive management, 25, 49, 161, 229, 236

Additive mortality, 116, 139, 140, 144

Aggression, 107, 198–200, 242, 243. *See also* Competition; Threat responses

Alces alces. See Moose

Allozymes, 37, 249

American colonization, effect on cougars, 7–8

Anderson, Allen, 21

Anderson, Charles R., Jr., 41, 46, 73, 184, 230

Animal Damage Control Program (ADC). *See* Wildlife Services

Animal Protection of New Mexico, 224, 228

Antilocapra americana. See Pronghorn

Antipredator behavior, animal, 158, 159, 171. *See also* Prey

Apparent competition, 54, 129, 139, 145–146, 152, 153

Argos satellites, 196

Arizona Game and Fish Department, 217–218, 230

Attacks
  assessing risk of, 196, 197, 199
  behaviors that influence, 200
  on humans, 52, 87, 179, 193, 200, 205
  number of, 190, 191, 197
  reducing risk, 201–203
  seasonal and activity patterns of, 198

Aversive conditioning, 86, 204, 205, 243

Bailey, James A., 222

Ballot initiatives, 14, 49, 179, 214, 216, 226–228, 264–265

Behavior modification, 240, 242–243

Beier, Paul, 72, 75, 179, 188, 200, 244

Big game, cougars and, 51, 237

Bighorn sheep, 254
  mortality of, 151, 155
  as prey, 122, 129, 154, 171

Black bears, 60, 138, 163, 167

Boundaries, related to gene flow in cougar populations, 35, 39, 61

Bounties, 8, 9, 42–43, 95–96, 208, 235. *See also* Hunting of cougars

Boyce, Mark S., 41

Breeding
  behavior, 107, 108, 111, 113
  populations, 50, 51, 53, 54, 185, 186

Bureau of Biological Survey, 9, 11, 42

Bureau of Land Management (BLM), 11, 223, 224

Cache sites, 166, 195, 202

California. *See also* Cuyamaca Rancho State Park (CRSP)
  home ranges of cougars, 181, 202
  hunting regulations, 14, 115, 216, 227–228
  prey of cougars, 169

California Department of Fish and Game (CDFG), 226, 228, 230

Camera traps, 17, 23, 25, 51, 81, 84

Canada, cougars in, 80, 87, 88, 125, 129
  breeding populations of, 50, 51, 53
  management and conservation, 51–53, 208
  range expansion, 50, 53–54

Canis latrans. See Coyotes

Canis lupus. See Wolves

Carnivores. *See also* Multicarnivore systems; Predation
  conservation of, 15, 222, 229, 244
  dietary overlap, 167–169, 171
  dispersal capability, 69, 72
  exploitation competition among, 164, 167–170, 171
  habitat area, 179
  interference competition among, 164, 165–167, 178
  management of, 233
  predation, 129, 159, 161, 169
  protection of, 14
  vocalizations, 113

Castration, 242

Characteristics, physical
  adults, 18–19, 119–120
  cubs, 107–108

Clark, Susan G., 216

Climate change. *See* Threats to cougars

Communication
  between cougars, methods of, 113–114
  sounds, 19

Compensatory predation, 141–143, 144, 145

Competition. *See also* Threats to cougars
  among carnivores, 164–170
    for food, 167–169, 171
    influencing factors, 170–171
  apparent, 54, 129, 139, 145–146, 152, 153
  exploitation, 164, 167–170, 171
  interference, 164, 165–167, 169, 170, 178
  in Latin America, 82

Connectedness, 6

Conservation
  actions advancing, 228–229
  in Canada, 51–53
  collaborative efforts, 231–232
  considerations for, 239
  of cores, 188
  definition of, 222
  of female cougars, 116, 225, 228
  forensic genetic techniques in, 38–39
  of habitat, 16, 89, 180–181
  of isolated cougar populations, 181–185
  in Latin America, 81, 82, 87–88, 232
  in Patagonia and Southern Andes, 98–99
  preventing overkill, 227
  relevance of genetics to, 30, 39–40
  of source populations, 117
  in United States, 73, 222, 223

Conservation movement, 11–12. *See also* Preservation movement

Conservation planning, cougar, 177–178
  as a flagship species in, 179, 188–189
  as a focal species for, 178–179, 187–188
  use of habitat, 180–181

Convention on International Trade in Endangered Species (CITES), 78, 244

Corridors, 15, 72, 89, 188–189, 231–232

Cougar characteristics. *See* Characteristics, physical

Cougar Fund, 224, 229

Cougar habitat. *See* Habitat, cougar

Cougar–human encounters. *See also* Attacks
  conflicts with people, 238–239
  cougar response to humans, 191–193, 194
  Cuyamaca Rancho State Park (CRSP), 197, 201, 203
    daily human and cougar activity patterns in, 196
    research on cougar–human interactions in, 193–196
  human attributes and behaviors affecting, 200
  New Mexico research on, 196, 200, 204
  probability of, 194, 197
  protocols for responding to, 203, 204
  reducing risk of, 197, 201–203

Cougar–human interactions. *See* Cougar–human encounters

Cougar management. *See* Management, cougar

Cougar Management Guidelines Working
    Group, 170, 190, 202
    Cougar Management Guidelines, 25, 26, 49,
        228, 230, 245
Cougar Network, 185, 186
Cougar–prey relationships, 138
    factors affecting, 139
    limiting and compensatory effects of, 141–143
Cougar status
    in Latin America, 78, 87–88
    in United States, 42, 44, 50, 191, 210
Coyotes
    dietary overlap, 167
    effect on cougars, 42, 52
    as predator, 108, 152, 156
    as prey, 122, 134, 158, 163
    surgical sterilization in, 240
Cubs, 62, 110, 113
    birth weight, 107
    characteristics, physical, 107–108
    coat color, 107, 108
    independence, 111
    infanticide and cannibalism, 108, 111
    mortality and survival rate, 64, 107
    number at birth, 19, 107
    play behavior, 108
Culpeo, 91–93, 96, 97, 156
Culver, Melanie, 20, 37, 77
Cumulative effects model, 226
Cuyamaca Rancho State Park (CRSP),
    193–196, 197, 201, 203
Cypress Hills Interprovincial Park, 50, 51

Data analysis units, 43
Defensive responses, 198, 199
Dens, 62, 63, 74
Depredation, 42–43, 45, 84–85, 94. *See also*
    Livestock; Predation
    factors influencing, 158
    trends in, 48–49
Diet, 119, 125–128, 132, 133
    biomass in, 81, 94, 124, 130, 134
    in Canada, 81, 125
    of individual cougars, 133
    in Latin America, 81–82
    overlap. *See* Carnivores
    seasonal changes in, 129–130
    in the United States, 81, 125, 126, 127, 129
    variations in, 130
Dietary switching, 171
Disease. *See* Threats to cougars
Dispersal, 48, 50, 53, 116, 117, 181
    behavior, 38, 109
    documenting, 66–69
    in subadult stage, 64, 108–109
Distinct population segments (DPS), 39, 187
DNA
    fingerprint, 250
    microsatellite, 32, 33, 37, 250
    mitochondrial (mtDNA), 31–34, 77, 249–250
    nuclear, 31, 250
    sequencing, 249, 250
    viral sequencing, 251

Eastern cougar, 8, 51, 89, 178
    recolonization of, 39, 185–187
Ecological role of cougars, 159, 178–179,
    233, 244

Endangered Species Act (ESA), 13, 14
    protection of cougar, 186–187
Environmentalism, 12–16
Environmental Protection Agency (EPA), 13, 42
European colonization, effects on cougars,
    93–94
Evolution of cougars, 27–28, 106
Exploitation competition, 164, 167–170, 171
Extirpation, 20, 32–33, 178, 219

Feeding behavior, 123–124, 136, 146
Feline immunodeficiency virus (FIV), 34–35, 37
Feline microsatellites, 32–34
Female cougars. *See also* Conservation
    behavior of, 108, 110
    home ranges, 110–111
    philopatry in, 38, 64, 66–67, 72, 108–110
    reproductive success of, 110
    sterilization, 242
Fine-scale movement, 181
FIV. *See* Feline immunodeficiency virus
Florida panther
    genetics, 30, 34, 39, 182, 250
    habitat, 180
    home range, 110, 111
    population, 13, 36–38, 182–183
    reintroduction of, 182–183, 187, 240
Focal species. *See* Conservation planning
Forensic genetic techniques, 38–39
Functional response, 140, 144
Fund for Animals, 207, 227

Gabrielson, Ira N., 222
Game depletion, effect on cougars, 8, 84
Genetic markers, 23
    DNA-based, 249
    molecular, 31–32, 35, 248–249
    neutral, 32, 248–249
Genetics
    and cougar behavior, 242–243
    diversity, 30, 31, 37, 184, 185, 248
    relevant to cougar conservation, 30, 39–40
    restoration, 37, 183
    tools, 32, 35, 248
    variation, 30, 31, 33, 39
Gestation, 5, 13, 62, 107
Gill, R. Bruce, 221, 241
Global positioning systems (GPS), 23, 62, 66,
    193–194
    location, 59, 74, 119, 195, 196
Goldman, Edward, 17, 20, 21, 42, 70
GPS. *See* Global positioning systems
Greater Yellowstone Ecosystem, 152, 169, 171,
    172, 225
Grizzly bears, 60, 82, 169
    habitat areas, 179
    protection of, 14
    recovery, 225, 246
Guanacos, 91–94, 98, 156
    as prey, 82, 91–93, 155–156

Habitat, cougar
    conservation of, 16, 89, 180–181
    dispersal, 108–109
    fragmentation of, 54, 77, 81, 88, 230
    human effect on, 74–75
    management of, 202, 204, 223
Habitat loss. *See* Threats to cougars

Habitat Suitability Map, 228, 231
Habituation, 181, 196, 198, 199
Harvest methods, 252–253. *See also* Hunting
    of cougars
    general seasons, 43, 47, 48
    limited entry, 43
    quota, 22, 43, 44, 46
Hernández, Lucina, 121–123, 230, 232, 238
Hibben, Frank C., 20
Home ranges, 180–181
    females, 75, 80, 92, 110–112
    males, 75, 80, 92
    philopatry, 66, 108
    subadults, 64, 108
Hornocker, Maurice, 21–23, 57, 124, 130,
    132, 161
Human population growth, effect on cougars,
    88–89, 231, 244
Hunting behavior of cougars, 121–123
Hunting of cougars, 9, 153, 243–244
    bounty, 50, 95, 226, 235
    changing management climate, 208–214
    early techniques, 6–8
    effects on livestock, 96
    effects on prey, 86, 88, 164
    harvest trends, 47–48, 65, 252–253. *See also*
        Harvest methods
    with hounds, 43, 45, 49, 256
    hunter numbers and distribution, 43–44
    impact on cougar populations, 239
    to increase safety, 95
    influence on management strategy, 219–220
    limitations, 239
    limited-entry programs, 43
    males, effects on, 116–117, 138
    methods of, 45
        bag limits/permits, 45
        pursuit seasons, 45
    to reduce depredation, 95
    regulations, 14, 42–44, 227, 230
    selective, 96
        v. nonselective, 238
    sport hunting, 43, 49, 87, 96, 117, 243
        numbers harvested, 47, 231
Husbandry practices, 86
    guard dogs, 96–97

Illegal trade of pumas, 96
Infanticide. *See* Threats to cougars
Intensive capture-recapture method, 23, 70
Interference competition, 164, 165–167, 169,
    170, 178
Intraguild predation, 163, 164, 167, 170, 171
Isolated populations, 15, 66, 181–185
    landscape-scale permeability, 185

Jaguars
    conservation, 88–89
    dietary overlap between cougar and, 82–84,
        130, 169, 170
    habitats, 84
Jalkotzy, Martin G., 41, 64, 130

Kellert, Steven, 207, 212, 255
Kinship levels for cougar populations, 38
Kittens. *See* Cubs
Kleptoparasitism, 164, 171
Knopff, Kyle H., 41

Lama guanicoe. See Guanacos
Land ethic, 12, 16, 222
Latin American pumas
  activity level, 81
  conflict between human and, 84–87
  conservation, 78, 82, 87–88
  diet, 81–83, 84
  distribution, 77–80
  ecology of, 80–81
  effect on livestock, 84–85, 158
  future of, 89–90
  genetics, 32–33
  habitat, 80–82, 84
  home ranges of, 80
  impact on prey populations, 82, 84
  predation rate, 82
  predator-prey interactions, 81
  public education, 85–87, 88, 158
  taxonomic status of, 77
  threats, 88–89
Laundré, John W., 23, 82, 121–123, 146,
  152, 238
Leopold, Aldo, 12, 222, 246
Limiting factors, on cougar populations,
  139–140
  disease, 31, 65, 139
  human persecution, 158, 177
  predation, 151, 153, 159
  territorialism, 238, 241
Lindzey, Frederick, 41, 46, 64, 68, 73, 155
Lineage, cougar, 27–30
Litigation, 214, 215–216, 225 227, 260–263
Livestock
  compensation for loss of, 86, 87
  conflict in Latin America, 84, 94
  cougar predation on, 129, 158–161
  effect on cougar populations, 9, 84
  grazing habits, 156–158, 160
  loss in Latin America, 85
Logan, Kenneth A., 21, 23, 26, 64–67,
  72, 151

Male cougars
  dispersal, 38
  effects of hunting, 116–117
  infanticide, 108
  mating behavior of, 106–107
  sterilization, 242
  territorialism, 61, 71, 112–113, 117, 236
Management, cougar
  ballot initiatives, 49, 214, 216, 226, 228,
    264–265
  behavior and social organization, 115–117
  conflicts, 218–219
  decision making in, 208–214
  examinations of, 207, 220
  groups participating in, 208–210, 219,
    254–259
  litigation in, 215–216
  in North America, 51–53, 208–210, 211
    evaluation of, 44, 47–49
    jurisdiction, 223–224
    phases of, 41
    plans, 46–47
    status changes in, 49–50
  policy, changing, 224–230
  programs, 44
Mark-recapture theory, 23

Mating behavior, 30, 60, 63, 109. *See also*
  Reproduction; Sexual selection
Mattson, David J., 124, 134, 149, 197,
  199, 200
Mesopredator release, 158
Molecular genetics
  approaches, 248–249
  case studies
    individual-level, 38–40
    population-level, 35–38
    species-level, 31–32
    subspecies-level, 32–35
Molecular markers, 248
Moose, 53, 122, 130, 153, 169
Mortality, 61, 63–65
  cannibalism, 108
  infanticide, 108, 111, 113
Mountain caribou, 153–154
Mountain Lion Foundation (MLF), 16, 224,
  226, 228–230
Muir, John, 10, 222
Mule deer
  habitat, 122, 132, 157, 158
  population, 145, 152–155, 167, 178
  as prey, 71, 129–132
Multicarnivore systems, 140, 171–172
Murphy, Kerry, 74, 132, 133, 149, 165, 239

National Environmental Policy Act (NEPA), 13,
  215, 227
National Park Service (NPS), 11, 196, 223
National Rifle Association (NRA), 14
Native Americans, 5–7
Negri, Sharon, 228, 235, 246
Novaro, Andrés, 156, 161
Numerical response, 139–140, 145

Odocoileus hemionus. See Mule deer
Olfaction, 120
Olfactory signals, 113–114
Ontario puma, 53
Ovis canadensis. See Bighorn sheep

Panthera onca. See Jaguars
Patagonia, cougars in
  management, 95–99
  population density, 79, 91
  prey selection, 91–93
Paternity studies, 38
Philopatry, 64, 66–69, 72, 108–110, 238. *See*
  also Female cougars
Phylogenetic analyses, 32, 33, 38, 249
Pinchot, Gifford, 11, 222
Polymerase chain reaction (PCR), 249
Polymorphism, 249–251
Population components, 61, 68
Population densities, 70, 71–72, 78, 156
Population status, 44, 46, 50–51
Population trends
  in Canada, 44, 50
  in Latin America, 78–80
  in the United States, 44
Populations, types of
  isolated, 15, 66, 181–185. *See also* Isolated
    populations
  metapopulations, 61, 66, 72, 73
  source and sink populations, 72, 73
Precautionary principle, 225

Predation. *See also* Carnivores
  behavior, 121–122
  compensatory v. additive effects, 141–145
  effect on livestock in Patagonia, 94
  habitat influences, 122, 157, 161
  indirect effects of, 157–158
  influence on prey, 171–172
  intraguild, 163, 164, 171, 172
  limiting v. regulating effects, 139
  livestock grazing effects on, 161
  in multicarnivore systems, 171–172
  patterns in, 156–157
  rate, 82, 146–151
  risk, 157, 158
  seasonal, 129
  theory, 130
Predation, cougar on. *See also* Carnivores
  bighorn sheep, 130, 154–155, 157, 178
  cattle, 85, 160
  deer, 151–153
  domestic prey, 158–161
  elk, 151–153
  guanacos, 82, 94, 155–156
  llanos, 160
  moose, 153
  mountain caribou, 153–154
  pronghorn, 155
  small mammals, 156
  vizacacha, 82
  wild horses, 153
Predator–prey interactions, 81–82, 139, 162
Predator–prey theory, 139
Preservation movement, 10–11, 15–16, 222
Prey. *See also* Carnivores
  animals, 125–128
  antipredator behavior, animal, 158, 159, 171
  consumption, 123–124
  density, 139, 140, 144, 166, 171
  populations
    ecological carrying capacity of, 140
    effects of habitat loss, 8
    regulated by predation, 140
  response to predation risk, 157–158
  selection, 124, 130–135, 169
  switching, 145–146, 154, 169
  vulnerability, 122, 130, 131, 135
Project CAT, 229
Pronghorn, 122, 129, 155, 171, 178
Pseudalopex culpaeus. See Culpeo
Public attitudes
  about reintroduction, 187
  nature-views, 5, 84, 187, 207
  toward predators, 10, 12, 14, 86, 190
Public safety, 243
Puma concolor coryi. See Florida panther

Quigley, Howard, 71, 237, 246

Ranchers, 9, 11, 12, 93
  conflict between pumas and, 94–95
  in cougar management, 254
  selective killing of pumas by, 96
Ranges, cougar
  current, vii, 19
  expansion of, 50, 53–54, 69, 180
  historic, vii, 19, 222
Rangifer tarandus caribou. See Mountain
  caribou

Recolonization, cougar, 53, 94, 98, 185–187
Refugia, 94, 188, 223
Regulating factors, 140
Reintroduction
    cougar, 187. *See also* Recolonization
    wolf, 14, 53, 87
Remote camera. *See* Camera traps
Removal, cougar, 52, 73
Reproduction, 107, 111, 187, 238. *See also*
    Mating behavior
    age, 19
    birth intervals, 111
    characteristics, 33
    communication, 106, 113–115
    competition among males, 106, 113
    copulation rates, 106
    effect of mortality factors on, 240
    fecundity rates, 59–60, 62, 72, 75
    polyestrous, 107
    prey influence on, 140
    sterilization effect on, 240–242
    success, 105–106
        females, 106, 110
        males, 113
    timing of births, 62–63
Research techniques
    camera traps, 23, 25, 51, 81, 84
    capture-recapture, 23, 236
    DNA analysis, 23, 32, 37, 77, 249–251
    GPS, 63, 84, 193–194, 241
        location, 59, 74, 119, 195, 196
        radio-tracking methodology, 23
    hair analysis, 25
    radio collars, 22, 23, 146, 191
    radiotelemetry, 22, 23, 68, 70
Reserve design, 187–189, 231
Roosevelt, Theodore, 11
Ruth, Toni K., 68, 167, 170, 238, 239, 244

Seidensticker, John, 22, 23, 71
Self-limiting hypothesis, 71, 105, 113, 237,
    241, 242
Sexual selection, 106, 107, 113
Shaw, Harley, 130, 136, 160, 210
Sinapu, now WildEarth Guardians, 224, 228
Social organization of cougars, 115–117
    dispersal, 108–109. *See also* Dispersal
    in home ranges, 110, 111. *See also* Home
        ranges
    sexual selection, 107, 116. *See also* Reproduc-
        tion; Sexual selection
    solitary living, 106

Source-sink populations, 48, 72–73. *See also*
    Populations, types of
South Coast Missing Linkages Project
    (SCMLP), 231
Stalking, 110, 121, 122, 131, 180–181
Stegner, Wallace, 246
Sterilization, 237, 239–242
Subadult cougars, 196
    dispersal of, 64, 66, 106, 108–110, 113, 203
    prey selection of, 133, 136, 159–160
    survival rate, 64, 72
Subspecies, 29, 30, 37, 77, 182, 187
    phylogenetic analyses of, 32, 34
Sweanor, Linda L., 21, 23, 25, 64–67, 151, 181

Taxonomy, 17, 28–30, 33, 39
Territorialism
    in male cougars, 71, 109, 112–114,
        117, 236
    for self-regulation of cougar numbers, 71,
        237–238, 242
Threat management, 230–232
Threat responses, 192–193
Threats to cougars
    climate change, 15, 231
    competition among carnivores, 164–165
    cumulative impacts, 221, 225–226, 231, 233
    disease, 65, 139, 231
    habitat loss, 78, 88, 90, 230, 231, 245
    human population growth, 88, 97, 230
    infanticide, 108, 113, 117
    in Latin America, 87–88
    loss of genetic variation, 182. *See also* Isolated
        populations
    overhunting, 229–230, 243–244, 245
    poisoning, 11, 42, 96
    and puma conservation, 87–88
    roads, 231
    urbanization. *See* Urbanization, effects of
Total predation rate, 144, 151
    limitations of, 149
Transients, 109
Translocation, 154, 203
Trophic cascades, 139, 158–159, 163, 178

Ungulates
    habitats, 132, 136
    population, 50, 53
    as prey, 128, 129, 157, 167–168, 178
United States
    cougar ballot initiatives in, 264–265
    cougar harvest in, 216, 252–253

Urbanization, effects of, 12–13, 97, 157, 185,
    232, 255
Ursus americanus. *See* Black bears
Ursus arctos horribilis. *See* Grizzly bears
U.S. Fish and Wildlife Service (USFWS), 183,
    187, 223
U.S. Forest Service (USFS), 224

Viral DNA sequencing, 251
Visual signals, 113–114
Vocalizations, 113–114, 192, 198

Walker, Susan, 156, 161
WildFutures, 228
Wildlands Network, 231
Wildlife agencies, 12, 52, 215, 217, 223
    management policies, 217–218
    mandates, 223
    in Oregon and Washington, 14
    personnel, 228, 256–257
    with regulatory responsibility, 16, 42
Wildlife commissions, 210–212, 218, 220,
    223
Wildlife Protection Act, 227
Wildlife Services, 42, 45, 215, 226
Williams, Jim, 246
Wise use movement, 14, 15
Wolves
    competition between cougars and,
        166–167, 171
    control of, 9, 41
    dietary overlap between cougars and,
        170, 171
    effects on cougars, 42, 52, 75, 244
    as a predator, 108, 151, 158, 167, 169
    reintroduction of, 14, 53, 87
Wyoming cougars, 35
Wyoming Game and Fish Commission,
    42, 229

X- and Y-chromosome linked genes, 31

Y-chromosome, 251
Yellowstone National Park
    cougar reproduction in, 62, 64
    public attitudes toward predators in,
        10–11
    wolf reintroduction in, 14, 53, 87, 167
Young, Stanley P., 12, 20, 42, 70
Yuma puma (Colorado River cougar), 30

Zone management, 117